Exploring Ocean Science

Second Edition

Exploring Ocean Science

Second Edition

KEITH STOWE
California Polytechnic State University

JOHN WILEY & SONS, INC.
New York • Chichester • Brisbane • Toronto • Singapore

Acquistions Editor Chris Rogers
Director of Development Johnna Barto
Marketing Manager Catherine Faduska
Production Director Pam Kennedy
Senior Production Editor Bonnie Cabot
Designer Laura Nicholls
Manufacturing Manager Mark Cirillo
Photo Editor Hilary Newman
Illustration Coordinator Edward Starr

This book was set in 10/12 ITC Cheltenham Light by General Graphic Services and printed and bound by Von Hoffmann Press. The cover was printed by Phoenix Color.

Recognizing the importance of preserving what has been written, it is a policy of John Wiley & Sons, Inc. to have books of enduring value published in the United States printed on acid-free paper, and we exert our best efforts to that end.

John Wiley & Sons, Inc., places great value on the environment and is actively involved in efforts to preserve it. Currently, paper of high enough quality to reproduce full-color art effectively contains a maximum of 10% recovered and recycled post-consumer fiber. Wherever possible, Wiley uses paper containing the maximum amount of recycled fibers. In addition, the paper in this book was manufactured by a mill whose forest management programs include sustained yield harvesting of its timberlands. Sustained yield harvesting principles ensure that the number of trees cut each year does not exceed the amount of new growth.

Library of Congress Cataloging in Publication Data:
Stowe, Keith S., 1943 -
Exploring ocean science/Keith Stowe. -- 2nd ed.
p. cm.
Rev. ed. of: Essentials of ocean science. c1987.
Includes index.
ISBN 0-471-54376-4 (pbk. : alk. paper)
1. Oceanography. I. Stowe, Keith S., 1943- Essentials of ocean science. II. Title.
GC16.S75 1995
551.46 -- dc20 95-32932
CIP

Printed in the United States of America

10 9 8 7 6 5 4 3 2 1

Preface

The oceans are the single most dominating and distinctive feature of this planet. They control our climate, modify and mold all Earth's surface features, and nurture the life that was born from them. This life now has a myriad of forms, some of which have left the oceans and colonized the land. Among these forms are we humans who have the curiosity to study this unique and fascinating feature of the Earth.

This second edition of *Exploring Ocean Science* is intended for a one semester introductory or survey course in oceanography for university students with little or no background in the sciences. This course provides an opportunity for curious students to question the world around them. Why are the oceans the way they are? How did they get that way? How are they changing, and why? What are their many important behaviors and how do these affect us? What is the spectrum of organisms and lifestyles that they support? What are the underlying principles and causes that tie the many and varied details together into a unified and harmonious picture? It is hoped that this inquisitive attitude will stay with the student long after this course is over.

GOALS

I want students to experience the excitement of learning how nature works, and I hope students will find this experience so rewarding that they continue to question and learn about the world around them for the rest of their lives. This overall objective has nurtured several more specific goals.

TO INSTILL CURIOSITY AND CONFIDENCE

My first goal is to stimulate the students' curiosity by drawing their attention to the most interesting and important features of this planet, virtually all of which involve the oceans. I wish to expose the students to the pleasure and beauty in understanding these features. Interesting analogies and everyday examples are used throughout the text to help students realize that this understanding is simple and easy to acquire.

TO USE AN INQUISITIVE AND EXPLANATORY APPROACH

Secondly, I try to employ a lively and interesting writing style so students will look forward to reading and learning this material. Unlike the more encyclopedic or descriptive approaches of other texts, this book takes an inquisitive and explanatory approach. It answers not only *what,* but also *how* and *why.* By providing the *how* and *why,* we create a framework that provides students with an understanding of the underlying causes. Without this information some students may see the subject matter merely as a group of unrelated and confusing details. They may view science as simply a search for data, rather than see that these data are merely tools that we use as we strive for a better understanding of nature.

TO SHOW THAT NATURE IS EASILY UNDERSTOOD

The theme throughout the book is that nature is neither random nor irrational. There are good reasons for things being the way they are, and most of these reasons are easily understood through everyday experiences and common sense. This text uses the oceans as the medium for conveying this greater message. It is my hope that the students will learn that most natural phenomena have simple explanations that can be understood easily without scientific sophistication. With this attitude, students will be more eager to probe the world around them and continue to develop a first-hand appreciation for science long after they have finished this course.

IMPROVEMENTS IN THE SECOND EDITION

Feedback from students and colleagues who have used and reviewed the first edition of *Exploring Ocean Science* indicated that the writing style was a strong point. I tried to maintain a similar lively writing style throughout this edition. All chapters have been completely rewritten and reordered to increase efficiency and clarity of presentation. Topics are included within the text at the appropriate places in the conceptual development; there are no separate boxes to interrupt the presentation.

NEW CHAPTERS

In response to numerous suggestion for expanded topical coverage, three new chapters have been added: one on the history of oceanography and the tools used by oceanographers (Chapter 1), one on materials and motion (Chapter 2), and an additional chapter on marine biology (Chapter 12).

SOCIETAL AND ENVIRONMENTAL ISSUES

This edition includes an increased emphasis on environmental and societal issues. The last two chapters (Chapter 14, "Coastal Development, Pollution, and Food," and Chapter 15, "Ocean Resources and Law") deal entirely with these subjects, and additional topics of special societal and environmental concern are discussed at appropriate places within the earlier chapters, as denoted by an environmental icon.

ENHANCED VISUAL APPEAL

In an effort to make the student's study as pleasant, interesting, and informative as possible, we have improved the layout and visual appeal of this edition. Full color is used to increase the quality and attractiveness of the photos and illustrations throughout the book. In addition, topics of special environmental interest are identified within the text by a special blue environmental icon. We have also highlighted quotes from the narrative to help focus the student on analogies and examples that illustrate the central ideas.

ORGANIZATION

INTRODUCTORY CHAPTERS

The book begins with two introductory chapters. Chapter 1 gives the student a quick overview of the history of the Earth and ocean exploration, and introduces some of the common tools used in present ocean research. Chapter 2 reviews the basic underlying scientific concepts that are common to many different topics in the remainder of the book. This review includes atomic structure, chemical bonds, the forces that are important in governing the motion of fluids, and the concept of energy and its transformation from one form to another. Students who are familiar with this material may choose to skim it quickly. Those who are not should master it. This chapter serves as an "equalizer" for students with weaker scientific backgrounds.

GEOLOGICAL, CHEMICAL, PHYSICAL, AND BIOLOGICAL OCEANOGRAPHY

The next eleven chapters address the conventional areas of geological, chemical, physical, and biological oceanography. Chapters 3–5 investigate geological oceanography—the origin and evolution of the Earth and its oceans, the features of the ocean floor, and the marine sediments that blanket the Earth's surface. In Chapter 6 we examine chemical oceanography—the many exceptional properties of water, the materials that are dissolved and suspended within it, and we question why the Earth's hydrosphere should be so different from those of its planetary neighbors. Chapters 7–10 deal with physical oceanography, including a study of the ocean's surface waters and their influence on our climate, global warming, global wind patterns, surface and deep currents, waves, tides, shoreline processes, and the distinctive properties of coastal waters. Marine biology is the focus of Chapters 11–13. Here we examine factors that influence food production by marine plants, the resulting distribution patterns for organisms of all kinds, the food web, the various kinds of marine organisms, the spectrum of environments they inhabit, and how marine organisms cope with the special challenges of these environments.

SOCIETAL AND ENVIRONMENTAL ISSUES

Although environmental and societal issues are discussed throughout the text, the final two chapters (Chapters 14 and 15) are devoted entirely to this subject. Topics include the problems associated with coastal development, pollution, fishery harvests, various important mineral resources, the search for renewable energy resources, the extraction of fossil fuels and other resources from the ocean, and the international concern over pollution and the ownership and extraction of ocean resources.

PEDAGOGICAL STRUCTURE

Each chapter is organized with a consistent pedagogical structure to reinforce the main ideas and to encourage effective study habits.

- Each chapter begins with an **outline**, which provides a quick preview of the topics included in the chapter, and an **introduction,** which gives a quick overview of how these topics fit into the larger picture of the study of oceanography.

- **Highlighted quotes** from the narrative focus on analogies that help to illuminate key ideas.
- **Environmental icons** alert students to areas of particularly sensitive societal and environmental concern.
- **Key terms** are emphasized with bold face type within the text. They are defined when first introduced and then are listed in order at the end of each chapter as a study aid.
- Each chapter concludes with a **summary** which includes the chapter subheadings for organized review.
- **Study questions** help students to check mastery of topics discussed in the chapter.
- **Critical thinking questions** ask students to apply the information learned by going beyond mere recall and use synthesis, analysis, and extrapolation.
- **Suggestions for further reading** include popular articles in nontechnical journals for interested students to pursue.
- A **glossary** at the end of the book includes all key terms and additional words that may be helpful to students with particularly weak scientific backgrounds.
- An extensive **index** permits easy reference to any topic.

I strongly encourage the student to augment the passive reading with more aggressive activities, such as highlighting, outlining the material within each chapter and section, and using the key terms, study questions, and critical thinking questions at the end of the chapters for self quizzes. Many students also find it helpful to verbally summarize the material under each heading and describe the important features of each figure while paging through a chapter.

SUPPLEMENTS

Instructional aids that accompany this text include an ***Instructor's Manual/Test Bank,*** which includes the following:

- The *answers* to all the study questions and critical thinking questions appearing in the book.
- *Matching-the-answers vocabulary test* for all the key terms of each chapter.
- A bank of approximately *2500 multiple choice questions* broken up according to chapter and sections.
- Brief descriptions of simple and inexpensive *coastal field trips.*

The test questions are also available in **computerized** format, allowing professors to choose and modify questions, create and print their own tests, and save for later use or modification (IBM and Macintosh®).

There are approximately **75 overhead transparencies**, taken from textbook illustrations. to support lecture presentations.

In addition to the pedagogical aids included in the text, a ***Study Guide*** by John F. Looney, Jr. (University of Massachussetts, Boston) is offered. This student aid provides chapter summaries and analogy sections, as well as matching, multiple choice, true/false, and "thought and research" questions to reinforce and expand key concepts and ideas.

ACKNOWLEDGMENTS

The development of this edition involved the effort of a large number of people. For their careful reading of the manuscript and their many valuable suggestions regarding its content, I thank my academic colleagues:

Claude E. Bolze
Tulsa Jr. College-Metro

Laurie L. Brown
University of Massachusetts

Richard A. Crooker
Kutztown University

David B. Eggleston
North Carolina State University

Jack C. Hall
University of North Carolina

Ronald E. Johnson
Old Dominion University

Charles Ernest Knowles
North Carolina State University

Lawrence A. Krissek
Ohio State University

Glenn C. Kroeger
Trinity University

John F. Looney
University of Massachusetts

James E. Mackin
SUNY Stony Brook

Robert A. Radulski
Southern Connecticut State University

Joanne R. Shelley
Pierce College

Keith A. Sverdrup
University of Wisconsin

Scott W. Synder
East Carolina University

Terri L. Woods
East Carolina University

John H. Wormuth
Texas A & M University

Thanks also to Larry Balthaser for answers to innumerable inquiries on geological topics, and to Howard Georgi and Harvard University for hosting me during the final stages of this endeavor.

I would also like to thank the staff at John Wiley & Sons both for extensive help in the preparation of this book and for producing and marketing it: Chris Rogers, executive editor; Bonnie Cabot, senior production editor; Johnna Barto, director of development; Laura Nicholls, senior designer; Edward Starr, senior illustration coordinator; Hilary Newman, photo editor; Pam Kennedy, production director, and Catherine Faduska, senior marketing manager. Special thanks go to Mary Konstant and Patti Brecht for their exceptionally careful editing of the manuscript and their many excellent suggestions.

Finally, I express my love and appreciation to Marianne, Adam, and Bryan for their support and understanding of my chronic obsession for writing.

Keith Stowe
Cambridge, MA

Brief Contents

Contents

Exploring Ocean Science

Second Edition

The world is a masterpiece. Every scene is more beautiful than the finest painting, and every piece displays an intricacy and harmony finer than Beethoven's best.

HMS *Beagle,* on which Charles Darwin sailed as a naturalist (off Rio de Janeiro, July 5, 1832).

One

HISTORY AND TOOLS OF OCEANOGRAPHY

The world is a masterpiece. Every scene is more beautiful than the finest painting, and every piece displays an intricacy and harmony finer than Beethoven's best. The more we ask how something is made, how it works, or how it evolved, the more we are in awe of the magnificence of nature. This is the beauty that science teachers would like their students to appreciate. It is everywhere, and its enjoyment requires no scientific sophistication. The inquisitiveness of a child is all it takes to turn our everyday world into a magical kingdom.

The primary objective of this book is to foster an inquisitive attitude that will remain long after many details are forgotten. As a course in history may allow us a deeper appreciation of the various social institutions, or a course in art or music may allow us new enjoyment of these expressive forms, a course in the sciences should offer us a new way of looking at our world for greater appreciation (Figure 1.1). In this book, the study of the oceans is the vehicle for experiencing how personally rewarding this inquisitive attitude can be.

Figure 1.1
A course in science offers a deeper appreciation of Nature's beauty. In this course, we explore the oceans.

1.1 OVERVIEW

The most unique feature of the Earth is its oceans. These huge bodies of water lie between widely separated continents and cover more than twice as much area as all continents combined. They are responsible for life, and they most decisively distinguish us from our planetary neighbors (Figure 1.2). Our human curiosity quite naturally leads us to study this unique and dominating feature of Earth. This field of study is called *oceanography*.

Oceanographic research requires imagination to decide what information would be most interesting and revealing, and to design the equipment needed to collect this information. Due to the interdependence among the various components of the ocean environment, most research requires familiarity with several different scientific disciplines. Examples include studies in the areas of geophysics, biophysics, nutrition, petrology, anthropology, meteorology, geography, and pharmacology. Some relevant ocean-related disciplines are outside the realm of the pure sciences, such as history, law, or sociology. The scientists who study the oceans are motivated by curiosity about the world around them, and their desire to find answers to their questions leads them through many fields of study.

To provide a basic framework for the systematic study of oceans, it is customary to divide the discipline into four broad areas: geological oceanography, chemical oceanography, physical oceanography, and biological oceanography. There is also a large and growing interest in studies of human interactions with the oceans, which involve all four areas.

These four categories, along with the study of human–ocean interactions, form the basis for the organization of this book. Some examples of the types of questions that will be investigated in each section are listed next.

INTRODUCTION (CHAPTERS 1 AND 2)

How and when did the Earth and its oceans first form? What time scales are involved in the birth and evolution of oceans, and what are some of the events in the Earth's history that affect what we see today? How do humans fit in, and what have been some of the milestones in human exploration of the oceans? What are some of the tools we have used in our investigations? What is the basic atomic and molecular structure of the materials on Earth, such as the oceans, atmosphere, and sediments? What basic forces are responsible for the evolution and motions of these materials?

Mercury

Earth

Mars

Moon

Figure 1.2

Earth, Mercury, Mars, and our Moon. The ocean is the feature that distinguishes us most from our planetary neighbors.

GEOLOGICAL OCEANOGRAPHY (CHAPTERS 3–5)

What lies beneath the oceans? How have the oceans evolved to their present form, and how is the evidence of this evolution recorded on the ocean floor? Why are there continents, and how does their underlying structure differ from that of the ocean basins? What are the sources of the sediments that cover the ocean floor, and why do they cover the continents as well?

CHEMICAL OCEANOGRAPHY (CHAPTER 6)

What makes water such a unique material? Why is it so much more abundant on Earth than neighboring planets? What other substances are found in ocean water? Where did they come from, and what important processes are they involved in?

PHYSICAL OCEANOGRAPHY (CHAPTERS 7–10)

How do oceans control the Earth's climate? What becomes of the sunlight that strikes the Earth? What drives the winds and ocean currents? How are the various types of waves created, and how do they travel? Why are there tides, and how are they related to the phases of the Moon? What causes hurricanes, and why are they so damaging to coastal communities? What causes narrow swift surface currents, like the Gulf Stream? How do the deep waters move? What processes are at work in coastal waters, and why are they so sensitive to human activities? What causes the various types of coastal features?

Figure 1.3

The Trifid Nebula in Sagittarius is a typical cloud of interstellar gas and dust that is trillions of kilometers across. The Sun, Earth, and all the other planets and moons of our Solar System were born of a region such as this.

MARINE BIOLOGY (CHAPTERS 11–13)

How do plants produce food? Why are some oceanic regions much more productive than others? How does life in the ocean compare to that on land? What is the spectrum of environments and lifestyles of marine organisms? What are some of the problems they encounter, and how do they overcome them? What are some of the more prominent classes of organisms in the oceans, and what role does each play in the web of life there?

OUR USE OF THE OCEANS (CHAPTERS 14 AND 15)

What types of coastal construction are particularly vulnerable to storm damage, and why? In what ways are we polluting our coastal waters, and what are the dangers associated with that pollution? What role can the oceans play in feeding the growing world population? How might we increase ocean harvests? What are possible sources of energy from the oceans, and how might each be tapped? What other ocean resources are being harvested? How effectively can we regulate the use of oceans for commerce and the extraction of resources?

1.2 PERSPECTIVE IN TIME

The Earth condensed from a cloud of interstellar gas and dust (Figure 1.3) about 4.6 billion years ago, along with the Sun and the rest of our Solar System. Since that time, the Earth and its oceans have evolved through a long and interesting history, virtually none of which was witnessed by humans. We appeared only very recently. With curiosity appropriate to our "newborn" status, we have been vigorously probing this fascinating world.

What we see today is a result of what has happened in the past. To understand our present world, we must probe its history. For very short-term changes, this is an easy task. For example, we are familiar with the disappearance of beach sand in the winter and its reappearance in the summer. Similarly, the slow extension of a river delta can be noticed within a lifetime. With regard to long-term changes, we are looking at a snapshot in time. In that snapshot, we search for clues regarding the evolution and behavior of the Earth and its oceans.

With regard to long-term changes, we are looking at a snapshot in time. In that snapshot, we search for clues regarding the evolution and behavior of the Earth and its oceans.

For example, it is relatively easy to demonstrate that the surface of the solid Earth is changing.[1] Dating a typical rock from the ocean bottom yields an age of a few hundred million years or less. A glance at Figure 1.4 will convince you that this age is only a

[1]The study of the solid Earth, including materials of the surface and those of the interior, is called *geology*. So movement or changes in the solid Earth is referred to as *geological activity*.

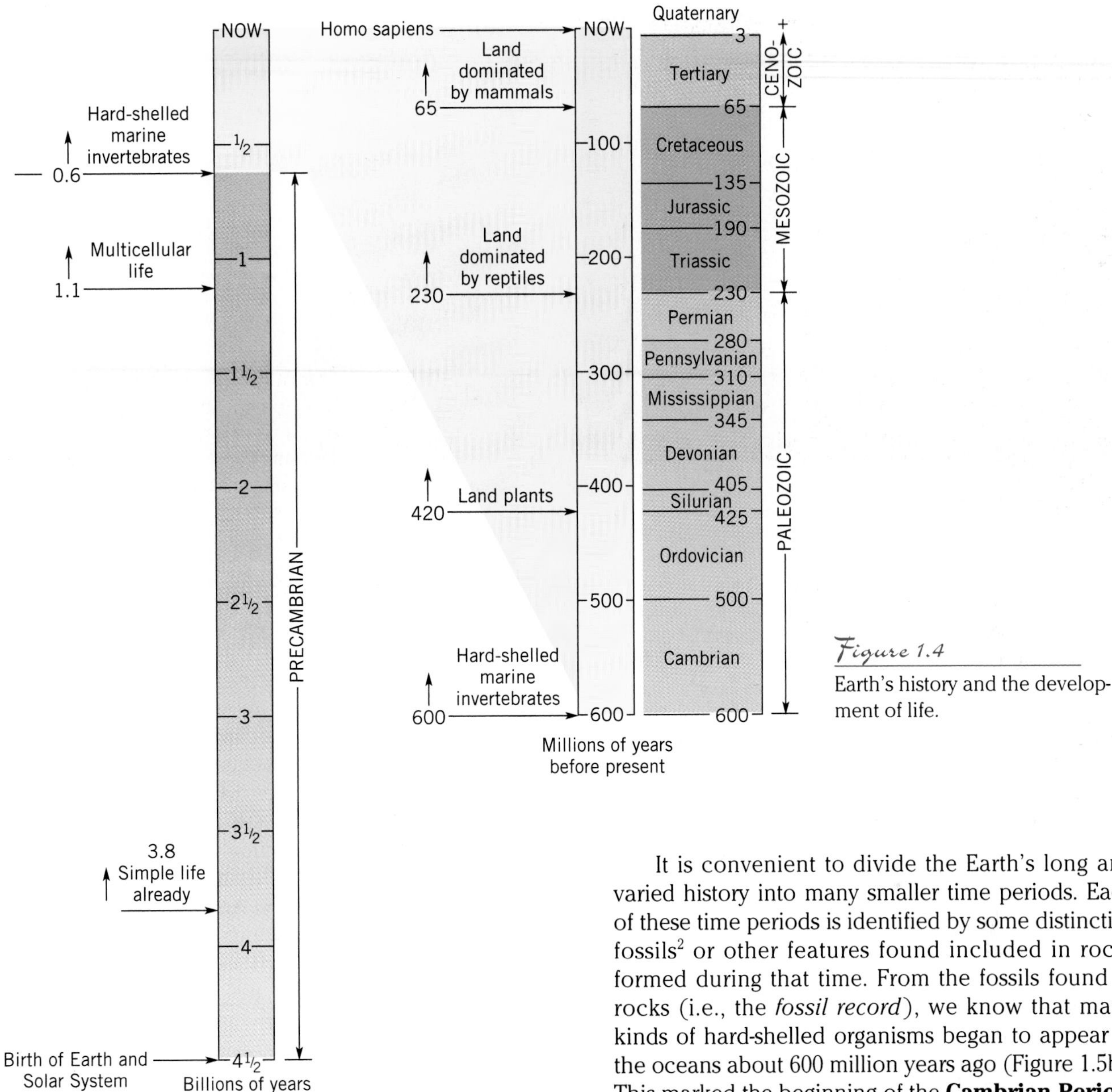

Figure 1.4
Earth's history and the development of life.

few percent of the Earth's lifetime, which suggests the rate at which oceans are being created and destroyed. The Atlantic Ocean, for example, has been around much longer than most species of mammals, but not as long as the reptiles. If the history of the Earth were compressed into 24 hours, the Atlantic Ocean first appeared less than an hour ago.

For the first 90% of the Earth's history, nearly all life was oceanic. Little, if any, was on land. Scientists have found evidence of single-celled marine organisms from 3.8 billion years ago, so life on Earth must be at least as old as that. Blue-green algae (Figure 1.5a) existed as early as 2.7 billion years ago.

It is convenient to divide the Earth's long and varied history into many smaller time periods. Each of these time periods is identified by some distinctive fossils[2] or other features found included in rocks formed during that time. From the fossils found in rocks (i.e., the *fossil record*), we know that many kinds of hard-shelled organisms began to appear in the oceans about 600 million years ago (Figure 1.5b). This marked the beginning of the **Cambrian Period**; the entire 4 billion years preceding it is called the **Precambrian Period**.

As can be seen in Figure 1.4, the identified time periods since the beginning of the Cambrian Period are smaller and more numerous than those earlier in Earth's history. There are two reasons for this finer slicing of recent history. They both have to do with our ability to detect changes, rather than with nature's inclination to produce them. First, more recent history is more easily discovered, because less of its remains have been destroyed or irrecoverably buried by subsequent geological processes. Consequently,

[2]Fossils are evidence of earlier life that is found in rocks. This evidence can take many forms, such as imprints, mineralized skeletons, carbon deposits, footprints, and so on.

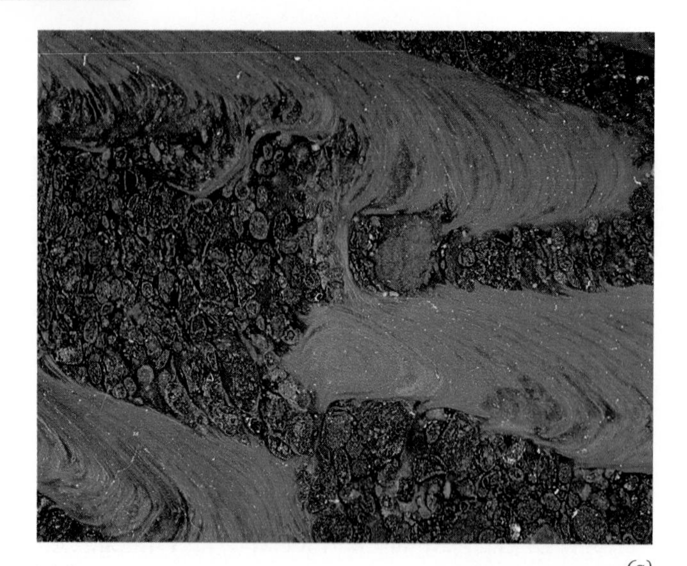

(a)

(b)

Figure 1.5

(a) Precambrian fossil. A rock slab that has been sawn to reveal the red-colored fossilized filamentous blue-green algae ("cyanobacteria"). Blue-green algae first appeared in our oceans nearly 3 billion years ago. (b) Cambrian trilobite fossil. Trilobites had external skeletal plates that provide good fossils. They first appeared during the Cambrian Period, and died out about 200 million years later.

we know recent history in much greater detail, which facilitates the more elaborate classification scheme.

Second, organisms with hard shells or skeletons make better fossils (Figure 1.5b). They preserve better and are easier to identify. We can identify the ages of many rocks by the species of organisms whose fossils are found in them. The scarcity of well-preserved fossils in Precambrian rocks means that more inaccurate and unreliable techniques must be employed to date geological events of these earlier times. For this reason, our understanding of Precambrian events is not as accurate, and dividing the Precambrian into detailed time slots would be inappropriate.

In addition to the appearance of hard skeletons, some more recent developments were also significant. One of these was the appearance of simple land plants about 420 million years ago. Soon thereafter, land animals began to appear. Notice that these were relatively recent events (Figure 1.4). Throughout most of the Earth's history, life occurred almost entirely in the oceans, and the land was barren.

Although reptiles appeared first during the late **Paleozoic Era** (600–230 million years ago), they became the dominant form of land animal during the **Mesozoic Era** (230–65 million years ago). The **Cenozoic Era** (65 million years ago–the present) is marked by the dominance of mammals. Erect, two-legged primates that might be considered precursors to humans first appeared about 5 million years ago, and the earliest stable, organized, agrarian societies seem to have emerged about 6000 years ago.

On a 24-hour scale for the history of Earth, life evolved and remained in the oceans for the first 22 hours. Land plants and animals first appeared about 2 hours ago, dinosaurs ruled about 1 hour ago, the Atlantic Ocean opened as Africa and Eurasia tore away from the Americas about 50 minutes ago, mammals became dominant on land about 20 minutes ago, erect two-legged primates first appeared about 2 minutes ago, and the great Egyptian pyramids were built less than a tenth of a second ago.

> On a 24-hour scale life evolved and remained in the oceans for the first 22 hours. Land plants and animals first appeared about 2 hours ago, dinosaurs ruled about 1 hour ago, the Atlantic Ocean opened about 50 minutes ago

1.3 HISTORY OF OCEANOGRAPHY

Much of the history of humanity has gone unrecorded, and of those portions that were recorded, many records have been lost. After the Middle Ages,

Westerners had to relearn much of what had been available to them centuries earlier. When the Europeans began long voyages of exploration in the late fifteenth and early sixteenth centuries, they found that the "newly discovered" lands had already been discovered and populated ages earlier by people of different cultural backgrounds.

This section presents a brief overview of some of the highlights of oceanographic exploration from the point of view of Western societies. It is done with apologies to those enlightened past cultures whose records have not survived.

Early studies of the oceans were incidental to other more practical and urgent interests, such as harvesting food, expanding trade, or protecting trade and other interests. Of primary concern in these endeavors were the ocean's surface and coastal features, along with wind and weather patterns. Today, along with environmental concerns, these same practical interests provide considerable motivation and financial backing for oceanographic studies. Today, however, most of the research is carried out by scientists. Furthermore, both our interests and abilities have expanded to include all aspects of the oceans.

ANCIENT CIVILIZATIONS

Early humans used the oceans primarily as a source of food. As more stable societies developed, their interests in the oceans expanded to include trade, national security, and more recently, exploration. More advanced societies relied on waterways for the shipment of heavy cargoes, including food, dry goods, building materials, troops, and military hardware. Water was an important lifeline to great ancient civilizations such as the Egyptians, Greeks, and Romans.

The earliest boats were made of materials that have long since decomposed (Figure 1.6). Our knowledge of them comes from such sources as cave drawings and artifacts found in tombs. Reeds, log dugouts, and skins or bark stretched over wooden frames formed the structures of these early boats. Although most were probably used on inland and protected coastal waters, we know that the Micronesians were sailing long distances over the Pacific Ocean thousands of years ago in boats made of these materials. In the West, people were crossing the Straits of Gibraltar at least 10,000 years ago. The bones of deep sea fish on ancient refuse piles indicate that people have been using boats on the ocean to harvest food for thousands of years.

Some of the earliest great exploratory voyages were made by Phoenicians, who lived in the eastern Mediterranean, roughly where Lebanon, Syria, and Israel now stand. Their cultural influence peaked about 1000 BC, but their seafaring exploits began well before that. They colonized many islands and coastal regions throughout the Mediterranean. As early as 2000 BC they were exploring the Red Sea and Indian Ocean, and by 1500 BC they were trading with people in the Persian Gulf area.

Phoenicians also sailed westward beyond the Straits of Gibraltar, reaching Great Britain to the north and the Canary Islands to the south. In 590 BC, Phoenician sailors under Egyptian employ sailed completely around Africa, leaving by way of the Red Sea and returning through the Straits of Gibraltar.

In the following century, Carthaginian sailors explored the West African Coast nearly as far down as the Congo River. In 325 BC, the Greek explorer Pytheas sailed to Great Britain and Iceland, and recognized that tides are related to the phases of the Moon. But for the most part, the Western societies turned their attention toward the East, where trading became a thriving business. Eastern spices, silks, and gems were especially prized during Roman times.

EARLY NAVIGATION

For two millennia, people have been using **latitude** to describe north–south positions on Earth. Latitude is measured in degrees north or south of the equator, with the equator being 0° latitude, and the poles being 90° north or south latitude, respectively (Figure 1.7). For the past three centuries, a similar measure, called **longitude**, has been used for determining east–west positions on Earth. Longitude is measured in degrees east or west of an imaginary reference line, called the *prime meridian*, which runs from the North Pole to the South Pole through the Royal Observatory in Greenwich, England. The Eastern and Western Hemispheres are those to the east and west of this meridian, respectively, just as the Northern and Southern Hemispheres are north and south of the equator, respectively.

Early sailors usually stayed within sight of land, because otherwise, it was difficult to determine the ship's position. In the third century BC, Pytheas demonstrated how one's latitude could be determined by the elevation of the North Star above the horizon (Figure 1.8).[3] But because of the Earth's spin, stars cannot be used to determine one's longitude without a good clock. Stars are continually moving from east to west across our sky. Therefore, the east-west position of a star in the sky depends on not

[3]One reason that early European sailors were reluctant to sail south of the equator was the disappearance of the North Star below the horizon, making it impossible to use this method to determine latitude.

Figure 1.6
A few of the boats and ships used by various ancient civilizations.

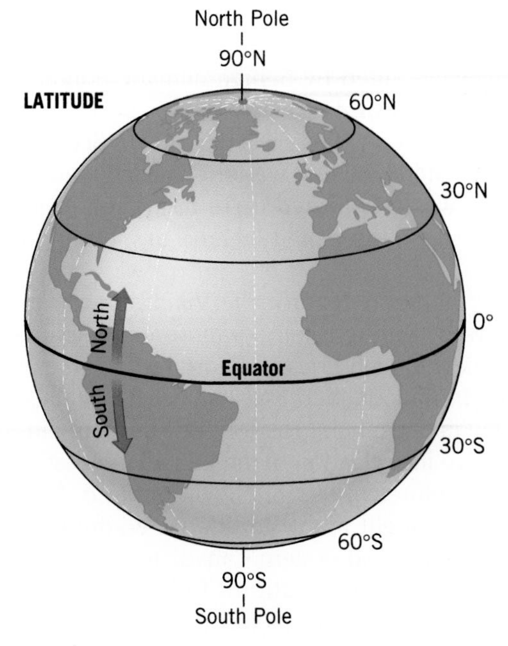

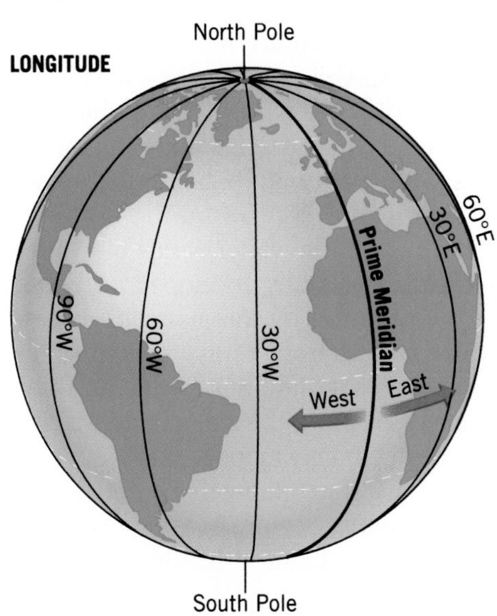

Figure 1.7

Latitude is measured in degrees north or south of the equator, with the equator being 0°, and the poles being at 90° North and South, respectively. Longitude is measured in degrees east or west of the prime meridian, which runs from the North Pole to the South Pole through the Royal Observatory in Greenwich, England. (Where do you suppose the east and west meet on the other side of the globe?)

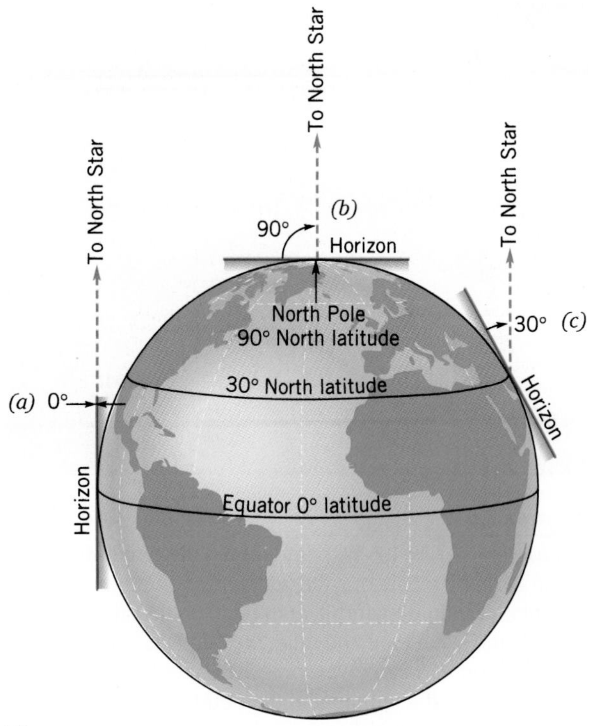

Figure 1.8

As Pytheas demonstrated, a person's latitude is equal to the angle of the North Star above the northward horizon. Note in particular that (a) on the equator (0° latitude), it is on the horizon (0° above the horizon), (b) at the North Pole (90° N latitude), it is directly overhead (90° above the horizon), and (c) at 30° N latitude, it is 30° above the horizon.

only where you are (e.g., Canary Islands, Boston, or Seattle), but also the time of day. It wasn't until the 1770s that clocks were developed to the point that they could remain accurate on extended voyages, so it wasn't until this time that longitudinal positions were accurately known. You will notice longitudinal distortions on maps that were drawn before 1770, but not after that.

THE GREEKS AND ROMANS

An Alexandrian librarian, Eratosthenes (276–195 BC) made a remarkably simple, elegant, and accurate measurement of the circumference of the Earth (Figure 1.9). This measurement, combined with the writings of earlier Greeks, demonstrated a remarkable knowledge of the world and its relationships to the Sun and Moon. The map of Figure 1.10a illustrates Greek knowledge of Earth geography at that time.

The Romans were aware that the Earth was active geologically. Strabo's (63 BC–24 AD) observations of volcanism, erosion, sediment transport by rivers, and marine sediments on land, led him to appreciate the geological dynamics of land masses. He also recorded data on tides and made measurements of ocean depths down to 2 kilometers.

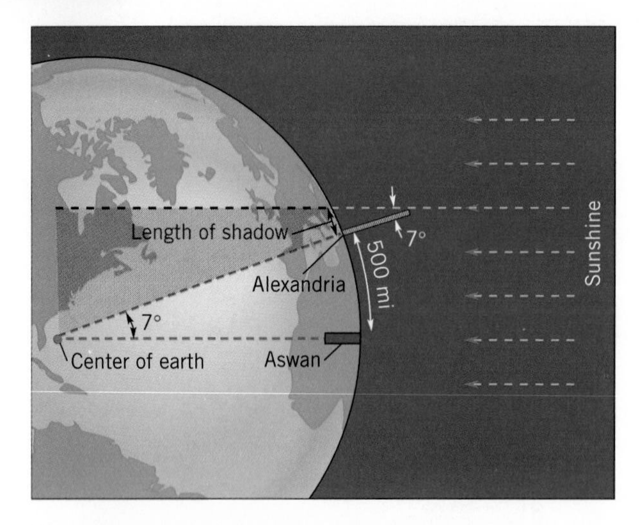

Figure 1.9

Eratosthenes (276–195 BC) determined the circumference of the Earth by measuring the length of the shadow cast by a vertical wall in Alexandria at midday on the summer solstice. At the same date and time, the Sun was known to be directly overhead in a town some 500 miles to the south, since the sunshine illuminated the bottom of vertical wells there. In Alexandria, the Sun was 7° south of being overhead, which meant that the 500-mile distance between the two amounted to 7/360 of the Earth's circumference.

Both Aristotle (384–322 BC) and Seneca (54 BC–39 AD) observed that the flow of rivers does not make the oceans overflow, and evaporation does not make them go dry. The water is involved in a never-ending **hydrologic cycle**,[4] evaporating from the oceans, and returning as precipitation. Some of the water that falls on land returns to the oceans via rivers and streams, and some is evaporated again, continuing the cycle.

A map by Ptolemy of about 150 AD (Figure 1.10b) summarizes geographical knowledge in Roman times. Although later Arabic maps display similar precision, knowledge in the West deteriorated greatly after that time (Figure 1.10c).

THE DARK AND MIDDLE AGES

In Europe, the periods from about 500 to 900 and 900 to 1200 are referred to as the **Dark Ages** and **Middle Ages**, respectively. The Dark Ages were characterized by lack of inquisitiveness and deterioration of knowledge in many fields. Europeans did develop larger, more seaworthy ships during this time, which became important when ocean explorations resumed in the 1400s.

[4]The prefix "hydro" comes from Greek and means "water."

In other parts of the world, inquisitive attitudes were alive and exploration continued. Northern Europeans were aided by benign climatic conditions in their voyages of discovery. In the seventh century, Iceland was visited by Picts of northern Scotland and by Celts of Ireland and western England, and a Celtic colony was established there in the following century. Irish priests visited Iceland in the period after 750.

In the tenth century, the Vikings began moving westward. During a temporary forced expatriation, an outlaw named Eric the Red sailed as far as Baffin Island in Canada, and 3 years later he established a colony in Greenland. His son, Leif Ericson, did more extensive investigation of the Canadian coast, spending the winter of 995 in Newfoundland. The next decade brought many additional Viking explorers into northern and eastern Canadian waters. But as a result of the deterioration of climatic conditions around 1200, the Vikings lost touch with these western lands.

Meanwhile, the Arabs were actively sailing the Red Sea, Persian Gulf, and Indian Ocean, primarily for purposes of trade with Africa, India, Southeast Asia, and China. They took advantage of the seasonal monsoon wind patterns, to sail eastward with the summer winds blowing toward the northeast, and sail westward with the winter winds blowing toward the southwest. The Arabs developed accurate instruments for navigation by the stars and brought the magnetic compass from China. It is an interesting quirk of history that although the Indian Ocean was the first major ocean to be sailed extensively, it was the last to be accurately mapped.

Although we don't know many details, we know that some extremely impressive navigational feats were being accomplished in the Far East. In the Western Pacific Ocean,[5] Micronesians and Polynesians were sailing large distances between small islands in small and primitive double-hulled sailing canoes, sometimes spending many days without sight of land. The navigational accuracy needed to sail such large distance between such tiny islands is impressive, as is the courage needed to undertake such voyages.

The Chinese sailed extensive areas of the Indian and Pacific Oceans. At least some of the time, the motivation was for communication with neighbors, rather than for exploitation or expansion. Their ships were very large by European standards, some carry-

[5]Remember that the Eastern Hemisphere is the half of the world extending eastward from the Prime Meridian. This includes the Western Pacific Ocean.

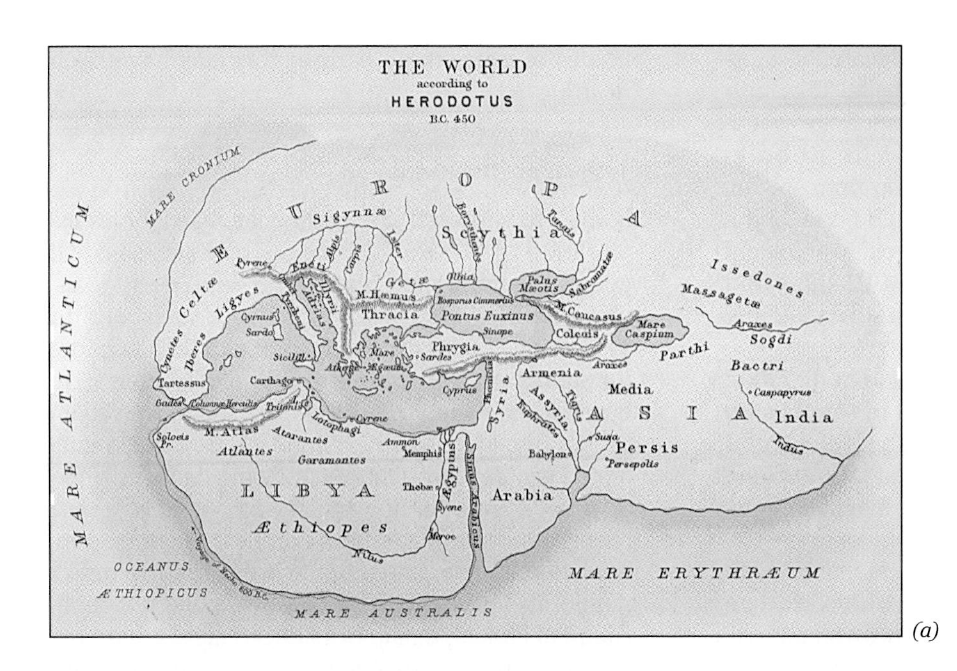

(a)

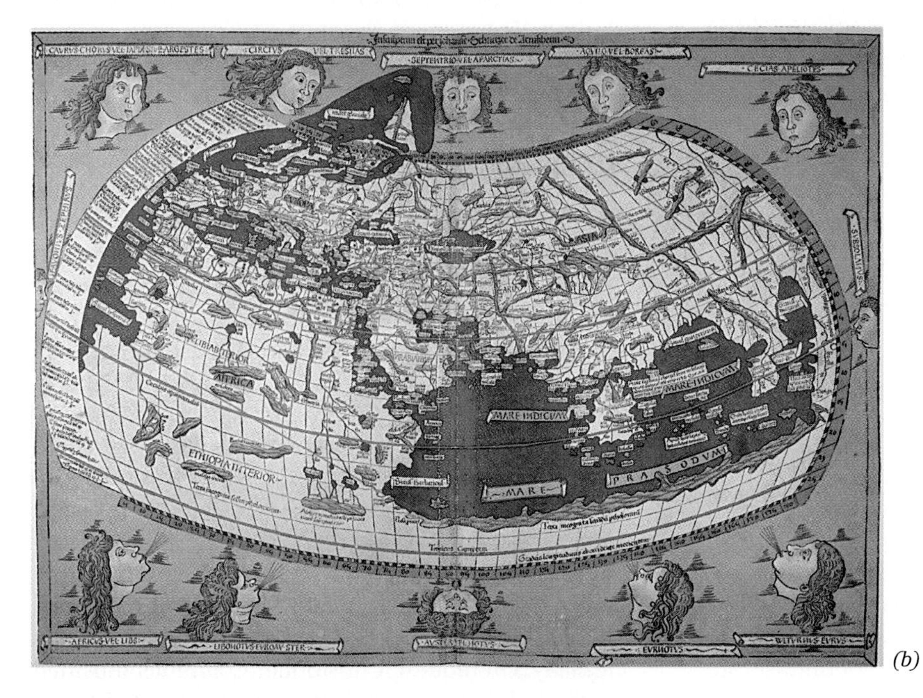

(b)

(c)

Figure 1.10

Early western knowledge of world geography. (a)The state of Greek knowledge of the world around 450 BC is represented in this map by Herodotus. By 250 BC, England and Ireland (identified as Bretannica and Hierne) were added. (b) The world according to Ptolemy, approximately 140 AD (Taken from Ptolemy's Geography, and published in Ulm in 1482.) (c) A 6th century European map of the world. Note the considerable loss of geographical information compared to maps of earlier times.

ing hundreds of sailors each. They developed magnetic compasses and navigational charts. They also divided the large ships into several separate watertight compartments with vertical walls, or **bulkheads**. This innovation prevents a ship from sinking even if one of the compartments should flood, and it is used in most modern ships. One relatively recent Chinese voyage began in 1405, used 62 ships and 27,800 sailors, and sailed as far as East Africa and the Arabian Peninsula.

During the **Renaissance** that began in the1200s, Europeans became more interested in scholarship and the arts, and more inquisitive about the world around them. Trade routes developed and expanded. Returning Crusaders brought the magnetic compass with them in the 1200s. Its usefulness in navigation outweighed superstitious fears associated with magical stones and needles. The oldest known Western navigational chart that includes compass directions is dated 1275.

Because early sailors remained mostly within sight of land, little or nothing was known of vast unexplored regions of the world's major oceans. From a sailor's perspective, there was little difference between sailing along the coast of the Mediterranean Sea and that of the Atlantic Ocean, for example. There was little or no appreciation of the size difference between the huge oceans, and the smaller seas.

Since the late 1700's, however, the major oceans have been accurately mapped, and we have an appreciation of their true sizes. We now use the word "ocean" to describe the major bodies of water that separate continents and cover much more of the Earth's area than the continents. We divide the world's oceans into three: the Atlantic, Pacific, and Indian. Although all three are connected at their southern ends, we run an imaginary line southward from the Southern tip of South America to separate the Atlantic and Pacific, and a similar imaginary line southward from the southern tip of Africa to separate the Indian from the Atlantic. **Seas**, by contrast, are much smaller bodies of water, smaller than continents, and are to a large extent separated from the major oceans by land forms. Examples would include the Mediterranean Sea, Baltic Sea, Gulf of Mexico, and Caribbean Sea.

The language used by early sailors is still embedded in some of the vocabulary we use today. Many words such as "seafloor," "sea water," "sea breeze," "seafaring," and so on should really have the prefix "ocean," rather than "sea." But sailors developed this jargon long before the oceans were explored and the difference between oceans and seas was known. You may have heard the expression "sailing the seven seas" and may wish to speculate which seven bodies of water these early sailors may have been referring to.

THE AGE OF DISCOVERY

In the 1400s, several events stimulated a vigorous exploration of the world's oceans by Europeans. Of course, many of their discoveries were really rediscoveries of places known to earlier civilizations. In 1420, Prince Henry the Navigator established a school for sailors at Sagres, on the Portuguese coast. He hired skilled Italian navigators and map makers to instruct Portuguese sailors, who previously were reluctant to sail beyond sight of land. Simultaneously, the Spaniards were forcing the Moors out of southern Spain and capturing Islamic libraries, which held information long forgotten in the West. The expansion of information in the early 1400s included the publication of Ptolemy's map of 150 AD (Figure 1.10b).

Motivation for exploration was increased when Sultan Mohammed II captured Constantinople in 1453. With this, the Turks had cut off the last of the overland trade routes between West and East, thereby providing economic incentive for the search for alternate routes. Greeks expelled from Constantinople at this time brought with them a knowledge of the world that had long been forgotten in the West.

The Portuguese discovered the Canary Islands in 1416 and the Azores around 1430. Students of Prince Henry's school sailed further south. Bartholomew Diaz rounded the Cape of Good Hope in a voyage of 1487 to 1488, and Vasco da Gama sailed all the way to India in 1498.

Meanwhile, some famous voyages were made under the Spanish flag. In 1474, a Florentine astronomer named Toscanelli wrote to the King of Portugal, suggesting a route to the East Indies by sailing west around the globe. He attached a map that greatly underestimated the distance, placing the East Indies just slightly west of where the unknown Americas lay. On request, Christopher Columbus obtained a copy of that letter and persuaded Spain to finance the suggested voyage. He left the Canary Islands with 88 men and 3 ships on August 3, 1492, and first arrived at a small island in the Bahamas on October 12.

In 1497, an Englishman, John Cabot, sailed a northern route to the northeastern coast of what is now the United States. From 1500 to 1520, there were numerous Spanish and Portuguese expeditions to the "New World," exploring most of the Atlantic Coast of South America, Central America, and up through Florida. Balboa was the first European to see the Pacific Ocean, when in 1513 he spotted it from a mountaintop in the Isthmus of Panama. In these times, many exploratory voyages were being financed by wealthy individuals with hopes of new trade opportunities and financial gain. Increased knowledge of the world and its oceans was a byproduct.

This exploratory period culminated in the historic circumnavigation of the globe by the **Magellan Expedition** in the years 1519 to 1522. It started from the southern tip of Spain with five ships and 230 sailors, sailing southward at first, through a passage that we now call the Straits of Magellan at the southern tip of South America, and then westward across the Pacific and Indian Oceans, rounding the southern tip of Africa, and then sailing northward back to Spain. The expedition endured many hardships along the way. While in the Straits of Magellan, the sailors on one ship mutinied and escaped with their ship. Other ships had to be cannibalized to keep the remaining ships going, and when the expedition arrived back at Seville on September 8, 1522, only one ship and 18 bedraggled sailors remained. Magellan was not among them, having been killed while participating in a dispute among tribes in the Philippines.

In those days, the life expectancy of a sailor was very short, mainly due to what we now recognize as dietary problems. The lack of fresh fruit and vegetables on ship led to a deficiency of certain needed vitamins. For the Magellan expedition, the trip north and west across the Pacific turned out to be exceptionally slow and tedious, with considerable cost to the health and lives of the sailors. The trip from the East Indies home should have been fairly routine, due to the pioneering work of Portuguese sailors in the previous decades. However, the remaining ship

(a)

Figure 1.11

(a) Engraving showing Drake's ship on the right at Callao, (Peru). The Golden Hind was a relatively small and fast ship, about 23 meters long and carrying 18 cannons. (b) The route Drake took around the world in 1577–1580.

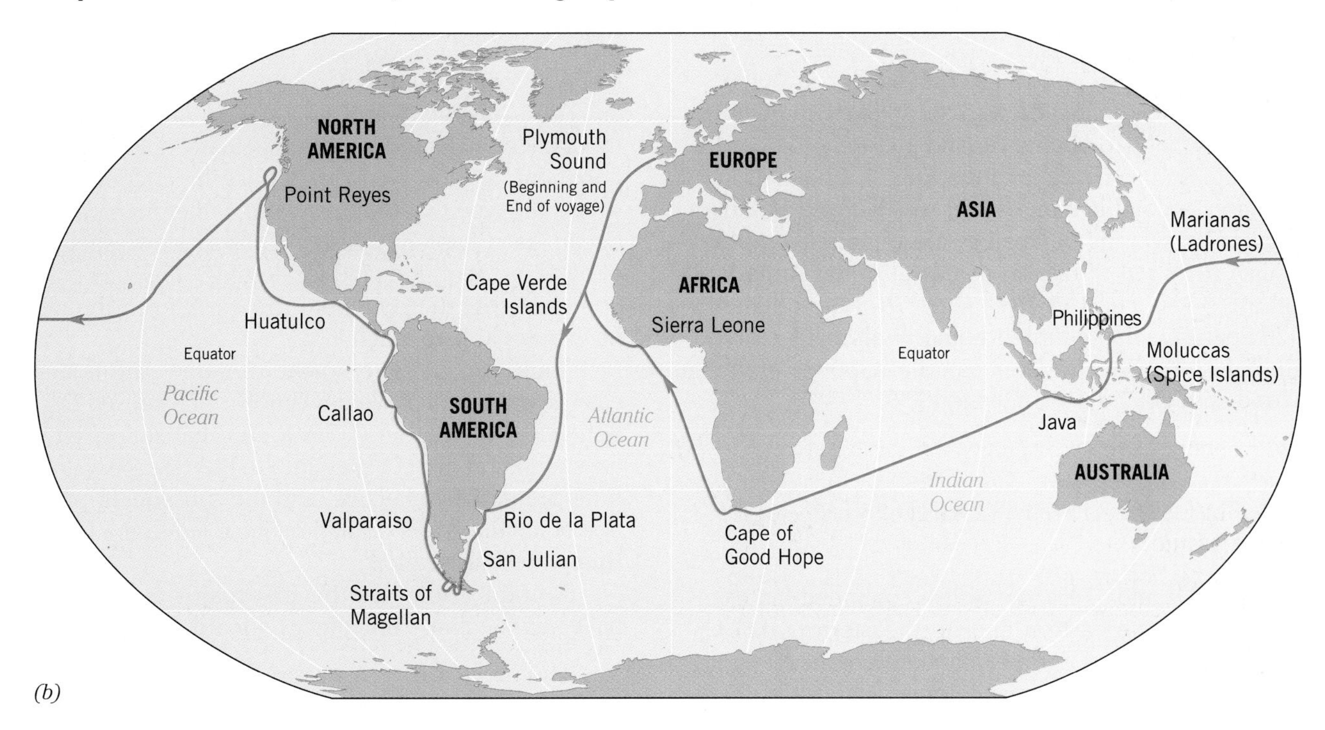

(b)

was hardly seaworthy, the remaining crew was sickly, and they had to sail a route that would keep them out of sight of Portuguese ships. Portugal jealously guarded its monopoly on the southern route to the East Indies.

The 1500s marked fierce national rivalries in the exploitative use of the oceans. Shipboard cannons were installed routinely. The Vatican had officially divided the world's oceans between Portugal and Spain, which didn't please other maritime nations, such as England and Holland. It even caused troubles for Portugal and Spain when one wanted to colonize or establish trade routes in the other's waters. In the New World the Portuguese and Spanish plundered the Aztecs and Incas, and the British and Dutch plundered the Portuguese and Spanish.

England's Queen Elizabeth I (1533–1603) encouraged both exploration and raiding of Spanish ships. One of her favorite cavalier and daring "sea dogs," Sir Francis Drake (1540–1596), circumnavigated the world in 1577 to 1580 (Figure 1.11). Drake's journey began as a voyage to raid Portuguese ships and Spanish settlements on both coasts of South America, but he continued westward across the Pacific rather than risk reprisal by the Spanish on a return trip. King Philip II of Spain demanded that he be punished; instead, Queen Elizabeth knighted him. Infuriated, King Philip initiated the formation of a large fleet of warships, called the **Spanish Armada**, with which to raid and punish England. In 1587, while the fleet was still being assembled, Drake made a preemptive surprise attack on it in the Spanish port of Cadiz and reported sinking about 30 ships.

Nevertheless, the Armada was completed and headed for the British Isles, but it was beset by further misfortune. Due in part to obstruction by the Dutch, it never made connection with the soldiers and fighting crews that had come overland and were to board the ships somewhere along the English Channel. In addition, a combination of foul weather and skillful attacks by swarms of smaller, more maneuverable British ships led to disastrous defeat of the Armada off the British coast in 1588. Only 67 of the original 130 Spanish ships returned home. With this victory, the British became the dominant naval power in the world and remained so until the 1900s.

THE SEVENTEENTH AND EIGHTEENTH CENTURIES

In the 1600s and 1700s, the British continued the exploration of the eastern and northeastern parts of North America, and European nations colonized and set up trade with their new colonies. Interest in the sciences increased, and societies formed to fund scientific research and publication. The motivation for oceanographic exploration was changing. There was a need for some systematic studies of the oceans to improve speed and safety along trade routes.

In 1700, W. Dampier published *A Discourse of the Winds*, a compilation of information intended to aid ocean travel. In the 1770s and 1780s, Benjamin Franklin mapped and published various charts of the **Gulf Stream** (Figure 1.12), using both his knowledge of travel times of mail ships from his days as Postmaster General for the Colonies, and the experiences of his cousin, a Nantucket whaler. Because the Gulf Stream flows eastward toward Europe, the sea captains wanted to stay within the Gulf Stream while sailing east and avoid it while sailing west. Because the Gulf Stream waters are warm, sea captains employed thermometers to locate it.

The first recorded voyages with primarily scientific objectives were those under Captain James Cook of the British Royal Navy in the period 1768 to 1779. He had previously been commissioned to map the area of the St. Lawrence River and the Newfoundland Coast. From this work, he gained a reputation for being skillful in mapping, mathematics, and astronomy. At the request of the Royal Society (a private organization that supported what it considered to be worthy projects), the Navy sent a scientific expedition to Tahiti in 1768 to observe the transit of the planet Venus.[6] Cook was chosen to captain that voyage, and he was sent back to the Pacific twice more before being killed by Hawaiians in 1779 in a skirmish over a stolen boat. The Pacific was largely uncharted before that time, but Cook changed that.

Cook is credited with many major accomplishments during these voyages. He established the existence of a great southern continent, then referred to as "Terra Australis," and now called Antarctica. Although ice prevented him from actually reaching this continent, he observed that the ice he encountered was of the type that forms on land, rather than on the ocean (see Chapter 7). Also, he observed two species of birds that required land for nesting. Thus, he could conclude that the continent did exist, even though he was not able to reach it.

Cook also conquered the dreaded sailors disease, **scurvy**, which we now know to be caused by a deficiency of vitamin C. He found that sailors who ate sauerkraut did not succumb to this disease. Citrus fruits are equally effective, and the subsequent practice of having crews drink lime juice led to the name **limey** for British sailors.

Captain Cook was the first major explorer to have very accurate timepieces aboard ship. This en-

[6]That is, they watched it as it crossed in front of the Sun.

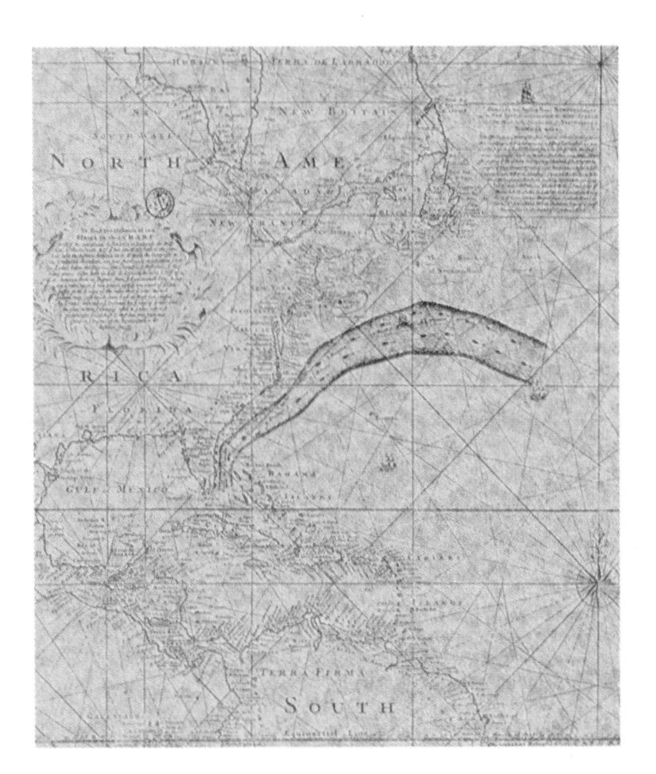

Figure 1.12

Benjamin Franklin's chart of the Gulf Stream, printed ca. 1869-1870.

abled him to determine longitude accurately by timing the rising, setting, or overhead passage of the Sun and stars. The maps he made were excellent. By the end of his voyages, the general geography of the world was fairly well known, except for Antarctica and other regions beyond 70° North or South. From then on, the arrangement of the continents, coastlines, and major islands on maps has remained pretty much the same as they are today.

THE EARLY NINETEENTH CENTURY

In the 1800s, there was considerable incentive for more scientific ventures into the oceans. Concerns for trade, fisheries, passenger safety, bulk transport, transoceanic cables, and national defense created a demand for organized data on winds, currents, seasonal weather patterns, depth measurements, ocean floor environments, marine organisms, ice hazards, and so on. Much of the research was sponsored by governments, but some was initiated by scientific societies, wealthy individuals, and private businesses.

Until the 1800s, no one was able to make reliable measurements of the oceans' depths, except in shallow coastal waters. For millennia, these depth measurements, called **soundings**, had been made by tying a weight to the end of a rope, lowering it into the water, and measuring how much rope must be let out before the weight strikes the bottom. This process works well in shallow water. But unfortunately, much of the ocean is so deep and the rope is so long and heavy that it is impossible to feel when the weight strikes the bottom. Sir John Ross overcame these problems and began making reliable soundings of less than 2 kilometers in Baffin Bay in 1817 to 1818. In the late 1840s, his nephew, Sir James Clark Ross, made soundings at more typical deep ocean depths (4–5 kilometers) in the Southern Atlantic. He also found organisms on the ocean floor that were identical to those his uncle had found in the Northern Atlantic. He reasoned that because these organisms were very sensitive to changes in environments, the entire deep ocean bottom must have water of similar characteristics. Otherwise, the migration of these species would not have been possible.

The need for more accurate sailing and navigational information was addressed by an American naval officer, Matthew Fontaine Maury (1806–1873). Having been crippled by a stagecoach accident early in his career, he was assigned to the Depot of Naval Charts and Instruments in Washington, D. C., where he systematically analyzed the log books and other records of thousands of voyages. Based on this information, he published charts and sailing instructions, detailing patterns of wind, currents, and waves throughout the well-traveled oceans. He organized the International Meteorological Conference in Brussels in 1853, where conventions were agreed on for keeping records of the state of the air and sea during all oceanic voyages. With some modifications, these procedures are still used today. In 1855, he published what is considered the first textbook in oceanography, entitled *The Physical Geography of the Sea*.

Erratic catches of European fisheries in the mid–1800s stimulated interest in studying the ocean's biological resources. This interest has continued since that time and spread to all oceans, as the various national fisheries have expanded and erratic catches have continued to be a problem.

Christian Gottfried Ehrenberg's (1795–1876) finding that some sedimentary deposits were made of myriads of skeletons of microscopic organisms, called **plankton**, established that these tiny creatures could be very important in the ecology of the oceans. Later, researchers found that the size of the catch of some species could be traced back to their survival as youth, when they fed on the microscopic plankton. Therefore, the planktonic community in one year influences the fish harvest a few years later.

During the early nineteenth century, considerable interest developed in the question of life on the deep ocean floor. A noted Scottish biologist, Edward Forbes (1815–1854), observed that the abundance of

(a)

(b)

Figure 1.13

(a) The *H.M.S. Challenger* carried a research staff of 6 scientists on a voyage that lasted from December, 1872 to May, 1876. (b) The *Fram* left Oslo in June, 1893, froze in the ice North of Siberia in September of that year, and remained icebound for 3 years, drifting with the ice at an average speed of about 2 kilometers per day.

life decreased with depth in the oceans. He conjectured that since the entire biological community is ultimately dependent on the plants which live in the upper sunlit surface waters only, then probably no life exists below a depth of about 600 meters. This viewpoint had significant following, even though Sir John Ross had pulled up worms and a brittle star from 1800 meters in the North Atlantic many years earlier. The issue was settled in the 1860s when transoceanic cables brought up from the ocean floor for repair had a variety of living organisms on them.

In 1831, the *H.M.S.* ***Beagle*** (Chapter opening photo) set sail for a 5-year voyage around the world that included a stretch up the western coast of South America as far as the Galapagos Islands. On board was a young naturalist named Charles Darwin (1809–1882). His observations of such things as basic similarities in mammal skeletons and specialized adaptations formed the basis of his ideas on evolution and natural selection. These were published in his famous *Origin of the Species* in 1859. He also postulated what has proven to be the correct explanation for the formation of atoll reefs (Chapter 4).

THE LATER NINETEENTH CENTURY

The study of ocean chemistry was advanced by Georg Forchhammer, a Dane who analyzed sea water samples from a wide range of ocean areas. In 1865, he confirmed that the major salts are always present in the same ratios. Only the relative amount of freshwater in the mixture changes. This **rule of constant proportions** was further confirmed by samples taken on the extensive *Challenger* Expedition described below.

In the United States, interest in oceanographic studies increased after the Civil War. A rather comprehensive study of the U.S. coastal waters was undertaken by the U.S. Coast and Geodetic Survey (now the National Ocean Survey). The Fish Commission (later the National Marine Fisheries Service) was established in 1871, to help strengthen the national fishery industry. In 1874 to 1875, Americans were making systematic measurements in the Pacific Ocean aboard the *Tuscarora* in preparation for laying a transpacific cable to Japan. Among other things, the greatest depth yet known was measured in the Kurile Trench just east of Japan. From 1877 to 1880, an American expedition studied the Caribbean region aboard the ship *Blake*.

One of the most ambitious voyages of all times was made by the British ship, *H.M.S.* ***Challenger***, in the years 1872 to 1876. This 2300-ton corvette[7] was roughly one-fifth the dimensions of modern tankers and similar in size to several modern research ships (Figure 1.13). It was a relatively small, fast naval vessel that could sail under either wind or steam. Its cannons were removed and it was refitted for its scientific mission. It had a crew of 243 and a scientific party of 6.

The voyage was sponsored by the British Admiralty and the Royal Society of London. Its assignment was to probe all physical, chemical, and biological aspects of the oceans. It traveled 109,000 kilome-

[7]A corvette was a kind of ship that had both sails and an engine.

ters in $3\frac{1}{2}$ years through the Atlantic, Pacific, and "Southern" Oceans. The data gathered took an additional 19 years for scientists to organize and edit, and eventually made up 50 large volumes—29,500 pages in all. Significant findings included 4717 new species of marine organisms and a new record depth of 8.2 kilometers in the Mariana Trench.

The successful *Challenger* Expedition stimulated international interest and resulted in expanded oceanographic exploration throughout the world that continues to this day. Some of the more interesting or noteworthy of those that followed the *Challenger* are described next.

During the years 1893 to 1896, a Norwegian crew of 13 under the command of Fridtjof Nansen explored the Arctic Ocean aboard the ***Fram*** (Figure 1.13b), a three-masted schooner[8] with a 1.3-m-thick reinforced hull, designed so that ice could not crush it. The voyage was inspired partly by the discovery of wreckage near Greenland from an American ship that had been crushed by ice north of Siberia some years earlier. This suggested currents and ice flow from the Pacific side toward the Atlantic side of the Arctic Ocean.

After eventually securing the financial backing for their ship's construction, Nansen and his crew sailed it into the Arctic Ocean, entering from the Pacific Ocean through the Bering Straits. They allowed it to become frozen in the ice north of Siberia, so that they could study the motion of the Arctic ice pack. Their slow journey with the ice took 3 years, and they did not pass directly over the North Pole as Nansen had hoped. Consequently, Nansen and a companion set out for that destination by dog sled. Unfortunately, they did not quite succeed in reaching the North Pole and suffered some hardships on the way back, including missing their ship and having to spend the winter on one of the Franz Joseph Islands. From there, they eventually were picked up by a British ship, which returned them home 1 week before their ship, the *Fram*, arrived.

The voyage of the *Fram* established that there was no continent at the North Pole, and soundings from the *Fram* showed that the Arctic Ocean was deep like the oceans. Deep water samples were retrieved with a *Nansen bottle*, a device that was used for many subsequent decades of oceanographic study. Careful records revealed that the ice had drifted in a direction to the right of the wind, which was later explained by V. W. Ekman as being caused by the Earth's rotation.

[8] A schooner is a kind of sailing ship that has the familiar triangular fore and aft sails on at least the rear mast, and perhaps on the other masts as well.

THE TWENTIETH CENTURY

In the late 1800s and early 1900s, a significant portion of oceanographic research was supported by private individuals and foundations, such as the Carnegie Institution, the Rockefeller Foundation, and the Royal Society. The first two major oceanographic research centers in the United States began as private institutions, although both are now associated with major universities and receive most of their funding from state and federal sources. These are the Scripps Institution of Oceanography on the West Coast, started in 1903, and the Woods Hole Oceanographic Institute on the East Coast, started in 1930.

During World War II, federal funding for oceanographic research increased significantly in the United States. At first, this research was directed toward immediate practical applications, such as currents, bottom topography, harbors, the transportation of troops and materials, troop landings, and locating submarines. After the war, the research interests broadened, and the federal funding tended to return to the universities.

Although many of these universities also receive funding from state and private sources, the major support for oceanographic research in the United States still comes from federal programs. These include the U.S. Navy Oceanographic Office, Office of Naval Research, U.S. Coast Guard, National Science Foundation, and National Oceanic and Atmospheric Administration (NOAA). Major divisions within NOAA include the National Sea Grant Program, National Marine Fisheries Service, and the National Ocean Survey.

During the most recent three decades, there has been an increase in large-scale research projects that involve collaboration among many universities and funding agencies, and even among nations. For example, a coordinated international effort produced maps of the ocean floor during the International Geophysical Year (1957–1958). Similar international cooperation involved research into currents, food, fuel, and minerals during the International Decade of Oceanography of the 1970s. Such cooperation has continued more recently in large deep sea drilling programs, which involve major equipment expenses and extended periods at sea.

Some particularly interesting oceanographic voyages early in this century included that of the *USS Stewart* in 1922 and the German ship *Meteor* in 1925 to 1927. These ships were the first to use **echo sounding** equipment (Figure 1.14), which allows continuous mapping of the ocean floor while the ship is moving. The depth is determined by the length of time it takes a sound signal sent from the ship to reach the ocean floor and return. The *Me-*

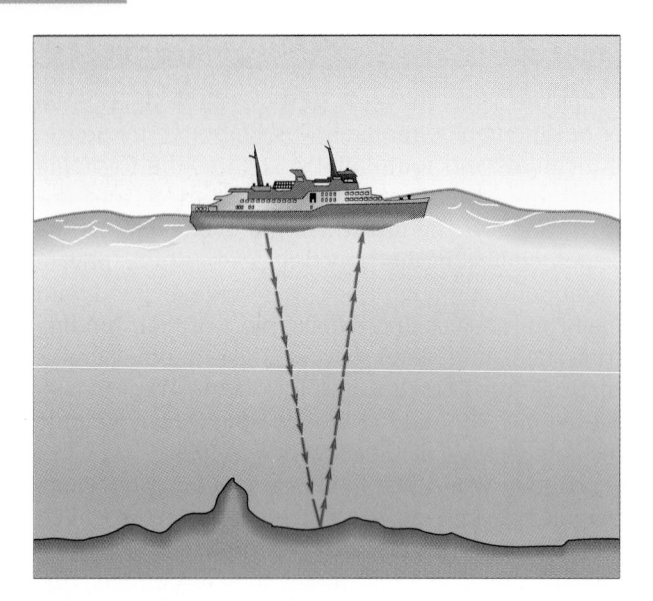

Figure 1.14
Illustration of how echoes are used to map the ocean floor. Echoes may return not only from the ocean floor, but also from schools of fish above the floor, and from layers of sediments and rocks beneath the ocean floor, which give us even more information.

teor's mapping of the Atlantic Ocean bottom displayed how rugged the ocean floor can be.

Another interesting research ship was used in the Deep Sea Drilling Project from 1968 to 1983. The *Glomar Challenger* (Figure 1.15) was a specially outfitted drill ship that could operate in water up to 6 kilometers deep, retrieving sediment and rock samples, called *cores*, from drill holes that penetrated as far as 1.6 kilometers into the ocean floor. Despite wind and waves, this ship could maintain steady position above the drill hole, retrieve and replace the drill bit, and then relocate the drill hole when the drill string was lowered again. This is a remarkable technological feat for a ship tossing in the surface several kilometers above that small hole. The *Glomar Challenger* made hundreds of drill holes, and the layers of ancient sediments and bedrock it retrieved have revealed valuable information about the ocean's history, movement of the ocean floor, past climates, and biological organisms.

The *Glomar Challenger* was retired and replaced by the better-equipped and slightly larger *JOIDES Resolution*, which requires an operating crew of 65 persons and can accommodate up to 50 researchers as well. A program called the Ocean Drilling Project is run by Texas A & M University on behalf of a collaboration of 10 U.S. universities (called JOIDES). Several foreign countries also contribute both researchers and funds to the project.

(b)

Figure 1.15
The drilling ship, *Glomar Challenger (a)*, and its replacement, *JOIDES Resolution (b)*.

In recent years, interest in coastal research has been increasing. One reason for this is the tightening of budgets, which encourages the use of smaller, more fuel-efficient boats in coastal waters, rather than the larger oceangoing research ships that are more expensive to operate. But there are other reasons as well. Water covering the submerged edges of continents, and waters of the marginal seas, such as the Baltic Sea, Mediterranean Sea, Gulf of Mexico, and Caribbean Sea, constitute 21% of total ocean area. Because of the proximity and accessibility to coastal populations, increasing attention is being directed toward the extraction of food, oil, and minerals from these coastal waters, and the problems of coastal construction and pollution. Another incentive for local coastal studies is that increased international awareness, jealousy, mistrust, and the lack of a uniform law of the sea have made it increasingly difficult to do extensive research in large areas of the oceans without provoking protests from neighboring coastal nations.

1.4 TOOLS OF OCEANOGRAPHY

Throughout the history of oceanographic exploration, the spectrum of tools used by oceanographers has been as broad as their imaginations. Part of the fun of working in this field is creating devices that can record the data that are wanted. Here we will consider only briefly some of the tools now in use.

PLATFORMS

For centuries, ships have been the main platform from which oceanographic measurements are made. A typical large modern research ship is roughly 60 meters long, displaces 2000 tons,[9] requires a crew of 20 to 25, and can accommodate a scientific party of 20 to 25 (Figure 1.16a). It costs about 2.5 million dollars per year to operate such a ship, which gives research institutions incentive to use it efficiently. Because of the large expense, multipurpose voyages and research collaborations among several institutions are the rule.

Water and sediment samples can be collected from any depth by lowering the appropriate equipment from surface ships. But when more precision or detail is needed, it is sometimes more appropriate to use **submersibles**, which are crafts that can dive and operate beneath the surface (Figure 1.16b). Some submersibles are manned, but for many purposes, unmanned remote-operated vehicles (**ROVs**) are preferred. Video cameras can do the "seeing," and researchers can better control the navigation and mechanical operations from a surface ship than cramped quarters within a submersible. In waters shallower than 90 meters, divers in SCUBA gear continue to provide a relatively inexpensive and safe way to perform subsurface operations.

Many oceanographic research projects are also carried out on unpowered platforms, either manned or unmanned, and either floating or fixed (Figure 1.17). Manned platforms include free-floating boats,

[9]According to Archimedes' principle, a floating object displaces an amount of water that is equal to its own weight. So a ship that displaces 2000 tons of water must itself also weigh 2000 tons.

(a)

(b)

Figure 1.16

(a) Woods Hole Oceanographic Institution with research ships in dock. (b) Research ship Atlantus II along with the submersible, Alvin. (WHOI)

Figure 1.17

A research buoy being carried to its launch site aboard the research ship Oceanus. (WHOI)

barges, and ice, and fixed platforms include oil rig platforms or anchored boats and barges. Unmanned platforms include buoys of various sorts, some free-floating and some anchored. Some are on the surface, and some are below. Research buoys have a strong inclination to become lost at sea, so it is important to include a reliable mechanism for having them pop to the surface and broadcast their position after they have performed their data-collecting function. Some surface buoys are made to broadcast their data as they are taking it, and some store the data for later retrieval.

REMOTE SENSING

Remote sensing technology is also becoming increasingly sophisticated. In some cases, this technology involves the use of unmanned oceanographic instruments that are "remote" from their operators, such as unmanned submersibles and buoys that take and broadcast data to a remote receiver. In other cases, remote sensing involves the use of instruments that are remote from the ocean they are observing, such as those carried in airplanes and artificial satellites orbiting the Earth. Observations from aircraft provide routine day-to-day updates on sailing conditions, wind and wave activity, weather patterns, and ice movement. Earth-orbiting satellites (Figure 1.18) provide nearly complete, continuous coverage of all oceans. Recent advances in remote sensing technologies have greatly accelerated our exploration and understanding of the oceans.

Figure 1.18
Painting of the SEASAT-A spacecraft studying the Earth's oceans from orbit.

SATELLITES

Satellite measurements have the advantage of viewing large areas of the world simultaneously. Therefore, they are particularly useful for measuring large-scale features, and overall patterns and changes. Satellite capabilities that are particularly useful to oceanographers include the abilities to measure wind velocity and direction, surface currents, ice and ice movement, wave activity, ocean surface temperatures, plant productivities (through chlorophyll), the distribution of fish stocks, and to monitor the progress of oil spills or other pollutants. They cannot "see" directly very far below the water's surface, but are able to identify shoals shallower than about 30 meters.

Satellites can measure sea floor topography using indirect methods. The ups and downs of the sea floor cause slight variations in gravity that are reflected in very slight changes in the mean level of the ocean surface. Satellites can average over all the surface waves to find this mean sea level, and from this map the overall topography of the ocean floor (Figure 1.19). Finally, satellites are useful for communicating with research buoys around the world, relaying instructions being sent to the buoys or data being sent from the buoys.

SUBMERSIBLES

Whereas satellites are particularly well suited to observe overall features and changes, submersibles are useful for detailed observations of isolated local features. Advances in remote-controlled submersible technologies have enabled us to engage in deep ocean studies using submersibles that are much smaller, cheaper, safer, more maneuverable, more versatile, and more efficient than those that carry human passengers.

Submersibles are being used to locate and study a variety of sea floor treasures of economic, scientific, and cultural interest. Examples include mineral deposits, deep ocean organisms, regions of seafloor spreading and other interesting geological and tec-

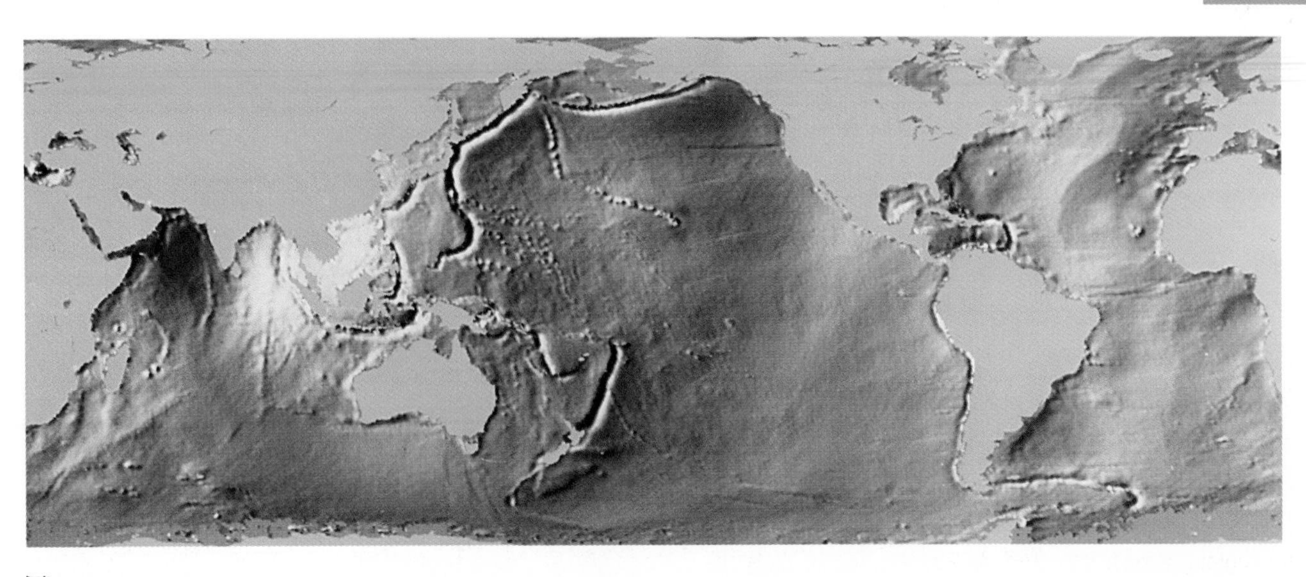

Figure 1.19
Topography of the ocean floor, as inferred from altimeter data measured by Earth orbiting satellites.

Figure 1.20
(a) On April 15, 1912, during its maiden voyage, the "unsinkable" *R.M.S. Titanic* struck an iceberg and sank in the North Atlantic, with the loss of 1517 of the 2222 lives on board. It was discovered on September 1, 1985, using a remote submersible system. In this photo the smaller submersible is tethered to a larger submersible, which is taking this picture. (b) Community of organisms growing near a hydrothermal vent on the ocean floor.

(a)

(b)

tonic processes, submersed archaeological sites, and ancient shipwrecks. ROVs have received front-page coverage because of some spectacular successes, such as the recovery of a lost thermonuclear bomb off the coast of Spain, and the discoveries of the wrecks of the *Monitor*, *Bismarck*, and *Titanic*. Also important are the finding and study of underwater geysers called **hydrothermal vents** and the communities of organisms associated with these vents, called **vent communities** (Figure 1.20).

Figure 1.21
Photo of scientists preparing to lower a cluster of water sampling bottles into the ocean near Antarctica.

DEVICES

Oceanographers employ a diverse array of mechanical devices to sample winds, water, sea floor sediments, and organisms, and to study the movements and interactions of these (Figure 1.21). Electrical and electronic devices are also common, especially for measurements that are made in place and instantly (in contrast to measurements that are made on retrieved samples, and for which the data are available at a later date). Standard electrical devices include thermistors for measuring water temperatures and instruments that determine the abundance of the salts (**salinity**) of a sample by its electrical conductivity. Ocean currents can also be measured electronically. But since it is helpful to gather ocean current information over a longer period of time, it is often done mechanically, either by tracking free-floating buoys or measuring the speed of the water passing an anchored meter.

Echo sounding is a standard method for mapping the sea floor (Figure 1.14). It evolved quickly during World War II to help locate submarines and shallow waters. Since that time, the technology has advanced considerably. Now both the direction that the sound is sent, and the direction from which the echoes are received, can be monitored very precisely, which enables us to get much clearer pictures of the subsurface features from which the sound reflects.

Continuing developments in electronics and computers are making it possible to collect and analyze larger amounts of data with greater accuracy, and with equipment that is more reliable. Modern electronics also allows ship navigators to know their positions anywhere on Earth to within 100 meters. One such method is based on the difference in arrival times of signals from different land-based transmitters, and another uses signals from Earth-orbiting satellites.

Summary

Oceanography is a diverse field that involves many different disciplines. This book will focus on the scientific aspects of oceanography and its primary objective is to foster an inquisitive attitude toward the world. The book's organization is based on the four broad areas of geological, chemical, physical, and biological oceanography, along with the growing field of human–ocean interactions.

Perspective in Time

The Earth is extremely old by human standards. Life began here nearly 4 billion years ago, but only recently has there been any life on land. The beginning of the Cambrian Period was marked by the development of hard shells on some marine organisms about 600 million years ago. About 420 million years ago, life moved onto land, but not until 65 million years ago did mammals become dominant. Erect, two-legged primates appeared about 4 million years ago, and the first records of stable agrarian civilizations are only about 6,000 years old. To understand the Earth and the development of its oceans, we must look for clues left by ancient events.

History of Oceanography

Early uses of the oceans involved food, transportation, trade, and national defense. Waterways were a lifeline to great ancient civilizations in the Mediterranean area. The Phoenicians were an early civilization with accomplished sailors. The Greeks had a remarkably accurate picture of the world, and the Romans were quite aware of the Earth's geological activity and the hydrologic cycle. Maps of explored areas were fairly accurate.

During the Middle Ages in Europe, the Vikings were exploring the North Atlantic, the Micronesians and Polynesians were accomplished navigators of the Western Pacific, the Chinese undertook some impressively large expeditions, and the Arabs were actively trading throughout the Persian Gulf and Indian Oceans.

In the fifteenth century, several events occurred that motivated Europe to explore the oceans. One of these was the capture of Constantinople, which severed conventional trade routes between the West and the East. The Portuguese sailed around the Cape of Good Hope, and the Spaniards discovered the "New World." This period was culminated by the circumnavigation of the world by the Magellan expedition in 1519 to 1522. The drive for riches caused the Spaniards and Portuguese to plunder the New World cultures, and the British and Dutch to plunder the Spanish and Portuguese. The defeat of the Spanish Armada in 1588 was a turning point in world naval history.

Colonization, exploitation, and trade were great motivations for increasing understanding of the oceans in the seventeenth and eighteenth centuries. Captain James Cook made several scientific expeditions during which he charted much of the oceans, confirmed the existence of Antarctica, and conquered scurvy.

In the nineteenth century, ocean science received increased attention for several reasons, including the laying of transoceanic cables, flourishing trade, interest in improving fisheries, scientific curiosity, and the question of whether there was life on the deep ocean floor. Maury published data to aid navigation, Darwin published his findings on natural selection, and the rule of constant proportions was established. The United States began investigations of its own coastal waters and the floor of the Pacific Ocean. The British *Challenger* expedition set the standards for subsequent oceanographic studies, and the Norwegian *Fram* expedition was the first to explore Arctic waters.

In the twentieth century, governments became more involved in the funding of oceanographic research, especially during and after World War II. Collaboration among many groups—government and private, national and international—has become increasingly common on large research projects. Deep sea drilling, echo soundings, and remote sensing are among the important technologies developed in this century.

Tools of Oceanography

Oceanographic data are collected from ships, submersibles, platforms, buoys, airplanes, and satellites. Standard equipment includes sampling bottles, current meters, echo sounders, and electrical devices for determining water properties. Electronics and remote sensing techniques are heavily used in present-day research.

Key Terms

Cambrian Period
Precambrian Period
Paleozoic Era
Mesozoic Era
Cenozoic Era
latitude
longitude
hydrologic cycle
Dark Ages
Middle Ages
bulkhead
Renaissance
sea
Magellan Expedition
Spanish Armada
Gulf Stream
scurvy
limey
sounding
plankton
Beagle
rule of constant proportions
Challenger
Fram
echo sounding
submersibles
ROV
remote sensing
hydrothermal vents
vent communities
salinity

Study Questions

1. What are the conventional four broad subdivisions of the science of oceanography?
2. In what way is our present investigation of the Earth a "snapshot"?
3. When did the Cambrian Period begin, and what development distinguishes it from the Precambrian Period?
4. What time spans are included by the Paleozoic, Mesozoic, and Cenozoic Eras? Give one major development of life on land during each of these eras.
5. What were some of the accomplishments of the Phoenician sailors?
6. Why are accurate clocks needed to determine longitude, but not latitude?
7. How did each of the following contribute to our understanding of our world? Pytheas, Eratosthenes, Strabo, Seneca, Aristotle.
8. Name at least one noteworthy accomplishment for each of the following persons: Eric the Red, Leif Ericson, Prince Henry the Navigator, Sultan Mohammed II, Bartholomew Diaz, Vasco da Gama, Christopher Columbus, John Cabot, Balboa, Magellan, Sir Francis Drake.
9. How did each of the following contribute to our understanding of the oceans? Dampier, Benjamin Franklin, James Cook, James Clark Ross, Matthew Fontaine Maury, Christian Gottfried Ehrenberg, Edward Forbes, Charles Darwin, Georg Forchhammer, C. Wyville Thompson, Fridtjof Nansen.
10. During the Middle Ages in Europe, what were some of the interesting oceanographic exploits being carried out elsewhere in the world?
11. What were some of the events of the 1400s that gave Europeans increased motivation for exploratory voyages?
12. How did Captain Cook deduce the existence of Antarctica? What device enabled Captain Cook to be the first to make accurate determinations of longitude?
13. Explain why making deep sea soundings with a weighted line was difficult. What evidence made James Clark Ross reason that the deep sea environment must be fairly uniform the world over?
14. Discuss the *rule of constant proportions*.
15. What were some of the discoveries of the *Challenger* and *Fram* Expeditions?
16. How does echo sounding work?
17. Describe some of the tools used by oceanographers.
18. In what ways can remote sensing be "remote"?
19. What are some of the reasons for the recent increased interest in coastal research?

Critical Thinking Questions

1. Why are Precambrian fossils rare compared to post–Cambrian fossils? (See if you can think of several reasons.)
2. Why are there so many more smaller time divisions for the Earth's recent history (e.g., the last 600 million years) than for its earlier history?
3. The Atlantic Ocean opened up about 150 million years ago. What percent of the Earth's lifetime is that? Estimate how long ago various notable events would have occurred if the Earth's history were condensed to a 24-hour scale.
4. At the Greenwich Royal Observatory, a certain very accurate clock is set so that it reads 12:00 noon when the Sun is at its highest point in the sky, half-way between sunrise and sunset. This clock is then taken on a ship that cruises the world. At a certain point in the journey, the clock registers 8:00 PM when the Sun is at its highest point in the sky. What is the longitude of the ship?
5. If you were aboard the *Challenger*, how would you get water samples from great depths? How would you design the hull of the *Fram* so that it would not be crushed by the movement of ice around it?

Suggestions for Further Reading

BAILEY, H. S. 1953. The Voyage of the *Challenger*. *Scientific American* 188:5, 88.

BAKER, D. JAMES. 1989. The New Wave of Ocean Studies. *Oceanus* 32:2, 10.

BAKER, D. JAMES. 1991. Toward a Global Ocean Observing System. *Oceanus* 34:4, 76.

BASS, G. F. 1987. Oldest Known Shipwreck Reveals Splendors of the Bronze Age. *National Geographic* 172:6, 693.

BOXHALL, PETER. 1989. Arabian Seafarers in the Indian Ocean. *Asian Affairs* 20:3, 287.

HARTWIG, ERIC O. 1990. Trends in Ocean Science. *Oceanus* 33:4, 96.

MCCLINTOCK, JACK. 1987. Remote Sensing: Adding to Our Knowledge of the Oceans and Earth. *Sea Frontiers* 33:2, 105.

NELSON, STEWART B. 1990. Naval Oceanography: A Look Back. *Oceanus* 33:4, 10.

O'REGAN, TIPENE. 1991. Pacific Pioneers. *UNESCO Courier* Aug.–Sept., 28.

SAYLES, FRED L. 1992. Benthic Landers: Taking the Laboratory to the Seafloor. *Oceanus* 35:1, 8.

Because of the many heavier elements here on Earth, we know we are made of materials that have been cycled through the interiors of stars at least once before.

In the battle between water and rock, which one wins, and why? The answer requires a basic understanding of forces, motion, and the microscopic makeup of matter.

two

MATERIALS AND MOTION

In this book, we study the Earth's oceans—their origins, evolution, properties, chemical makeup, and various motions and interactions that affect us. To understand all these aspects of oceans, a certain amount of scientific background is necessary. Particularly relevant is a basic understanding of the atomic and molecular structure of the materials from which Earth is made and of the forces that are responsible for the formation, evolution, and present behaviors of the oceans.

This chapter provides this very basic yet important scientific background. Readers who are familiar with this information may decide to proceed directly into the next chapter. Others may find this background material an essential first step. For either reader, this chapter serves as a reference for concepts introduced in subsequent chapters.

2.1 ATOMS

STRUCTURE

At a very fundamental level, all matter is made up of myriads of tiny **atoms**. Each atom is so small that it would take about 100,000 of them laid side by side to span the width of a hair. Each atom is composed of an extremely small and dense nucleus, surrounded by a much more nebulous cloud of electrons moving at very high speeds (Figure 2.1). In the atomic nucleus are neutrons and protons, similar particles except that the proton carries a positive electrical charge, and the neutrons are neutral. By contrast, the very light electrons carry negative electrical charge.

Each atom is so small that it would take about 100,000 of them laid side by side to span the width of a hair.

The different kinds of atoms are distinguished by the number of protons in the nucleus. Under normal conditions, atoms are electrically neutral, so the number of positively charged protons in the nucleus is equal to the number of negatively charged electrons in the electron cloud. For example, hydrogen has one of each, helium two of each, lithium three of each, and so on. The list continues up through uranium, which has 92 of each. As a result, there are 92 different kinds of atoms, or **elements**, from which materials can be made. A few elements heavier than uranium (i.e., they have more than 92 protons and electrons per atom) can be artificially produced, but they are unstable and not found in nature. The number of protons (or electrons in the neutral atom) is called the **atomic number**, and is what distinguishes one type of atom from another.

ABUNDANCES OF THE ELEMENTS

The lightest two elements, hydrogen and helium (atomic numbers 1 and 2, respectively), were produced in what is called the *Big Bang* birth of our Universe. All other elements were produced later, as stars began to form from these clouds of hydrogen and helium. These *heavier elements* (i.e., heavier than hydrogen and helium) were produced by the nuclear reactions in the centers of these stars, and so their relative abundances reflect the complicated details of these nuclear reactions. Many stars go through eruptive convulsions late in their lives, which

Figure 2.1
An atom is composed of an extremely small and dense nucleus, with a much more nebulous cloud of electrons orbiting it. In the nucleus are positively charged protons and neutral neutrons, each of which is nearly 2000 times more massive than a negatively charged electron, so essentially all the mass of an atom is in its nucleus.

Figure 2.2

Hydrogen and most of the helium were formed in the "Big Bang" origin of the Universe. All heavier elements were made in the cores of stars, and released in spectacular stellar explosions. The Crab Nebula is the remnant of such an explosion, which was observed in the year 1051.

Table 2.1 Estimated Relative Abundances of the 20 Most Common Elements in the Universe, in Terms of the Number of Atoms of That Element per Million Hydrogen Atoms

Other than hydrogen and helium, the abundances of the elements are a reflection of nuclear reactions in stars. It is seen that hydrogen and helium comprise the vast majority of all materials. Presumably, the Earth and inner planets are made of the small residue left over when the hydrogen and helium escaped.

Element	Atomic Number[a]	Atomic Weight[b]	Abundance Relative to 10^6 Hydrogen Atoms
Hydrogen (H)	1	1	1,000,000
Helium (He)	2	4	80,000
Oxygen (O)	8	16	690
Carbon (C)	6	12	420
Nitrogen (N)	7	14	87
Silicon (Si)	14	28	40
Neon (Ne)	10	20	37
Magnesium (Mg)	12	24	32
Iron (Fe)	26	56	25
Sulphur (S)	16	32	16
Aluminum (Al)	13	27	3.3
Calcium (Ca)	20	40	2.5
Nickel (Ni)	28	59	2.1
Sodium (Na)	11	23	1.9
Argon (Ar)	18	40	1.0
Chromium (Cr)	24	62	0.69
Phosphorus (P)	15	31	0.39
Manganese (Mn)	25	55	0.26
Chlorine (Cl)	17	35	0.22
Potassium (K)	19	39	0.12

[a]Atomic number is the number of protons in the nucleus or, equivalently, the number of electrons in the electron cloud.

[b]Atomic weight is the total number of nucleons (i.e., protons and neutrons) in the nucleus, given here for the most common isotope of each.

spew these stellar materials back out into space (Figure 2.2). Subsequent gravitational collapse of these clouds of regurgitated material can result in new second-generation stars and solar systems. Because of the many heavier elements we find in our Earth and Solar System, we know we are made of materials that have been cycled through the interior of stars at least once before. (Can you believe that all the materials in your own body once helped make a star shine?)

Because of the many heavier elements we find in our Earth and Solar System, we know we are made of materials that have been cycled through the interior of stars at least once before.

You can see from Table 2.1 that the heavier elements (i.e., heavier than hydrogen and helium) altogether represent only about 0.1% of all the atoms in the universe. Even among these, there is still a tendency for the lighter ones to be more plentiful, with oxygen and carbon being the most common, followed by nitrogen, silicon, and so on. Although the Earth is formed of these same materials, there are some interesting differences between the relative abundances of the various elements found on Earth, and those found in the Universe as a whole.

When the Earth was forming, its gravity was too weak to be able to hold onto the very light and elusive hydrogen and helium, so these gases are nearly absent from Earth. A few of the other lighter elements are somewhat depleted as well. A very small amount of helium is still trapped beneath the surface, some of which is found when we drill for natural gas. Some

Table 2.2 Relative Abundances of Various Elements in the Earth's Crust, by Weight

Element	%
Oxygen	47.0%
Silicon	28.0%
Aluminum	8.0%
Iron	5.0%
Calcium	3.6%
Sodium	2.8%
Potassium	2.6%
Magnesium	2.1%
All others together	0.9%

Source: After E. Tarbuck and F. Lutgens, 1991, *Earth Science*, 6th ed., p. 20, Macmillan, New York.

hydrogen atoms are bound to heavier elements, such as oxygen, which helps prevent their escape from Earth. But for the most part, the abundance of these two elements on Earth is far below that which might be expected given their relative abundance in the Universe as a whole.

Consequently, the overall internal composition of the Earth is similar to that of the Universe as a whole, with the notable exception of the light gases, hydrogen and helium, and some relatively smaller depletion of a few other light elements.

In studying oceanography, however, we are particularly concerned with the materials of the Earth's surface, and here we find further differences. As we will soon learn, there has been some sorting of the materials within the Earth. Denser materials have sunk toward the center, and lighter materials have risen to the surface. In the rocks near the Earth's surface, the most abundant elements are those listed in Table 2.2. Above these rocks, we find the even lighter materials of our oceans and atmosphere. Most prominent among these are the water and various dissolved materials found in our oceans, and the nitrogen, oxygen, water vapor, and carbon dioxide of our atmosphere.

2.2 CHEMICAL BONDS

Most atoms bind together in groups. If each group contains a definite and relatively small number of atoms, it is called a **molecule**. Examples include the 2 hydrogen and 1 oxygen atoms that combine to form a water molecule (H_2O), and the 2 nitrogen atoms that combine to form a nitrogen molecule (N_2). In other materials, the groups of atoms may grow to extremely large sizes as additional atoms bond to those that are already there. For example, a typical grain of table salt contains about a billion billion sodium atoms, and an equal number of chlorine atoms. These atoms form a periodic framework in which the sodium and chlorine atoms alternate in all three dimensions. Each sodium atom is surrounded by 6 chlorines (left, right, front, back, top, bottom), and likewise, each chlorine atom is surrounded by 6 sodiums. We identify this material as "NaCl" to indicate the ratio of sodium (Na) to chlorine (Cl) atoms is 1 to 1. Similarly, a small grain of a common yellowish beach sand contains billions of billions of silicon atoms, and twice as many oxygen atoms (SiO_2), bonded together in repeating regular arrangements.

ELECTRON CONFIGURATIONS

Atoms bind together by sharing electrons. The electron configurations of the individual atoms determine what combinations can form. Within individual atoms, certain symmetric electron configurations are preferred. One preferred arrangement has two electrons that occupy the same **shell**, which is the name we give a region of space surrounding the atomic nucleus. Preferred arrangements for larger atoms include an additional shell with eight more electrons (10 altogether), a third shell with still eight more electrons (18 altogether), or additional larger completed shells for the very large atoms (Figure 2.3).

We use the idea of shells for conceptual ease and clarity of representation, although it is an oversimplification. The electrons are so unpredictable in their behaviors that we should only talk about them in terms of probabilities. These probabilities indicate that the distribution of electrons in the electron cloud is more like "fuzzy blurs" than shells (Figure 2.4).

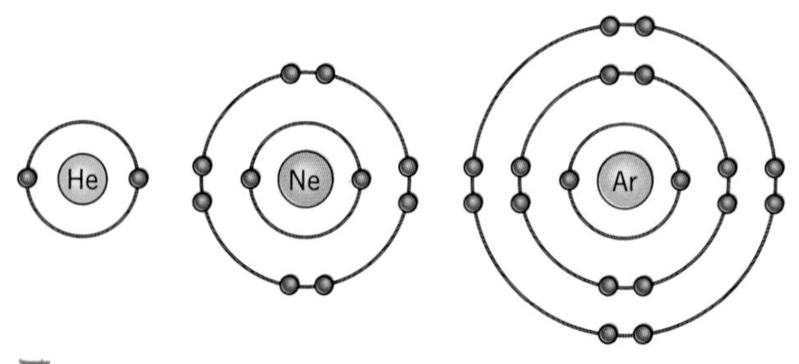

Figure 2.3

Idealized illustration of the preferred symmetric electronic configurations, which are those of the inert gases. The above illustrations are for helium, neon, and argon, which have one, two, and three completed electron shells, respectively. These diagrams serve as conceptual aids only. In real atoms, the electrons are moving in all directions at very high speeds.

Figure 2.4

The electrons in atoms move extremely quickly, and electron *shells* are actually more like "blurs." The electrons that orbit atoms are so fast and elusive that we can only describe their locations with probabilities. The above figure for various possible configurations of electrons in the first three shells, represents the relative probabilities of where electrons might be found.

With this understanding, we now investigate how the preference for certain symmetric electronic configurations determines the chemical behaviors of the various elements, and the combinations they may form. We begin by examining some of the more common materials of the Universe. From Table 2.1, you can see that there is a general tendency for lighter atoms to be more plentiful. (There are some interesting exceptions, however, such as the elements with atomic numbers 3, 4, and 5.) Consequently, molecules made of lighter atoms are much more common than those containing heavier ones. For example, you should expect molecules made of hydrogen and oxygen, near the top of Table 2.1, to be much more common in the Universe than those made of chlorine and potassium, near the bottom of that list.

THE BEHAVIORS OF HYDROGEN AND HELIUM

We begin by examining the simplest and most common element in the Universe, hydrogen, which has only one electron (Table 2.1). This is one electron *less* than the preferred arrangement of two electrons and is also one *more* than the very symmetric situation of zero electrons. Consequently, hydrogen combines with other atoms to share an electron, in order that some of the time it may enjoy the preferred states of either zero or two electrons. We say that the hydrogen atom has a **valence** of 1, meaning that it shares one electron.

For example, hydrogen may combine with another hydrogen atom, to form the H_2 molecule. Each hydrogen atom shares its electron, so there are two shared electrons altogether. In this situation, each of the two hydrogen atoms can some of the time enjoy the preferred state of having either zero or two electrons nearby.

The second simplest atom, helium, already has its preferred arrangement of a completed two-electron shell (Figure 2.3a). Consequently, it doesn't share any electrons with other atoms, and so it doesn't interact chemically at all. Even the mutual interactions of helium atoms among themselves are so weak, that it doesn't condense into liquid form until temperatures very close to absolute zero. For these reason, helium is called an **inert gas**.[1] Neon is another inert gas, with a full compliment of 10 electrons: two in the first shell and eight in the next. Other still heavier inert gases include argon, with 18 electrons in three shells, and much rarer inert gases with four, five, and six completed shells (krypton, xenon, and radon).

MOLECULES WITH CARBON, NITROGEN, AND OXYGEN

We now look at the chemistry of carbon, nitrogen, and oxygen (Figure 2.5), which are the fourth, fifth, and third most abundant materials in the universe, respectively (Table 2.1). Carbon has six electrons: two in the first shell and four in the next. This is four electrons short of the preferred arrangement of 10 electrons in two closed shells, and it is also four electrons in excess of the one closed-shell arrangement. Consequently, it is quite eager to share four electrons

[1] It is *inert* because it doesn't interact chemically with other atoms, and it is a *gas* because its own self-interactions are so weak that it doesn't condense into a liquid state until temperatures very near absolute zero.

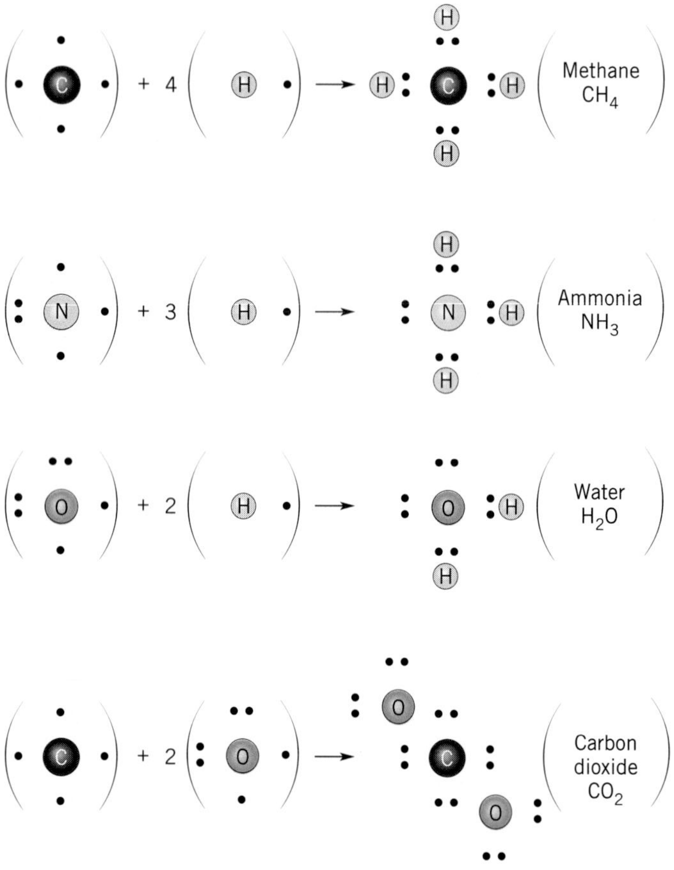

Figure 2.5

Simplified illustration of how the chemistry of an atom is determined by the preference for the electron configurations of the inert gases, which are very symmetrical. Carbon is four electrons short of the neon configuration and four in excess of the helium configuration. So it eagerly shares its four with four others. Similarly, nitrogen shares three, and oxygen two. Hydrogen is one electron short of the helium configuration, and one electron in excess of the symmetric configuration with no electrons. Therefore, hydrogen eagerly shares its one electron with others.

with other atoms. In other words, it has a valence of 4. This enables it to form the basis of quite complicated molecules, such as those in living tissues (Chapter 11). But it also forms some very simple molecules, such as when one carbon (valence of 4) combines with four hydrogens (valence of 1 each) to form the CH_4 molecule, called **methane**. Because of the abundance of hydrogen and carbon, this is a fairly common molecule in the universe, although not nearly as common as H_2, of course.

Nitrogen has one more electron than carbon, with seven electrons altogether: two in the first shell and five in the next. This is three short of the preferred arrangement of 10 electrons. So nitrogen combines with other atoms to share three electrons. For example, **ammonia** (NH_3) is a fairly common molecule in the universe. Oxygen has one more electron than nitrogen, with eight electrons altogether: two in the first shell and six in the next. Therefore, it needs two more for a completed outer shell. Consequently, **water** (H_2O) is a common molecule.

Because of hydrogen's abundance, molecules that include hydrogen are most common in the universe. Of course, hydrogen gas, H_2, is by far the most abundant of these. But water, methane, and ammonia are also quite common. These three materials often form tiny frozen crystals in the vast frozen regions of outer space, and they will also be important materials in our study of the evolution of planetary atmospheres and oceans in Chapter 6. Water is the most abundant of these, and it should be no surprise that much of the Earth is covered by it. The more interesting question is why the other planets are not.

But large numbers of other kinds of molecules are also important, many of which do not involve hydrogen at all. For example, carbon and oxygen may form carbon dioxide (CO_2) molecules. Since carbon has a valence of 4 and each oxygen a valence of 2, the two oxygens and one carbon provide mutual satisfaction in electron sharing. Later in this book, we will learn that large amounts of carbon dioxide gas are dissolved in our oceans, where it is involved in many important chemical reactions, including those that support life.

IONIC AND COVALENT BONDS

Although atoms join together by sharing one or more of their electrons, the "shared" electrons are often not shared equally. Some atoms are generous with electrons, and others are greedy. If the electrons are shared roughly equally between the bonded atoms, the chemical bond between them is called **covalent**. Examples of covalent bonds include those in hydrogen (H_2) and oxygen (O_2) molecules. At the other extreme are the **ionic** bonds, in which the shared electron or electrons spend much more time with one of the bonded atoms than the other. Because an electron carries a negative electrical charge, the atom with the extra electron carries a net negative charge, and the one lacking the electron carries a net positive charge. The electrostatic attraction between the two oppositely charged parts, called **ions**, is what holds atoms together in ionic bonds. Later in this book, we will see that the most abundant materials dissolved in the oceans have ionic bonds, because the separation of the electrical charges between the ions makes it particularly easy for water to dissolve them.

METALS, RADICALS, AND OTHER COMMON MOLECULAR INGREDIENTS

In our first examples of chemical bonding, we used the light elements hydrogen, carbon, nitrogen, and oxygen. Although these materials are very important and abundant in the Universe as a whole, we have seen that hydrogen is heavily depleted on Earth. A considerable fraction of the important materials on Earth involve heavier elements. We now examine the chemical behavior of some of these materials.

Examples of heavier elements that are generous with their electrons include the metals, which is one reason they are good electrical conductors. The most common metals near the Earth's surface are listed below in order of decreasing abundance, along with the customary notation for how many electrons each is willing to give up:

Aluminum	Al^{3+}
Iron	$Fe^{2+ \text{ or } 3+}$
Calcium	Ca^{2+}
Sodium	Na^{+}
Potassium	K^{+}
Magnesium	Mg^{2+}

Elements that could accept the electrons given up by these metals include chlorine, which accepts one extra electron (symbolically, Cl^-), and oxygen, which accepts two (symbolically, O^{2-}). You can see that a suitable combination would be sodium and chlorine, because the sodium easily gives up an electron, and the chlorine eagerly accepts it. Together, they form the compound sodium chloride (NaCl), or common table salt. Similarly, you can see that calcium oxide (CaO) would be a stable compound for similar reasons. Water (H_2O) is a stable molecule, because the two hydrogens willingly share one electron each, and the oxygen eagerly accepts both. Silicon shares four electrons, as does carbon. Consequently, these two elements are commonly found in combinations with oxygen, such as *silica* or *quartz* (SiO_2) or carbon dioxide (CO_2).

Oxygen also combines in excess with some elements to form groups called **radicals** which are covalently bonded groups that function as ions. Because of the excess of oxygen, radicals accept extra electrons. For example, oxygen combines with silicon to form the **silicate** (SiO_4^{4-}) radical. The four oxygens would accept eight electrons altogether (two each), but silicon has only four to give. So the radical needs four more, as is indicated by the superscript 4–. Similarly, oxygen combines in excess with carbon to form **carbonate** (CO_3^{2-}). Other common radicals that you may have heard of include **sulfate** (SO_4^{2-}) and **phosphate** (PO_4^{3-}). Below are listed some of the more common electron-accepting elements and radicals, along with the number of electrons they readily accept:

Chlorine	Cl^-
Oxygen	O^{2-}
Silicate	SiO_4^{4-}
Carbonate	CO_3^{2-}
Sulfate	SO_4^{2-}
Phosphate	PO_4^{3-}

From the abundances of silicon and oxygen listed in Table 2.2, you might correctly guess that the most common materials near the Earth's surface are silicates, such as the silicates of aluminum, iron, calcium, and so on. Likewise, another very common sediment material is **silica** (SiO_2), which forms much of the familiar yellow quartz sands. The carbonates are also common, especially calcium carbonate (limestone). Other fairly common materials include ordinary salt (NaCl), and various metal oxides and sulfates.

With a little thought, you can see that a very large number of different atomic combinations are possible (Figure 2.6). This variety is reflected in the large number of tiny flecks of different minerals that are often revealed when we examine a rock or some sediments under a magnifying glass.

This brief overview of the atomic nature of matter gives us insight into the materials from which the Earth and its oceans are made. This is important background knowledge for the following chapters. In addition, we will need to have a basic understanding of the forces that have given birth to the oceans, caused their evolution over time, and govern the important oceanic processes going on today.

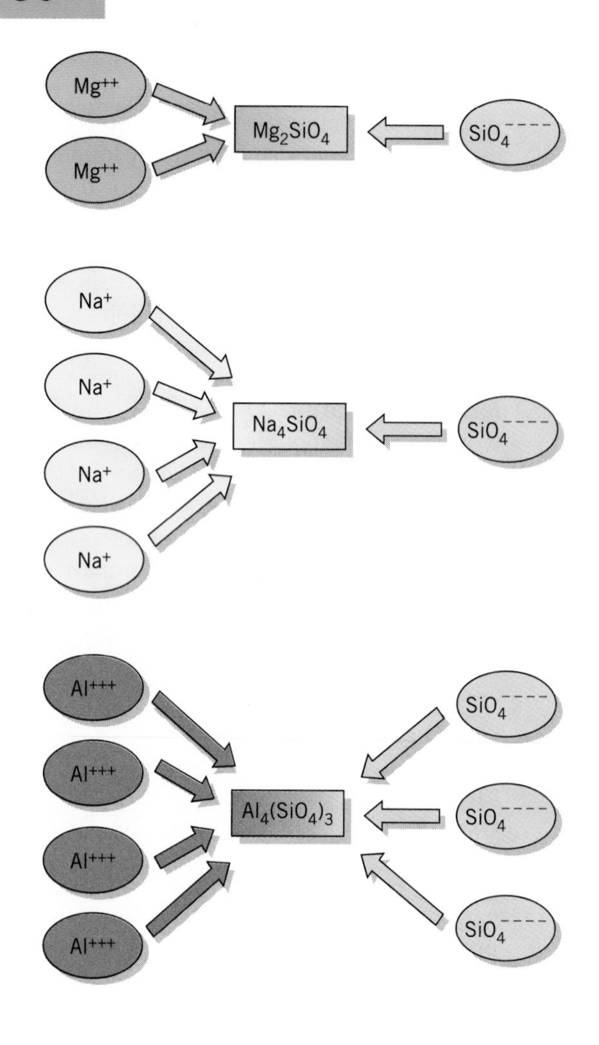

Figure 2.6

Atoms combine in such a way that the total number of electrons furnished by some equals the total number accepted by others. This is illustrated here for three possible combinations involving the silicate (SiO_4^{4-}) radical.

2.3 FORCES

We now investigate motion. According to Newton's first law of motion, an object at rest tends to stay at rest and an object in motion tends to continue moving in a straight line unless acted upon by a force. Forces, then, cause resting objects to move and moving objects to change their motion.

On the most fundamental level, there are only three types of forces at work in nature. The **strong force** holds together the atomic nuclei. The **electrical force**[2] binds the clouds of swarming electrons to the atomic nuclei and binds neighboring atoms together in molecules and larger groups. Finally, the gravitational force, or **gravity**, pulls masses together and causes stars and planets to form from clouds of interstellar gas and dust.

[2]More properly, they are called *electroweak*, because they have been shown to be related to a force operating within the nucleus that is weaker than the strong nuclear force.

Although electrical and strong forces operate primarily at very tiny atomic and subatomic distances, their ramifications can sometimes extend to much larger distances. Extremely violent motion caused by the strong force within atomic nuclei may result in radiation that shoots outward. For example, the sunlight that supports life on Earth comes ultimately from nuclear reactions in the deep interior of our Sun (Chapter 7). Similarly, atomic electrical forces often have ramifications at larger distances. When you push on a door, the electrical forces between billions of atoms in your hand against billions of atoms on the door cause the door to move. Electrical forces between the atoms within the door hold them in place and make the door rigid, so that the force you exert on some of them is transferred to all of them. In this book we will frequently be interested in forces in fluids. A force exerted at one point in a fluid, is quickly transferred throughout the fluid as neighboring atoms push against each other with electrical forces. Later in this same section we will investigate buoyancy and other pressure gradient forces which are of this nature.

ELECTRICAL FORCES

For reasons we do not fundamentally understand, electrical charges come in two types: positive and negative. Like charges repel, and unlike charges attract. For example, two positive charges repel each other, two negative charges repel each other, but a positive and a negative attract. Although these forces are dominant at an atomic scale, they are considerably less influential at larger distances, because matter as a whole is electrically neutral. For every positive charge, there is a negative one nearby; for every proton, there is an electron.

Within atoms and molecules, however, the various charged particles are found at different places, and the effects of this charge separation are extremely strong. For example, one side of a molecule may carry a net positive charge, and the other side a negative charge. The positive charge of one molecule will attract strongly the negative side of another. This bonding between neighboring molecules is extremely important in understanding the properties of materials, such as the ocean water that we study in Chapter 6.

GRAVITY

Also for reasons we do not yet fundamentally understand, any two objects in the universe attract each other with a force proportional to their masses. We call this mutual attractive force gravity. The force of the Earth's gravity on you is called your weight. Actually, all objects in the universe are pulling on you, including both the person sitting next to you, and the farthest star. Because of its size and closeness, Earth's gravity dominates, and it is the only one you notice.

The gravitational force between two objects depends on both their masses[3] and their separation (Figure 2.7). It is directly proportional to the mass of each (call these two masses m_1 and m_2) and inversely proportional to the square of their separation (r). The constant of proportionality is given the symbol G.

$$\text{Force} = G\frac{m_1 m_2}{r^2}$$

This is called the **law of gravity**. You can see that the force of attraction increases if either mass (m_1 or m_2) increases and decreases if their separation r increases.

For example, consider the force of the Earth's gravity on you. You would weigh more if there were an increase in either your mass m_1 or that of the Earth m_2. And you would weigh less if you increased your distance from the center of the Earth (r). You weigh slightly less on a mountain than in a valley, because the mountaintop is slightly farther from the Earth's center (Table 2.3). If you doubled your distance from the Earth's center, you would weigh one-fourth as much ($\frac{1}{2}^2 = \frac{1}{4}$). The fact that the force of gravity depends on distance means that the effect of the Moon's gravity on Earth varies slightly from one side of the Earth to the other. In Chapter 9, we will see that this difference is responsible for the tides.

BUOYANT FORCE

After having reviewed the three fundamental forces of nature, we now look at ways that these forces influence large scale motion. First, we examine the **buoyant force**, which is particularly important in the motion of **fluids**. The word *fluid* refers to anything that flows. It is normally associated with liquids

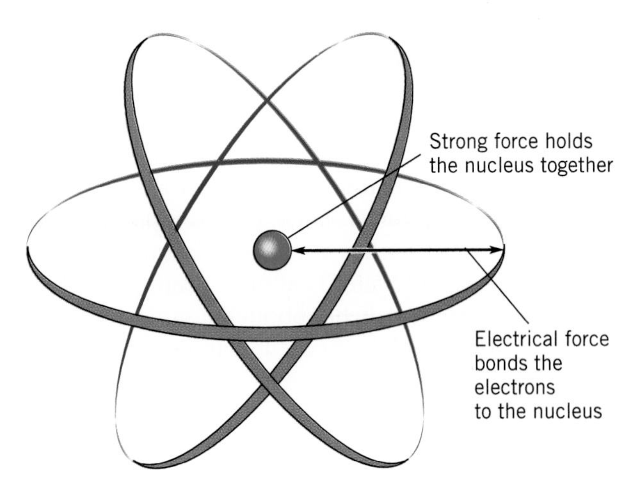

(a) Atom enlarged 100,000,000 times

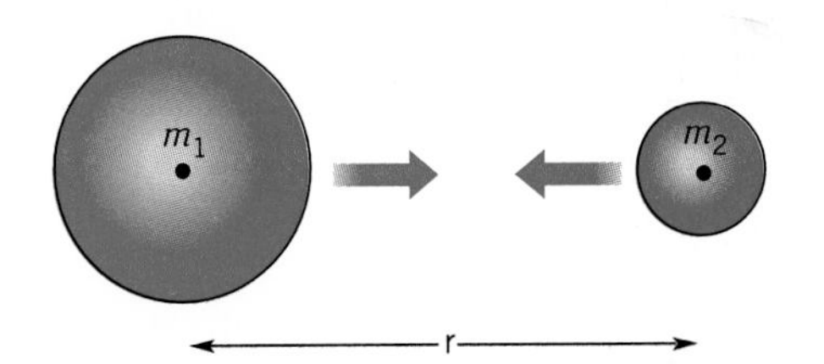

(b) Gravity: masses and distances could be anything from atomic to intergalactic.

Figure 2.7

(a) The strong force holds together the nucleus, and the electrical force keeps the electrons in orbit about the nucleus. Although both are extremely influential at these small atomic and subatomic distances and are responsible for the structure of matter, normally neither is very influential at larger distances. (b) Gravity dominates at larger distances. Any two objects attract each other with a force that is proportional to the mass of each (m_1 and m_2) and inversely proportional to their separation (r).

Table 2.3 Variation of the Earth's Gravitational Force with Altitude above Sea Level

The gravitational acceleration g measures how much speed, in meters per second, an object gains in 1 second of free fall.

Altitude (km)	g (m/sec^2)
0	9.806
1	9.803
10	9.777
100	9.521
1000	7.433
10,000	1.587
100,000	0.039

[3]Don't confuse *mass* with *weight*. Mass measures an object's inertia, or resistance to change in motion, whereas weight measures how hard the Earth pulls on it. For example, a bowling ball has a certain amount of mass no matter where it is. If you collided with a bowling ball in outer space, it would hurt just as much as if you collided with it on Earth, even though it doesn't weigh anything in outer space. The two concepts are often confused because the force of gravity is proportional to the mass. Therefore, on Earth more massive things also weigh more.

and gases. We will soon learn, however, that the "solid" rocks of the Earth's interior also deform and flow, albeit very slowly. So the buoyant force is also important in studying motion and evolution within the solid Earth.

The buoyant force is caused both by gravity, which gives the fluid weight, and by the electrical force between neighboring atoms, which allows an applied force to be transmitted from atom to atom throughout the fluid. The weight of a fluid is proportional to its **density**, which is a measure of average mass per unit volume. For example, 1 cubic centimeter of water has a mass of 1 gram, so we say that the density of water is about 1 gram per cubic centimeter. Your body has about the same average density. A typical rock is about three times as dense as water, so it would have a density of about 3 grams per cubic centimeter.

The fluid must support its own weight. At greater depths, the weight of the overlying layers is greater, and so correspondingly greater force is required to support this greater weight. A measure of this force is the **pressure**, which is the amount of force per unit area. In the metric system, force is measured in Newtons and area in square meters, so *Newtons per square meter* is a unit of pressure in the metric system.

At any point within a fluid, the pressure is the same in all directions (Figure 2.8). If it weren't, the fluid would be pushed away from the higher pressure, flowing toward the lower pressure until the pressure became equalized. If the fluid is not moving, then you know that the pressure at any point must be the same in all directions. The fluid at any point has nowhere to go.

Pressure increases with depth in a fluid, as your ears may have noticed when changing altitudes in an airplane, or when diving beneath the surface in a swimming pool. You might also have noticed that it doesn't help to turn your head; the pressure is the same in all directions. At each level, the pressure must be sufficient to support the weight of all the fluid above it. At greater depths, more of the fluid is overhead, so the pressure must be larger to support this greater weight.

When an object is submerged within a fluid, its bottom side is deeper and therefore experiences greater pressure. Therefore, the upward pressure from beneath is greater than the downward pressure from above (Figure 2.9). The result is a net upward force on the object, called a buoyant force.[4]

[4]Notice that the buoyant force is a consequence of gravity, because it is due to the weight of a fluid. If there were no gravity, there would be no weight and, therefore, no buoyant force.

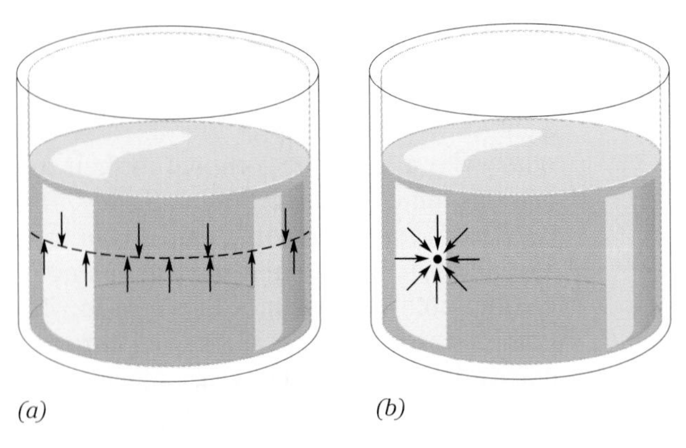

Figure 2.8

(a) The pressure at any depth in a fluid is caused by the weight of the overlying layers and is exactly sufficient to support this weight. (b) When in equilibrium (i.e., not moving), the pressure at any point in a fluid is the same in all directions. If it weren't, then the fluid would flow in the direction favored by the greater pressure.

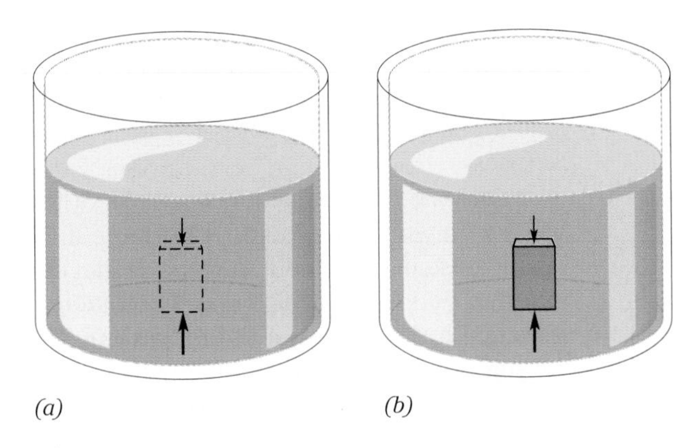

Figure 2.9

Pressure increases with depth in a fluid. (a) Consider the fluid inside an imaginary box (e.g., inside the dashed lines). The pressure on the bottom of this imaginary box is larger than that on the top, giving a net upward buoyant force. Since the fluid just sits there, we know that the net upward buoyant force is exactly sufficient to support its weight. (b) When the fluid is replaced by any other object, that object experiences the same upward force as the fluid that used to be there. That is, the object receives an upward buoyant force exactly equal to the weight of the fluid that it replaced (Archimedes' principle).

The magnitude of the upward buoyant force on an object is always equal to the weight of the fluid it displaces. This statement is known as **Archimedes' principle**. To understand Archimedes' principle, notice that the pressure in a fluid is always sufficient to support its own weight. When some of that fluid is replaced by another object, the pressure in the fluid around the object is still sufficient to support the weight of the fluid that used to be there (Figure 2.9).

That is, the fluid always pushes upward with a force that is sufficient to support its own weight. So that is the amount of upward buoyant force experienced by anything immersed in it.

As an example, a ship displaces an amount of water that is equal to the weight of the ship. The upward buoyant force on the ship has to support the ship's weight. If we add some heavy cargo to the ship, it sinks deeper into the water. It must displace more water to support the extra weight. Since the pressure increases with depth, the upward buoyant force increases as the ship sinks deeper.

As another example, notice that when you get into a swimming pool, you sink down until the weight of the water displaced is equal to your weight. Since you are almost the same density as water, virtually all your body must me submerged before you have displaced enough water. If you take a deep breath of air, your lungs expand, so you displace more water, and you float slightly higher. If you then let out all this air, your chest is smaller and you displace less water. So the buoyant force is not sufficient to support your weight, and you sink.

Although we have used water in our examples of buoyancy, the same applies to all fluids. For example, your body displaces some air. This air weighs a small fraction of a pound, so you receive an upward buoyant force of a fraction of a pound when standing in air. A helium-filled balloon actually weighs less than the air it displaces, so it receives a buoyant force larger than its weight, and it rises.

With regard to motions of air, water, and the evolution of our Earth, notice that things denser than the fluid they displace will sink; the buoyant force won't be sufficient to completely support their weight. They weigh more than the fluid they displace. Similarly, things that are less dense than the fluid in which they are immersed will rise toward the surface, because the upward buoyant force is greater than their weight. They weigh less than the fluid they displace.

Within the Earth, denser materials have gradually sunk toward the interior, whereas lighter materials have risen toward the surface. This concept will be important in Chapters 3 and 4. Similarly, within the ocean, denser waters sink toward the ocean floor, and lighter waters rise to the surface. This concept will be important in understanding the oceans' structure and motion in Chapters 6 to 8.

Other things being equal, materials usually become less dense when heated and more dense when cooled.[5] Hot air balloons rise because the hot air is less dense and therefore weighs less than the cooler air it displaces. Within the oceans, we will learn that warmer water tends to remain near the surface, and cooler water tends to sink toward the bottom.

PRESSURE GRADIENT FORCES

Fluids tend to flow from higher pressures toward lower pressures, because the extra pressure pushes them away. For example, when you puncture a tire, the air rushes out. Because the air pressure within the tire is greater than that outside, the air gets pushed out. When studying our atmospheric circulation in Chapter 7, we will learn that winds tend to blow from areas of higher pressures toward areas of low pressures, although some interesting things may happen along the way. A change in pressure is called a *pressure gradient*, so forces caused by changes in pressure are called **pressure gradient forces**.

We have already noticed that pressure increases with depth in any fluid, so there is always an upward buoyant force on objects immersed in a fluid due to this pressure gradient. That is, the buoyant force is really a pressure gradient force, where the pressure gradient is caused by the weight of the overlying layers of the fluid. However, it is customary to treat the upward buoyant force separately and reserve the term pressure gradient for forces other than that due to the weight of the fluid.

INERTIAL FORCES AND THE CORIOLIS EFFECT

If you are on a spinning merry-go-round, a car going in circles, or even just standing in one place and turning around, you notice that the people, trees, clouds, and so on around you all seem to be going in circles. Even a bird flying across the sky, or a ball sailing overhead seem to be going in circles. Of course, the trees, birds, balls, and so on are not really going in circles. It just looks that way because you are spinning.

We Earthbound observers are on a large merry-go-round that rotates once every 24 hours. Consequently, moving things seem to curve rather than going straight. Of course, it is not the object that is curving. Rather, it is we that are turning, along with our trees, houses, streets, light posts, clouds, and so on. This apparent curvature of trajectories due to the Earth's rotation is called the **Coriolis effect**.

We Earthbound observers are on a large merry-go-round that rotates once every 24 hours. Consequently, moving things seem to curve rather than going straight.

[5] An important exception is ice, which is less dense than the warmer water on which it floats.

The Earth spins in such a way that in the Northern Hemisphere, things seem to curve to the right as they go. This happens to snowballs, ballistic missiles, air currents, and water currents alike. In the Southern Hemisphere, things appear to curve to the left. A merry-go-round spinning counterclockwise as seen from the top is spinning clockwise as seen from the bottom. So the apparent deflections in the Southern Hemisphere are the opposite of those in the North. At the equator, there is no effect at all, but the closer one gets to the poles (i.e., the "top" or "bottom" of the merry-go-round), the more pronounced the curvature of trajectories becomes.

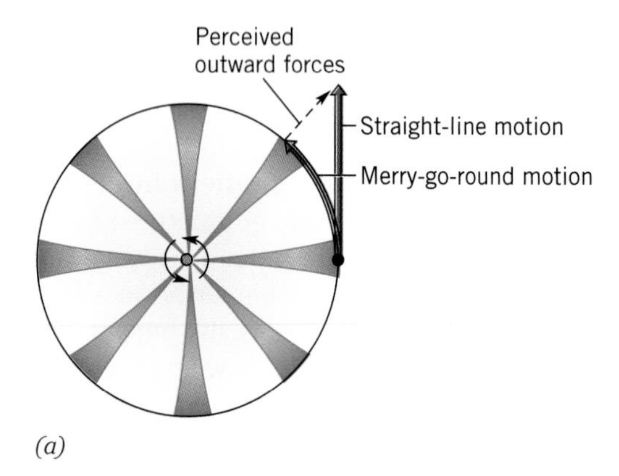

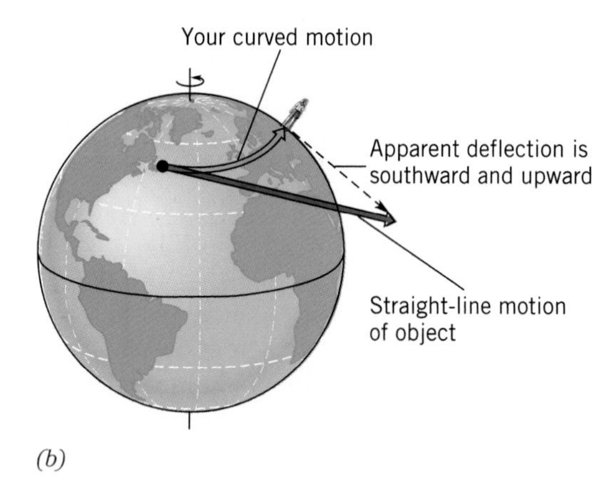

Figure 2.10
Inertial forces and the Coriolis effect. (a) On a spinning merry-go-round, your body is trying to follow a straight path, but the merry-go-round prevents it from doing so. You think that there is some invisible outward force on your body. In fact, however, there is no outward force; rather, the merry-go-round is pulling you in. Therefore, the perceived outward force is an *inertial force*, one that is not real. (b) An object moving initially eastward in the Northern Hemisphere appears to be deflected both southward and outward (i.e., up away from the Earth's surface), as we observers are carried with the spinning Earth. However, the "outward" vertical deflection of air or water is severely constrained by the thinness of the ocean and atmosphere. Therefore, their vertical deflections are less pronounced than the horizontal ones.

We would have an indication that our immediate environment is fooling us by watching "stationary" far-away things, such as the Sun, Moon, and stars, which also seem to whirl around us every day. Nonetheless, it is understandable that both our perceptions and scientific measurements are made relative to our immediate worldly environment rather than relative to the distant stars.

It is sometimes convenient to imagine that we are not rotating, and that some imaginary force is causing the perceived deflection of moving objects. We call these imaginary forces **inertial forces**, because they are simply the result of the object trying to maintain motion in a straight line while we are turning. The tendency of an object to continue in a straight line is called its **inertia**, which is measured in units of mass (e.g., kilograms). More massive objects have more inertia, and so correspondingly, larger forces are required to alter their motion.

Inertial forces can seem very real (Figure 2.10a). For example, when our car turns a corner, our bodies tend to tip over as if being shoved toward the outside of the curve. Of course, this "shove" is really an imaginary *inertial* force. Our bodies simply try to keep going in a straight line, as the car turns out from under us. We are fooled by our immediate environment—the interior of the car. Relative to that, we are tending to slide outward. For a more global view, we would have to look at things outside the car—the trees, clouds, buildings, and so on—and then we would see that our bodies are simply trying to go in a straight line.

When our car turns a corner, our bodies tend to tip over as if being shoved toward the outside of the curve. Of course, this "shove" is really an imaginary inertial force. Our bodies simply try to keep going in a straight line, as the car turns out from under us.

Because the Earth's surface is curved, rather than being flat like a merry-go-round, there is a vertical component to the apparent deflection of things traveling in straight lines, as illustrated in Figure 2.10b. But the ocean and atmosphere are very thin, and so the vertical motions of these fluids are highly constrained. Consequently, in this book we will be only concerned with the horizontal (i.e. parallel to the Earth's surface) part of the Coriolis effect.

As an example of the Coriolis effect, consider the flow of winds around high- and low-pressure centers in the Northern Hemisphere. Because of the pressure gradient, air tends to flow from high pressures to low

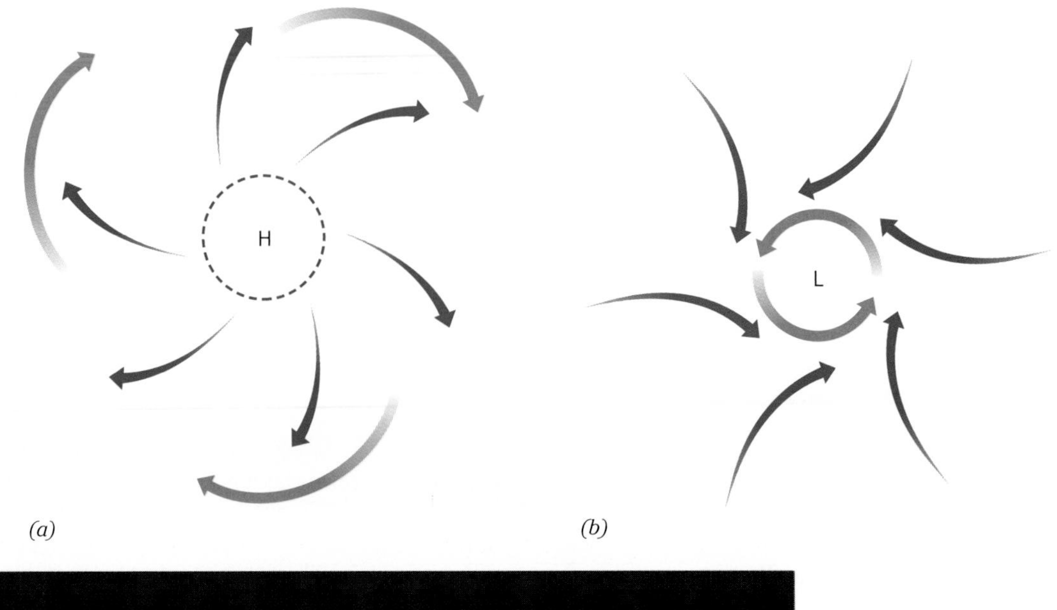

Figure 2.11
(a) Due to the Coriolis effect, winds leaving high-pressure centers get deflected to the right (Northern Hemisphere), causing clockwise motion around a high. (b) Winds coming toward a low-pressure center are deflected and miss the center to the right, causing counterclockwise motion of the air around a low. (c) Satellite photo that shows surface winds spiraling inward towards a low pressure center in the Northern Hemisphere.

pressures. Because of the Coriolis effect, winds leaving a high-pressure center in the Northern Hemisphere curve to the right as they go. Seen from above, they spiral out clockwise, as illustrated in Figure 2.11a. Similarly, winds flowing inwards toward a low-pressure center in the Northern Hemisphere curve to the right and miss their mark on that side. This causes counterclockwise circulation around such low-pressure centers, as shown in Figure 2.11b.

Winds circling around a low-pressure center are called **cyclonic winds**, and the wind direction around a high-pressure center is **anticyclonic**. In the Northern Hemisphere, cyclonic winds circle counterclockwise, and anticyclonic winds go clockwise. The motion is opposite in the Southern Hemisphere, because the Coriolis deflection is opposite.

In the oceans, many currents appear to follow looped paths, with the loops being caused by the Earth's spin (Figure 2.12). These loops are called **inertial currents**, being named after the *inertial forces* that make them seem to curve.[6] Because of the Earth's spin, all ocean currents tend to do this, but their response is often constrained by other forces. For example, we will learn in Chapter 8, that the boundaries of continents and slopes in the water's surface are also important factors that influence the direction in which the water may flow.

[6]Remember that this apparent curvature is because we are turning along with the Earth, while these currents are trying to flow in a straight line.

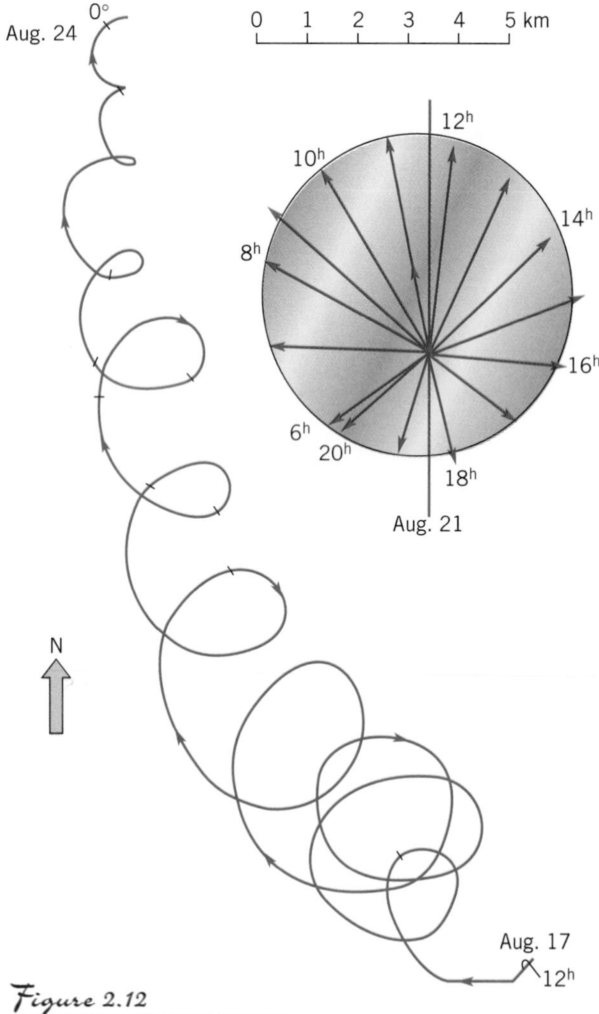

Figure 2.12
Chart of the progress of an inertial current in the Baltic Sea, showing the characteristic looped path. (From Sverdrup, Johnson, and Fleming, 1970, *The Oceans*, Prentice-Hall, Englewood Cliffs, NJ.)

2.4 ENERGY

Energy is the ability to do **work**. Work is defined as the product of force times distance. For example, if you push a chair across the floor or lift it into the air, the amount of work that you have done on the chair is the product of the force you exerted times the distance you pushed or lifted. In short, *energy* is anything that has the ability to move something.

FORMS OF ENERGY

Energy comes in a variety of forms. A car speeding down the highway has energy of motion, called **kinetic energy**. So does a ball flying through the air. The rapid random thermal motions of the individual atoms or molecules in a substance are called *heat*, or **thermal energy**, which is kinetic energy on an atomic scale. Notice that all have the ability to push something. A speeding car will obviously push just about anything it hits. So will a flying ball, although it may not push as hard or far as the car. The rapidly moving molecules of the hot gases in the cylinders of that car push the pistons that make the car go. The high pressure of the heated gases is due to the violent motions of the individual molecules. On a beach, the kinetic energy of the waves moves around the sediments.

> In short, energy is anything that has the ability to move something.

The energy of electrons moving through a wire is called *electrical energy*, and energy stored in a stretched rubber band or compressed spring is referred to as **potential energy**, because it is quickly converted into energy of motion when the spring or rubber band is released. **Chemical energy** is potential energy stored on an atomic scale in many substances, including fossil fuels and food. This stored chemical energy is released when fossil fuels or food are burned. That is, when the molecules combine with oxygen, the electrons fall into configurations of lower potential energy—similar to balls rolling downhill, but on an atomic scale. When this chemical energy is released, the molecules jiggle more violently, which we identify as the release of heat or, more appropriately, thermal energy.

All these forms of energy are familiar to us. All have the ability to do work—that is, to make something move. You can probably think of common examples where each of these does work.

The oceans carry energy in many ways. There is the kinetic energy of motion in the ocean currents and waves, and the thermal energy stored in the surface waters that are heated by the Sun. There is potential energy in elevated portions of the ocean surface, such as in the wave crests or slopes in the water surface caused by winds and currents. This potential energy is converted to kinetic energy of motion as gravity pulls the elevated regions back down, or water slides downhill off the slopes. Chemical energy is stored and released in a variety of oceanic processes, not the least of which is the production and consumption of food by marine plants and animals.

Energy can be converted from one form to another. For example, the chemical potential energy stored in gasoline becomes thermal energy when it is burned, and the heated gases expand against a piston in a cylinder of your car. Through a series of mechanical devices, these moving pistons push your car down the road, giving the car kinetic energy (energy of motion). In this book, we will learn how some of the Sun's thermal energy, through a series of conversions, ends up in the motions of winds, waves, ocean currents, and in living organisms. We will also learn how energy transformations in the Earth's interior cause portions of its surface to move, ocean basins to open, and continents to collide.

UNITS

The **temperature** of a material is a measure of its molecular motion. Scientists usually measure temperatures using the **Celsius scale**, using the notation °C for degrees Celsius. At atmospheric pressure, water boils at 100°C, freezes at 0°C, and the coldest temperature possible (*absolute zero*) is at –273°C. Higher temperatures mean that atoms and molecules are wiggling and jiggling more violently. They also collide more violently, causing them to spread out (usually). That is why most materials expand and become less dense when heated.

The amount of thermal energy held by an object depends on both its temperature and size. For example, water has more thermal energy when hot than when cold, and an ocean has more than a raindrop. A common unit for measuring thermal energy is the **calorie**, which is defined as the amount of thermal energy needed to raise the temperature of 1 gram of water by 1°C. In the metric system, a common unit of energy is the **joule**, which is the equivalent of 4.18 calories. Although any kind of energy may be measured in either unit, there is a tendency to use calories for thermal energy and joules for all other forms.

Many of the processes that occur on Earth are driven by energy received from the Sun. This energy is converted, stored, and utilized in a variety of ways, including those essential to the atmosphere, the oceans, and life. We will be studying these effects of solar energy throughout much of this book.

Summary

Atoms

Each atom has a tiny dense nucleus, surrounded by a cloud of much lighter electrons. Normally, an atom is electrically neutral, with the number of negatively charged electrons in the electron cloud being equal to the number of positively charged protons in the nucleus. The simplest and most abundant element in the Universe is hydrogen which has one proton in the nucleus and one electron in the electron cloud. Next comes helium, with two of each, and the list continues up to uranium, which has 92 of each.

Hydrogen and helium were the only materials present in the early Universe and are still by far the most abundant. All the other elements were produced in the nuclear reactions that go on in the interiors of stars.

Chemical Bonds

Atoms prefer certain symmetric electron configurations. These include one shell with two electrons, a second shell with eight more, a third shell with still eight more, and additional shells with still more electrons. Atoms combine by sharing electrons. They tend to combine in such ways that each atom is at least part of the time able to enjoy the preferred arrangement of completed shells.

Helium already has a completed shell of two electrons, so it doesn't share electrons. It is chemically inert. Other inert gases are those elements with two, three, four, and five completed shells, respectively. Hydrogen has only one electron and has a valence of 1. Carbon, nitrogen, and oxygen are fairly abundant elements that have valences of 4, 3, and 2, respectively, and may combine with hydrogen to form methane, ammonia, and water.

If the shared electrons are shared rather equally among the atoms, the bond is called covalent. If one bonding partner hoards the shared electron(s) at the expense of the other, the bond is said to be ionic, and the two parts are oppositely charged. Common materials that give up electrons are metals like aluminum, iron, calcium, sodium, potassium, and magnesium. Common materials that accept electrons include oxygen, chlorine, and many radicals such as silicate, carbonate, sulfate, and phosphate.

Forces

There are fundamentally only three kinds of forces at work in nature. The strong forces bind together the nucleus. The electrical forces are responsible for the structure of matter at the atomic and molecular level. Although both these types of forces are extremely strong and influential at these tiny distances, neither is very influential at larger distances, such as those that govern the motions of planets and oceans. Imposed forces may be transferred through a material via the electrical forces between neighboring atoms. The force of gravity between two objects is directly proportional to the mass of each and inversely proportional to the square of their separation. Although all objects pull on all others, the strongest gravitational forces come from those objects that are massive and close.

At each depth in a fluid, the pressure must be sufficient to support the weight of all the overlying fluid. Therefore, the pressure must increase with depth. This means that the pressure on the bottom of a submerged object is greater than that on its top, and so it receives a net upward force from the fluid, called the *buoyant force*. In equilibrium, a fluid supports its own weight. Therefore, when a submerged object displaces some of this fluid, it receives an upward force from the fluid that is equal to the weight of the fluid it displaced. This is Archimedes' principle. Fluids flow from high pressure to low pressures. A change in pressure is called a *pressure gradient*.

The Coriolis effect is caused by the Earth's spin. As we turn, things that are stationary or moving in straight lines appear to turn the opposite way. In the Northern Hemisphere, moving objects appear to curve to the right as they go, and in the Southern Hemisphere, they appear to curve to the left. The tendency of things to continue moving in a straight line is called *inertia*. Therefore, we sometime attribute the apparent curvature of things moving in a straight line while we observers are turning to *inertial forces*. The Coriolis effect and inertial forces can be experienced in any turning reference frame, such as a merry-go-round, or a car going around a corner.

Energy

Energy is the ability to do work or, equivalently, to push something over some distance. It comes in many forms, such as the kinetic energy of motion, thermal energy of motion on an atomic level, potential energy of position (which can become motion when it is released), electrical energy of electrons in a wire, or chemical energy that is held in electron configurations. It can be converted from one form to another, as happens in most important processes on Earth. Temperature measures atomic and molecular motion, with higher temperatures corresponding to more violent motions. Thermal energy is commonly measured in a unit called a calorie, which is the amount of energy needed to raise the temperature of 1 gram of water by 1°C. The metric unit for energy is the joule, which is the equivalent of 4.18 calories.

Key Terms

atoms
elements
atomic number
molecules
electron shell
valence
inert gas
methane
ammonia
water
covalent bond
ionic bond
ion
radical
silicate
carbonate
sulfate
phosphate
silica
strong force
electrical force
gravity
law of gravity
buoyant force
fluids
density
pressure
Archimedes' principle
pressure gradient force
Coriolis effect
inertial forces
inertia
cyclonic
anticyclonic
inertial currents
energy
work
kinetic energy
thermal energy
potential energy
chemical energy
temperature
Celsius scale
calorie
joule

Study Questions

1. Why do the abundances of the elements on and near the Earth's surface differ from their abundances in the Universe as a whole?
2. Describe the preferred electronic arrangement in atoms. Explain how this influences the chemical interactions of hydrogen, helium, carbon, nitrogen, and oxygen.
3. What is the difference between covalent and ionic bonds? Give an example of each.
4. What is the valence of an oxygen atom? Use this and the indicated valences of the silicate, sulfate, and phosphate radicals to decide the valence of silicon, sulfur, and phosphorus, respectively.
5. Because each aluminum atom gives three electrons and each silicate radical accepts four, it takes four aluminum atoms to give the same number of electrons that are accepted by three silicate radicals (12 electrons total, in each case). Therefore, a possible material is $Al_4(SiO_4)_3$. Using similar reasoning, construct some other possible metal-radical combinations.
6. What are the three basic kinds of forces, and at what distance is each important? Why are electrical forces usually not important at large distances?

7. State Archimedes' principle and explain why it is true.

8. Explain why a hot air balloon rises, and why cold water sinks. Would the hot air balloon rise if it were in a vacuum? Why does a ship sink deeper into the water as it is being loaded?

9. Explain why you tend to tip over when your car goes around a curve.

10. Why would it be difficult to engage in a snowball fight on a spinning merry-go-round?

11. Temperature measures what? What else does the thermal energy of an object depend on?

13. Give some common examples that illustrate energy being converted from one form to another.

Critical Thinking Questions

1. If the strong force can hold together nuclei with 10 or 50 protons, why can't it hold together 100? That is, why are the large nuclei unstable?

2. In the violent interiors of stars, nuclei engage in very forceful collisions. Once formed, some nuclei are stable, and some are unstable (i.e., fall apart). From the information given in this chapter, would you guess that nuclei of elements with atomic numbers 3, 4, and 5 are more or less stable than those with atomic numbers 6, 7, and 8? Defend your answer.

3. Why do you suppose the colder planets that are farther from the Sun contain large amounts of hydrogen and helium, whereas the warmer inner planets have very little?

4. Rather than flowing straight from a high-pressure center to a low-pressure center, winds tend to blow sideways between them. For example, if you are standing in a high-pressure center and looking directly toward a low-pressure center, the wind would be blowing from left to right (Northern Hemisphere) in front of you. Why?

5. Suppose a man weighs 180 pounds (800 Newtons). If he moved three times farther from the Earth's center, how much would he weigh? How much would he weigh on the surface of Mars which has one-tenth as much mass as Earth and one-half the radius?

Suggestions for Further Reading

CHANG, RAYMOND. 1991. *Chemistry*, 4th ed. McGraw-Hill, New York.

EBBING, DARRELL D. 1984. *General Chemistry*. Houghton Mifflin, Boston, MA.

GIANCOLI, DOUGLAS C. 1984. *General Physics*. Prentice-Hall, Englewood Cliffs, NJ.

HALLIDAY, DAVID, ROBERT RESNICK, AND JEARL WALKER. 1993. *Fundamentals of Physics*, 4th ed. John Wiley & Sons, New York.

MCQUARRIE, DONALD A. AND PETER A. ROCK. 1984. *General Chemistry*. W. H. Freeman, New York.

SEARS, ZEMANSKI, AND YOUNG. 1976. *University Physics*, 5th ed. Addison-Wesley, Reading, MA.

SIVARDIERE, J. AND A. P. FRENCH. 1988. Coriolis Deflection and Air Resistance. *American Journal of Physics* 56:1, 88.

The relative thickness of the hydrosphere is equivalent to that of a coat of paint on a bowling ball.

View of the ocean planet from outer space.

Three

PLANET EARTH

About 5 billion years ago, a huge cloud of interstellar gas and dust slowly began to collapse, being drawn together by gravitational forces among the myriad of microscopic components (Figure 3.1). As it collapsed, the cloud spun faster and faster, like a huge cosmic figure skater pulling in her limbs. The spinning caused it to flatten. Further gravitational collapse of smaller regions within this flattened cloud formed the members of our Solar System. The central massive portion became the Sun. Due to the Sun's immense mass and strong gravity, its deep interior regions experienced tremendous compression, igniting the nuclear reactions that make the Sun shine.

The entire system, including the Sun, planets, satellites, and asteroids, were all formed from the same massive mixture of materials in the original interstellar cloud. Some of the smaller and warmer objects were not able to hold onto the two lightest and most elusive gases: hydrogen and helium. But with this exception, all the members of the Solar System are made of the same materials.

Why, then, do the various planets and satellites look so different? The Earth sparkles in blue and silver, rather than carrying the drab reds or browns of its neighbors. (Refer to Figure 1.2 in Chapter 1.) It blossoms with life in myriads of forms—grass, trees, alligators, jellyfish, koala bears, and people who are able to study and appreciate these features.

The recognition of our own uniqueness raises the question, "To what do we owe this distinction?" The answer is that the single most important feature that sets Earth apart from its neighbors is its oceans. To them, we living creatures owe both our origin and our sustenance. Without them, Earth would be quite different, and we would not be here to study it.

3.1 VIEW FROM OUTER SPACE

To notice Earth's distinction does not require close observations of its life forms. In fact, its unique coloration could be spotted from distances of billions of kilometers—a point not overlooked by writers of science fiction space adventures. Interstellar travelers with any curiosity would certainly want to investigate.

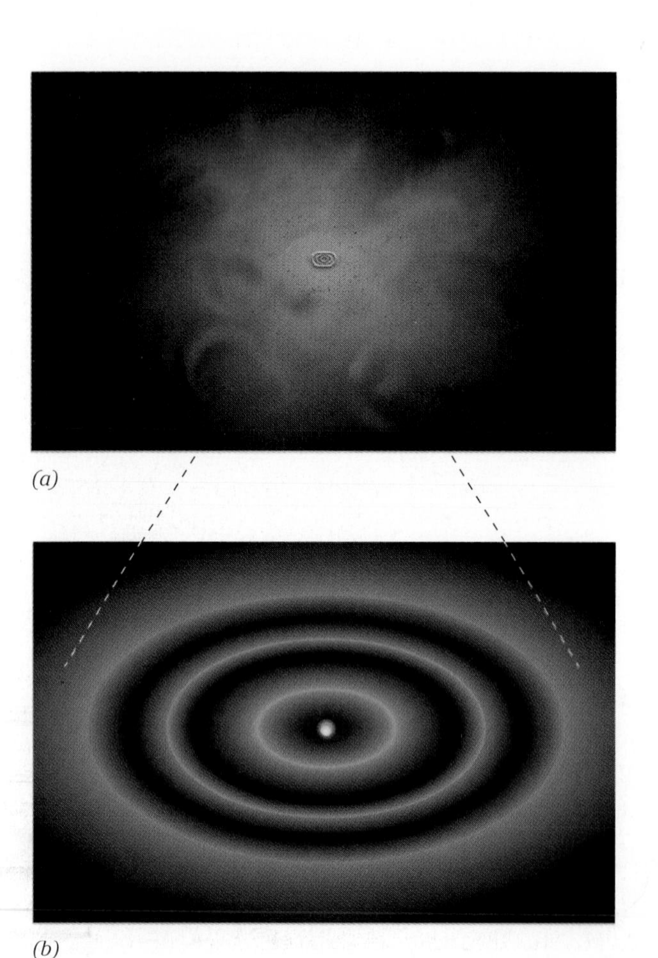

(a)

(b)

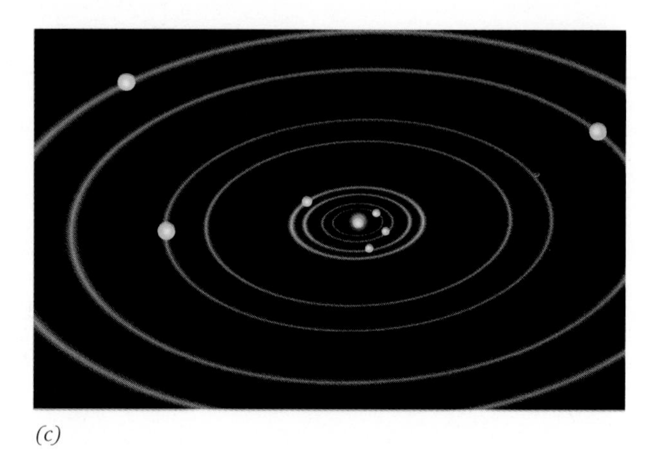

(c)

Figure 3.1
Formation and evolution of the Solar System: (a) Gravitational collapse of a very large cloud of interstellar gas and dust, forming a denser central region. As this region collapses further, it spins faster and flattens more. (b) Enlargement of the spinning central region. Further gravitational attraction causes accumulations of matter within the spinning disc. (c) Gravity continues to pull the matter together until it has nearly all accumulated into the individual massive objects that form the Sun, planets, asteroids, moons, and so on.

THE HYDROSPHERE

The Earth's distinctive deep blue and bright silvery coloration is due to the very thin outer coating, which includes the ocean and atmosphere. This thin fluid region is called the **hydrosphere**, because of the abundance and influence of water. Both the ocean and atmosphere are only a few kilometers thick. This makes the relative thickness of the hydrosphere equivalent to that of a coat of paint on a bowling ball. (You can see this in the chapter opener, or Figure 3.4, for example.) This fluid "coat of paint" is what makes Earth appear so distinctively different from the other **terrestrial** (i.e., hard-surfaced, or Earthlike) planets.

> The relative thickness of the hydrosphere equivalent to that of a coat of paint on a bowling ball.

Within the hydrosphere are several different kinds of water reservoirs, listed in Table 3.1. Because we land creatures are so familiar with our local freshwater sources, it may be surprising to learn that only 0.6% of the water is stored in ground water, and only 0.01% in lakes and rivers. Even more surprising might be that the atmosphere holds only a thousandth of a percent of the water. If all this water were to fall in one huge worldwide rainstorm, it would average only 3 centimeters of rainfall. Practically all the Earth's water is contained in the oceans, and most of the rest is stored in polar ice.

Table 3.1 Amounts of Water in the Various Reservoirs on Earth

Reservoir	Percent of Total	Sphere Depth[a] (m)
Oceans	97.41	2685
Polar caps and ice	1.94	53
Ground water	0.64	18
Lakes	0.01	0.3
Atmosphere	0.001	0.03

[a]This is the depth the water would have if the Earth's surface were perfectly smooth, and the water were spread uniformly over it. (Although the volume of the oceans is accurately known, estimates of ground water and Antarctic ice vary.)

In spite of its small water content, the atmosphere is vital in the transfer of water from one reservoir to another. The ocean loses water to the atmosphere via evaporation, but gains it back through precipitation, runoff from land, and melting ice. The atmosphere carries some of the water from the oceans to the polar caps, where it is deposited as snow. Some is deposited on the continents through rainfall, snow, and dew. The excess precipitation

Figure 3.2

The hydrologic cycle. Annual water transport via the various pathways are indicated in thousands of cubic kilometers of water per year. (For reference, 1000 cubic kilometers are about 0.00007% of the ocean's volume.) (After Maurits la Riveiere, *Scientific American* **261**:3, Sept. 1989.)

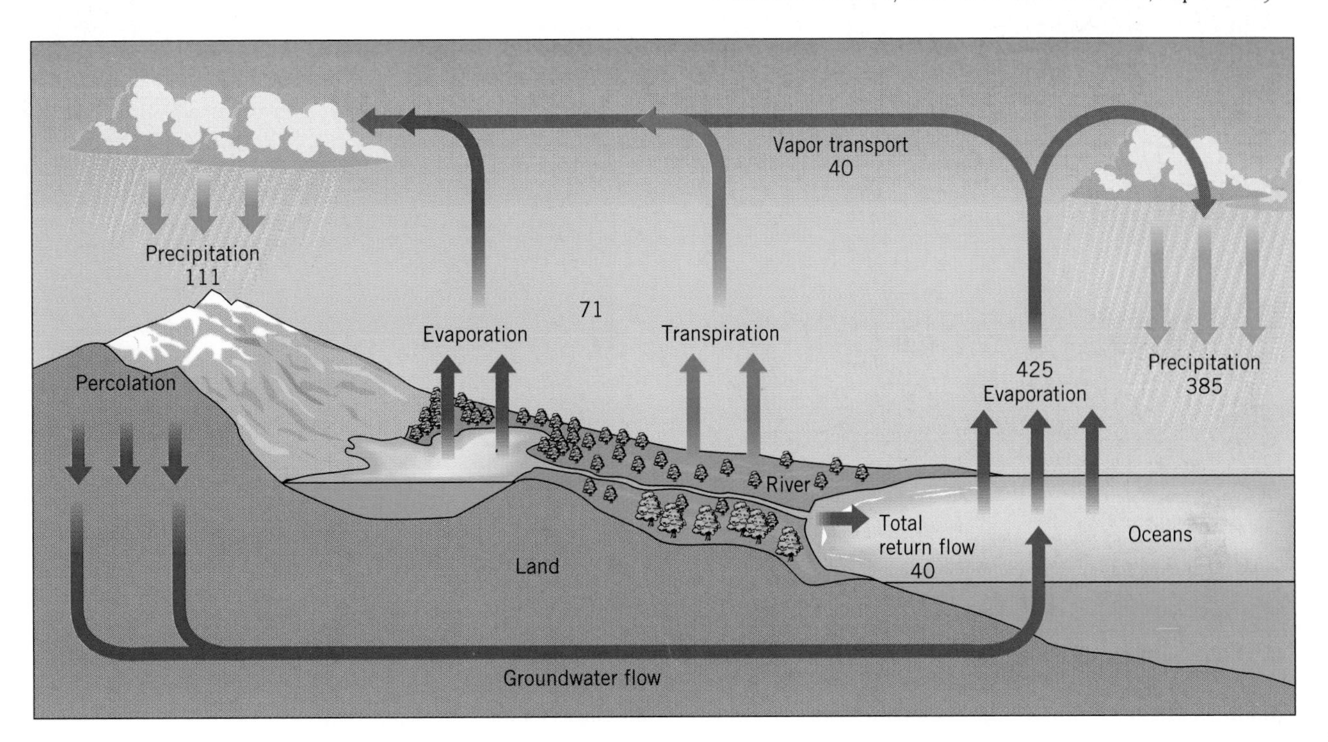

over the continents is returned to the oceans by rivers and underground flow. The elements of this *hydrologic cycle* are depicted in Figure 3.2.

OVERALL SHAPE

Unlike its coloration, the Earth's shape is similar to that of any other planet. Gravity has pulled it into a nearly spherical shape, with an average radius of 6371 kilometers. The reason for the spherical shape is that gravity pulls toward the Earth's center, pulling equally from all directions. Any deviations from a purely spherical shape would amount to huge mountains and valleys on the surface. Over time, these mountains would flatten and the valleys would fill in, because gravity causes materials to flow downhill, some more quickly than others. Geological features, such as mountains on the Earth's rocky surface, require millions of years for major deformation. Although this is slow on human time scales, it is fast compared to the 4.6-billion-year history of the Earth. So over time, the Earth and all large astronomical bodies pulled themselves into spherical shapes. Only small objects, whose gravity is too weak to deform rock over these long time scales, remain irregular today (Figure 3.3a).

Just as spinning pizza dough tends to be stretched and flattened, so does the Earth. Although the dominant force of the Earth's gravity keeps it nearly spherical, the spin-flattening causes a small **equatorial bulge**, or fattening near the equator. The Earth's equatorial radius is about 21 kilometers or 0.3% larger than the polar radius (Table 3.2). This is too small to be noticed by the naked eye. By contrast, the larger and faster spinning outer planets, such as Jupiter or Saturn, are sufficiently flattened so that their equatorial bulges are noticeable (Figure 3.3b).

EVIDENCE OF ACTIVITY

Any interstellar observers approaching Earth to examine it more closely would discover additional curious and unique features (Figure 3.4). The solid surface has two distinct elevations: one characteristic of the continents (0–1 km above sea level) and one characteristic of the ocean basins (4–5 km below sea level). Relatively little surface is at higher, lower, or intermediate elevations. What process could have caused these two distinct surface regions?

Our hypothetical interstellar observers might notice that one ocean (the Pacific Ocean) dominates

(a)

(b)

Figure 3.3

(a) Phobos. This small Martian Moon is only 22 km across. Its gravity is too weak to pull it into a smooth spherical shape. (b) Saturn. This planet is much larger than Earth, and spins much faster. Consequently, its spin flattening is more pronounced. You can see that it is fatter in the equatorial plane (defined by the rings) than it is from pole to pole.

Table 3.2 Variation of the Earth's Gravity and Radius with Latitude

Latitude	Radius (km)	Acceleration of Gravity[a] (cm/sec^2)
0°	6378	978.04
45°	6366	980.53
90°	6357	983.22
Average	6371	981.00

[a]This measures the rate of increase in velocity of a freely falling object.

Figure 3.4
Although the Earth's distinctive markings would be noticeable from great distances, closer inspection would reveal additional curious features, such as places where the continents are tearing apart.

half the world and is as large as the other two oceans (the Atlantic and Indian Oceans) combined (Figure 3.5). They might also notice that the edges of some continents fit together like the two halves of a torn sheet of paper (Figure 3.6). For example, the two sides of the Atlantic Ocean fit together as if Europe and Africa have somehow been torn away from Greenland and the Americas.

An even closer look would reveal curious long, thin features, such as mountain ranges and ocean trenches. Had these formed early in the Earth's history, and if there were no ongoing geological processes to create or maintain them, gravitational flow and erosion would long ago have erased them from the surface. Therefore, they must be relatively young features, resulting from recent or ongoing processes (Figure 3.7). Samples of rocks from the ocean bottoms and the continents would reveal that the ocean bottoms are relatively simple and young, typically being only 1 or 2% as old as the Earth. By contrast, the continents would be found to be old and complex, indicating a long and dynamic history. These two types of terrains must clearly have different origins and evolutions.

In summary, not only do our deep blue oceans and silvery clouds give Earth a unique appearance that can be noticed from great distances, but also, a closer inspection reveals that the Earth is geologically both intriguing and active. You might suspect that these peculiar features are related—that the unique appearance might be related to the geological structure and activity. Indeed they are, as we will see in the rest of this chapter and the next. The oceans owe both their origin and evolution to the Earth's geological activity, and this activity sets Earth apart from other planets in the Solar System.

But that is getting ahead of our story. For now, just notice that the oceans flow into the low spots in

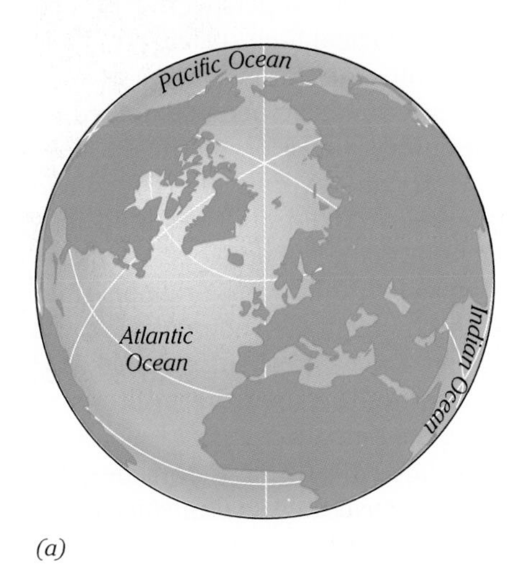

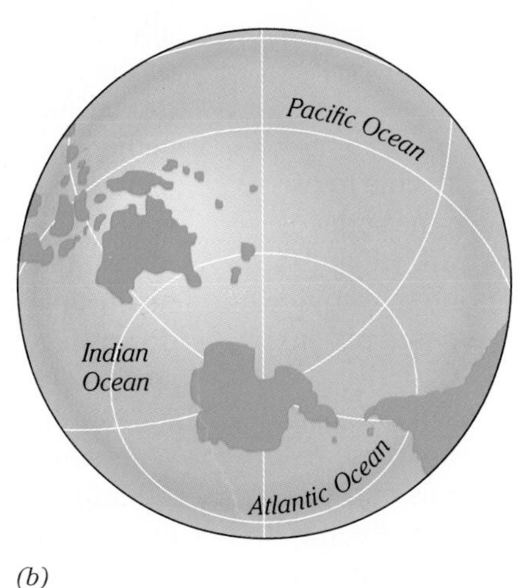

Figure 3.5
The distribution of continents and oceans is quite asymmetrical. One hemisphere (a) contains most of the land mass. The other hemisphere (b) is dominated by the Pacific Ocean.

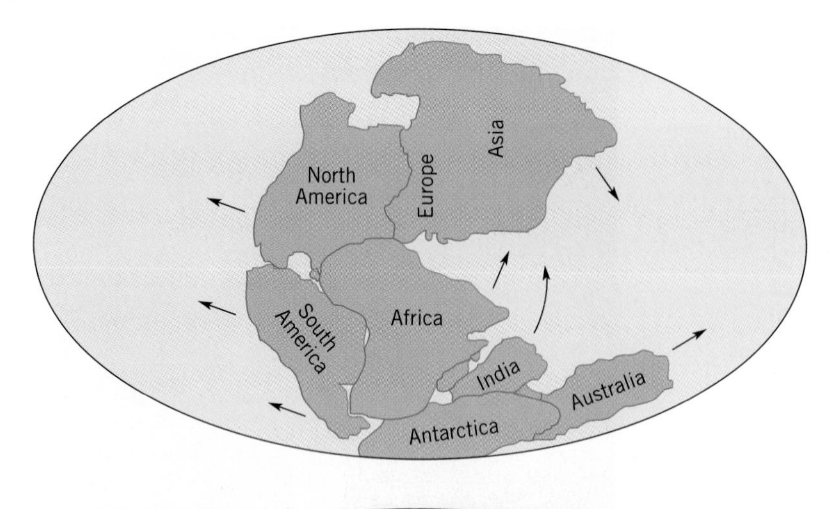

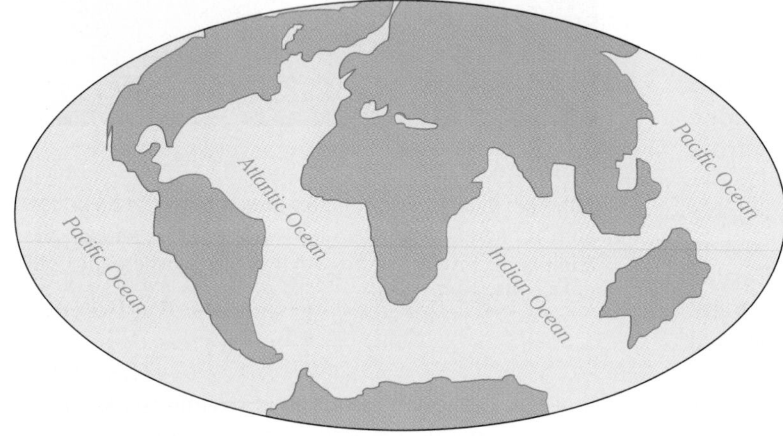

Figure 3.6

(top) The supercontinent, Pangaea, as it appeared 200 million years ago. The subsequent motion of the various continental fragments is indicated by the arrows. (bottom) The continents as they appear today (including submerged continental margins). North and South America did indeed tear away from Europe and Africa, which accounts for the similar shapes of the Atlantic boundaries of these continents today.

the Earth's surface, thereby taking on whatever configuration permitted them by the solid Earth beneath. Therefore, to understand the origin, evolution, and present general features of the oceans requires an understanding of the structure and processes in the "solid" Earth beneath them.

3.2 VIEW OF INNER SPACE

The Earth's inner space is much less accessible to us than its outer space. We can only scratch the surface with our direct probes, so for the most part, we must use indirect techniques to determine its internal composition. Some of these techniques are described in Section 3.3.

The Earth has several interior regions of different material composition, and the properties of these regions vary with both the types of materials and the physical environment they are in. In addition, there is increasing evidence of multiple layers within each of these major interior regions, with denser materials found in deeper layers.

TIME SCALES AND MOTION

Much of the Earth's interior is composed of what we might describe as "solid" rock. However, what appears to be solid from a human perspective is actually fairly "fluid" on geological time scales. You have probably noticed slanted or warped layers of rock formed from sediments that have been cemented together (i.e., **sedimentary rock**) on mountain faces or that have been exposed by cuts made for highways. These have been raised and wrinkled into forms quite different from the flat horizontal beds beneath the ocean in which they were originally laid down (Figure 3.7). Although these sedimentary rocks seem solid to us, there has clearly been considerable movement and deformation over the millions of years since their original formation.

Although the solid Earth displays considerable fluid motion over geological time scales, some of its interior regions flow more easily than others. This de-

Figure 3.7
Originally laid down in horizontal layers beneath the ocean and slowly hardened into rock, these sedimentary rocks were later gradually tilted and raised to create these tall mountains, which testify to the Earth's geological activity. (Rocky Mountains, British Columbia)

pends on not only the material, but also the temperature and pressure. Materials tend to flow more easily at higher temperatures, as you can demonstrate by heating butter or molasses in your kitchen. Since temperature increases with depth in the Earth, you might think that deeper materials are more fluid. However, pressure also increases with depth, and greater pressures tend to make things *less* fluid. These two opposing influences, coupled with variations in materials, make for interesting variations in the resistance to flow, or the **viscosity**, of the Earth's interior materials.

INTERNAL ENVIRONMENTS

As we learned in Chapter 2, pressure increases with depth, because each layer must support the weight of the layers above. At greater depth, there is more material above to be supported. Temperature also increases with depth, because of various sources of internal heating. The primary heat source is the **radioactive decay** of certain types of atomic nuclei that are unstable and fall apart, releasing various kinds of energetic subatomic particles when they do.

But many other processes provide interior heating as well. One is the original heat released when our planet first formed. As gravity drew together the original materials, the collisions and compression of these materials produced heating. Some of that heat still remains trapped inside the Earth. Heat is still being released as denser materials sink slowly toward the Earth's center, and lighter materials rise toward the surface. This separation of materials of different densities is referred to as **gravitational differentiation**. As you can guess, there is a great deal of friction in the motion of these dense and viscous materials, and both friction and compression release heat. Other sources of internal heating include the heat released as materials slowly change their atomic arrangements, tidal friction due to deformations in the Earth caused by the Moon and Sun, and friction in the flow of the liquid portions of the Earth's core.

The heat released in the Earth's interior eventually makes its way to the surface, where it is radiated into outer space. The Earth is coolest on the outside and gets hotter toward the interior, because heat can escape from the outer surface only (Figure 3.8). For the same reason, a bowl of soup or cup of hot coffee are coolest on the surface and hotter below.

The Earth is coolest on the outside and gets hotter toward the interior, because heat can escape from the outer surface only. For the same reason, a bowl of soup or cup of hot coffee are coolest on the surface and hotter below.

Not only does temperature increase as we go deeper into the Earth, but the density of the materials also increases, and for two reasons. First, denser materials tend to sink toward the interior, and lighter materials tend to rise toward the surface. Second, increasing pressures compress materials more at greater depths.

The tendency of denser materials to sink has caused the formation of three general regions of distinct compositions within the Earth, called the core, mantle, and crust, respectively. These three regions are illustrated and described in Figure 3.9 and Table 3.3. Within these three basic divisions is additional layering, but this layering is due primarily to differences in structure (i.e., atomic arrangements) rather than differences in materials.

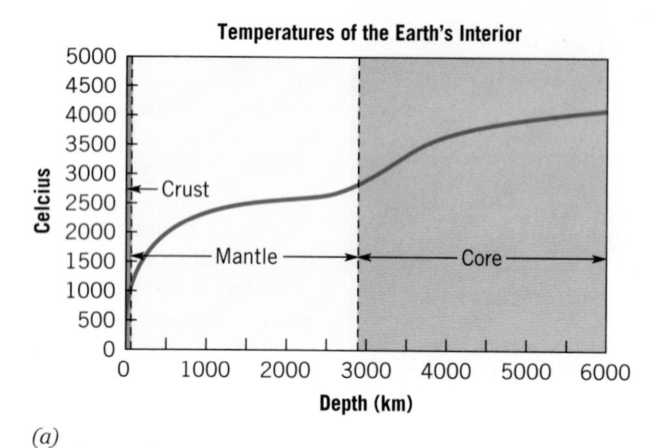

(a)

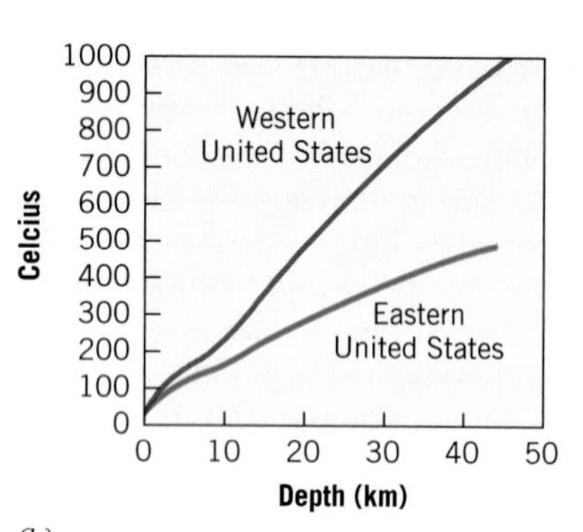

(b)

Figure 3.8

(a) Graph illustrating how the temperature increases with depth within the Earth. The sudden rise in temperature immediately beneath the surface indicates that the surface layers are good thermal insulators, trapping excess heat inside. At greater depths, the gradual rise in temperature is mostly due to the slow motion of the materials, with materials being compressed and heated as they sink, and expanded and cooled as they rise. Also indicated in this figure are the three general regions of the Earth's interior: the crust, mantle, and core. (b) Estimated temperature profiles for the first 50 kilometers beneath the surface (within the crust), typical of the western and eastern United States, respectively. Notice that the temperature rises more quickly with depth on the West Coast, where there is greater geological activity, and the crust tends to be thicker, providing both more heat sources and better overlying insulation. (After Lachenbruch and Sass, 1977, in Pecock (ed.), *The Earth's Crust* (Geophysical Monograph 20), pp. 626-675.)

THE CORE

The innermost region of the Earth is called the **core**. It extends outward from the Earth's center to a little more than half-way to the surface, encompassing 16% of the Earth's total volume.[1] The core can be further subdivided into a solid **inner core** and liquid **outer core**. The transition from solid to liquid at the interface between the inner and outer core is probably due primarily to differences in the pressure on the materials rather than differences in their composition or temperature.

How could we possibly know what is in the Earth's core? Although we cannot sample the core, we can make fairly reliable estimates of its composition from two different approaches. In the first approach, we estimate the density of the Earth's internal regions from the strength of its gravity. We then look for common materials that would have that particular density at the pressures and temperatures existing in these regions.

In the other approach, we compare the Earth's composition with that of other nearby objects in the Solar System, which formed from the same primordial cloud and therefore are made of the same materials. The most convenient of these objects are fallen meteorites. These comparisons reveal many differences. We conclude that those materials relatively lacking at the Earth's surface must have sunk deeper into the interior.

From both these approaches, we arrive at the same conclusion. The core is composed mostly of a mixture of the two metals iron and nickel, although other materials must also be present in smaller amounts. Sulfur may be one of these other important materials, although there is presently considerable variation among estimates of its relative abundance there.

The density of the core is estimated to range from 9.4 grams per cubic centimeter near the outer edge to 17.2 grams per cubic centimeter near the center. For comparison, the density of water is 1 gram per cubic centimeter, so the core ranges from

[1]The volume of an object is proportional to (width)×(height)×(thickness). If you double the dimensions of an object, its volume increases by a factor of 2×2×2 = 8. Therefore, the inner half of a sphere contains only one-eighth of its total volume, and the outer half contains the remaining seven-eighths of the volume.

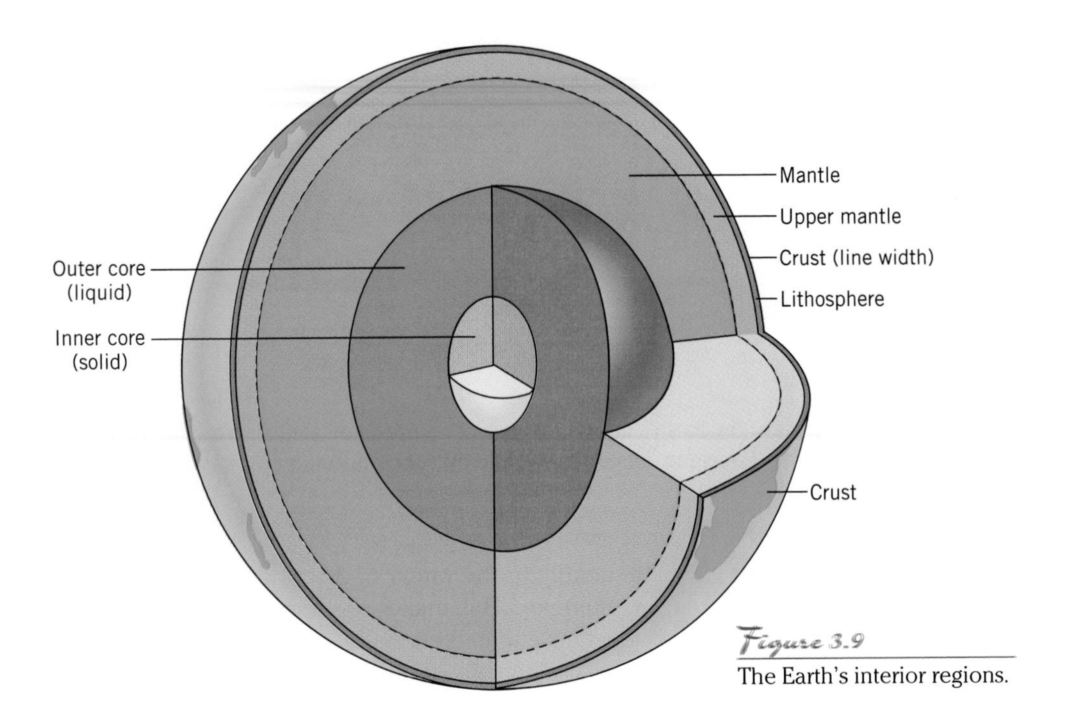

Figure 3.9
The Earth's interior regions.

Table 3.3 Interior Regions of the Earth

Region	Depth below the Earth's Surface (km)	Density (g/cm^3) Range	Average
Crust	0–30	2.5– 3.3	2.8
Mantle	30–2900	3.3– 5.7	4.5
Outer core	2900–5100	9.4–14.2	11.8
Inner core	5100–6370	16.8–17.2	17.0
Entire Earth	0–6370	2.5–17.2	5.5

9.4 to 17.2 times the density of water. Because of its high density, the core contains 33% of the Earth's total mass.

The outer core is responsible for the Earth's magnetic field, which makes navigation by magnetic compasses possible. Scientists are not sure of the exact mechanism by which the magnetic field is generated, but most believe it is related to the fact that the metallic core is a good electrical conductor and it is spinning.

Even more puzzling are the changes that occur in this magnetic field. It is not aligned exactly with the Earth's spin axis, and it moves around slowly. Compass needles today point to a slightly different magnetic North Pole position than they did in Columbus' time. Even more alarming is that at irregular intervals, averaging a few hundred thousand years apart, the magnetic field flips completely, so that a compass needle would turn around and point in the opposite direction. These phenomena must reflect interesting processes in the core, which we do not yet understand.

THE MANTLE

The **mantle** is the region of the Earth that extends from the outer edge of the core nearly to the surface. It is 2900 kilometers thick. It encompasses about 84% of the Earth's total volume and 67% of the total mass.[2] The mantle's density varies from about 5.7 grams per cubic centimeter near the bottom to about 3.3 grams per cubic centimeter at its outer edge.

Although there is still some uncertainty regarding its interior composition, we believe that at least the mantle's outer regions are composed primarily of metallic silicates and oxides. Iron and magnesium silicates and magnesium oxide are the main components, but various oxides of iron, aluminum, and calcium must also be present. These materials are more similar to the rocky materials we find on the Earth's surface than the iron-nickel alloy of the core (Figure 3.10).

As studies become more sophisticated and thorough, there is increasing evidence that the upper 670 kilometers of the mantle may be quite complex. Within this **upper mantle**, there is considerable vertical and horizontal motion (called *convection*), as well as regions where blocks of former surface materials are slowly sinking. All these motions require time scales of millions of years, but are rather fast for

[2]If you add up the numbers for the core and mantle, you will see that we have accounted for 100% of the Earth's volume and 100% of its mass. That doesn't leave much for the crust, which we discuss in the next section. The crust is extremely thin and accounts for much less than 1% of the Earth's mass and volume.

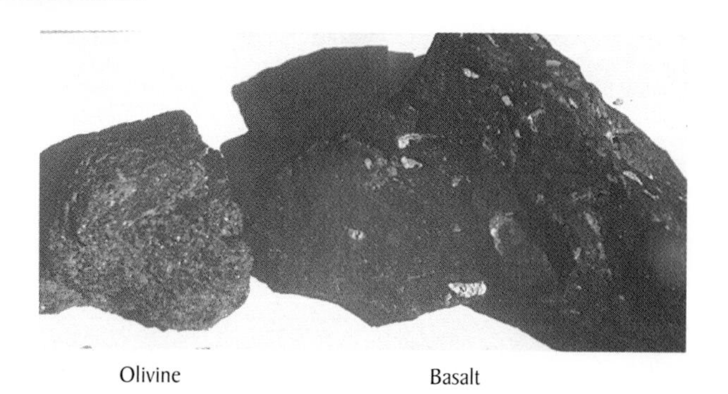

Granite

Diorite

Figure 3.10

Some examples of rocky materials from various regions of the Earth's interior. A rock of composition similar to the materials of the upper mantle (olivene), one of composition typical of oceanic crust (basalt), and two of composition typical of continental crust (granite and diorite).

geological processes. Since the materials that form the continents and ocean floor ride above it, the motions of this upper mantle are closely coupled to those of the continents and to the changing shapes of our oceans.

Our models for deeper portions of the mantle are more uncertain, because these deeper regions are further removed from our probing sensors. The interface between the upper and lower mantle at a depth of 670 kilometers (Figure 3.9) probably represents a distinct change in density. Like other less pronounced interfaces and layers within the mantle, this one may represent a difference in the atomic arrangements within the materials, rather than a change in composition.[3] Materials sufficiently dense to sink through the upper mantle may stop when they encounter this interface.[4] Occasionally, materials may cross this boundary as the Earth continues its gravitational differentiation.

THE CRUST

The very thin outer shell of the Earth is called the **crust**. Like the ice that forms on ponds and streams in the winter, the materials of the Earth's crust tend to be lighter, cooler, more rigid, and more brittle than those beneath them. They have floated to the surface and cooled, to form a very thin, light, cool rigid layer, which "floats" atop the denser, warmer, less rigid materials beneath. The densities of crustal materials are generally in the range of 2.5 to 3.3 grams per cubic centimeter. This makes them denser than water but less dense than the mantle, so water floats atop these crustal materials, and these crustal materials, in turn, float atop the mantle.

Like the ice that forms on ponds and streams in the winter, the materials of the Earth's crust tend to be lighter, cooler, more rigid, and more brittle than those beneath them.

Crustal Thickness The dividing line between the crust and mantle is called the **Moho** (Figure 3.11a), which is short for *Mohorovicic discontinuity*, named after the scientist who discovered it. The Moho shows up as a reflecting boundary for vibrational waves, called **seismic waves**, traveling toward the Earth's interior. Seismic waves can be created by such things as explosives or earthquakes.

By using the reflection of these seismic waves to find the depth of the Moho at thousands of different locations, scientists have made two rather interesting discoveries. First, the crust is very thin, averaging only about 15 to 20 kilometers in thickness. Compared to its lateral dimensions of tens of thousands of kilometers, this makes the crust's relative thickness roughly that of the rubber walls of an inflated balloon. Consequently, in order to make the crust's distinctive features visible when we sketch it in cross sections, we must use a great deal of vertical exaggeration.

The second curious feature is that the Moho is deepest beneath the highest mountain ranges and shallower beneath regions of lower surface elevations, as is illustrated in Figure 3.11a. For example,

[3]Both temperature and pressure increase with greater depth, and these environmental factors influence the preferred atomic arrangements within materials.

[4]This idea is still controversial. The gradual heating of materials as they sink through the upper mantle increases their buoyancy. Some suggest that this increased buoyancy is more important in preventing further sinking than a change in density of the mantle materials.

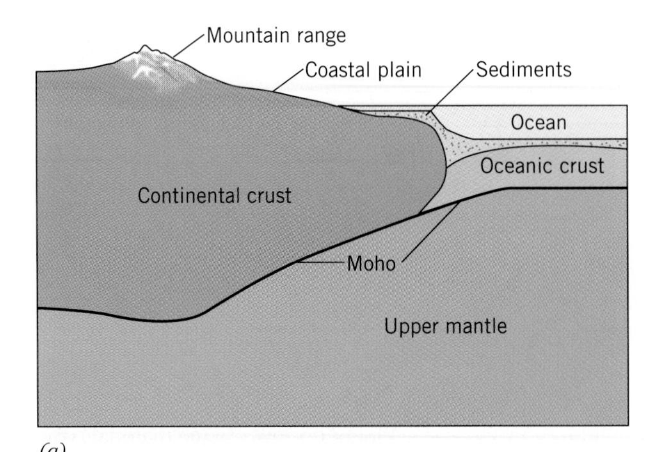

(a)

(b)

Figure 3.11

(a) Various thicknesses of the Earth's crust. The oceanic crust is relatively thin, 6 kilometers being typical. The crust thickens beneath the continents, typically 40 kilometers beneath coastal plains, and up to 80 kilometers beneath mountain ranges. The higher surface elevations have thicker underlying crust, which sinks deeper into the mantle due to its greater weight. (Vertical exaggeration about 10 times.) (b) Thicker objects sink deeper to gain the buoyancy needed to support their greater weight. For example, the much thicker iceberg in the background of this photo must sink much deeper into the ocean in order to support its weight than the thinner ice in the foreground. (For ice on water, there is about 9 times as much below the surface as there is above.)

the Moho reaches depths of about 70 kilometers beneath large mountain ranges, 40 kilometers beneath more typical continental regions, and only about 6 kilometers beneath the ocean bottom.

This is further evidence that the crust is "floating" on the mantle, with more massive portions reaching deeper to obtain the needed **buoyancy** (i.e., vertical lift, or support). Not only do more massive objects sink deeper, but their top surfaces also reach greater heights (Figure 3.11b). By analogy, the iceberg that sank the Titanic extended both deeper beneath the water and higher above the water than would a thin sheet of ice or an ice cube.

Just as a boat sinks when a hippopotamus steps in, so must the Earth's crust begin to sink when a large mass is placed on it. This is what happens when a volcano grows, for example. Conversely, as weight is removed from the crust, as happens as mountains erode, the crust rises. The adjustments that are required to obtain buoyant equilibrium[5] are referred to as **isostatic adjustments**. The condition of buoyant equilibrium, where the buoyant force is sufficient to support the weight, is called **isostasy**. More massive portions of the crust must extend deeper into the mantle to obtain the required buoyancy.

Crustal Materials There are two distinct types of crustal materials: one characteristic of continents and one characteristic of ocean bottoms. The materials of the **continental crust** tend to be less dense, lighter in color, thicker, and considerably older than those of the **oceanic crust**. Both types of crust are made primarily of the silicates of various metals. But a comparison of the two reveals that the oceanic crust has somewhat more iron, magnesium, and calcium, whereas the continental crust has comparatively more aluminum, potassium, and sodium. These two distinctive types of crustal materials are reflected in two distinct ranges of surface elevations observed on the Earth. As is illustrated in Figure 3.12, most of the Earth's surface elevations are either near sea level (i.e., the continents and their submerged outer edges) or around 4 or 5 kilometers below sea level (i.e., the deep ocean floor).

Given the differences in the two types of crustal materials, we can easily understand why the continents reach higher elevations than the sea floor. Both float on the denser materials below. But compared to the oceanic crust,

1. Continents are thicker.
2. Continents are less dense.

Being thicker, continents extend both deeper into the mantle and higher into the air. Being less dense, continents would float higher even if they were the same thickness.

[5]For a discussion of buoyancy, refer to Chapter 2.

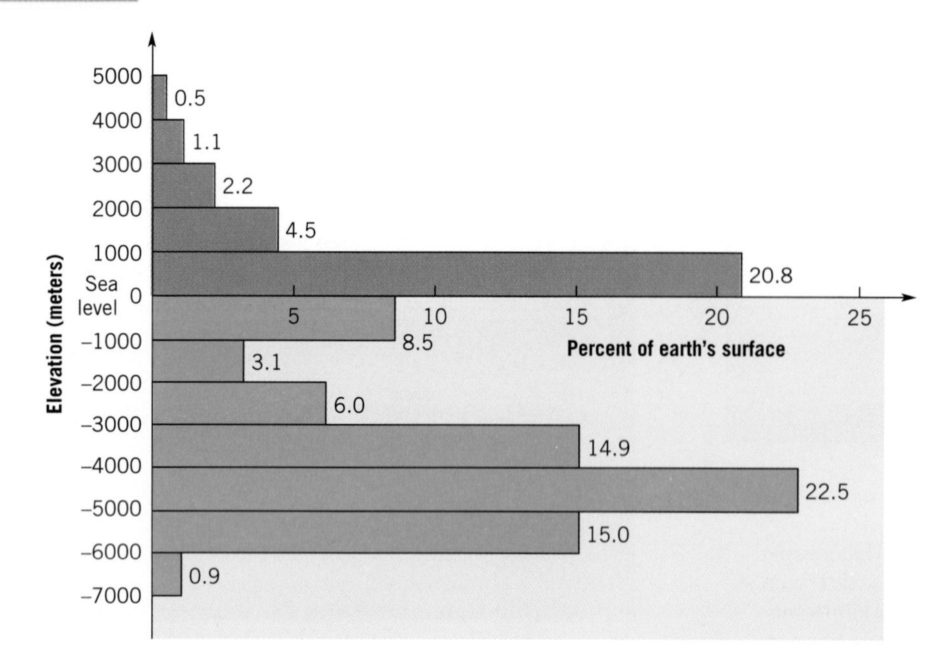

Figure 3.12

Plot of the percentage of crustal surface located at various elevations, in 1 kilometer intervals. Note the concentrations at two elevations: one just above sea level and the other around 4 to 5 kilometers below.

What is not so evident, however, is *why* there are the two distinct types of crustal materials in the first place. We will answer this question in the next chapter.

LITHOSPHERE AND ASTHENOSPHERE

Because they are so close to the cool outer surface of the Earth, the upper 80 to 100 kilometers of the upper mantle also are rather cool and rigid like the crust (see Figure 3.8a) and move along with the crust as a unit (Figure 3.13). This combined cool rigid outer region, which includes both the crust and top layer of the upper mantle, is called the **lithosphere**. Although we use considerable vertical exaggeration in our drawings, the lithosphere is still extremely thin in comparison to its width. Its relative dimensions are thinner than those of an egg shell, for example. The lithosphere is divided into several large fragments called **plates**, and the motion of these surface plates over the years is referred to as **plate tectonics**. The lithosphere appears to be quite complex, especially beneath the continents, where it sometimes includes fragments of former crustal plates that have been thrust downward during collisions.

Immediately beneath the lithosphere, at a depth extending from about 100 to 200 kilometers, is the **low velocity zone**. The name is derived from the relatively slow speed of seismic waves traveling through this region. Compared to the lithosphere above, this region is more "plastic" or "less viscous"; that is, it is softer, more pliable, and capable of bending or deforming without breaking. The higher temperatures at these depths cause some **partial melting** of the materials, which means that some materials melt or soften and lubricate the motion of others that are more rigid or resilient. Beneath the low velocity zone is no sudden or decisive change in physical properties like occurs at the top where it abuts the lithosphere. The pressure increases with depth, which causes the speed of seismic waves to increase as well.

> Descriptive words like "soft," "plastic," or "movement" are written from a geological perspective, not a human one. Although the asthenosphere is soft compared to the lithosphere, it is still much more rigid than steel.

The entire relatively plastic region of the upper mantle, including the low velocity zone and some of the material beneath it, is called the **asthenosphere** (Figure 3.13). Scientists do not completely agree on the lower boundary of this region, but many now view it as extending down to the 670-kilometer-deep base of the upper mantle. From this viewpoint, we could divide the upper mantle into two regions: the upper 80 to 100 kilometers of cool rigid material that together with the crust make the lithosphere, and the

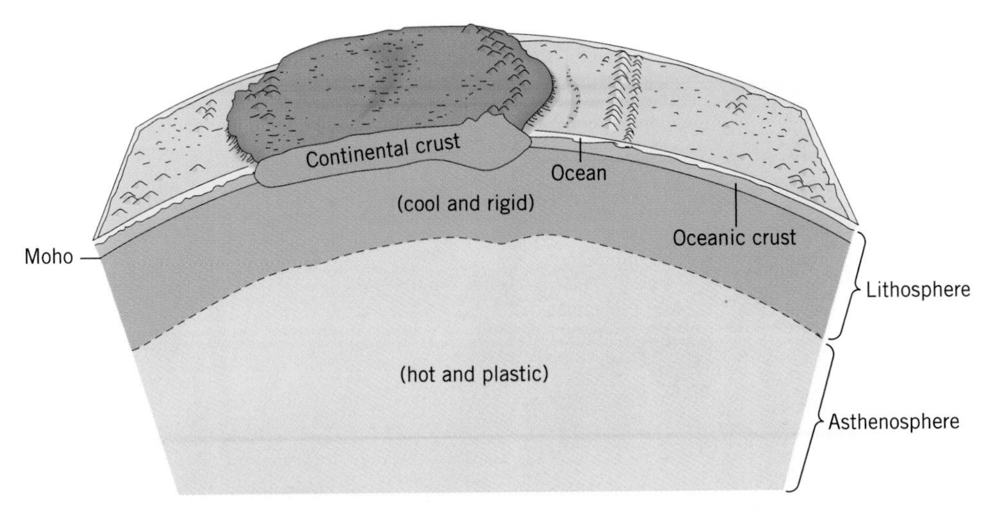

Figure 3.13

The upper 80 to 100 kilometers of the mantle are relatively cool and rigid. Together with the crust, they ride on the hotter, denser, and more plastic portions of the mantle beneath. (Vertical exaggeration about 10 times.)

lower 570 to 590 kilometers of warmer and more plastic material that make up the asthenosphere.

It is important to remember that descriptive words like "soft," "plastic," or "movement" are written from a geological perspective, not a human one. Although the asthenosphere is soft compared to the lithosphere, it is still much more rigid than steel, for example. Furthermore, the movement we refer to is on a time scale of millions of years, not human lifetimes. Things that bend or flow over millions of years can be quite resilient to the blow of a hammer.

SUBSIDING PLATES AND THE UPPER MANTLE

Because it is closest to us, the upper mantle is more easily studied than deeper regions, and from these studies, an interesting picture is beginning to emerge.

As we have seen, the overall internal structure within the Earth resulted from gravitational differentiation. Denser materials sank toward the center, and lighter materials rose toward the surface. You would think, then, that any layer should be lighter than that beneath it. When materials reach the Earth's surface, however, they cool off, and cooling makes them denser. As a result of this cooling, parts of the lithosphere have actually become slightly denser than the warmer material beneath.

The lithosphere has two components: the thin light crust and a second thicker layer of material from the upper mantle that has cooled and become more rigid. Compared to the asthenosphere beneath it, this second layer is made of similar materials, but is cooler and therefore denser. The question of whether or not the lithosphere can "float" depends on whether the crust is sufficiently buoyant to compensate for the extra weight of the layer of cool mantle material on its underside.

Beneath the continents, the crust is relatively thick and light, and can support this added weight. Thus, continental plates tend to remain "afloat." Oceanic crust, however, is thinner, denser, and less buoyant. Therefore, beneath the oceans, there are large regions where the crust is not sufficiently buoyant to support the weight of the cool mantle material on its underside. In these regions, the lithosphere would sink if it were not covering the asthenosphere and thereby preventing the asthenosphere from rising to replace it. In some places, however, the oceanic lithosphere slips edgewise into the asthenosphere below.

Although pressure increases are transmitted immediately, it takes millions of years for significant heating to occur, because the subsiding plates are both huge and poor conductors of heat. Consequently, as surface plates sink into the upper mantle, they are compressed by the greater pressures, but they remain relatively cool for millions of years after their entry. In this compressed, cool condition, they tend to be denser than the materials around them, so they continue to sink, often dragging the adjacent lithosphere with them.

> To humans, the upper mantle would seem as solid and rigid as rock. But on geological time scales, it may well resemble a broad, shallow cauldron of hot simmering stew.

The slow heating of these sinking fragments of surface plates makes them less dense, which may eventually arrest their sinking. The heating also causes melting or partial melting of some of the ma-

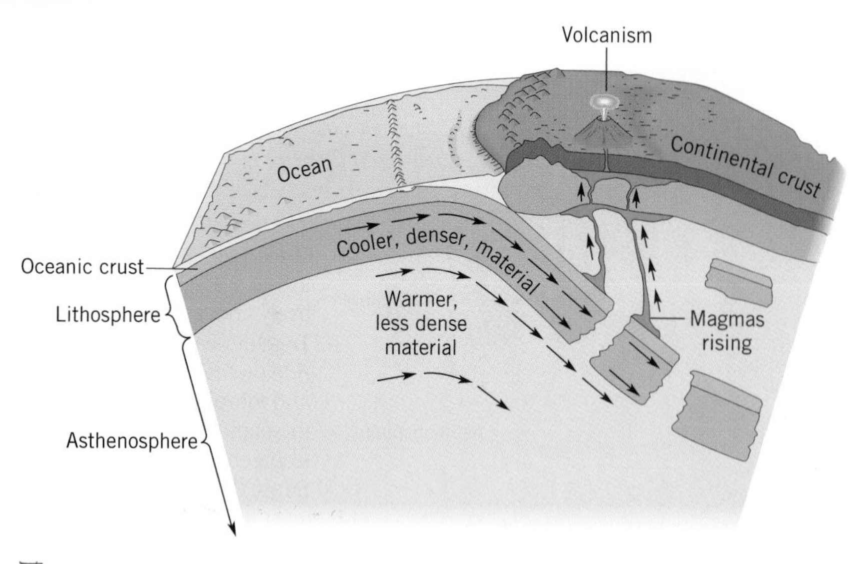

Figure 3.14

Because it is on the Earth's outer surface, the lithosphere is cooler and more rigid than the materials beneath it. The lithosphere has two layers: the crust on top and a thicker layer of cool mantle material on the underside. This second layer has a composition similar to that of the mantle beneath it, but it is cooler and therefore denser. Although the crust itself is light and buoyant, beneath the oceans it is sometimes too thin to support the weight of the denser cool mantle material on its underside. Consequently, there are large regions beneath the oceans where the lithosphere as a whole is actually slightly denser than the material beneath it. In some places, this lithosphere slides edgewise into the asthenosphere. The descending plates are heated very slowly and release magmas that rise toward the surface and cause uplifting (pushing up on the lithosphere) and volcanism. The upper mantle holds many ancient plate fragments, which are slowly moving, as is the material of the upper mantle itself.

terials in these plate fragments. Rising plumes of these molten materials, or **magmas**, may cause volcanism and mountain building when they reach the surface (Figure 3.14).

Of course, we must remember that we are considering motion over long time periods here. To humans, the upper mantle would seem as solid and rigid as rock. But on geological time scales, it may well resemble a broad, shallow cauldron of hot simmering stew. Fragments of underthrust lithosphere are churning slowly in the more fluid bulk of the asthenosphere, which is driven in its motion by the heating from beneath. To make the analogy complete, we would have to have some floating lid on our stew pot, to represent the present surface plates on which we walk and our oceans float.

3.3. STUDYING THE EARTH'S INTERIOR

Our knowledge of the Earth's internal structure, as presented in the preceding section, is based primarily on studies that use indirect techniques. Mine shafts and boreholes extend no more than a few kilometers downward, which only scratches the surface of the crust. So we have no direct observations of the deeper crust, let alone the mantle or deeper regions. Even where the crust is very thin, such as beneath the oceans, we have not yet been able to drill entirely through it.

Scientists sometimes try to study the properties of materials in environments similar to those of the Earth's interior by subjecting them to very high temperatures and pressures, using special presses or explosives. The results are not completely satisfactory for two reasons. First, we are not quite sure what materials we should be studying, because we have no direct knowledge of the precise makeup of the Earth's deeper materials. Second, laboratory environments do not last long enough to adequately simulate the environment of the inner Earth. Many materials require extended periods of time, some greater than a million years, to complete the transformations into forms that are stable under the extreme conditions of the Earth's interior.

Because of limitations such as these, scientists must get much of their knowledge regarding the structure and properties of the Earth's interior by using indirect techniques—observations made without actually touching or seeing the materials being studied. Foremost among these techniques are studies of the Earth's gravitational field and the seismic waves that propagate through the Earth.

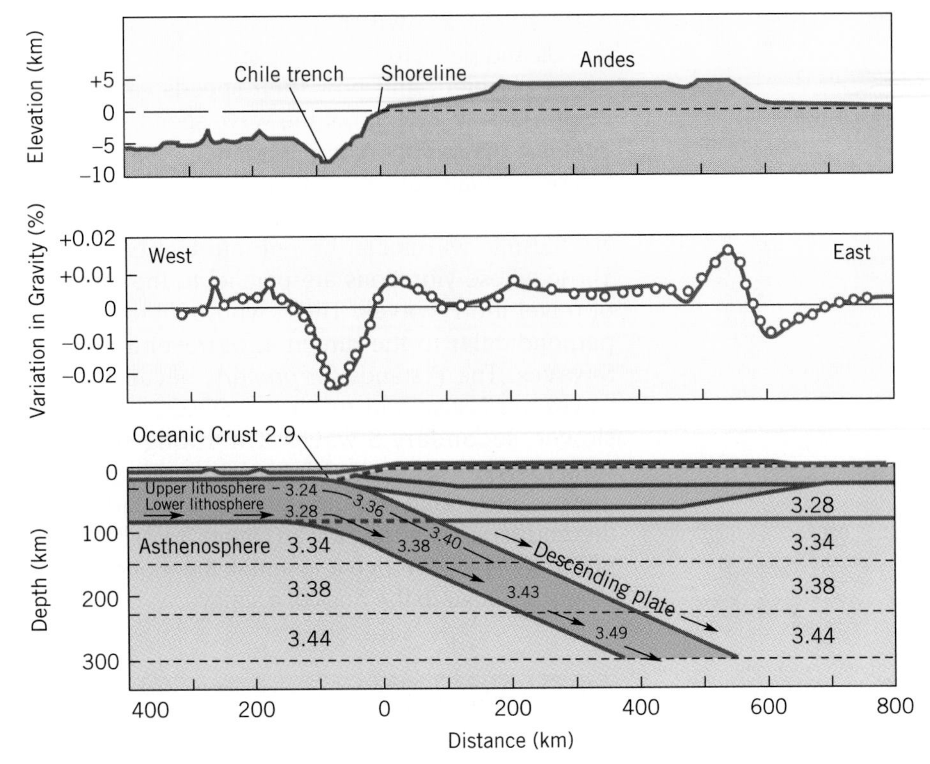

Figure 3.15

The Chile Trench and Andes Mountains in South America. The drawing at the top shows surface elevations relative to sea level. The middle drawing shows gravity variations in hundredths of a percent. The bottom drawing shows subsurface structure as inferred from seismic studies. In the bottom figure, the densities are given in grams per cubic centimeter. Notice that the descending plate initially is *not* denser than the materials beneath it, but as it sinks, it gets compressed and its density increases. In fact, it becomes denser than the materials surrounding it. Thus, it tends to continue to sink, and to drag the rest of the plate with it. (After Grow and Bowin, 1975, *Journal of Geophysical Research* **80**, 1449–1458.)

GRAVITY STUDIES

As we learned in Chapter 2, the gravitational force between two objects depends on their masses and separation. It is directly proportional to the mass of each and inversely proportional to the square of their separation. Therefore, careful studies of the Earth's gravity can yield a great deal of information about subsurface masses and their distributions (Figure 3.15). This is particularly true of mass distributions that differ from the purely spherical shapes, such as the Earth's equatorial bulge, and deformations deep within the crust or upper mantle.

Very sensitive measurements of the strength of the Earth's gravity can be made with sensitive instruments called **gravimeters**. We can also use Earth-orbiting satellites. Variations in mass distribution within the Earth cause small changes in a satellite's orbit and orbital speed, which can be easily detected.

SEISMIC STUDIES

Seismic waves are also an important tool used to probe the Earth's interior. These waves are usually created by disturbances within the Earth's outer crust, such as an Earthquake, or an underground nuclear explosion. The waves travel through the Earth and are detected at seismographic stations on the Earth's surface in various other parts of the world.

Seismic waves have two properties that make them particularly useful in revealing interior structures. One is that a portion is reflected when striking an interface between two materials, much like some light is reflected when it strikes glass. This means, for example, that if you were operating a seismograph and some seismic disturbance occurred beneath you, then you would record a series of echoes as seismic waves bounced off various interfaces deep within the Earth (Figure 3.16a). You could even estimate the depths of the various interfaces by the amount of time it takes the echoes to reach you. Later echoes must come from greater distances, therefore revealing deeper interfaces.

The second very helpful property of seismic waves is that they tend to bend, or **refract**, toward regions where they go more slowly. This property is shared by all waves. For example, ocean waves travel slower in shallow water and therefore always bend in toward shore. That is why waves are always coming in toward the beach and not going sideways along it. By analogy, if the wheels on one side of your car go slower than those on the other, the car will turn toward the side that is going slower.

When a wave crosses the interface between two materials, the amount by which it is bent depends on the difference in wave velocities between the two materials. The greater the change in speed, the more

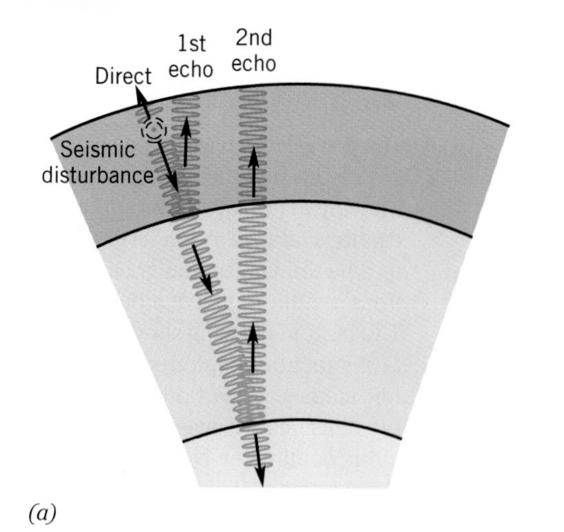

(a)

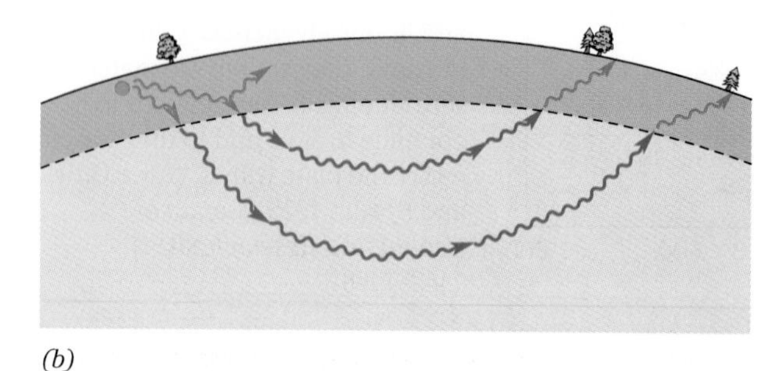
(b)

Figure 3.16

(a) Seismometers record echoes of seismic disturbances when they reflect from subsurface interfaces. Later echoes (such as the second echo indicated here) come from greater depths. (b) Seismic waves travel curved trajectories through the Earth, because the wave speed changes with depth. Changes in wave speeds are caused by changes in materials, temperature, and pressure.

it will bend. When the wave speed changes gradually within a material, the wave will also bend gradually. Consequently, when a seismic wave crosses an interface between two distinct regions, we expect it to change directions abruptly, and when it passes through regions where there are gradual changes in pressure, temperature, or composition, we expect it to bend gradually (Figure 3.16b).

We cannot follow a seismic wave through the Earth. We can only witness its creation near the surface and then detect its arrival at various other points on the surface. The time required for the wave to arrive at any detector depends on both the path of the wave through the Earth and the speed of the wave along that path. But the two are related. As we have seen, the path followed by the wave depends on the distribution of wave speeds within the Earth.

So from an analysis of arrival times for seismic waves at various points along the Earth's surface, we can use the known relationship between wave speeds and paths to discover exactly what routes the waves traveled, and how their speeds varied with depth. This knowledge of the wave speed at various depths gives us important information on the properties of the materials there, including how dense and rigid they are.

Seismic waves can be put into two categories. Those whose vibrations are parallel to the direction of travel are **P-waves**. Those whose vibrations are perpendicular to the direction of travel are called **S-waves**. The P stands for *primary*, because the P-waves are fastest and reach the sensors before the slower, secondary S waves. On a long spring or slinky, a P-wave can be demonstrated by quickly moving one end in and out. An S-wave is made when the end is wiggled sideways (Figure 3.17a). As seismic waves pass through a material, the motion of the atoms is similar to the rings of a slinky.

On a long spring or slinky, a P-wave can be demonstrated by quickly moving one end in and out. An S-wave is made when the end is wiggled sideways. As seismic waves pass through a material, the motion of the atoms is similar to the rings of a slinky.

The P-waves can cross an interface between a solid and a liquid, but the S-waves cannot. To illustrate this, consider a glass of water, and how the motion of the glass affects that of the water. If the glass is pushed forward and backward, the water goes with it. However, if the glass is rotated, the water won't move. In and out motion is transferred between the glass and the water, but sideways motion is not. The liquid water is not rigid enough to transfer such sideways motion. The same is true for seismic waves within the Earth. The in and out motion of the P-waves can travel through a liquid, but the sideways motion of S-waves cannot.

When there is a seismic disturbance somewhere, sensors on the opposite side of the Earth receive only P-waves directly. This indicates that the Earth has a liquid outer core, which prevents the direct arrival of S-waves through the Earth's center. Of course, some will eventually arrive by traveling through the crust all the way around, or by some other circuitous route avoiding the core, but these waves will arrive much later, having traveled farther. By determining where S-waves are and are not received on the far side from a disturbance, scientists can map the outer border of the liquid region (Figure 3.17b).

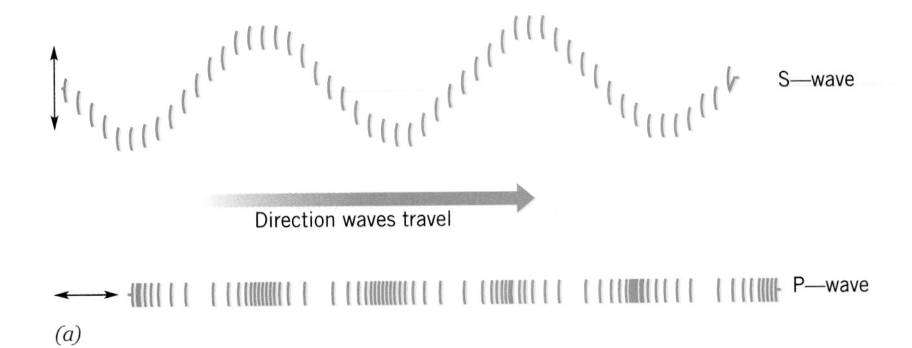

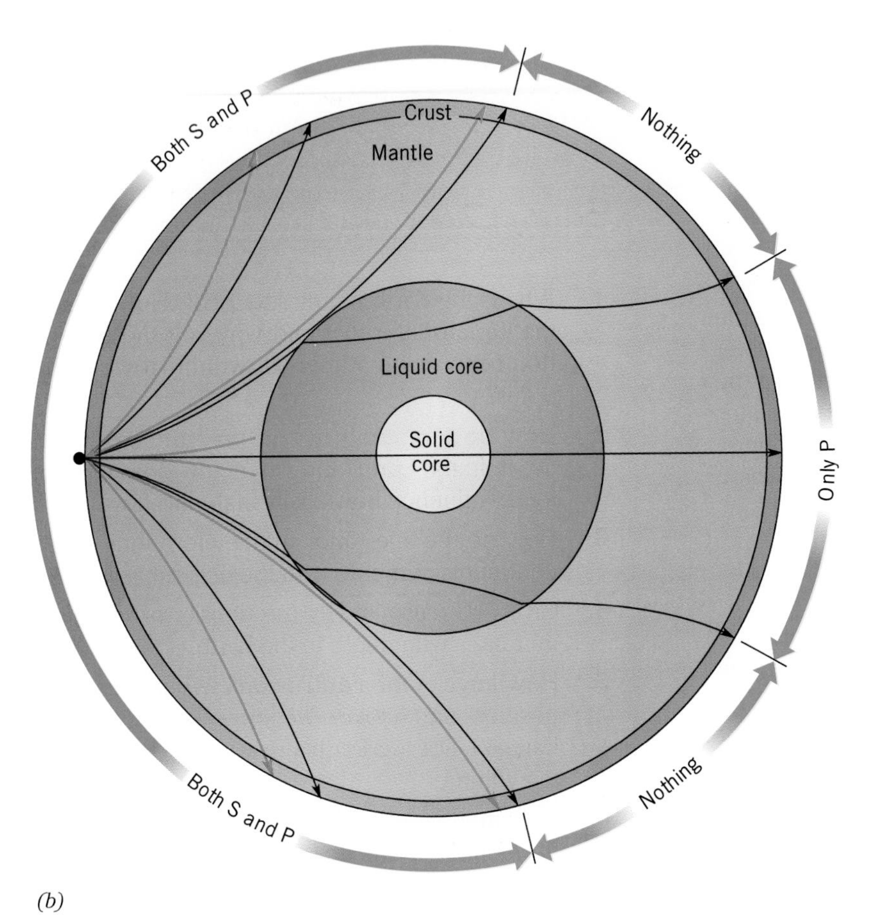

Figure 3.17

(a) Waves on a spring or slinky. When the motion of the individual links is perpendicular to the direction that the waves travel, they are S-waves. When the links move back and forth in the same dimension that the waves travel, they are P-waves. (b) Schematic diagram of the paths followed by P-waves (black lines) and S-waves (colored lines) through the Earth's interior, after being generated at the left side of the figure by some seismic disturbance. Notice that because the S-waves cannot cross the interface between the solid mantle and liquid core, they will not travel directly to the opposite side of the Earth. In addition, due to refraction of the waves, there will be shadow zones where neither type of wave will be received directly.

Summary

View from Outer Space

The Earth is a unique planet. The large abundance of water in its outer *hydrosphere* gives it a distinctive blue and silver coloration that would set it apart from other planets even when viewed from large distances. Earth is spherical, with a slight equatorial bulge due to its spin. Its surface has two distinct elevations: one characteristic of continents and one of ocean floors. The symmetry of coastlines on opposite sides of the Atlantic Ocean and other peculiar features also indicate ongoing geological activity.

View of Inner Space

Over time, denser materials have tended to sink toward the Earth's interior and lighter materials have tended to rise toward the surface. Increased temperature and pressure toward the interior also cause variation in the properties of materials within the Earth. There are several internal sources of heat, among which the decay of radioactive materials is probably presently most important. The surface is cooled as it radiates heat into outer space.

The Earth's interior has three distinct regions based on material composition. The innermost region, which extends about halfway to the surface, is called the *core*. The core is mostly an iron-nickel alloy with other materials present in smaller amounts. The inner portion of the core is solid, and the outer portion is liquid. Extending from the core nearly to the surface is the mantle. The outer portion of the mantle is composed mostly of "rocky" metallic silicates. The crust is very thin, cool, light, and brittle compared with the interior regions.

The crust is thicker beneath the continents and thin beneath the oceans, which implies that the crust is "floating" on the materials beneath. Massive continental mountain ranges require the crust to sink down deep into the mantle beneath them in order to gain the required support. The thinner oceanic crust tends to have slightly denser composition than the thicker continental crust. The outer 80 to 100 kilometers of the mantle are sufficiently cool and rigid that they tend to move with the crust, and together they form the lithosphere. The rigid plates of the lithosphere ride atop the more plastic region of the mantle beneath, which is called the *asthenosphere*. Cooler regions of oceanic plates are slightly denser than the asthenosphere beneath them. In some places, plates are slowly sliding edgewise into the asthenosphere, and the upper mantle in some ways resembles a "stew" that contains old plate fragments.

Studying the Earth's Interior

To study the Earth's interior requires indirect techniques, because we can barely scratch the surface with direct observations. Careful studies of the Earth's gravitational field, using either gravimeters or satellites, yield some information on the distribution of subsurface masses. Seismic waves are also used to infer internal structure. Seismic waves reflect from subsurface interfaces, so we can use echoes to locate these interfaces. Because seismic waves bend toward regions where they go more slowly, we can study the physical properties of the Earth's interior by analyzing the paths that these waves follow. In contrast to S-waves, P-waves travel more quickly and can cross a solid–liquid interface, such as the mantle-core boundary.

Key Terms

hydrosphere
terrestrial planet
equatorial bulge
sedimentary rock
viscosity
radioactive decay
gravitational differentiation
core
inner core
outer core
mantle
upper mantle
crust
Moho
seismic wave
buoyancy
isostatic adjustment
isostasy
continental crust
oceanic crust
lithosphere
plates
plate tectonics
low velocity zone
partial melting
asthenosphere
magmas
gravimeter
refraction
P-wave
S-wave

Study Questions

1. Why is the Earth nearly spherical? How large is the equatorial bulge and why is it there? Why don't the oceans flow "downhill" toward the poles?
2. How do temperature and pressure vary with depth in the Earth, and why? How does each of these influence how easily materials flow?
3. How do the ages and surface elevations of the continents compare with the ocean basins?
4. Give two reasons why the density of materials increases with depth in the Earth.
5. How large is the Earth's core? What is its composition, and how do we know this? How can it have 33% of the Earth's mass with only 16% of the volume?
6. How much of the Earth's volume does the mantle include? How does the mantle differ from the crust and core?
7. How does the depth of the Moho vary from oceanic to continental regions, and why? What does isostasy have to do with the depth of the Moho beneath mountain ranges?
8. What are some of the differences between oceanic crust and continental crust?
9. Why are the surfaces of continents at higher elevations than the sea floor?
10. Describe the lithosphere and asthenosphere. What is *partial melting*, and in which of these regions does it occur?
11. Why is it more likely for oceanic plates to slip into the asthenosphere than for continental plates to do so?
12. Describe the upper mantle.

13. What are seismic waves, and how are they generated? Do they travel faster or slower in more rigid materials?

14. Explain how the refraction of seismic waves gives us information on the Earth's interior? What do we learn from the reflection of seismic waves?

15. What are the two types of seismic waves? How do they differ? Which is faster? Which cannot cross a solid–liquid interface?

Critical Thinking Questions

1. What do you think would be some of the important differences on Earth, if there were no oceans?

2. Why must tall mountains, deep valleys, and ocean trenches be much younger than the Earth?

3. Estimate how many times wider a sheet of paper is than it is thick. (*Hint*: Estimate how many sheets would make a stack 8 or 9 inches tall.) Next, estimate how many times wider the crust is than it is thick. (*Hint*: The Earth is 40,000 kilometers around.) How do these two numbers compare?

4. Figure 3.15 indicates that the Earth's gravity is weaker over the Chile Trench than elsewhere. But the trench is not empty; it is filled with water. Why then is the gravity weaker?

5. How do we know that the Earth has a liquid outer core?

Suggestions for Further Reading

ANDERSON, D. L. 1989. Composition of the Earth. *Science* 243:4889, 367.

ANDERSON, D. L. 1989. Where on Earth is the Crust? *Physics Today* 42:3, 38.

BROWNE, MALCOLM W. 1993. China Nuclear Tests Reveal Details of the Earth's Interior. *The New York Times* 142 (Feb. 11), A18.

CARR, MICHAEL H. 1992. Earth and Mars: Water Inventories as Clues to Accretional Histories. *Icarus* 98:1, 61.

GAYOT, FRANCOIS. 1994. The Earth's Innermost Secrets. *Nature* 369:6479, 360.

JEANLOZ, R. 1984. Earth Science: Dynamics of a Stratified Mantle. *Nature* 308, 15.

MADDOX, J. 1987. What the Solar Nebula Was Like. *Nature* 330:6150, 691.

WACK, D. 1987. Meteorites: News from the Early Solar System. *Nature* 331:6155, 387.

1989. The Earth's Sisters. *The Economist* 313:7634-5, 105.

The remarkable similarity of the coastlines on opposite sides of the Atlantic suggests that the continents would fit nicely together like pieces of a jigsaw puzzle.

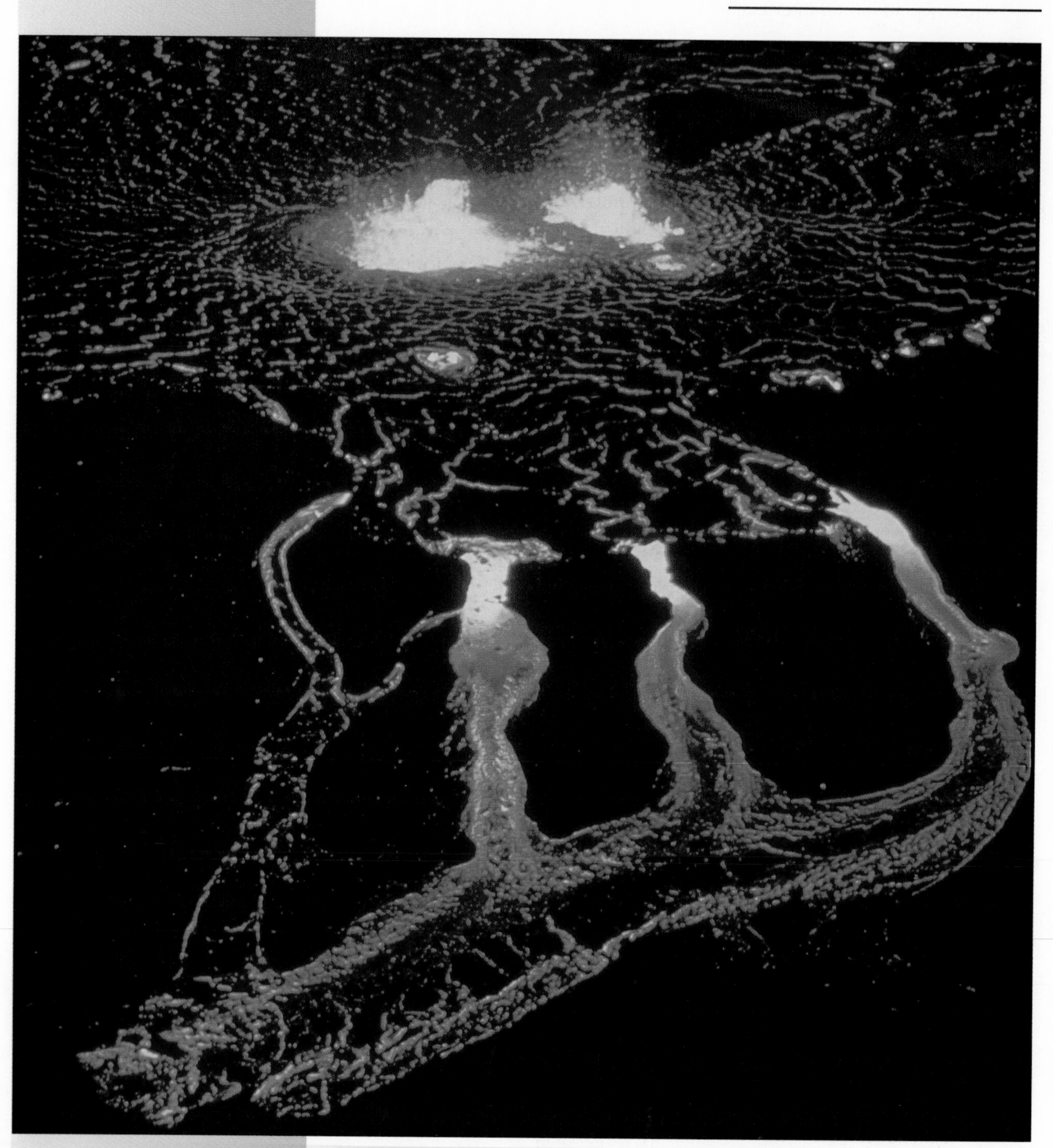

Evidence of the Earth's geological activity is all around us.

Four

PLATE MOTION AND THE OCEAN FLOOR

As we learned in Chapter 1, our observations of Earth are but a "snapshot" in its very long and dynamic history. From our perspective, the oceans seem eternal. Mountains seem immutable. "Solid as the Rock of Gibraltar" is a familiar expression. At first we may have difficulty accepting the idea that our snapshot captures only a moment in the history of a very dynamic Earth, a planet whose continents skitter across its surface, whose oceans open and close, and whose surface slips into depths and destruction in some places and is reborn in others. Yet, we may recall having seen sedimentary rocks that formed beneath the ocean floor, now folded and contorted and high in the mountains (Figure 3.7). We may have felt an earthquake or heard grumbling from a volcano, reminding us that things are changing. In this chapter, we will learn that the Earth is indeed very dynamic. The evidence of this motion is all around us and includes the dominant features of the ocean floor.

4.1 CONTINENTAL DRIFT

The remarkable similarity of the coastlines on opposite sides of the Atlantic suggests that the continents would fit nicely together like pieces of a jigsaw puzzle (Figure 3.6). We now know that at one time they did fit together. The continents were all part of one large supercontinent, which began tearing apart about 150 to 180 million years ago,[1] as the Atlantic Ocean opened up. This makes the Atlantic Ocean only about 3% as old as the Earth itself, so it is undoubtedly just one of the more recent events in the Earth's long history of changes. But the idea of continents splitting and new oceans opening up is the kind of sensationalistic theory toward which scientists are naturally skeptical. It took some time and convincing evidence before the theory was accepted.

One of the earliest and most influential proponents of the idea of continental drift was a German astronomer, meteorologist, and explorer named Alfred Wegener. He advanced good arguments to support the idea that at one time there was no Atlantic Ocean, and the Americas were welded with Europe and Africa into one large continent. He named this continent **Pangaea**.

[1] It didn't all "tear" at once, which is the reason for the time span given here.

One of Wegener's strong arguments was the remarkable similarity of the continental margins on opposite sides of the Atlantic. But he also looked at similarities in ancient geological structures, fossil records, and ancient climates to support his hypothesis. He pointed out that some unique geological structures in the Americas terminate abruptly in the Atlantic Ocean and then continue again on the other side of the ocean, as if the structures had formed when the continents were together and then were split as the continents tore apart. From fossils, he demonstrated that the species of ancient organisms on both sides of the Atlantic were similar up to about 150 million years ago and then diverged, indicating that the continents had become isolated from each other at that time. Finally, using fossils and sediments, Wegener was able to demonstrate that ancient climates on both sides of the Atlantic were similar, although quite different from those found presently in the respective regions. For example, New York and Spain had similar climates 150 million years ago, but both were on the equator then.

The remarkable similarity of the coastlines on opposite sides of the Atlantic suggests that the continents would fit nicely together like pieces of a jigsaw puzzle.

Although Wegener's ideas were interesting, they were far too revolutionary to be easily accepted or even seriously considered by most of the scientific community. Wegener died in 1930, more than two decades before further evidence would force reconsideration, and eventual acceptance, of his revolutionary way of looking at our world.

4.2 PALEOMAGNETISM

In the 1950s, convincing evidence for continental drift came from the field of **paleomagnetism**, which is the study of magnetism in rocks. Until this time, many Earth scientists thought paleomagnetism to be a rather esoteric study, of interest to only a few, and hardly capable of forcing any significant revolution in our ways of thinking. But it did. Ancient rocks contain a record of the Earth's magnetic field at the time when the rocks formed, and in this fossil magnetic record is contained undeniable proof that continents move.

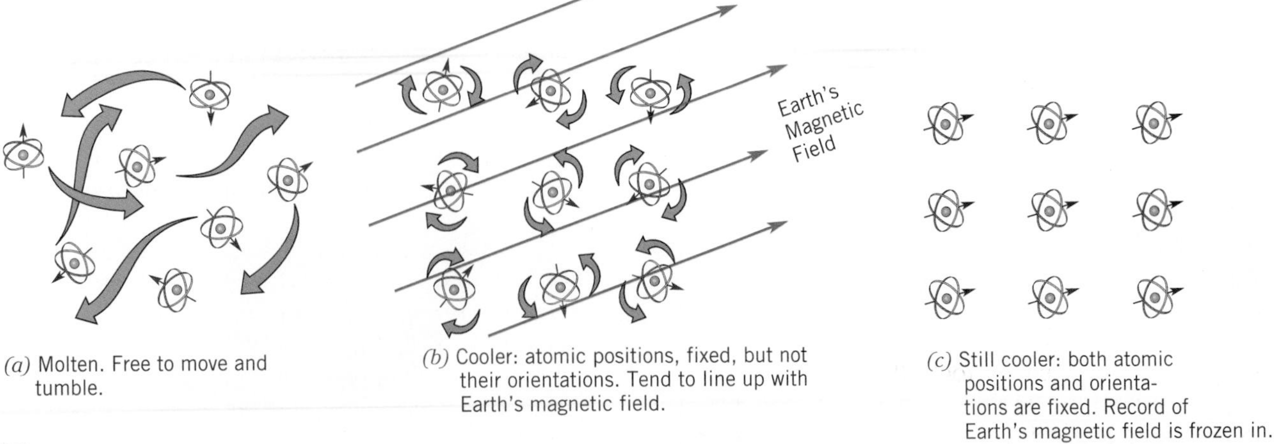

Figure 4.1

Like tiny compass needles, the atomic magnets in molten rock tend to line up with the Earth's magnetic field. (a) In molten rock the individual atoms are free to move and tumble about within the magma. Their tendency to align themselves with the Earth's magnetic field is reduced by the thermal agitation at these higher temperatures. (b) As the magma cools and solidifies, the atoms' positions become fixed, but they may still rotate. At these cooler temperatures, there is a stronger tendency for alignment with the Earth's magnetic field, which is especially pronounced in some iron-containing minerals. (c) At still lower temperature, the atomic orientations become fixed, and a record of the Earth's magnetic field is thereby frozen into the rock.

HOW ROCKS BECOME MAGNETIZED

All magnetism is caused by the motion of electrical charges, such as those moving about within atoms. Depending on the details of the electron orbits, the magnetism in some types of atoms may be particularly strong. Iron is one of these, so minerals containing iron are often exceptionally magnetic. For each mineral, there is a temperature (around 600°C for many common minerals) above which these **atomic magnets** are free to rotate and below which they are "frozen" in orientation (Figure 4.1). This temperature is normally below the melting point, so when hot molten magma cools, first the magma solidifies as the atoms become frozen in position, and then after further cooling, the atoms become frozen in orientation as well.

All crustal materials are derived ultimately from magmas (i.e., molten rock) that have risen toward the surface from the Earth's deeper interior. Rocks formed from magmas are called **igneous**. More particularly, they are **plutonic** if the magmas cool and solidify below the surface and **volcanic** if the magmas actually flow out onto the surface before freezing. Magma that flows onto the surface is called **lava**. In either case, while the materials are still molten, the tiny atomic magnets tend to line up with the Earth's magnetic field. As this magma cools and solidifies, these atomic magnets become both more strongly aligned and frozen in orientation. A record of the Earth's magnetic field at that time becomes permanently frozen in the rock.

THE AGES OF ROCKS

If the ancient magnetic record in a rock is to be helpful to our understanding of past events, we must be able to determine how long ago the rock was formed. One of the most common and accurate means of dating a igneous rocks is the **potassium-argon dating** technique. Other techniques, involving the radioactive decay of other atomic nuclei, are similar to this.

The atomic nuclei of potassium atoms come in several different forms, called *isotopes*.[2] One of these, called *potassium-40*, is unstable and decays radioactively into argon. Argon is a gas. When the rock is molten, gases are able to escape. Once the magma solidifies, however, gases can no longer escape. From that point on, any additional argon that is formed will be trapped inside. As an igneous rock ages, the continuing decay of potassium-40 creates more and more argon. So the abundance of argon within the rock increases, and that of potassium-40 decreases (Figure 4.2). Since the decay proceeds at a known rate,[3] the actual age of the solid rock can be determined by the relative abundances of these two materials.

[2]The isotopes of any given element differ from each other by the number of neutrons in the nucleus.

[3]Potassium-40 has a half-life of 1.3 billion years, which means that after any period of 1.3 billion years, half as much potassium-40 remains as was present at the beginning of that period.

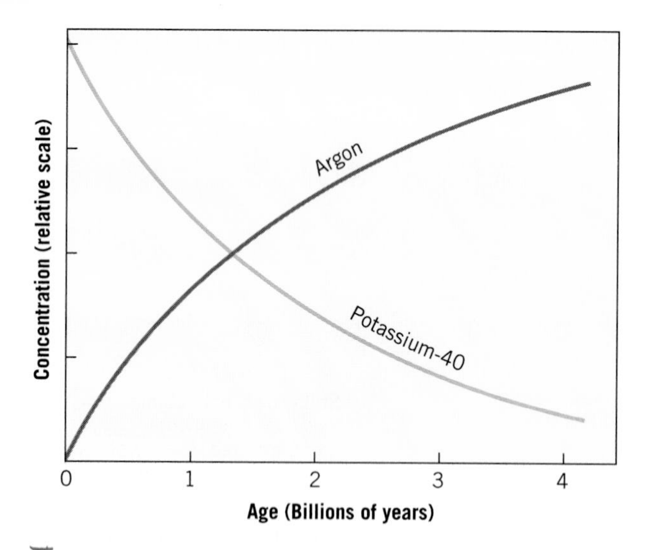

Figure 4.2

Potassium-40 decays into argon with a half-life of 1.3 billion years. Argon is a gas that escapes the molten magmas. Once this magma solidifies, however, any further argon released is trapped inside. As the potassium-40 decays, the argon concentration increases. By the relative concentrations of these two materials, scientists can tell how old the rock is. (If there were equal concentrations of these two materials, how old would the rock be?)

POLAR WANDERING

By taking samples of rocks of various ages from the Earth's crust and measuring the magnetization of each, we can find how the direction of the magnetization has changed over the years. But we are left with the question of whether it is the crust or the magnetic field that has moved. For example, suppose we examine a 10-million-year-old portion of the crust whose magnetism points east. Does this mean that the Earth's magnetic field pointed east back then, or did that portion of the crust rotate toward the east since its formation?

To answer this question, paleomagnetists in the 1950s made charts depicting **polar wandering**. For the sake of argument, they assumed that continents didn't move. Then, by studying the magnetization of ancient rocks from one continent over the years, they traced how the position of the magnetic North Pole must have shifted (Figure 4.3). Then they repeated the measurements for a different continent. Again, they traced the path of wandering for the magnetic pole according to the rocks of this second continent. The amazing result was that the two paths of polar wandering were not the same. Since the magnetic North Pole could not be in two different places at once, the original assumption must be false. The continents cannot be sitting still.

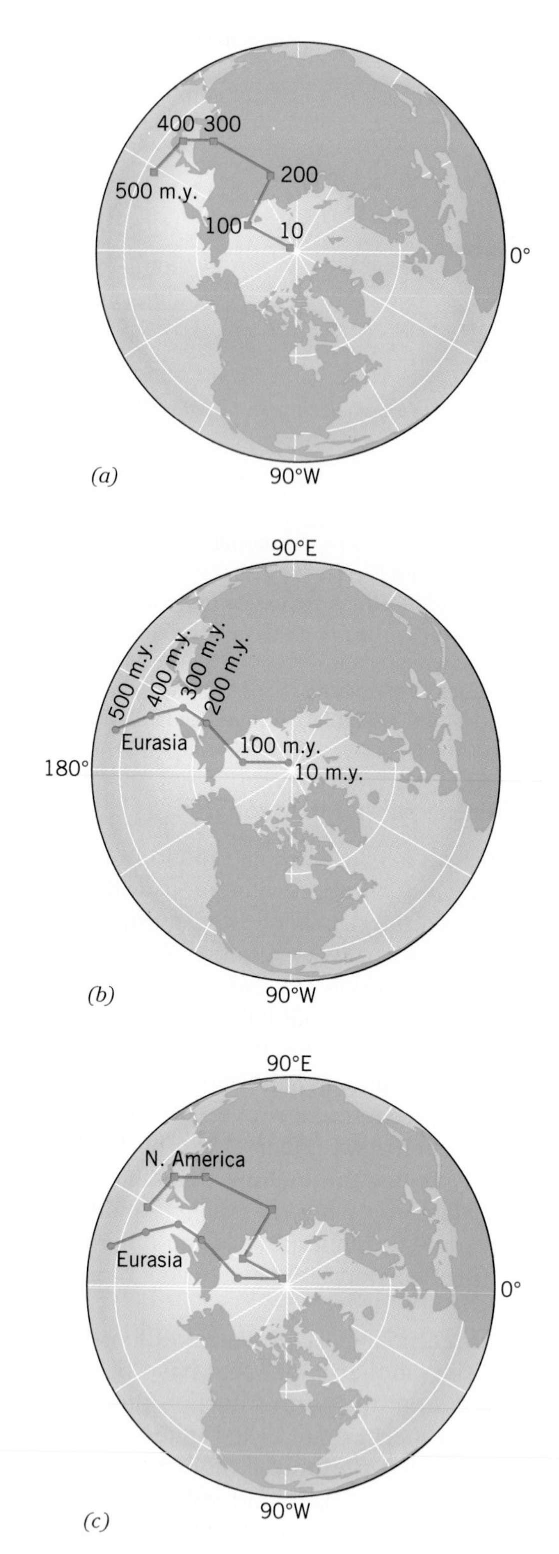

Figure 4.3

Paths of polar wandering over the past 500 million years, as determined from crustal rocks in (a) North America and (b) Eurasia. When combined as in (c), it is clear that they disagree as to where the magnetic North Pole has been. Thus, at least one of the continents must have moved. (From Peter J. Wyllie, 1976, *The Way the Earth Works*, John Wiley & Sons, New York.)

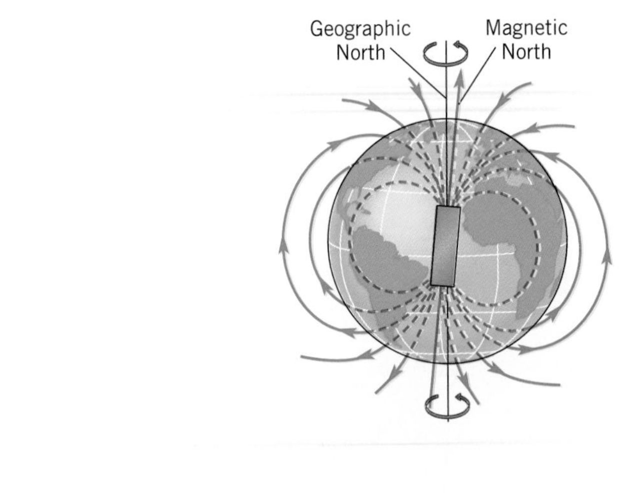

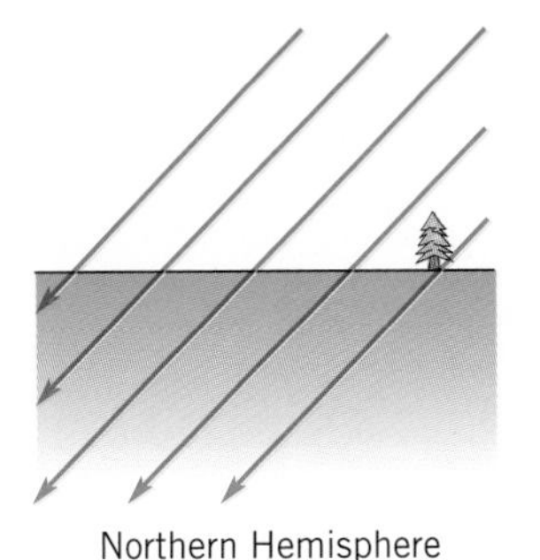

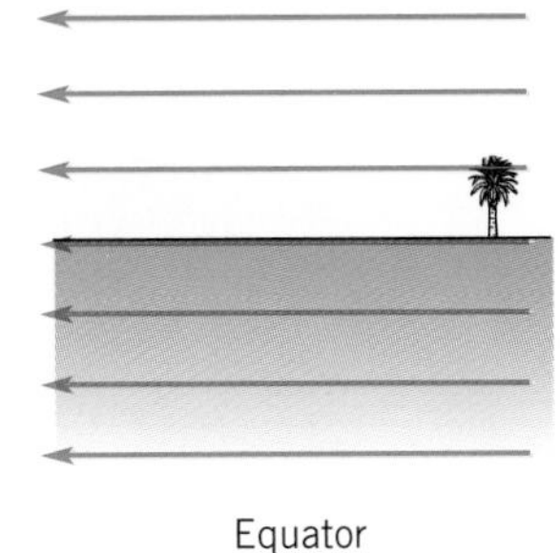

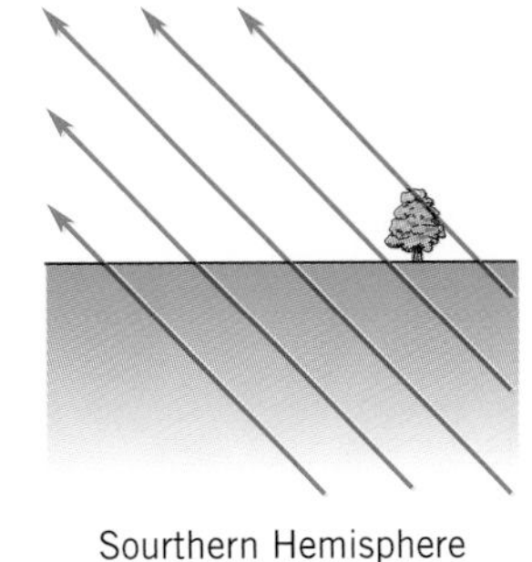

Figure 4.4

Illustration of how magnetic dip varies with latitude. In the Northern Hemisphere, the Earth's magnetic field points downward into the surface. In the Southern Hemisphere, it points upward out of the surface. By the direction and amount of dip, you can determine your magnetic latitude.

Once the paleomagnetists established that the continents had been moving, they then directed their efforts toward analyzing these motions in more detail. By taking the data on polar wandering and insisting that the rocks from all over the world agree on the past location of the magnetic North Pole, they traced the relative motions of the continents. For example, they identified the opening of the Atlantic Ocean during the period of 150 to 180 million years ago and traced the motions of the continents as the Atlantic continued to widen.

Fossil magnetism in rocks can also reveal the latitude of a continent at various times in its past. As is illustrated in Figure 4.4, although the Earth's magnetic field points generally northward, it intersects the surface at different angles at different latitudes. The magnetic field parallels the surface at the equator, points up and outward from the surface in the Southern Hemisphere, and points downward into the surface in the Northern Hemisphere. The angle it makes with the horizontal, called the **magnetic dip**, is zero at the equator, but increases with latitude and becomes vertical near the poles.

By measuring the dip of the fossil magnetism frozen into ancient lava flows, scientists can determine the latitude where each lava flow occurred. For example, New York was on the equator and India was near the South Pole about 180 million years ago. Figure 4.5 illustrates how the dip of fossil magnetism in rocks of various ages from India was used to trace the northward motion of that continent over the past 180 million years.

4.3 THE DRIVING FORCE

The knowledge that the continents are moving leads us to wonder what kind of force could be powerful enough to drive these huge massive structures. Of course, what seems "huge" and "massive" from our point of view may be rather small and inconsequential compared to the rest of the Earth. In fact, the entire lithosphere, including not only the continents but the ocean floor as well, is very thin and not nearly as massive as the material beneath it. Like leaves floating on a stream, the motions of the continents may simply reflect the motion of the asthenosphere beneath. Applying this leaf analogy to oceanic plates, we note that parts of the "oceanic" leaves are slightly denser than the water beneath, and the edges of some are starting to slide down into the water (Section 3.2).

We know that the Earth's deep interior is hot, for reasons mentioned in the previous chapter. This heat

180 m.y.	100 m.y.	65 m.y.	20 m.y.	Present	Age of rock
65°	60°	25°	17°	20°	Direction of fossil magnetization
65°	60°	25°	17°	20°	Angle of dip
46°S	41°S	12.5°S	9°N	12°N	Paleolatitude

(a)

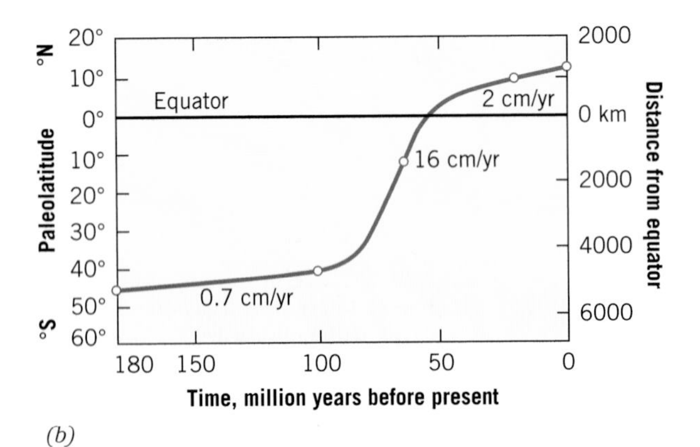

(b)

Figure 4.5

Measurements of the dip angle of the magnetism in ancient volcanic rocks in India have revealed the northward movement of that continent. (a) Illustration of how the magnetic dip in the rocks of various ages has determined the magnetic latitude of India in those ages. (b) Plot of the magnetic latitude of India vs. time for the last 180 million years. (From Peter J. Wyllie, 1976, *The Way the Earth Works*, John Wiley & Sons, New York.)

Figure 4.6

Like sheets of ice on water, the plates of the lithosphere are cool, rigid, and brittle compared to the asthenosphere on which they ride. They bump and grind together as they move. The thin oceanic plates can be likened to the thin sheets of ice in the above photo, and the thicker continental regions to the thicker iceberg on the horizon.

source beneath the asthenosphere causes motion, called **convection**. As the lower layers are heated, they expand and become less dense than the cooler layers above them. Therefore, they rise toward the surface. Likewise, as the surface cools off, it becomes denser and sinks. Think of the simmering motion of water in a pot on the stove. When fluids are heated from below, there are some regions where heated material rises toward the surface and then spreads outward. In other regions, cooler surface materials converge and sink.

4.4 THE PLATES

We learned in Chapter 3 that the lithosphere is cool, rigid, and brittle compared to the material on which it rides. In this way, the plates of the lithosphere are something like sheets of ice that are cool, rigid, and brittle compared to the water on which they float (Figure 4.6). As they move, the various plates push and grind against each other, creating earthquakes. Although the buildup of stress is sometimes removed through fractures and adjustments within the interior of a plate, most earthquakes occur along outer edges, called the **plate boundaries**. In fact, we can identify the edges of the plates by marking on a map the locations of recorded earthquakes, as is done in Figure 4.7. This technique allows us to identify 12 major plates. Although careful study of earthquakes helps scientists understand the present motions of the various plates, earthquakes leave no permanent records. Therefore, the study of past motions must rely on things that leave records, such as rock magnetism and radioactive dating.

The kinds of boundaries between adjacent plates fall into three general categories, according to whether the plates are spreading away from each other, are colliding with each other, or are moving laterally past each other (Figure 4.8). These are

1 = African Plate
2 = Arabian Plate
3 = Eurasian Plate
4 = Australian Plate
5 = Philippine Plate
6 = Pacific Plate
7 = Antarctic Plate
8 = Nazca Plate
9 = Cocos Plate
10 = Caribbean Plate
11 = North American Plate
12 = South American Plate

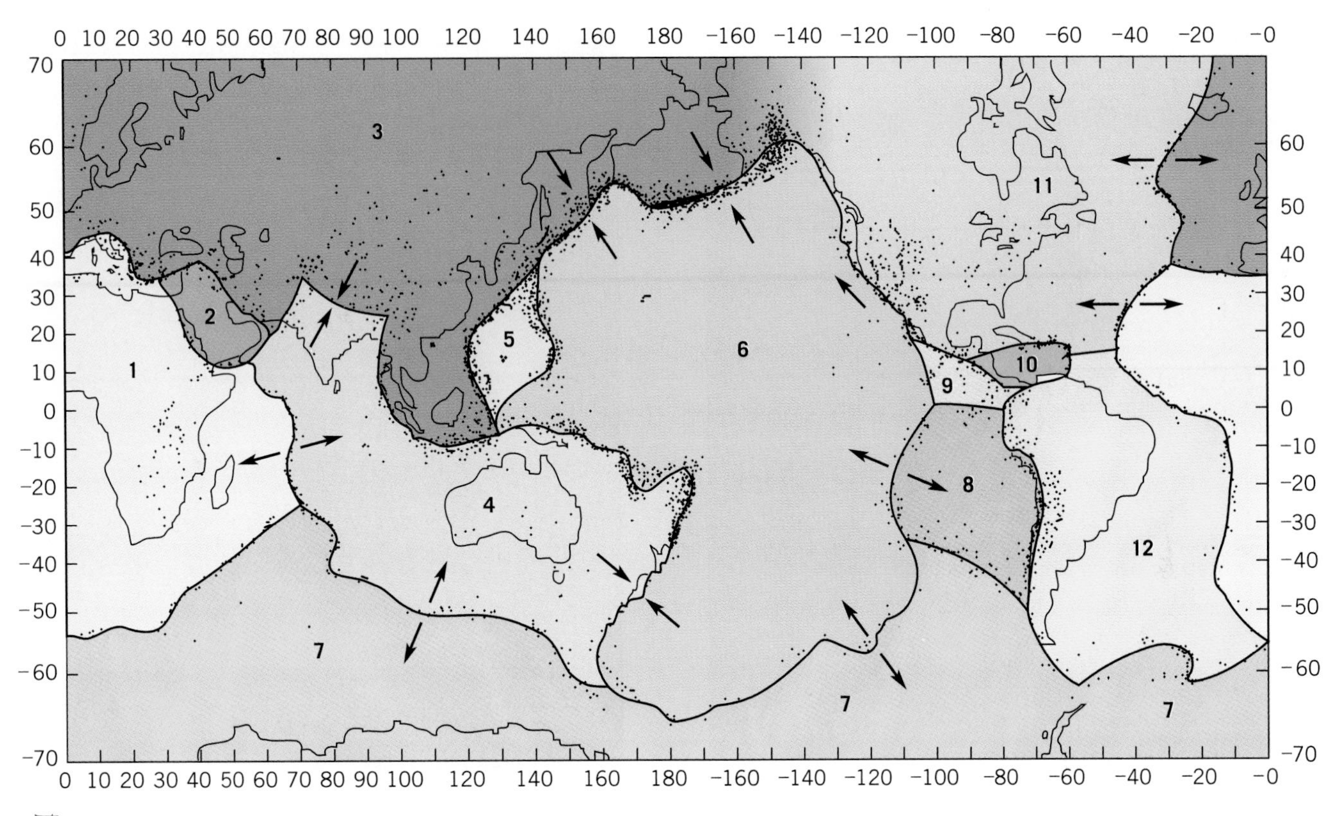

Figure 4.7

Distribution of earthquake epicenters for the years 1961 to 1967 and outline of the various plates of the lithosphere. Directions of the relative movement along plate boundaries are indicated. (After a plot by M. Barazangi and J. Dorman, Columbia University.)

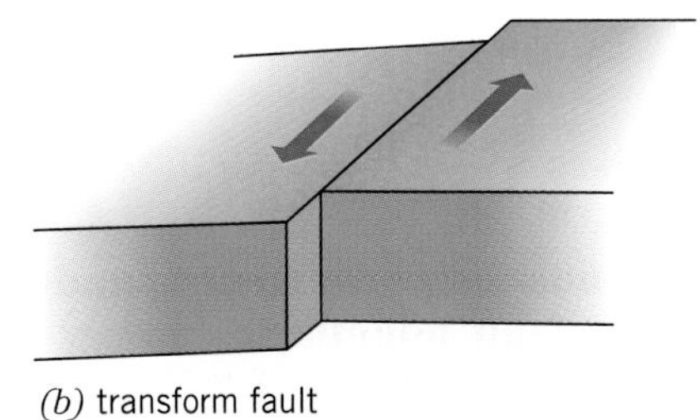

(*b*) transform fault

Figure 4.8

Illustration of the types of plate boundaries: (a) divergent, (b) transform fault, and (c) convergent.

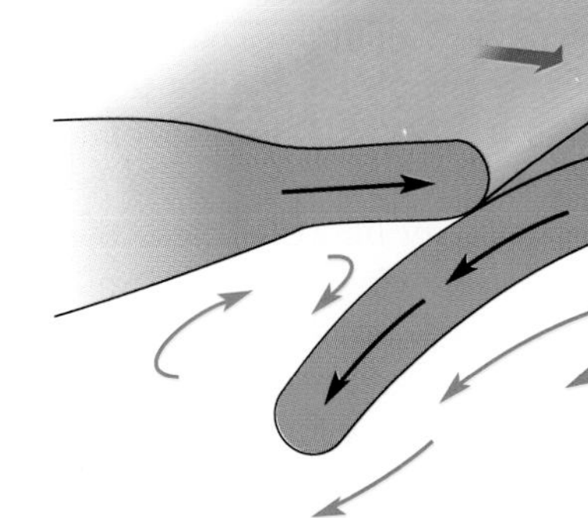

(*a*) divergent

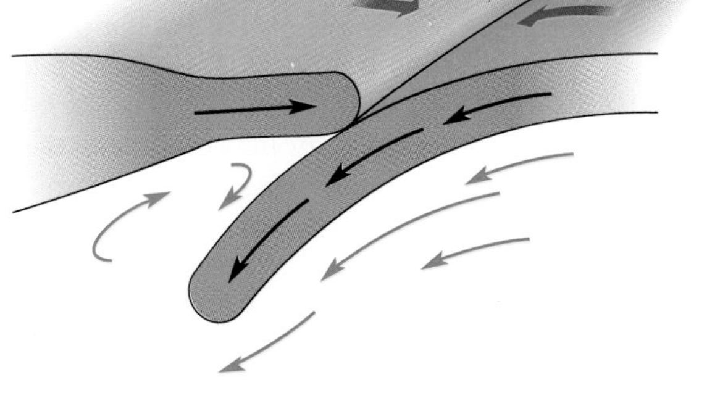

(*c*) convergent

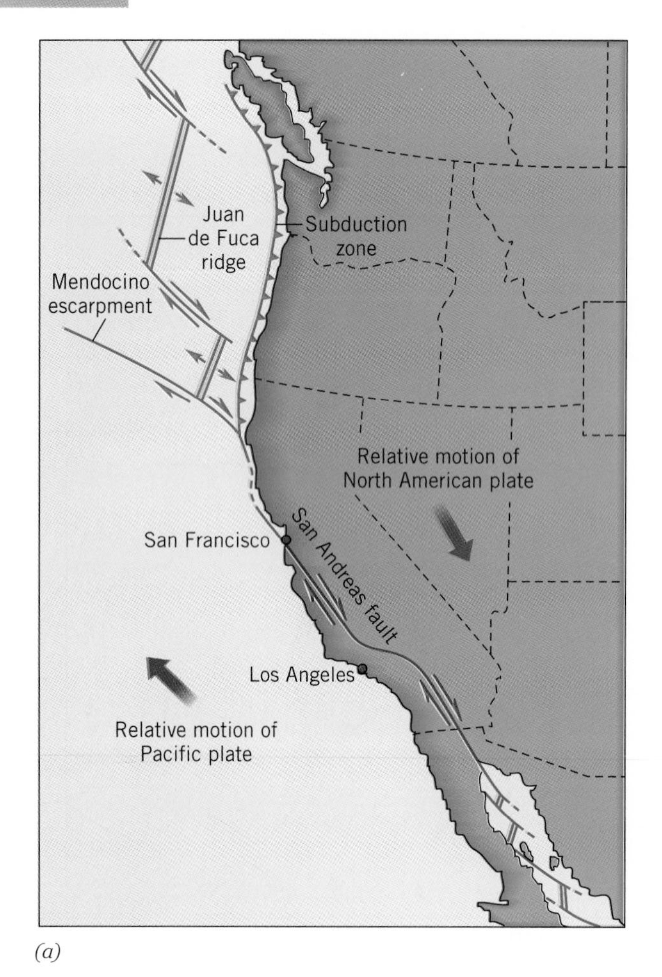

(a)

(b)

Figure 4.9

(a) Map indicating the location of the main San Andreas Fault, and (b) an aerial photo of one section along it.

called **divergent boundaries**, **convergent boundaries**, and **transform fault boundaries**, respectively. We will examine a transform fault boundary first and then study divergent boundaries and convergent boundaries in the following two sections.

A very famous transform fault boundary is the **San Andreas Fault** system in California (Figure 4.9). A small portion of California is on the Pacific plate, moving northwest, while the North American plate moves westward. This means that there is a net lateral motion along the boundary between these two plates, with the Pacific side moving northward relative to the North American side. Los Angeles, on the west (Pacific) side of the main fault, is slowly moving northward toward San Francisco, on the east (North American) side of the fault. If the present motion continues, the two cities should be neighbors in about 30 million years.

Along some sections of the San Andreas Fault, the motion is quite smooth. However, there are some regions where the two plates seem to stick together and do not slide smoothly past each other. Pressure builds, and when the two plates eventually break loose, the built-up pressure makes the resulting motion catastrophic. Unfortunately, the two most densely populated areas of California are near such regions. Although the San Andreas Fault is a famous example of a transform fault running through a continent, most transform faults are in oceanic plates, as we will learn in the following section.

4.5 THE OCEANIC RIDGE

We now investigate divergent plate boundaries. As we have seen, the asthenosphere is slowly moving, and this movement is reflected in motions of the plates of the lithosphere that ride above it. In particular, there are some regions where hot material rises through the asthenosphere, and upon reaching the surface, this hot material spreads outward, carrying the surface plates with it (Figure 4.10a). On the surface, then, we can recognize these regions by the hotter temperatures and the outward spreading plates—that is, the divergent plate boundaries. Furthermore,

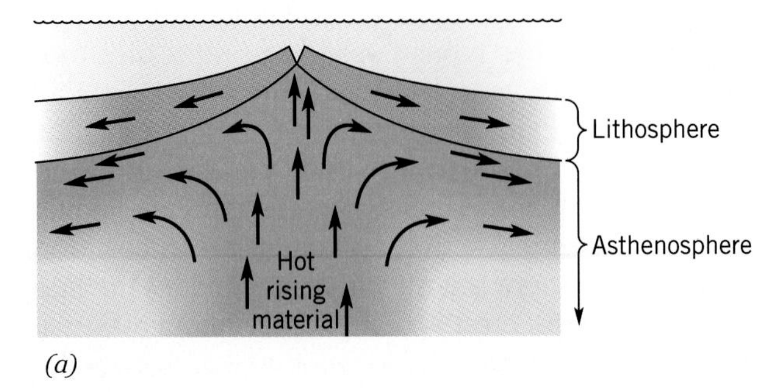

(a)

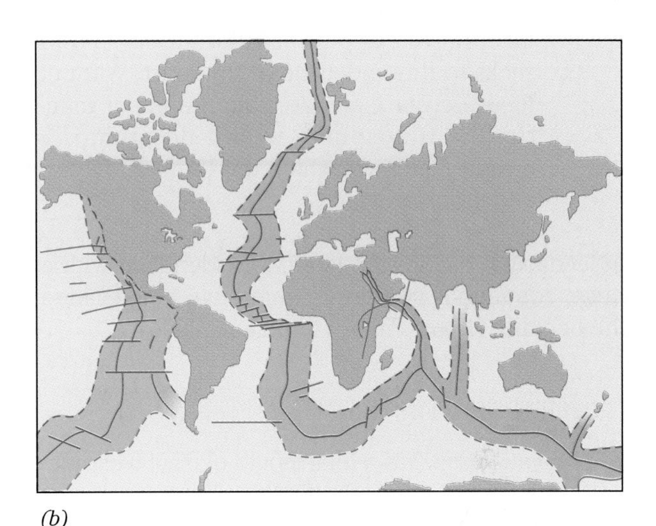

(b)

Figure 4.10

(a) Hot material rising through the mantle produces a ridge on the surface and drags the lithosphere away from the ridge as it flows to the side. (b) Extent of the oceanic ridge system. Boundaries are indicated by dashed lines, and the ridge crest and major fracture zones by the solid lines. (c) A detailed drawing of one section of the ridge, showing its relief.

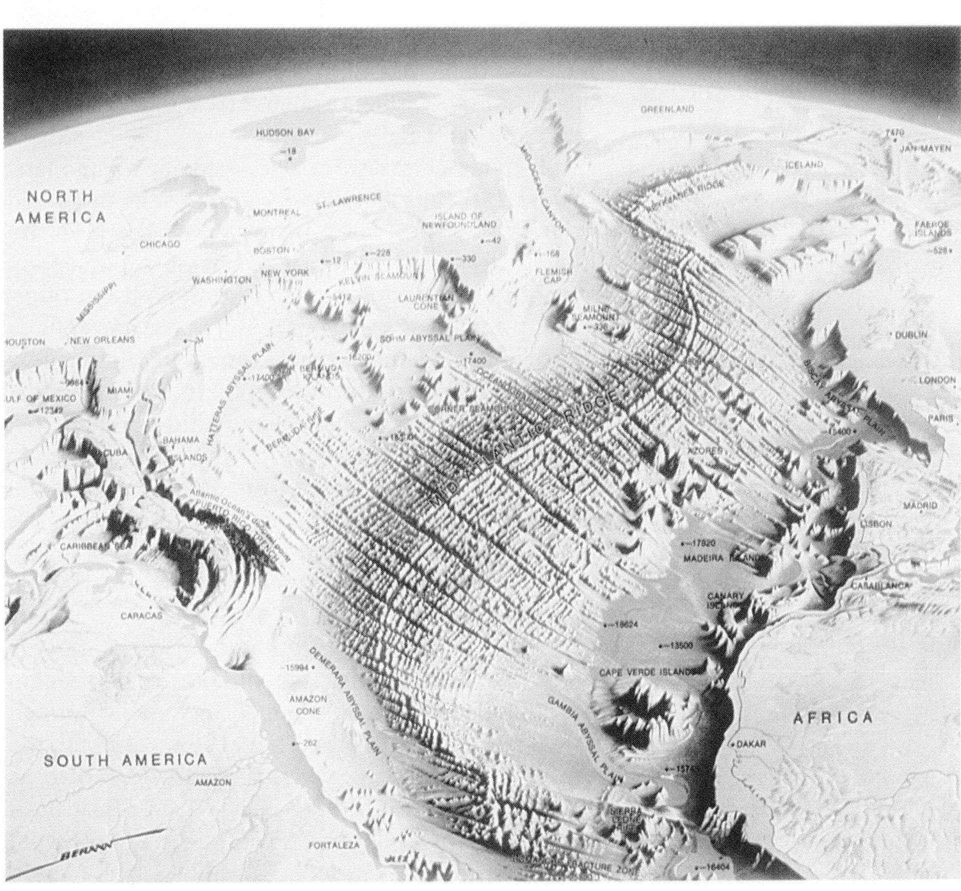

(c)

these regions should have higher surface elevations for three reasons:

1. The hotter lithosphere would be less dense and more buoyant.
2. The lithosphere would be pushed upward by the force of the material rising from beneath.
3. The heat would reduce the amount of material from the upper mantle that cools and sticks to the underside of the crust. Without these denser materials from the upper mantle, the lithosphere would be primarily crustal materials, which are thinner, lighter, and more buoyant.

Indeed, such regions are found and identified with a huge, continuous underwater mountain range called the **oceanic ridge** (Figures 4.10b and 4.10c).

GENERAL FEATURES

The oceanic ridge has a total length of 65,000 kilometers, or about 1.5 times the Earth's circumference. It has an average width of over 1000 kilometers, and it covers about 23% of the Earth's surface altogether—nearly as much as the continents. The ridge crest typically rises to an elevation of about 2 kilometers above the normal ocean floor. In some places, it rises much higher than this and even breeches the ocean surface occasionally. Iceland is a noteworthy example.

> The oceanic ridge looks like a huge zipper holding the Earth together. Some say that it resembles the seams on a baseball.

A large crack, called the **rift valley**, runs centrally along much of the length of the ridge, with hundreds of smaller faults crisscrossing it. The oceanic ridge looks like a huge zipper holding the Earth together. Some say that it resembles the seams on a baseball. Although this ridge system is mostly confined to the oceans, a branch of the system does invade the Red Sea and Eastern Africa.

AGE AND RELIEF

Radioactive dating of the oceanic crust reveals that not only is it all quite young, but also its age varies with distance from the ridge crest. At the crest, the crust is essentially "brand new," and it gets older at a rate of about a million years of age for every 20 to 40 kilometers of distance from the ridge crest. This suggests strongly, of course, that new crustal material is being born at the ridge crest and then is slowly moving away, getting farther from its birthplace as time goes on. Movement by 20 to 40 kilometers per million years implies typical spreading rates of 2 to 4 centimeters per year. If the oceanic ridge is a "zipper," then the Earth is continually being "*un*zipped."

The rate of spreading varies greatly along the oceanic ridge. Some areas, such as the East Pacific Rise, are extremely active, and others are almost dormant. Where there is a rift valley, it identifies the line along which the crest is being torn. Although its features vary greatly with the location and activity of the ridge, the rift valley is fairly steep-walled, typically 0.5 to 1.5 kilometers deep, and ranges from 12 to 48 kilometers wide. It is riddled with faults running both parallel and perpendicular to the rift valley.

As the plates spread outward from the ridge crest, they tear, and magma from below rises to fill in the new cracks that open up. This magma cools and solidifies, becoming part of the newly born oceanic crust. The new oceanic crust, then, comes from magmas that have risen to the top in the hotter parts of the asthenosphere and tend to be less dense than the materials that remained below. These characteristic materials of the oceanic crust, which have been "distilled" from the asthenosphere below, are commonly referred to as **basalts**.

The ridge crest is displaced laterally along many of the large faults that cross it. Between the displaced ridge crests, the plates separated by the fault are moving in opposite directions, making it a transform fault, and resulting in a great deal of earthquake activity in this region (Figure 4.11). The extensions of these faults, called **fracture zones**, frequently continue for hundreds or thousands of kilometers beyond the ridge crests, on down the ridge, and across the neighboring ocean floor. These extensions were once at the ridge crest, but were carried away as the plates spread outward. In these outer regions, the motion of plate fragments on both sides is in the same direction, so earthquake activity is considerably diminished.

Bordering the rift valley on both sides are very rugged **rift mountains**. The entire ridge crest region has frequent earthquakes and many cracks through which molten magma from below flows out onto the ocean floor. Such flows are often referred to as "toothpaste" or "pillow" basalts, because of their appearance as they ooze out of the cracks (Figure 4.12).

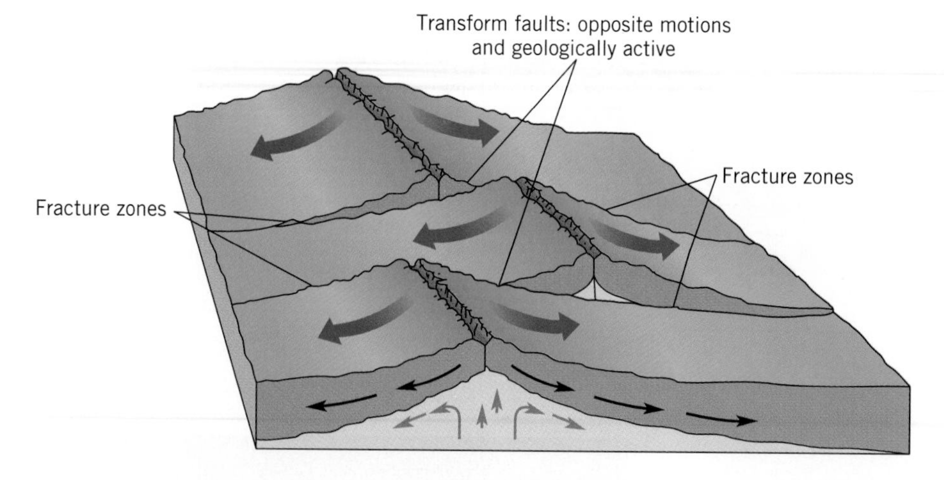

Figure 4.11

Faults cross the oceanic ridge. Transform faults between displaced ridge crests are seismically active due to opposite motions of the crust on opposite sides of the fault. The extensions of this outside the ridge crests are called *fracture zones*.

Going away from the rift valley, the volcanic and seismic activity decreases, the relief becomes less rugged, and the sediment cover thickens. All this reflects the increasing age of the underlying rock, because older crust is cooler, has undergone more flattening, and has had more time to accumulate sediment cover. This pattern continues on past the ridge and across the adjacent ocean basins as well.

MAGNETIC FIELD REVERSALS

Just as paleomagnetic studies of the continents established that the continents were moving, paleomagnetic studies of the ocean floor were also influential in verifying the spreading of the ocean floor from the oceanic ridge. But there was an unexpected surprise.

The spreading of the ocean floor outward from the ridge crest creates tears and cracks in the crust into which magma from below intrudes. As this magma cools and solidifies, a record of the Earth's magnetic field becomes frozen into the rock. Tens of millions of years later, after it has moved a considerable distance away from the ridge crest, it still carries with it a record of the Earth's magnetic field at the time when it formed.

When we measure the magnetization of the oceanic crust, we indeed find that very near the ridge crest, it has a magnetization that parallels the present Earth's magnetic field. However, a few tens of kilometers away from the ridge crest on either side, its magnetization is exactly reversed. As we continue on from the ridge crest, we find alternate regions of *normal* and *reversed* polarity of the crustal magnetization. These regions are in the form of long strips that parallel the ridge axis. Each strip is typically a few tens of kilometers wide, and the pattern is symmetric across the ridge axis (Figure 4.13).[4]

Figure 4.12

Hot fluid magmas flow up through cracks in the crust and onto the ocean floor, where they cool and harden to form "pillow basalts" or "toothpaste basalts," named for their appearance.

This curious pattern was explained by the discovery that the Earth's magnetic field undergoes intermittent **field reversals**. That is, the magnetic field points northward for awhile, then southward for awhile, then northward again, and so on. As each region of the sea floor was forming at the ridge axis, it

[4]This pattern is called the pattern of *magnetic anomalies*. The Earth's magnetic field strength in the region is either slightly strengthened or slightly reduced by the magnetization of these nearby crustal rocks, which is either parallel or backward to the present Earth's magnetic field. Deviation in the measured values of the Earth's magnetic field caused by the magnetization of the nearby rocks are called *anomalies*.

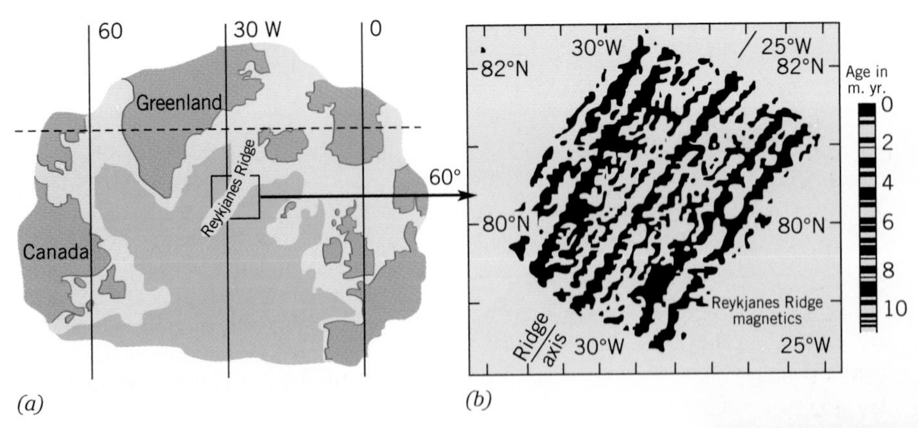

Figure 4.13

(a) The location of the region enlarged in (b) along the Reykjanes Ridge in the Northern Atlantic. (b) Results of measurements showing the symmetrical pattern of alternating strips of opposite magnetic polarity on both sides of the ridge axis. Black areas are regions of positive magnetic anomaly (in the direction of the Earth's present field), and negative anomalies are the regions in between. (From F. J. Vine, 1969, *Journal of Geological Education* **17**:1.)

froze into the solidifying magma a record of the Earth's magnetic field at that time, and then carried this record outward with it as it moved away from the ridge axis. Since the sea floor spreads in both directions from the ridge axis, the pattern is symmetrical across the ridge.

Verification that the Earth's magnetic field reverses direction came from studies of the magnetism in both ancient lava flows and ancient sedimentary deposits. These studies confirm that the field reversals occur at irregular intervals, and a period of one dominant polarity (i.e., north or south) typically lasts between several hundred thousand and a million years.

Although we now know that the Earth's magnetic field reverses itself intermittently, we do not know why. In fact, we don't even know why there is a magnetic field in the first place! We do know that magnetic fields are caused by electrical currents, and we do know that the Earth's core is an electrical conductor. But how all this fits together to produce electrical currents and magnetic fields and field reversals is still a big mystery to us.

HEAT FLOW

Because it lies above the region in the asthenosphere where heated material is rising, the oceanic ridge is hotter than any other region of the Earth's surface. It is characterized by an abundance of geysers, hot springs, and volcanism. The spreading opens tears and cracks in the crust into which cool ocean water flows from above, and hot magma from below. In many places, water that has trickled down through cracks in the crust gets heated in the deeper hotter regions and shoots back to the surface in undersea geysers called *hydrothermal vents* (Figure 4.14). The temperature of the water varies from vent to vent, depending on the depth within the rock that it reached and the amount of dilution by water that hasn't gone as deep. Water temperatures up to 350°C have been observed.[5]

Figure 4.14

A "black smoker" hydrothermal vent on the East Pacific Rise at a depth of 2.8 km.

The study of hydrothermal vents by submersible research vehicles has turned up some interesting surprises. One is that the chemical content of the water exiting these vents has been significantly altered by its high temperature and interaction with hot crustal rock. As the exiting water cools, some of the dissolved minerals precipitate out, giving deposits rich in the sulfides of metals such as iron, copper, and

[5]This is still less than the boiling point for water under the tremendous pressures at these depths.

zinc. Although the total flow rate from hydrothermal vents is only about 0.5% as large as the total flow rate of all continental rivers, the hydrothermal vents have nearly equal impact on the ocean's chemistry.[6] This has been good news to chemical oceanographers, who had previously been unable to explain some details of the ocean's chemical makeup with previously known natural processes.

Also bordering these vents are dense colonies of living organisms. The food that sustains these communities is produced by certain types of bacteria that use the water's temperature and dissolved materials, instead of sunlight. That is, right here on our very own Earth, in the dark depths of our oceans, we find colonies of living creatures that are *not* dependent on sunlight for their existence. We will study these special communities further in Chapter 13.

From heat flow studies, the extent of the sulfide deposits, and the ages of the biological communities, we find that hydrothermal vents are common, but relatively short-lived on geological time scales. Vents are found over large portions of the ridge, especially near the ridge crest. Going away from the ridge crest, increasing sediment cover and cooler rocks cause a reduction of hydrothermal activity.

4.6 CONVERGENT PLATE BOUNDARIES

The lithosphere is thin and buoyant beneath the oceanic ridge, because the heat prevents the mantle material from cooling and sticking to the underside of the crust. As it spreads away from the ridge, however, the combination of cooling and the addition of cool mantle material to its underside makes the oceanic plates not only thicker, but also denser. They may eventually become denser than the warm asthenosphere on which they ride. As we saw in the last chapter, these denser outer regions may slowly sink edgewise into the asthenosphere, dragging adjacent portions of the oceanic plate with them. Therefore, the motion of an oceanic plate is often a combination of being pushed away from the ridge by the flow of the asthenosphere beneath, and being pulled in the same direction by an outer edge that is sinking edgewise down into the asthenosphere. Collisions with neighboring plates add additional forces that influence its motion.

If plates are diverging in some places, then they must be converging in others. The regions where two massive plates are undergoing collision are not only geologically active and exciting, but also largely responsible for the features of this planet that make it so distinctive. From these zones of convergence are born our continents, oceans, and much of our atmosphere. The three types of collisions involving the continental and oceanic plates are illustrated in Figure 4.15.

Materials that make up the continental plates are sufficiently light and buoyant that when two continents collide, neither plunges downward. Both remain afloat atop the asthenosphere. Mountain ranges rise as the crust is compressed, wrinkled, and thickened in the region of collision. The Himalayas are still growing today as India presses on northward into Asia. Older mountain ranges, mellowed by time and erosion, still mark the margins of former continents that underwent collisions long ago (Figure 4.16).

The collisions of oceanic plates are different, however, because the outer edges of oceanic plates are thinner, denser, and might be inclined to sink down into the asthenosphere on their own, even without the forceful encouragement of a collision. Therefore, when two oceanic plates collide, one subsides, plunging downward into the asthenosphere beneath the other. When an oceanic plate collides with a continental plate, the denser oceanic plate subsides. Sometimes, we find fragments of old oceanic plates that were crumpled up against the outer edges of a continent, and in some places we find pieces of old oceanic plates that were squeezed between two colliding continents. In many places, especially where it is relatively young and buoyant, a portion of the ocean floor moves along with the adjacent continent as part of the same plate.

The areas where oceanic plates are slipping down into the asthenosphere are called **subduction zones**, and are marked by **ocean trenches** (Figure 4.17a). Trenches are deep creases in the ocean floor, whose explanation eluded oceanographers before the theory of plate tectonics was introduced. Scientists study earthquakes and seismic waves to trace the path of the cool rigid oceanic plate as it plunges through the trench and on into the asthenosphere below. Careful measurements indicate oceanic plates descend typically at inclinations of 35 to 45°.

As a crustal plate descends into a subduction zone, many of the minerals cannot tolerate the great heat and pressure encountered. Some partial melting

[6]From observed flow rates and heat studies, it is estimated that a water volume equal to that of the entire ocean flows through these vents in 8 million years. The flow rate from rivers is such that it could replace the entire ocean water content in about 40,000 years.

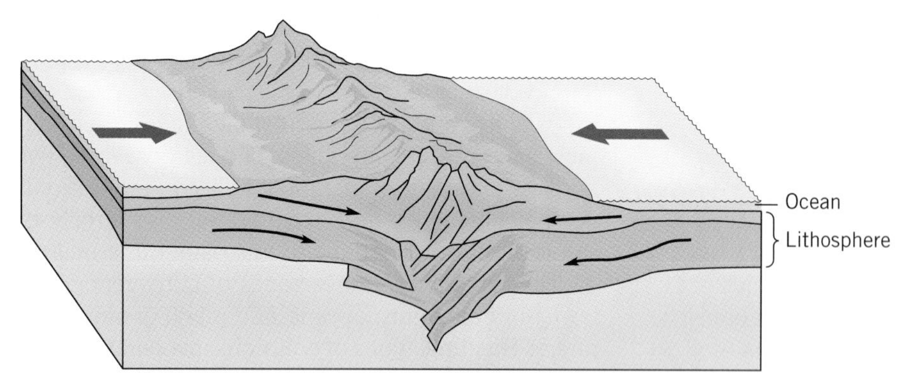

(a) Continent-Continent

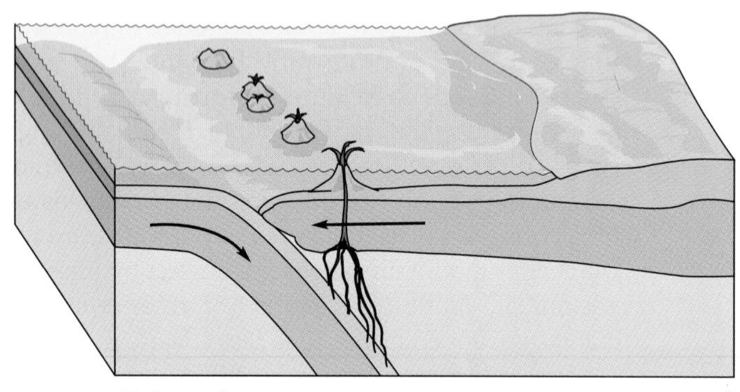

(b) Ocean-Ocean

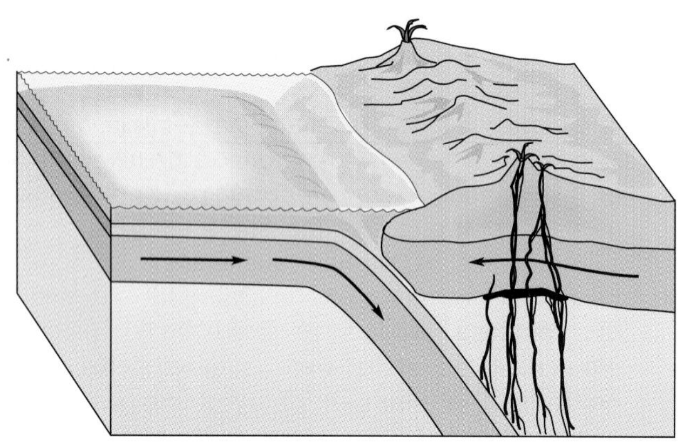

(c) Ocean-Continent

Figure 4.15

Illustration of the three kinds of collisions involving continental and oceanic plates.

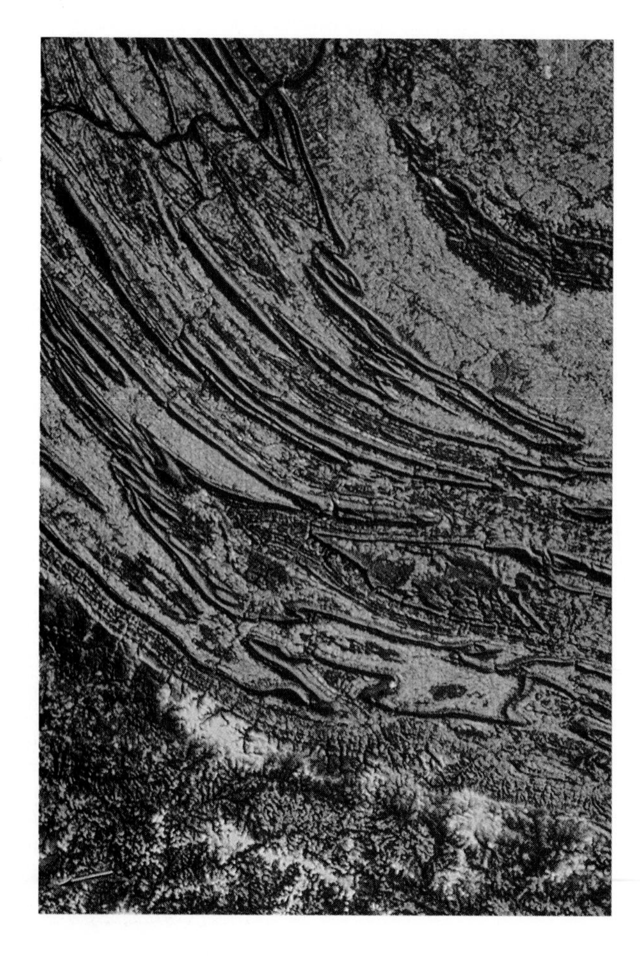

Figure 4.16

High altitude photo of the Appalachian Mountains, which were formed by the compression, lifting, and folding of plate boundaries, ocean floor sediments, and plate fragments as Africa collided with North America about 350 million years ago. Once towering and majestic, erosion has now reduced the Appalachians to a gentle rolling range. Some layers weathered and eroded more quickly than others, leaving the pattern of long parallel ridges and valleys.

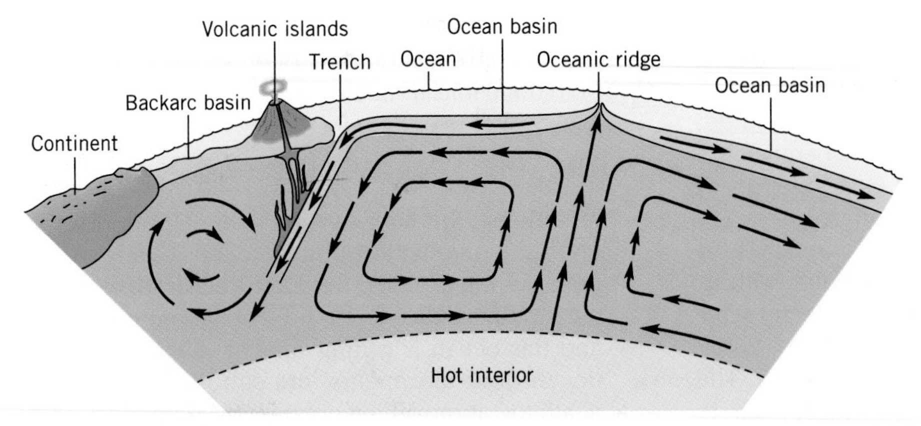

(a)

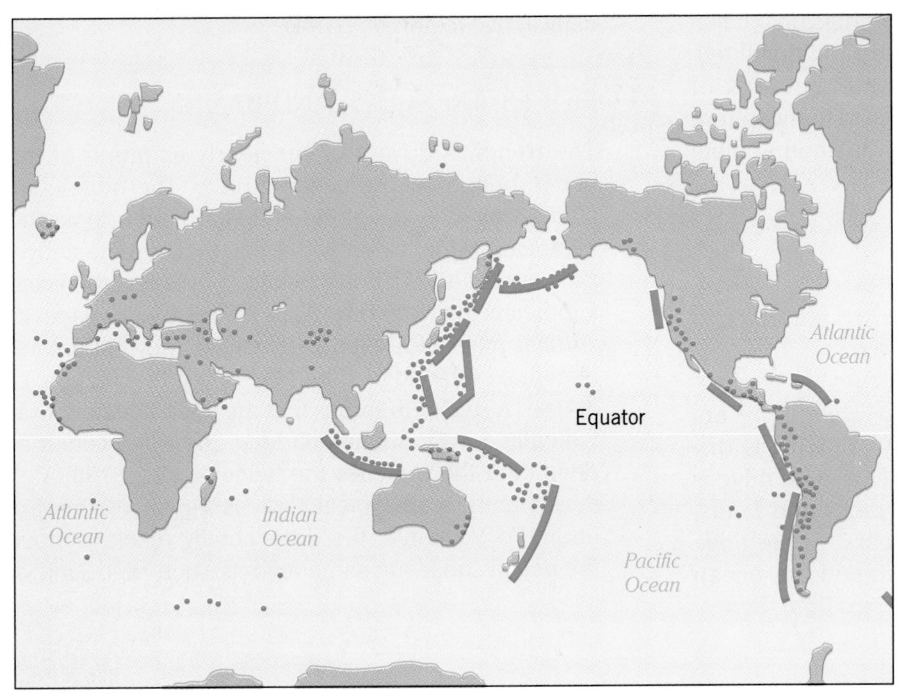

(b)

Figure 4.17
(a) Subduction zone (left). One plate is thrust under another. The increased temperature and pressure cause some of the lighter materials to rise back toward the surface, creating volcanism just outside the trench. (b) Location of the ocean trenches (heavy solid lines) and active volcanoes (dots). Notice that the trenches ring the Pacific and the majority of the Earth's volcanoes are concentrated just outside these trenches.

occurs, and some of the lighter and more volatile materials form magmas that rise back up toward the surface. This "distillation" of lighter materials from descending plates in subduction zones produces most of our atmosphere and the water and salt of our oceans. It also produces most of our continental materials and explains why they are lighter than those of the oceanic crust. It is part of the process of gravitational differentiation of materials referred to earlier, bringing lighter materials to the surface, where they give our planet its distinctive appearance.

As the lighter magmas approach the surface, they may push up beneath the overlying lithosphere, causing uplifting and mountain building. In some places these magmas may break through the surface in the form of volcanoes (Figure 3.14). The Cascade Range of the Pacific Northwest and the Andes Mountains of South America are places where this uplifting and volcanism are happening right now. Materials exiting these volcanoes include nitrogen, chlorine, carbon dioxide, and water, which leave as gases, as well as some of the lighter minerals, which solidify in lava flows.

Nearly all ocean trenches are found around the perimeter of the Pacific Ocean (Figure 4.17b). The Pacific Ocean bottom disappears beneath the North and South American plates on the north and east, and beneath the Asian and Australian plates on the west. As a result of the distillation of magmas from the descending oceanic plate in these subduction zones, the majority of all the Earth's active volcanoes are found just outside this perimeter of trenches. Among others, this includes the Aleutian Islands, Kamchatka, the Japanese Islands, the Philippines, and the Andes Mountains of South America. In fact, the perimeter of the Pacific Ocean is sometimes called the **ring of fire**, because of the volcanic activity there.

As can be seen in Figure 4.17b, the trenches occur in long gentle arcs. A subduction zone, with its trench and associated volcanoes, is frequently separated from the neighboring continent by a sizable shallow ocean basin, called a **backarc basin**. The crust beneath backarc basins frequently appears to be stretched, rather than compressed. This indicates that the subducted plate is going under voluntarily, rather than being forced under by collision. (Collision would cause compression.) It is additional evidence that the cooler denser outer regions of oceanic plates tend to sink into the asthenosphere on their own, although a collision with another plate may be needed to start the process.

4.7 THE OCEAN FLOOR

The dominant features of the deep ocean floor are those that result from the tectonic activity as discussed in the previous sections. The oceanic ridge is very rugged, being riddled with cracks and faults, steep cliffs, and sharp mountain peaks. At the ridge crest, new crustal material is born, and the ocean floor spreads outward in both directions from there.

OCEANS THAT ARE GROWING

In the Atlantic and Indian Oceans, the spreading sea floor drives the adjacent continents ahead of it as it spreads outward from the ridge (Figure 4.18). Therefore, these two oceans are growing. In the Indian Ocean, the oceanic ridge runs through the Gulf of Aden and Red Sea, so these narrow waterways are widening. We are witnessing the birth of a new ocean there (Figures 3.4 and 4.10b). Someday, Africa may be separated from the Middle East by a vast ocean that is just now beginning to grow.

In the Atlantic, the Americas and Greenland are getting farther from Europe and Africa, as spreading continues from the **Mid-Atlantic Ridge**. If we could run the clock backward, we would find that these continents were once joined into one large supercontinent. About 180 million years ago, Pangaea began splitting, as the ridge appeared and the Atlantic Ocean began to open up (Figure 3.6). The ridge still runs exactly down the center of the Atlantic Ocean, indicating that there has been equal spreading in both directions from the ridge.

THE SHRINKING PACIFIC

In contrast, the floor of the Pacific Ocean does not carry the adjacent continents with it as it spreads from the ridge, called the **East Pacific Rise**. In fact, the continents bordering the Pacific are converging, and this ocean is getting smaller (Figure 4.18). The ocean floor disappears into subduction zones beneath the approaching continental plates. From a geological viewpoint, it is far from "pacific". In fact, the perimeter of the Pacific Ocean is the most geologically active region on Earth.

RIDGES, TRENCHES, AND BASINS

The trenches do not cover nearly as much of the ocean bottom as the oceanic ridge. The ridge completely encircles the globe and is at least a thousand kilometers wide, being wider where it is more active. By contrast, a trench is typically only a few thousand kilometers long and only 50 to 100 kilometers wide. Its vertical relief, however, is usually somewhat greater than that of the ridge, but in the opposite direction, of course. A rule of thumb is that the deep ocean floor is typically 4 or 5 kilometers deep, the ridge is half as deep, and the trenches are twice as deep. Although these numbers are typical, there is a great deal of variation. For example, the ridge actally reaches above the ocean surface in some regions, such as Iceland or

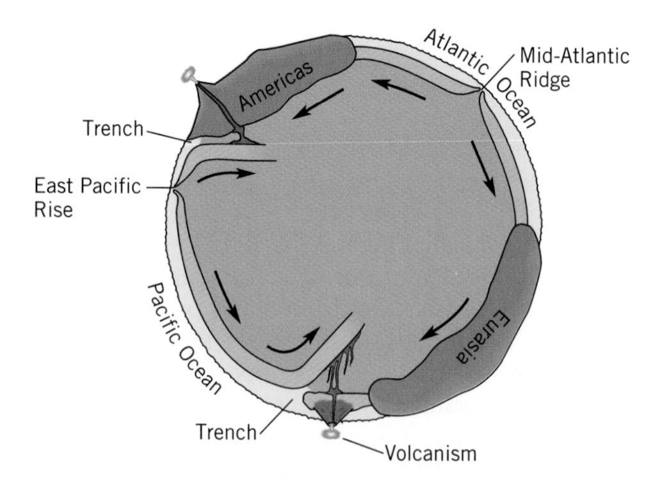

Figure 4.18
Cross section through the Earth. As the Atlantic Ocean floor spreads from the Mid-Atlantic Ridge, it pushes the continents ahead of it. The advancing continents override the Pacific Ocean floor, creating trenches on their Pacific margins. (Vertical exaggeration of plate thickness about 100 to 1.)

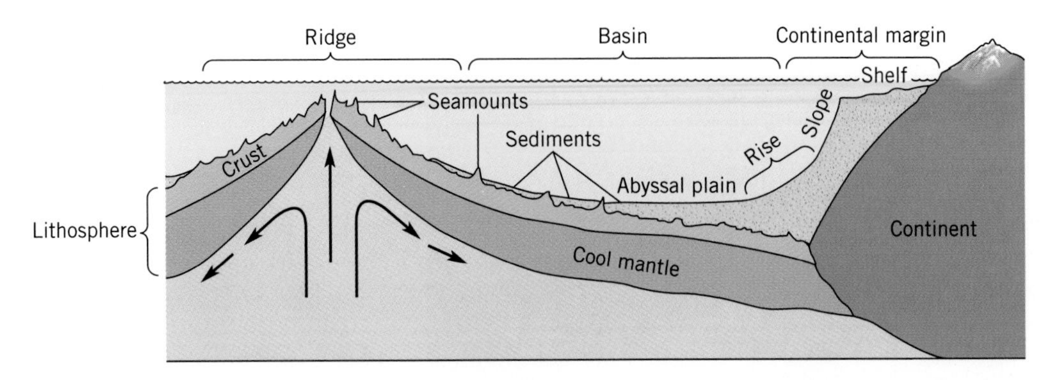

Figure 4.19
Illustration of the common ocean floor features, including those of the continental margins. (Vertical exaggeration about 400 to 1.)

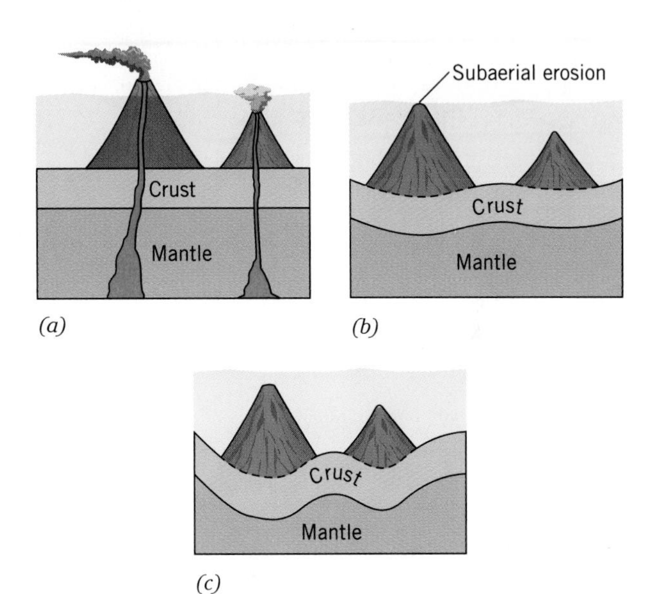

Figure 4.20
(a) Volcanoes. (b) Those protruding through the ocean surface are subjected to weather and wave erosion, flattening the tops. (c) Heavy volcanoes on thin crust subside until isostasy is reobtained, leaving flat-topped undersea mountains, called tablemounts or guyots.

the Tristan da Cuhna group, and trenches reach as deep as 11 kilometers in some places.

The portion of ocean floor extending from the ridge on one side to the trench or continental margin on the other is called an **ocean basin** (Figure 4.19). Ocean basins cover 30% of the Earth's surface altogether, more than either the continents or the oceanic ridge, and have depths in the range of 4 to 5 kilometers. Near the oceanic ridge, the ocean floor is still quite young and rugged, but because of increasing age and sediment cover, it tends to become smoother and more gentle with increasing distance from the ridge. Small hills are called **abyssal hills** or **seaknolls**, and larger mountains are referred to as **seamounts**. Near continental margins, there are frequently vast regions of the ocean basins that are extremely flat due to thick beds of sediments derived from the erosion of the adjoining continent. These regions are called **abyssal plains**.

Putting a massive volcano atop a relatively thin plate of oceanic crust causes this plate to sink, just as a floating air mattress sinks a bit when a person lays on it.

VOLCANIC FEATURES

Volcanoes are also common on the ocean floor. As we have seen, most of the Earth's large prominent volcanoes are found in subduction zones, where one oceanic plate is going under another. There is also considerable volcanism on the oceanic ridge, although much of this goes unnoticed because it is hidden beneath the ocean's surface. These magmas also tend to be comparatively hot, fluid, and low in water content, so that the lava flows on the oceanic ridge tend to be smooth compared with the more explosive and spectacular volcanism that sometimes occurs near subduction zones.

There are also a few very interesting volcanic regions, which are neither associated with subduction zones nor the oceanic ridge, but rather are caused by isolated regions in the mantle with elevated temperatures. Plumes of magmas rising from these **hot spots** create volcanism on the crust overhead.

Putting a massive volcano atop a relatively thin plate of oceanic crust causes this plate to sink (Figure 4.20), just as a floating air mattress sinks a bit when a person lays on it. The sinking continues until buoyant equilibrium is reestablished. This takes only

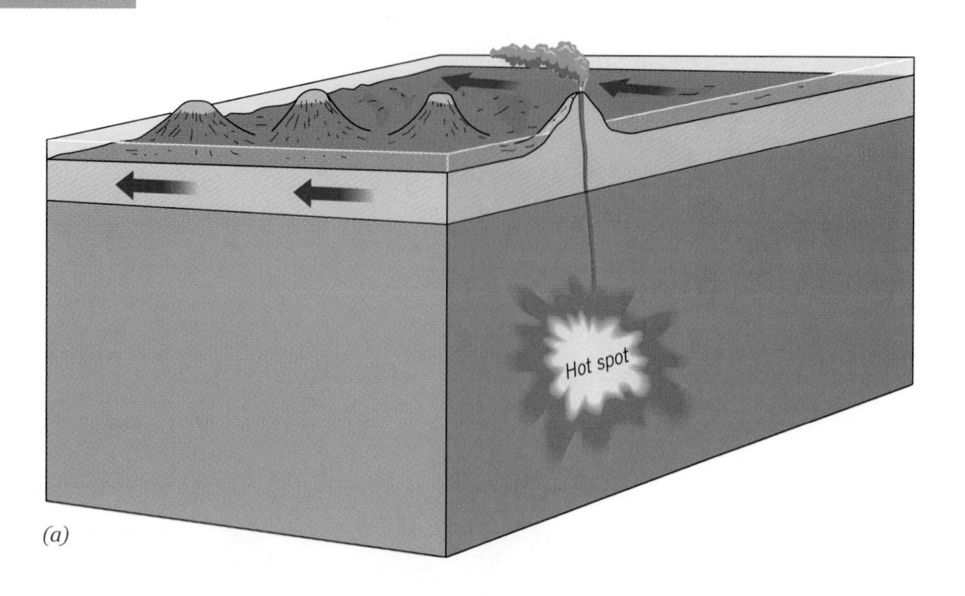

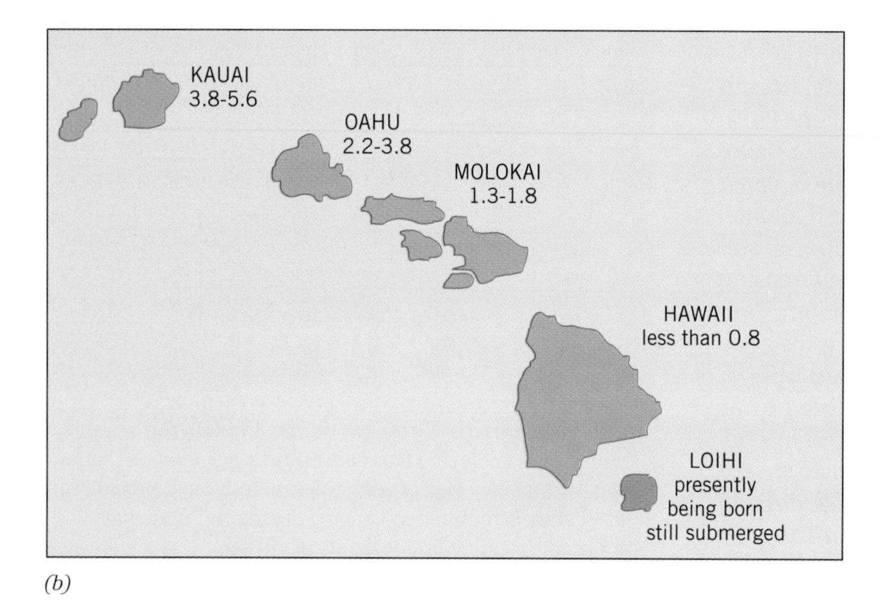

Figure 4.21
(a) Chains of volcanic activity are caused by the passage of the lithosphere over some sort of hot spot anchored in the mantle below. A chain of ancient volcanoes traces the path that the lithosphere has followed over the hot spot. (b) The Hawaiian Islands are at the southeastern end of a long chain of volcanic activity, most of which has sunk below the surface. The islands are born at the southeast end and carried northwest by the plate on which they ride. Ages are indicated in millions of years. (After F. Lutgens and E. Tarbuck, 1986, *Essentials of Geology,* 2nd ed., Merrill.)

a fraction of a second for a person on an air mattress, but it takes millions of years for a volcano on the ocean floor.

If the volcano extends above the sea surface, it becomes an island. Present volcanic islands, such as Hawaii, are slowly sinking and will someday be completely submerged. Because the exposure to waves and weather flatten the exposed portions of these volcanic islands, submerged former volcanic islands can be identified by their characteristic flat tops. They are called **guyots** or **tablemounts**.

As the oceanic plate passes over a stationary hot spot in the mantle below, long **volcanic chains** are created (Figure 4.21a). The rising magmas leave a string of volcanoes, like an upside-down sewing machine, whose needle pokes a row of holes into the cloth passing by. The youngest volcanoes in any chain may still protrude above the ocean surface and be identified as volcanic islands, whereas the older ones have by now sunk below sea level.

This type of volcanism is particularly evident in the central and western Pacific Ocean. As the sea floor moves toward the northwest from the East Pacific Rise, long chains of volcanoes are formed, with the older volcanoes on each chain extending to the northwest from the presently active volcanoes.

The Hawaiian Islands, Line Islands, French Polynesia, Marshall Islands, and many others are simply the youngest members in a long chain of volcanoes, whose older submerged members extend thousands of kilometers to the northwest (Figure 4.21b). In the distant future, we can expect these islands to move with the sea floor toward the northwest and to slowly sink beneath the sea surface. But new islands will appear to replace them, as the hot spot anchored in the mantle below continues its activity.

CORAL REEFS AND ATOLLS

Coral reefs are rigid, porous, wave-resistant structures made primarily of the interwoven and cemented skeletons of generations of corals, which are a kind of animal (Section 13.6). Shells of shellfish, along with skeletons of certain types of algae and other members of the intricate reef community, also make significant contributions to these structures. Most reef-building corals grow best in warm and salty water, which is most prevalent in the western tropical regions of oceans. Although coral reefs cover only a very minor portion of the ocean floor, they grow in shallow water in regions of the world that are often heavily populated. Consequently, they have an important impact, especially as a navigational hazard.

The primary food producer in most reef communities is an algae that lives together in a mutually beneficial arrangement with the coral. The algae must be close enough to the ocean surface to receive sunlight. The corals are stationary and require water motion to bring them the small particles of nutriments they need. Therefore, shallow water exposure to wave motion and sunlight are primary needs for living reefs. Corals do not grow well in the turbid coastal waters bordering large land masses, because the load of inorganic sediments carried by rivers and streams reduces the sunlight available to the alga and dilutes the nutritional value of the suspended particles captured by the corals. In summary, coral grows best in the western tropical oceans, away from river mouths and other sources of freshwater runoff from land masses, and exposed to heavy wave activity.

When conditions are right, **fringing reefs** first appear in the shallow waters bordering a land mass. The reef grows best on the outer edge, where the wave action and water motion are greatest, and the coral is farthest from the land's freshwater runoff. As it grows outward, it leaves a lagoon behind, and in this more mature stage it is called a **barrier reef**. The accumulating skeletal debris extends the platform, enabling continued seaward growth of the barrier reef, with a resulting widening of the lagoon. A famous example of this type of reef is the 2000-kilometer-long Great Barrier Reef of the northeastern Australian coast.

Because corals rely on water motion to bring them their food particles, they do not grow as well in the protected waters of the lagoons behind the barrier reefs. Although there is enough food to support occasional *patch reefs* that grow upward from the lagoon floor, life is not as vigorous as on the reef's outer edges, and much of the lagoon floor is covered with coralline skeletal debris.

Sometimes, small ring-shaped reefs are found growing in what would otherwise be deep ocean waters (Figure 4.22). These are called **atolls**, or *atoll reefs*. The coral atolls of the South Pacific were among the things mapped by Captain Cook and studied by Charles Darwin in his voyage on the *H. M. S. Beagle* in 1831 to 1836, and he formulated the basis of the correct explanation for their formation. In balmy climates, coral grows rapidly. As a volcanic island slowly sinks, a coral fringing reef starts growing around its flanks. As it continues to subside, the reef grows upward at a matching rate, leaving a lagoon between it and the remaining emerged part of the volcano. Finally, after the entire volcano has subsided below sea level, only the ring-shaped reef remains, and we call it an *atoll*.

Most of an atoll is submerged, but sometimes waves pile up the skeletal debris, forming low and rather unstable islands. The rim of the atoll tends to grow thickest on the windward side, because the heavier wave activity brings more nutriments to the feeding corals. A typical cross section of an atoll is given in Figure 4.22e.

PLATEAUS

There are numerous elevated portions of the ocean floor that rise 1 or 2 kilometers, but do not reach the ocean surface and are not part of the oceanic ridge. These **plateaus** are made of thicker, more buoyant materials than is typical for the ocean floor. Some of these may be continental fragments, and others the result of volcanism. (Remember, volcanism tends to bring lighter materials to the surface, which eventually aggregate onto continents.) As plateaus are carried with the ocean floor into a subduction zone, they tend to remain "afloat." The plateau collides into the adjacent continental margin, and the trench reappears behind the plateau where the oceanic plate resumes its descent (Figure 4.23). The collision causes folding, faulting, and uplifting of the colliding blocks, along with the sediments that were trapped between them. Taiwan is on one such plateau that is heading for a collision with the Asian continent in just a few million years.

(a)

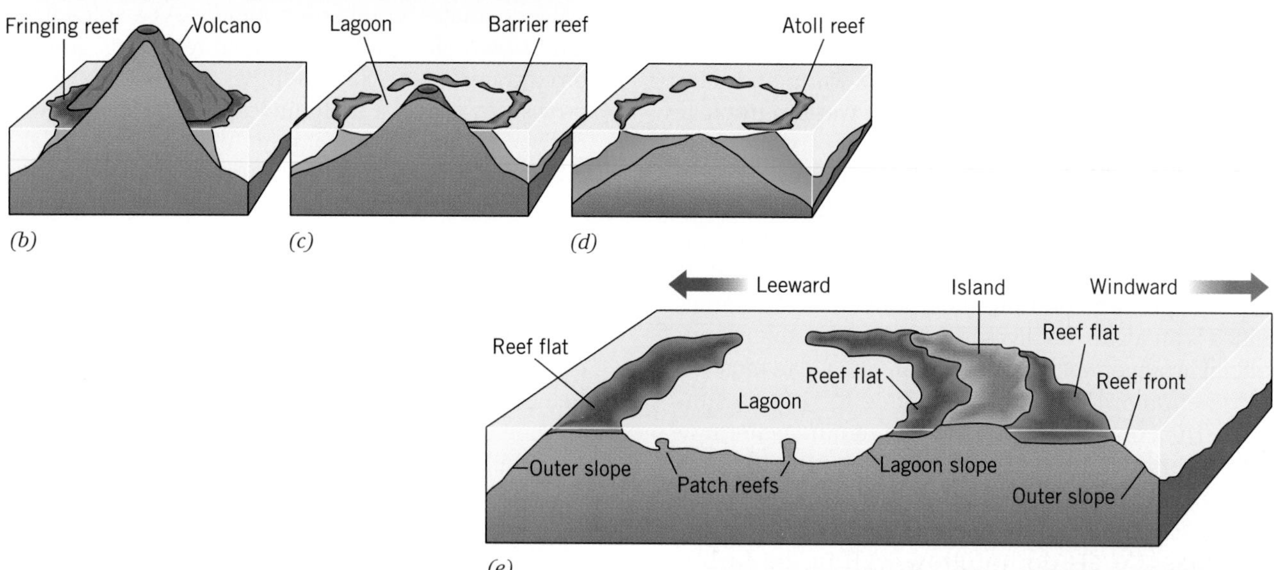

(f)

Figure 4.22

(a) Atolls, looking southeast across Tuamotu Archipelago in the South Pacific, as seen from Apollo 7. (b) The formation of an atoll begins as coral fringing reef grows around the edge of a volcanic island. (c) As a volcano subsides, coral grows upward at a matching rate, leaving a lagoon between the barrier reef and volcanic island. (d) Eventually, the volcano sinks entirely below sea level, leaving only a ring-shaped atoll reef with a central lagoon. (e) Cross section of a mature atoll. (f) Underwater photo of a reef.

4.8 THE CONTINENTAL MARGINS

The deep ocean floor ends abruptly at the continents. The edges of the continents rise rather steeply from the ocean depths and clearly identify where the deep ocean basins end and the shallow waters of the continental margins begin. We now turn our study to these areas (Figure 4.19 and Table 4.1).

OVERVIEW

Going seaward from shore, the first submerged region is the **continental shelf**. It is an extension of the continent that just happens to be under water at

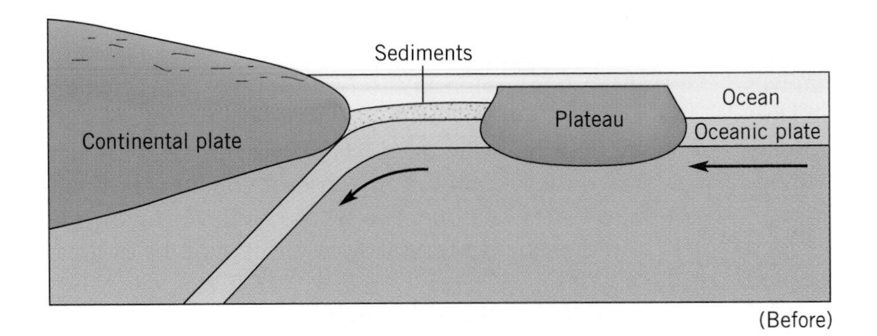

(Before)

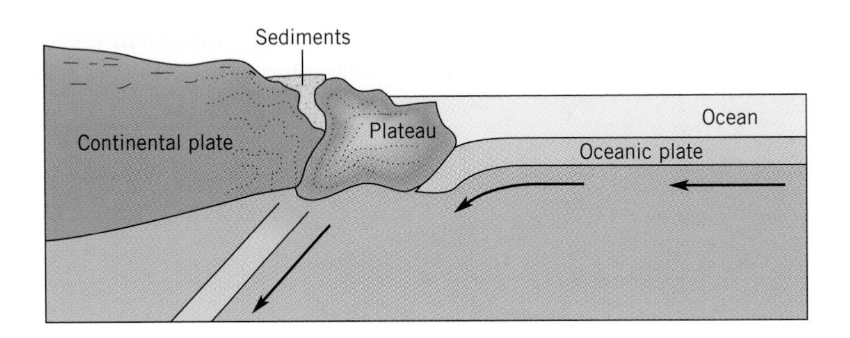

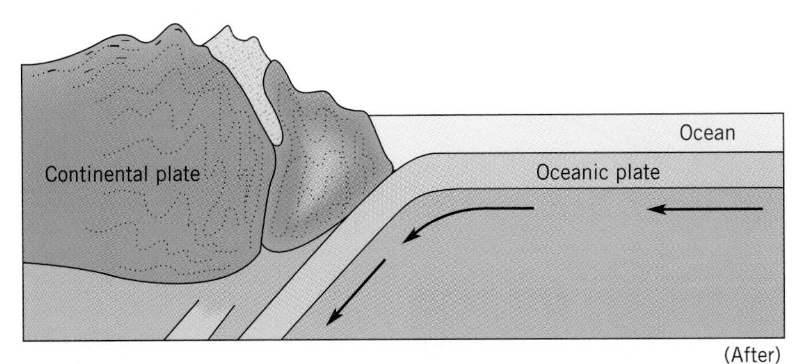

Figure 4.23

Oceanic plateaus do not descend with the oceanic plate in subduction zones. Rather, they get "scraped off," causing some uplifting and mountain building as they collide with the continent. The trench reappears behind the plateau after it has been plastered up against the continent.

Table 4.1 Ocean Areas and the Fraction of each Ocean's Area in the Various Provinces: Continental Shelf and Slope, Continental Rise, Deep Ocean Basins, Oceanic Ridge, and the Relatively Small Miscellaneous Areas Involved in Trenches, Volcanic Island Chains, and Other Minor Features

Ocean	Shelf and Slope (%)	Rise (%)	Basin (%)	Ridge (%)	Misc. (%)	Total Area ($10^6 km^2$)
Pacific and adjacent seas	13	3	43	36	5	181
Atlantic and adjacent seas	18	8	39	32	3	94
Indian and adjacent seas	9	6	49	30	6	74
Arctic and adjacent seas[a]	68	21	0	4	7	12

[a]Although the Arctic Ocean has some basins, their depths are more comparable to the rises of the three major oceans, so we include them under "rises."

Source: Menard and Smith, 1966, *Journal of Geophysical Research*. **71**, 4305.

this particular time. It hasn't always been submerged and will reemerge from time to time in the future in response to minor changes in sea level. The shelves are fairly flat, with an average depth of about 100 meters. This is barely enough to cover a 20-story building. Shelf widths vary greatly, but the average width is about 75 kilometers.

The outer edge of the continental shelf is called the **shelf break**. It marks the beginning of a drop-off in the underwater landscape, called the **continental slope**. The continental slope falls downward toward the adjoining ocean basin at an average angle of slightly over 4°, which is comparable to the steepest parts of a modern highway entering or leaving a range of mountains.

At the base of the slope may be a wedge of sediments that make up the **continental rise**. These sediments washed down the slope and were deposited at the base, often in the form of large "fans," where canyons and gullies in the slope empty out onto the sea floor below (Figure 4.24). These deposits are similar to those found at the base of continental mountains, where sediments wash down through canyons and deposit in **alluvial fans** at the mouths of these canyons.

In fact, if the water were removed and you could stand at the edge of an ocean basin and look upward at the slope, it would resemble a very high continental mountain range. One big difference, of course, is that if you were to climb over this range to see the other side, there wouldn't be one. You would find a flat continental shelf up there instead.

(a)

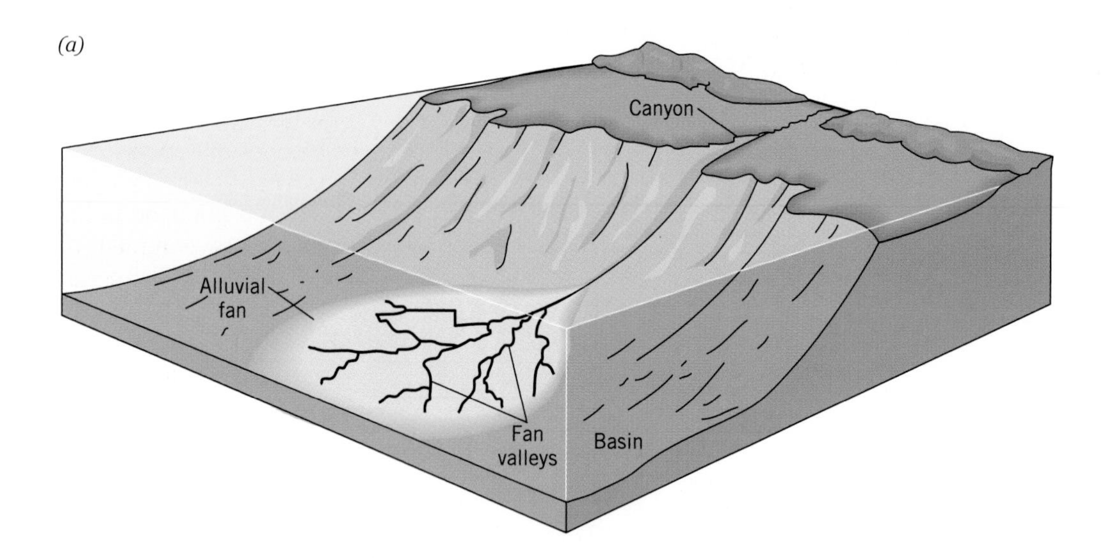

(b)

Figure 4.24

(a) Sediment slides down through canyons in the continental slope in muddy flows called *turbidity currents* and is deposited at the base of the slope in large fan-shaped deposits. Channels called *fan valleys* are cut across the fans by the muddy currents. (b) An alluvial fan at the base of a mountain in California.

> If the water were removed and you could stand at the edge of an ocean basin and look upward at the slope, it would resemble a very high continental mountain range. One big difference, of course, is that if you were to climb over this range to see the other side, there wouldn't be one.

THE SHELF

The outer margins of some continents are geologically active. This is the case when the continental crust and adjacent oceanic crust are part of separate plates, with the oceanic plate sliding past or plunging beneath the continental plate. In other cases, the continental crust and oceanic crust lie adjacent to each other on the same plate as a result of the splitting of a continent by the birth and spreading of an ocean basin. In this case, the continental margin is not a plate boundary and thus the tectonic and seismic activity associated with plate boundaries is absent. Because the features of the continental margins reflect the degree of geological activity, it is convenient to classify continental margins as being **active** or **passive**, according to this criterion. The continental shelves on passive margins tend to be broad and well developed, reflecting long periods of relative stability. On active margins, however, the shelves are rather narrow and poorly developed. For example, the continental shelves of the Atlantic Ocean tend to be very broad and well developed because the Americas, Europe, and Africa move ahead of and along with the oceanic crust as they spread from the Mid-Atlantic Ridge. By contrast, the West Coast shelves of the Americas tend to be narrow, because these continental margins are undergoing collision with the plates of the Pacific Ocean floor.

The outer edge of the continental shelf is marked by an abrupt change in slope. A typical depth at this point is 130 meters. Shelves are generally quite flat, having been smoothed by numerous processes, the most important being the movement of sediments by large waves. Most have also been exposed to erosion by wind and weather during recent ice ages when sea level was lower, and some were excavated by the ice itself.

The underlying cause of some of the broad shelves of the passive edges can be traced to the time when the oceanic ridge appeared and spreading began. The continental crust stretches and thins before tearing open, much like potters clay does as you pull it apart (Figure 4.25). This is what is going on in East Africa at this moment. Eventually, the continent begins to tear and a sea opens up, such as is happening in the Red Sea today, and as happened in the new Atlantic Ocean around 180 to 150 million years ago. As the newly torn and stretched continental margins leave the spreading center, where they were being heated and uplifted by rising hot material in the asthenosphere below, they begin subsiding. In buoyant equilibrium, these stretched thinner portions of the crust have lower elevations than the unstretched, thicker regions. The added weight of accumulating sediments also pushes them downward. These stretched continental margins become the continental shelves and low coastal plains of the new continental margins.

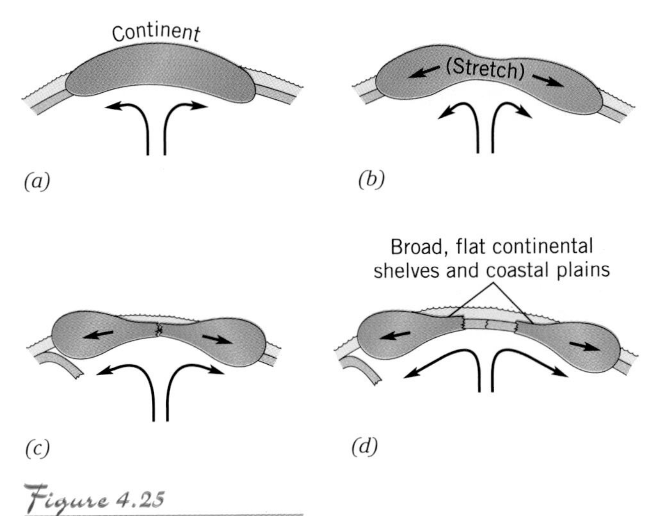

Figure 4.25

As a new ocean opens, the continent is stretched before it breaks apart. This creates broad flat continental shelves and coastal plains on the trailing edges of continents, such as we find on the Atlantic margins of the Americas, Europe, and Africa. (Vertical exaggeration of about 400 to 1.)

> The continental crust stretches and thins before tearing open, much like potters clay does as you pull it apart.

Although stretching and thinning are important mechanisms for the formation of the broad continental shelves on passive continental margins, waves are also important in the grooming of these shelves. In fact, waves would generate small shelves all by themselves in the absence of other processes. You can demonstrate this in a puddle of water (Figure 4.26). If you put a pile of sand in the middle and then make

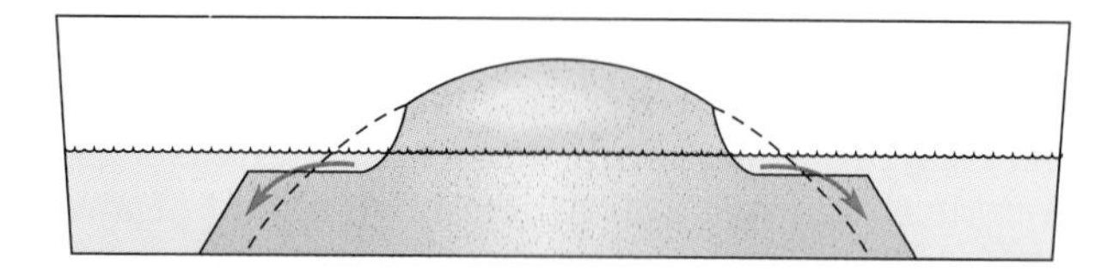

Figure 4.26
If you put a pile of sand in a tub of water and then make waves, the waves eat away at the sand, forming a sea cliff at the exposed face of the sand pile and a terrace extending outward below water level.

some waves, you will find that the waves cut into the sand, leaving a terrace. The sand removed from the exposed areas will be carried out into deeper water and deposited, extending the terrace further out.

In a similar fashion, waves cut into the edge of the continent and carry the debris out to deeper water. Although the incessant smaller waves play a role in beach erosion and the shaping of shallow beach features, the more infrequent but very forceful heavy storm waves actually do most of the excavation. Beneath these storm waves, water motion to a depth of about 100 meters is sufficient to transport sandy sediments, so we say that they "reach" to a depth of about 100 meters. Consequently, the terraces excavated by these waves are typically at a depth of about 100 meters. In summary, the exceptional width of some continental shelves on passive margins is due to the stretching and thinning of the continental plate as it tore open. Storm waves provide further excavation and grooming of these wide shelves, and they also carve narrower shelves on other continental margins.

Sea level changes by a hundred meters or so over periods of tens of thousands of years, as ice ages come and go. Wave-cut terraces are frequently found both far out on the continental shelf, and also on land above present coastlines, indicating past positions of sea level. Large portions of the shelves are sometimes exposed during ice ages, and the shelves show evidence of erosion by the weather, streams, and the ice sheets themselves. However, the shelves are typically over 100-million-years-old and therefore much more stable than the location of sea level. The overall wave-carved features, then, represent some sort of average water depth over this long period of time.

THE SLOPE

Although the continental shelves cover about 6% of the world's surface and are painted blue on maps, they are very shallow compared to the other regions

Figure 4.27
Chart of the Monterey and Carmel Submarine Canyons off the central California coast.

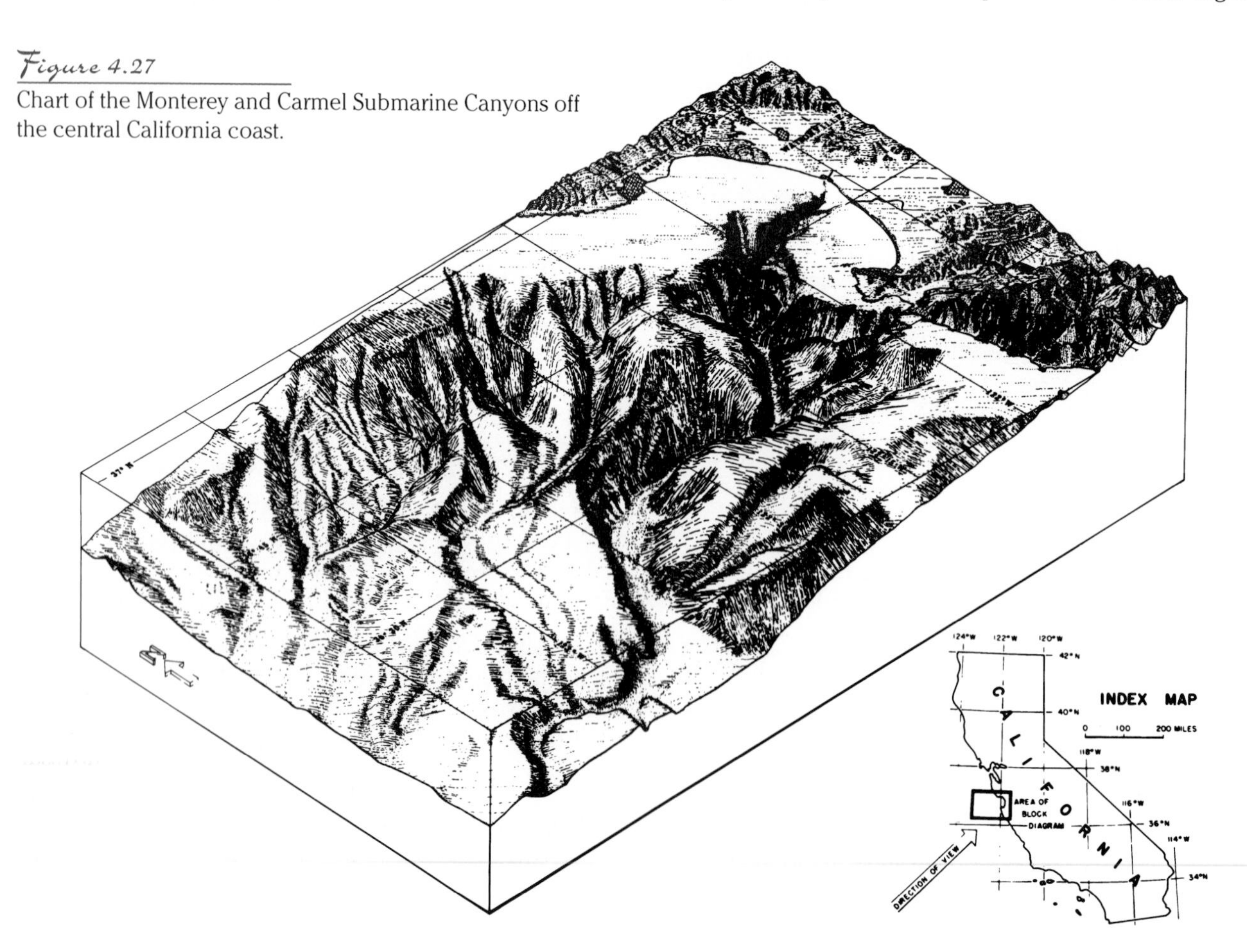

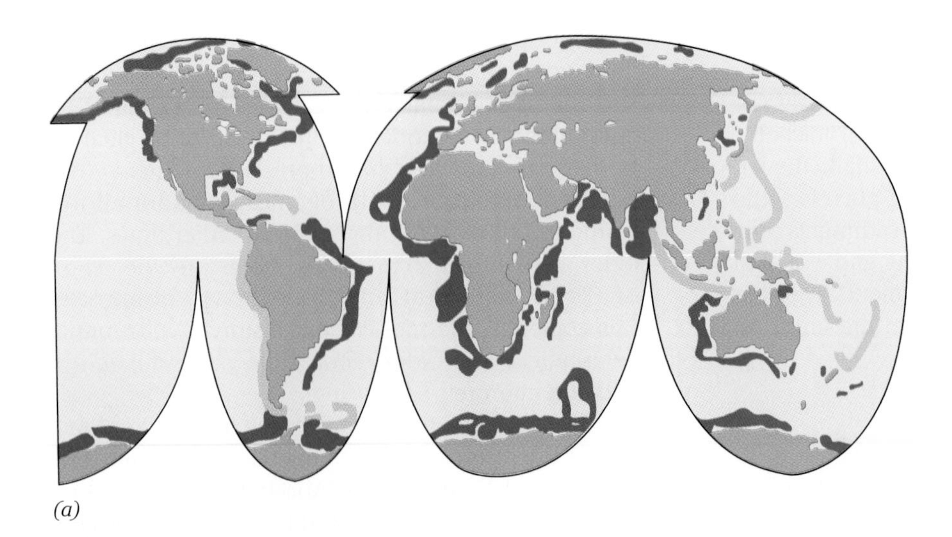

(a)

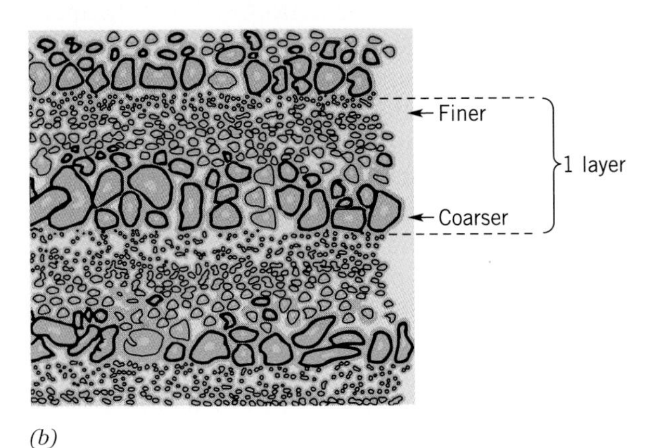

(b)

Figure 4.28

(a) The world distribution of continental rises shows their scarcity along the margins of the Pacific. The thick shaded lines indicate the locations of the trenches. Notice the lack of rises on margins bordered by trenches. Can you suggest an explanation? (From K. O. Emery, 1969, *Oil and Gas Journal* **167**:19, 231–243.) (b) Graded bedding as found in turbidite deposits. In any one layer, the coarser sediment is found at the bottom and the finer sediment near the top. The pattern repeats in each layer.

of the ocean. As we have seen, they are really extensions of the continents, whose outer edges are marked by the continental slope. In some ways, the "real" oceans begin at the continental slope, not at the beach.

Like the edges of large continental mountain ranges, the slope has large assortments of gouges and gullies running down its face (Figure 4.27). Some of these are huge and called **submarine canyons**. Many have relief comparable to that of the Grand Canyon, and some are even larger. Cores of sediments retrieved from fans at the base of canyons show that these deposits have been building up for many millions of years. This indicates that many present canyons may have already been forming when the ocean was still young, and some may have been started when the original rifting occurred. Some may have begun as river valleys cut across the shelf at a time when much of the shelf was exposed above sea level.

However these canyons started, they are further excavated by muddy murky avalanche-like slides of sediments, called **turbidity currents**. Sediments accumulate in the upper end of a canyon, either being deposited there by rivers flowing into the ocean, or by the longshore flow of sediments down the coast. As these sediments pile up, eventually their weight becomes too great, and they slide down through the canyons and out onto the sea floor below in these turbid flows. Repeated excavation by turbidity currents carves the canyons into increasingly larger and deeper features.

THE RISE

The sediments deposited at the base of the slope make up the continental rise (Figure 4.28). Some of these deposits are the accumulation of sediments that flow slowly across the shelf and down the slope. But most are the deposits of turbidity currents. In some regions where turbidity currents are particularly prominent, they flow on out across the continental rise and onto the deep ocean floor, forming large abyssal plains. As you might guess, it is impossible to determine where the rise stops and the abyssal plain begins, so the distinction between the two regions is a bit arbitrary.

The sediment deposits of the rise and abyssal plains are in layers, displaying **graded bedding**. Each layer has the coarsest sediments on the bottom and finest on top (Figure 4.28b). This tells us that the sediments arrived in occasional turbid flows, rather than being deposited slowly and continuously. To observe graded bedding, put a mixture of miscellaneous sediments into a tall jar of water, shake the jar, and then set it down. The stones and gravels settle out first, followed by sands. The finest sediments may take several minutes to settle out. In the end, you will see that you have sorted out your sediments, with the coarsest sediments on the bottom, and the finest on top.

To observe graded bedding, put a mixture of miscellaneous sediments into a tall jar of water, shake the jar, and then set it down. The stones and gravels settle out first, followed by sands. The finest sediments may take several minutes to settle out.

4.9 THE PAST

The evolution of the ocean floor, with its spreading from the oceanic ridge and eventual destruction in a subduction zone, is now fairly well understood. As it spreads, it carries a record of Earth history. Studies of the ocean floor—both the crustal rock and its sediment cover—can be interpreted rather directly to tell us details of the Earth's recent history. Unfortunately, ocean floors do not last more than a few hundred million years before disappearing into some trench, so studies of the ocean floor cannot be used to reveal elements of the Earth's history older than that. For insight into earlier history, scientists look to the continents or develop theoretical models. From very basic information, models can be constructed that produce general features of the Earth's evolution, although these models cannot provide the detailed history of any particular continent.

From studying such models, scientists believe that heat production inside the Earth was originally about three or four times its present value. As a consequence, the early lithosphere was hotter, thinner, more flexible, and more buoyant than it is now. As the heat production diminished, the lithosphere cooled, became less buoyant, and subduction zones began to form. Plateaus and other fragments of volcanically produced continental materials began to aggregate at these convergent plate boundaries, forming larger blocks of continents. Plate motion caused repeated collision and occasional division of these continental blocks, which is continuing today.

The actual details of these processes cannot be determined from models, but rather must be gleaned from the records held in the present continents. The ocean floors are no help because they are all too young to tell us anything of these earlier times. The older the continental materials, the greater the extent of the transformations they have undergone. Consequently, the most recent history is most easily deciphered, and older records are more difficult to find and interpret.

Scientists are gradually forming a fairly good picture of the history of the continents over the most recent 10% of Earth history. About 350 million years ago, a block of continents including what is now Africa collided with North America, forming Pangaea. The Appalachian Mountains mark this collision (Figure 4.16). Since that time, erosion has made the Appalachians into a gentle, rolling range, which is a far cry from the formidable rough range they were back then. About 180 million years ago, a ridge appeared, which began to separate the continents and open the Atlantic Ocean (Figure 3.6). This cleavage was east of the previous continental margins, so as Africa departed, it left North America with the land to the east of the Appalachians. This now forms the eastern coastal plain and continental shelf.

If you look at a map of the Earth, you find it covered with clues to previous and present tectonic activity. Why are the Ural Mountains in such a conspicuously straight line? What is going on in the Red Sea and the Gulf of Aden, with their peculiar geometry and an oceanic ridge running into them? Why the Rocky Mountains? Why the geysers and hot springs of Yellowstone National Park? To find the answers to questions such as these, we must remember that with the rapid rates of erosion and sedimentation, features more than a billion years old will be completely disguised, leveled, and buried, and their history will be very difficult to decipher. But even the history of the last billion years will keep us busy for awhile.

4.10 PLANETARY EVOLUTION

CONTINENTAL AND OCEANIC MATERIALS

We have seen that oceanic crust is born at the oceanic ridge, as magmas from the mantle rise to fill in the cracks that form in the crust as it spreads. Because they have risen to the top, these magmas tend to contain materials that are less dense than those of the mantle through which they rose. This "distillation" of materials from the mantle explains why the oceanic crust is less dense than the mantle.

The oceanic crust spreads away from the oceanic ridge and eventually reaches a subduction zone, where these materials undergo a second "distillation." Magmas of lighter, more volatile oceanic materials tend to be driven from it as it sinks into the hotter regions below. These lighter magmas then rise back up toward the surface in volcanic and plutonic flows, cooling and solidifying as they near the cooler surface regions. These are the materials from which the continents are made. This distillation in subduction zones explains why the continental crust is less dense than oceanic crust, which itself was previously distilled from the mantle.

As the plates move about, they undergo collisions. Unlike the oceanic plates, which tend to subduct into the interior, the lighter continental materials tend to remain afloat. Upon collision they are compressed and folded, making them thicker. Like soap bubbles on water, these more buoyant continental materials tend to aggregate into larger thicker masses, which become continents.

NEIGHBORING PLANETS

Now that we are beginning to understand the cause of the prominent features of our planet, such as its oceanic ridge, trenches, continents, and oceans, we must next wonder why we are unique. Why haven't the other terrestrial planets undergone as much geological evolution as the Earth?

The answer is that the Earth's greater geological activity is a result of its greater size. You may have noticed that soup in a teaspoon cools more quickly than soup in a bowl. Smaller objects have more surface area in comparison to their volume and can therefore cool more quickly. Planets having larger volumes have correspondingly more radioactive materials and other heat sources within them, and they also have greater difficulty cooling off. The heat tends to stay trapped inside, and so the interiors of larger terrestrial planets are hotter. Being the largest of the terrestrial planets, we should expect the Earth's interior to be hottest, and therefore it should be the most active geologically. It is.

Mercury and our Moon were apparently too small and cooled too quickly for any appreciable amount of plate motion to have occurred. Venus and Mars have clearly experienced some plate motion in the past, as is evident from the presence of "continents" on their surfaces. They are not as active as Earth, and of course, they do not have liquid water to fill in their ocean basins, as do we.

Summary

Continental Drift and Paleomagnetism

Early in this century, Alfred Wegener suggested that the continents now separated by the Atlantic Ocean were once united. In support of this, he pointed to the symmetry of the Atlantic coastlines and evidence from ancient fossils, climates, and geological features. In the 1950s, analysis of the fossil magnetic record in continental rocks proved that the continents were indeed moving.

The Driving Force and the Plates

Convection in the asthenosphere and mantle is driven by heating from the Earth's interior. This, along with sinking outer edges of oceanic plates, causes considerable motion of the lithosphere. The plate boundaries can be identified by the earthquakes and are of three general kinds: divergent, convergent, and transform fault boundaries. Along transform fault boundaries, there is relative sideways motion between the two plates.

The Oceanic Ridge

Divergent boundaries are identified by the oceanic ridge, a place where hot material in the asthenosphere is rising from beneath. As it rises, it heats and pushes up the ridge, and then drags the oceanic plate with it as it spreads outward. Evidence that supports this includes the age and sediment cover of the sea floor, both of which increase with increasing distance from the ridge crest, and the elevated temperatures of the ridge. The strips of sea floor of alternate magnetic polarity are a result of occasional reversals in the direction of the Earth's magnetic field, each strip freezing in it a record of the polarity at the time it was forming at the ridge crest.

Convergent Boundaries

Convergent boundaries result in mountain building when two continents collide, or in a subduction zone when one of the converging plates is oceanic. Subduction zones are identified by ocean trenches with neighboring volcanoes that result from the distillation of more volatile materials from the descending oceanic plate. Through this process, we gain the water and salts of our oceans, many of the gases of our atmosphere, and additional continental materials.

The Ocean Floor

The primary features of the ocean floor are the oceanic ridge, trenches, and ocean basins that lie between the ridge and the continental margins. Also conspicuous are the faults and fracture zones crisscrossing the ridge. Ocean floor volcanism includes chains of volcanoes caused by the lithosphere passing over stationary *hot spots* in the mantle below. Coral reefs, atolls, and plateaus are also found in certain oceanic regions.

The Continental Margins

The undersea continental margins include a shallow, flat continental shelf. Seaward from this is a drop-off toward the deep ocean floor, called the *continental slope*. Sediments wash down the continental slope, often through huge submarine canyons, and form wedges of deposits at the base of the slope, called the *continental rise*.

The Past, and Planetary Evolution

The continents are formed through the accumulation of lighter crustal materials, generally formed through the "distillation" of oceanic crust in subduction zones. The oceanic crust has a relatively short lifetime between its formation on the ridge and its destruction in a subduction zone. Therefore, information on the Earth's earlier history must come from either the continents or theoretical models.

Because of the geological activity, continents display a complex, contorted, and often confusing history. But some of that history is beginning to be understood, including both the formation of Pangaea and its breakup.

The Earth's geological activity makes it unique among the terrestrial planets. The primary reason for its exceptional activity is its greater size, which provides more internal heat sources and causes greater difficulty in cooling.

Key Terms

Pangaea
paleomagnetism
atomic magnets
igneous
plutonic
volcanic
lava
potassium-argon dating
polar wandering
magnetic dip
convection
plate boundaries
divergent boundaries
convergent boundaries
transform faults
San Andreas Fault
oceanic ridge
rift valley
basalt
fracture zone
rift mountains
field reversals
subduction zone
ocean trench
ring of fire
backarc basins
Mid-Atlantic Ridge
East Pacific Rise
ocean basin
abyssal hills
seaknolls
seamounts
abyssal plains
hot spots
guyots
tablemounts
volcanic chains
coral reefs
fringing reef
barrier reef
atoll
plateaus
continental shelf
shelf break
continental slope
continental rise
alluvial fans
active margins
passive margins
submarine canyons
turbidity currents
graded bedding

Study Questions

1. What evidence did Alfred Wegener use to support the idea of continental drift?
2. How do rocks become magnetized? Explain the difference between the freezing point and the temperature at which the magnetism becomes permanently frozen into a rock.
3. Discuss how magnetism in rocks can be used to verify the motions of continents. In particular, describe the significance of paths of polar wandering and how the north–south motion of a continent is determined.
4. What causes the motion of the plates of the lithosphere, and what evidence is there to support this explanation?
5. How can we determine the present locations of plate boundaries? What are the three general types of plate boundaries? Can you give an example of each?
6. Describe the oceanic ridge. What features identify it as a spreading center? Explain how the ocean floor changes with increasing distance from the ridge crest, and why.
7. Where are hydrothermal vents found, and what causes them? How does their flow rate and impact on ocean chemistry compare with that of rivers? What is especially interesting about the communities of organisms found around them?
8. What happens when two continental plates collide? What happens when one or both of the colliding plates is oceanic?
9. Discuss what happens in and around a subduction zone. How are these responsible for much

of the water of our oceans, the nitrogen of our atmosphere, and the materials of our continents?

10. Where are most of the world's trenches located? What are typical dimensions (i.e., sizes and depths) of trenches? How does this compare with the dimensions of the oceanic ridge?

11. How deep are ocean basins, and where are they found? How does the ruggedness of their relief change as you move away from the ridge? Why?

12. What are tablemounts, and how are they formed?

13. Describe how chains of volcanic islands are produced. Discuss the evolution of an individual volcano in a chain.

14. What conditions are preferred by reef-building corals? Why is there a lagoon between a mature reef and the neighboring land mass? What regions of the oceans are most likely to have coral reefs, and why?

15. How are atolls formed?

16. Describe two important mechanisms involved in the formation of continental shelves?

17. What are turbidity currents, and how are they related to submarine canyons?

18. What does graded bedding tell us about the manner in which a deposit of sediments was formed?

19. Explain why the continental crust is thicker and less dense than oceanic crust, and why oceanic crust is less dense than the materials of the mantle.

20. How did the early lithosphere differ from today's?

21. Why is the Earth more active geologically than other terrestrial planets?

Critical Thinking Questions

1. How can the ages of rocks be determined? Why do you suppose this wouldn't work for determining the ages of sedimentary rocks?

2. The Earth's magnetic field switched polarity about 700,000 years ago. How do you suppose we know this?

3. Why do you suppose scientists were at first surprised to find that some backarc basins are being stretched? Can you propose an explanation?

4. Why do you think that there is an overall tendency for plateaus and other fragments of continental materials to aggregate into larger continental masses, rather than for existing continental blocks to fragment into smaller pieces?

Suggestions for Further Reading

ACHACHE, J. ET AL. 1988. The Magnetic Anomalies of the Earth's Crust. *Endeavor* 12:4, 154.

BLOXHAM, J. AND D. COUBBINS. 1989. The Evolution of the Earth's Magnetic Field. *Scientific American* 261:6, 68.

BONATTI, E. 1987. The Rifting of Continents. *Scientific American* 256:3, 97.

COURTILLOT, V. AND G. E. VINK. 1983. How Continents Break Up. *Scientific American* 249:1, 43.

ELDERFIELD, HARRY. 1992. Iron Fountains on the Seabed. *New Scientist* 134:1825, 31.

GOLDEN, FREDERICK. 1991. Birth of the Caribbean: Tracing the Islands' Most Basic Roots. *Sea Frontiers* 37:5, 20.

HARLEY, S. L. 1987. Origin and Growth of Continents. *Nature* 329:6135, 108.

KUNZIG, ROBERT. 1992. Time Zero. *Discover* 13:12, 32.

LONSDALE, P. AND C. SMALL. 1991. Ridges and Rises: A Global View. *Oceanus* 34:4, 26.

MACDONALD, K. C. AND P. J. FOX. 1990. The Mid-Ocean Ridge. *Scientific American* 262:6, 72.

MOORES, E., ED. 1990. *Shaping the Earth—Tectonics of Continents and Oceans.* W. H. Freeman, New York.

MORGAN, JASON PHIPPS. 1992. Flattening of the Sea-Floor Depth-Age Curve as a Response to Asthenospheric Flow. *Nature* 359:6395, 524.

Open University. 1989. *The Ocean Basins: Their Structure and Evolution.* Pergamon Press.

NANCE, R. D. ET AL. 1988. Planetary Systems: Body Building in the Solar Nebula. *Nature* 334:6182, 474.

VAN DOVER, CINDY LEE. 1993. Depths of Ignorance. *Discover* 14:9, 37.

WATTERS, THOMAS R. 1992. Systems of Tectonic Features Common to Earth, Mars, and Venus. *Geology* 20:7, 609.

WEIJERMARS, R. 1989. Global Tectonics Since the Breakup of Pangaea 180 Million Years Ago. *Earth-Science Review* 26:2, 113.

Most of the sediments beneath you right now will someday be in an ocean, washing off the continent and collecting mostly on the continental margin.

The Grand Canyon of the Colorado River displays sediments that were deposited beneath an ocean long ago.

Five

SEDIMENTS

In the preceding chapter, we learned how the process of differentiation brings lighter materials to the Earth's surface. These include the materials of the crust, as well as its oceans and atmosphere. Most crustal materials begin as magmas that cool and solidify to form igneous rock. But rocks on the very surface of the crust are exposed to water, weather, and living organisms, which break them up and create a blanket of sediments covering the Earth.

Much of our information on deep sediment layers is gained from drilling and **seismic echo profiling**. This later technique is similar to the echo sounding of the ocean's depths, in that it involves producing a loud sound and analyzing the echoes from subsurface sediment layers (Figure 5.1).

Using these techniques, scientists estimate the average thickness of the sediments covering the Earth to be slightly over 1 kilometer, but with great variations. As seen in Table 5.1, the sediments tend to be thinnest on the deep ocean bottom, primarily due to the youth of the underlying crust, and thickest on the continental slopes and rises, where sediments washed off the continents tend to collect. There is little or no sediment cover in parts of some young rugged mountain ranges. At the other extreme, the relative stability and gradual subsidence of passive continental margins have enabled very thick layers of sediments to collect in these regions, sometimes up to 15 kilometers thick.

Even though the continents are old and the continental rocks are broken up by exposure to wind and weather, the majority of the sediments do not collect on the continents. They wash off, first filling the lowlands and valleys, and then being carried to the oceans by rivers and streams. Of course, the process takes millions of years, and many interesting things can happen along the way. Nonetheless, most of the sediments beneath you right now will someday be in an ocean, washing off the continent and collecting mostly on the continental margin.

Nonetheless, most of the sediments beneath you right now will someday be in an ocean, washing off the continent and collecting mostly on the continental margin.

Together with bits of organic matter and nutriments, sediments form the soils that nourish our land plants and agriculture. They also hold many mineral resources, as well as bits of organic matter that may some day turn into deposits of petroleum, coal, and natural gas. Look around yourself right now—at the room, walls, windows, furniture, your clothing, and this book. Everything you see was somehow produced or grown from sediments.

In this chapter, we will study the composition of the various kinds of sediments and the range of properties they display. We will learn of the processes through which the different types of sediments are created and distributed, and we will see how oceanographers retrieve samples from the sea floor for study.

5.1 SEDIMENT MATERIALS

SOURCES

Sediments derived from rock are called **lithogenous**, or sometimes **terrigenous**.[1] Another important source of sediments is the ocean's biological community. Sediments with this origin are called **biogenous**. The skeletal remains of myriads of microscopic organisms dominate large areas of the ocean floor, and in some regions, organic matter in various stages of decomposition is also important. The skeletal remains of larger organisms are important in a few areas, such as the coralline debris that forms the beach sands in some balmy areas. The third primary source of ocean sediments is the materials in solution, which sometimes encounter conditions that force them back out of the solution—a process called **precipitation**. Sediments derived from the materials in solution are called **hydrogenous**, or **authigenic**. Examples include salt and limestone deposits.

A fourth and extremely minor component of the sediments comes from the meteoritic debris that continually bombards the Earth. These sediments are called **cosmogenous**. Some meteors are sufficiently large (i.e., the size of a grain of sand, or larger) to heat up the air and leave visible streaks as they collide with our atmosphere. But the great majority are tiny **micrometeors**, which stop like tiny feathers immediately as they strike the outer atmosphere, and they gently float down to the Earth's surface as dust. The Earth gains several thousand tons of this meteoritic material each day, and most of this lands in the oceans. But several thousand tons spread over the

[1]The prefix "litho" comes from Greek for "stone," "terra" is latin for "land," and "genous" refers to origin.

Figure 5.1

Seismic echo profile of sea floor sediments in the Gulf of Mexico. (Horizontal distance is about 50 km.)

entire surface of the Earth is a minuscule contribution to the total sediment cover, amounting to the equivalent of one additional small grain of sand per square centimeter every few years.

TEXTURES

The various sediments display a variety of textures and compositions. As seen in Table 5.2, grain sizes vary greatly; from finest to coarsest, the categories are *clays*, *silts*, *sands*, *granules*, *pebbles*, *cobbles*, and *boulders*. (Colloids are so fine that they remain in suspension in the water and are therefore of only minor importance in the sediments.) Colors vary from white, lime, or yellow sands to dark red or brown muds. If you take the time to look carefully at a handful of sand or dirt under a magnifying glass, you will be amazed at the variety of sizes, shapes, and colors of the various tiny grains (Figure 5.2).

In many regions, natural processes sort sediments according to grain size and composition, so that vast areas of the ocean floor are covered by similar sediments. Sorting also occurs on beaches, where constant agitation by the waves continually winnows the beach materials. We tend to find rocky beaches where wave activity is heavy, such as on headlands

Table 5.1 Distribution and Thickness of the World's Sediments

Region	% of World Area	% of Total Sediments	Average Sediment Thickness (km)
Continents	29	8	0.3
Shelves	6	14	2.5
Slopes	4	38	9
Rises	4	28	8
Deep sea floor	56	12	0.2
(Average)			(1.0)

Source: After James Kennett, 1982, *Marine Geology*, Prentice-Hall, Englewood Cliffs, NJ.

Table 5.2 Classification of Beach Material According to Diameter

Classification		Sediment Diameter (mm)
Boulder		$256 = 2^8$
		$128 = 2^7$
Cobble		$64 = 2^6$
		$32 = 2^5$
Pebble		$16 = 2^4$
		$8 = 2^3$
		$4 = 2^2$
Granule		$2 = 2^1$
Sand	Very Coarse	$1 = 2^0$
	Coarse	$\frac{1}{2} = 2^{-1}$
	Medium	$\frac{1}{4} = 2^{-2}$
	Fine	$\frac{1}{8} = 2^{-3}$
	Very Fine	$\frac{1}{16} = 2^{-4}$
Silt	Coarse	$\frac{1}{16} = 2^{-4}$
	Medium	$\frac{1}{64} = 2^{-6}$
	Fine	$\frac{1}{128} = 2^{-7}$
	Very Fine	$\frac{1}{256} = 2^{-8}$
Clay	Coarse	$\frac{1}{512} = 2^{-9}$
	Medium	$\frac{1}{1024} = 2^{-10}$
	Fine	$\frac{1}{2048} = 2^{-11}$
	Very Fine	$\frac{1}{4096} = 2^{-12}$
Colloid		

(rocky points sticking out into the ocean) or at high latitudes where beaches are battered by winter storms. By contrast, we find muddy deposits of fine silts or clays in very protected bays and lagoons. Where wave activity is moderate, such as on many beaches in temperate latitudes, we find sands of intermediate grain sizes. Because different beaches have different wave exposures, the grain size may vary considerably from one beach to the next. Fine sediments carried away from a beach exposed to heavy waves may be deposited on a neighboring, more protected beach. If you collect sand from two different beaches and then compare the two samples side by side, you will be amazed at the difference.

In addition to grain size and degree of sorting, sediments may also be characterized by their porosity and permeability. **Porosity** is a measure of the amount of space between grains. For example, a porosity of 20% would mean that 80% of the space is occupied by sediment particles, and 20% is not. Usually, more poorly sorted deposits are also less porous, because smaller sediment grains tend to fill in the space between larger ones.

The **permeability** of a sediment deposit refers to the ability of water to flow through it. Usually, sediments of larger grain size are more permeable, because it is easier for water to flow through the larger gaps between grains. In fact, the tiny grain size of clays makes it almost impossible for water to flow through these deposits at all. Clay pots have been used to carry water for ages.

Figure 5.2
Closeup, sediments display an interesting variety of shapes and colors.

CHEMICAL COMPOSITION

Even among sediments of similar textures, a close-up examination reveals a variety of colors and shapes, reflecting a diversity of materials (Figure 5.3). Each material is distinguished by the kinds of atomic elements from which it is made, and by the way these groups of atoms are arranged. That is, different materials are characterized by their *chemical composition* and **crystal structure**.

Since sediments form only on the Earth's outer surface, their ingredients come from the crust and hydrosphere. The eight most common elements in the Earth's crust are listed in Table 2.2. The hydrosphere's most important contributions to sediments are carbon and water, which enter the hydrosphere via volcanic outgassing from the Earth's interior. Most carbon arrives as carbon dioxide gas and forms carbonates after dissolving in the water. The water molecule often attaches to other molecules to form a *hydrate*. Consequently, sediment materials are formed primarily from the crustal materials of Table 2.2, plus carbonate and water.

In Section 2.2, we studied the more common chemical ingredients found in the Earth's crust and sedimentary blanket. These materials include elements that donate electrons, such as the metals aluminum, iron, calcium, sodium, potassium, or magnesium, in combination with elements or radicals that

Figure 5.3
Photo of four materials having similar textures, but quite different chemical compositions. (top left) A fine grained quartz beach sand (SiO_2). (top right) A beach sand made up of skeletal debris of some marine organisms ($CaCO_3$). (Bottom left) Table salt (NaCl). (bottom right) Cane sugar ($C_{12}H_{22}O_{11}$) is not an ocean sediment, but is included here to illustrate that different materials can have similar textures.

accept electrons, such as chlorine, oxygen, silicate, carbonate, sulfate, or phosphate. For example, calcium carbonate, iron oxide, or potassium-aluminum silicate would be among the many materials that could be found in various types of sediments on Earth. In Table 2.2, you can see that oxygen and silicon are the two most abundant elements in the Earth's crust. Consequently, the silicates are extremely prevalent, and silicon oxide (SiO_2, or *silica*) is also very common.

CRYSTAL STRUCTURE

The characteristics of a material are determined by not only its atomic constituents, but also the way that these constituents are arranged (Figure 5.4). If the atomic constituents are joined together fairly randomly, like bodies in a crowded room, the material is **amorphous**. If there is some recurring order to their arrangement, like bricks in a wall, the material is **crystalline**.

> If the atomic constituents are joined together fairly randomly, like bodies in a crowded room, the material is amorphous. If there is some recurring order to their arrangement, like bricks in a wall, the material is crystalline.

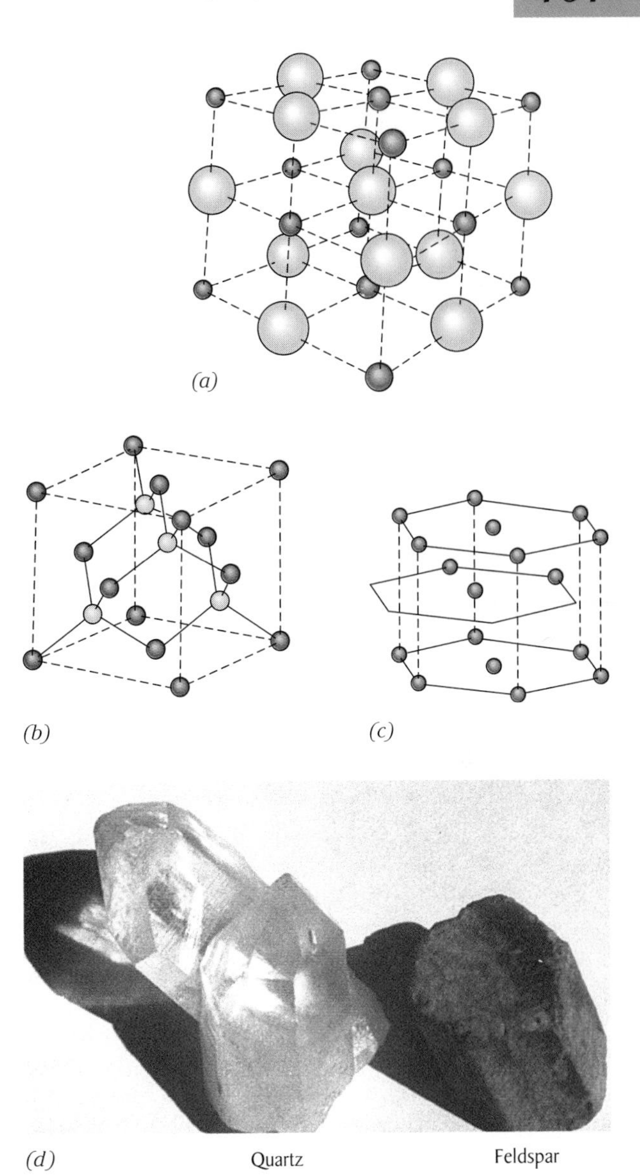

Figure 5.4
(a) The cubic structure of NaCl. The small red spheres represent Na^+ and the large spheres the Cl^-. Notice that each sodium is surrounded by several chlorines, and vice versa. (b) The cubic structure of diamond, or zinc sulfide. (c) One possible hexagonal structure. (d) Large crystals of two common minerals. The crystals in crustal rock are usually tiny or microscopic (see Figure 3.10).

The crystalline components of rocks are called **minerals**. Most large rocks contain many different minerals and are named according to their mineral content. Although the individual crystals may be as large as tennis balls, that is rare. More often, they are so small that you need a magnifying glass to see them, and they give the rock a speckled appearance and gritty texture.

A material of certain chemical composition may be able to have several different crystalline forms. Which of these forms is preferred may depend on the material's environment. For example, under heavy pressure, a denser arrangement may be preferred. When tectonic processes or erosion move a mineral from one environment to another, its crystalline structure may slowly transform to one that is more stable in the new environment. Some materials are amorphous under certain conditions and crystalline under others. Glass is a common material that can be amorphous (as in window glass) or crystalline (as in expensive crystal glassware).

Rocks that have formed deep within the Earth, or beneath kilometers of overlying sediments, are sometimes brought to the Earth's surface by tectonic activity or the erosion of overlying layers. Although stable in their birth environment, many of the materials in these rocks may be unstable in the exposed surface environment, with the result that portions of these rocks may crumble, dissolve, or erode (see chapter opening photo).

5.2 LITHOGENOUS SEDIMENTS

OVERVIEW

The majority of all sediments come from rocks of the continental surface. These rocks are exposed to large temperature changes, dissolution of materials by fresh water flowing over and through them, the freezing and thawing of water in their cracks and pores, desiccation from solar heat, abrasion by wind-carried dust, moving organisms, other rocks rubbing against them, and the probing organs and chemicals of plants and animals in search of sustenance. All these processes that cause rock to crack and crumble are referred to as **weathering**.

By contrast, the ocean floor is an extremely stable environment. Temperature and humidity are constant, there is very little abrasion from the slow ocean currents, and the crustal rock is protected by the layers of sediment covering it. Consequently, with the exception of sediments derived from occasional submarine volcanic and hydrothermal activity, the ocean floor makes very little contribution to its own sediment cover.

On the continents, sediments are usually removed from their place of formation by wind or water. Most sediments are then carried to the sea by rivers and streams, where they eventually deposit on the continental margins (Figure 5.5). The removal of sediments from the original place of formation is referred to as **erosion**, and their subsequent movement over a larger distance in streams, rivers, or ocean currents is called **transportation**.

The highest rates of erosion and transportation occur in regions with heavy rainfall. Heavier erosion leads to heavier weathering, because without the protection of overlying layers of sediments, underlying rocks experience greater exposure to weathering processes. Because of its heavy rainfall, southeast Asia is the world's largest sediment producer, followed by the southeast United States.

THE SKIN OF THE CONTINENTS

In Chapter 4, we learned that crustal materials have an igneous birth. Those of the continental crust are formed in volcanism and plutonism primarily associated with subduction zones. The very outer surface of the crust is blanketed by sediments, due to the harsh environmental conditions and consequent weathering that the outer surface has had to endure. The weathering of exposed continental rocks is constantly producing new sediments to replace those that are gradually working their way toward the oceans.

The long and active geological history of the Earth has brought most present continental regions below sea level at some time in their history, many regions more than once.[2] When submerged, sediment grains may slowly cement themselves together, forming sedimentary rock. This slow transformation generally occurs deep beneath overlying sediments. The weight of the overlying layers tends to squeeze out much of the water. The remaining water is permanently trapped between the sediment grains and slowly dissolves some of the minerals. This dissolution continues until the water becomes **saturated**, meaning that it is holding as much of these minerals as it can. When this point is reached, continued dissolution of minerals from some places is balanced by the precipitation of these minerals in others (Figure 5.6a). This precipitation of minerals tends to cement together neighboring grains and so the sedimentary deposit slowly becomes rigid (Figure 5.6b).

WEATHERING

If this sedimentary rock should ever become exposed at the Earth's surface, it would encounter an environ-

[2]The submergence of large regions of the continents is caused by changes in sea level relative to land. This can be caused by processes in and beneath the oceans as well as processes in and beneath the continents.

Figure 5.5

Aerial photo of the mouth of the Fraser River in British Columbia, showing the load of lithogenous sediments being fed into the Pacific Ocean.

Figure 5.6

(a) Magnified cross-sectional view of sediment grains in a deep sediment layer. The dark spots indicate the points of contact between neighboring sediment grains. Here, the stress due to the weight of the overlying sediment layers is greatest. Consequently, the sediment grains are weakened here. The entrapped water dissolves them at these pressure points and deposits the dissolved materials in the voids, cementing together the grains. (b) The sediments derived from the weathering of this ancient sedimentary rock formation find their way into rivers and streams that carry them to the ocean, where they will join in the formation of new sedimentary deposits.

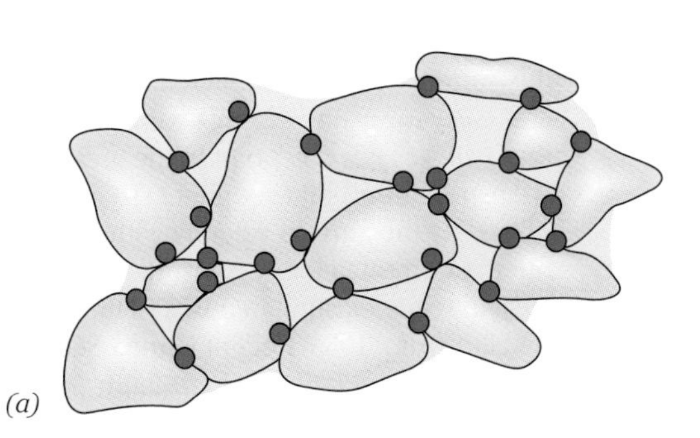

(b)

ment quite different from that of its birth. Relieved of the weight of the overlying layers, it would tend to expand and soak up water, rather than squeezing the water out. Furthermore, the water it absorbs would be fresh, rather than saturated in the various minerals. Consequently, this fresh water would slowly dissolve the minerals that cement together adjacent grains. The rock would start falling apart and the individual sediment grains would regain their freedom. In addition to the fresh water, other agents that would help break up these rocks include temperature changes, freezing and thawing, abrasion, solar heating, and the probing appendages and chemicals of animals and plants.

Although igneous rocks have quite different origins from sedimentary rocks, they weather when exposed at the surface for basically the same reason. Although stable in their protected birth environment, they are not stable under the very different conditions they experience at the surface. In their new location, they begin to break down, some mineral more quickly than others. Some of the tiny mineral grains dissolve completely, others break up into fine clays, and still others do not weather much at all.

As a rock falls apart, the more resilient materials remain pretty much unchanged, and the sizes of the tiny grains in the original rock determine the grain size of the sediments they form. The grains of common yellow sand are mostly tiny crystals of quartz, made of silica (SiO_2), which were originally formed as tiny flecks in igneous rocks. They are exceptionally resilient against weathering and can survive many times through the sedimentary cycle described next.

Although the continental crust is igneous in origin, the very surface is dominated by sediments. Nearly 100% of the continental surface is covered by sediments of some sort, and beneath more than 70% of these deposits lies sedimentary rock. If you look at the rocks exposed in the mountains, on river banks, roadway cuts, or building excavations, you will find the rock in most cases to be sedimentary. Since sedimentary rocks are so common on the continental surfaces, most new sediments are forming from the weathering of these previous deposits. Because today's sediments tend to form from the weathering of earlier ones, and future sedimentary deposits will result from the weathering of today's, many sediments will continue several times through this **sedimentary cycle** during their lifetimes (Figure 5.6b).

EROSION, TRANSPORTATION, AND DEPOSITION

After exposed rock has weathered, it still takes a long time for the sediments to reach the continental margins, and some never make it. Erosion is usually accomplished by wind or water, but sometimes it receives help from other agents, such as animals, gravity, or landslides.

Once the sediment grains reach a river or stream,

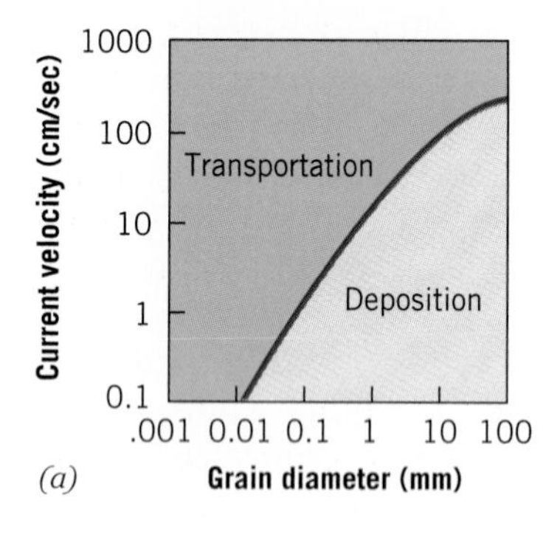

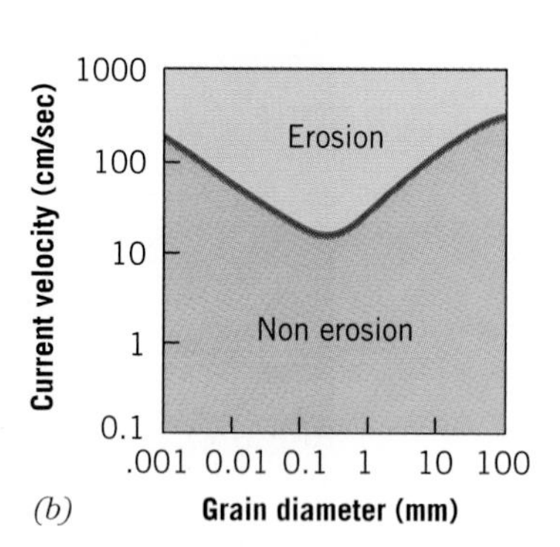

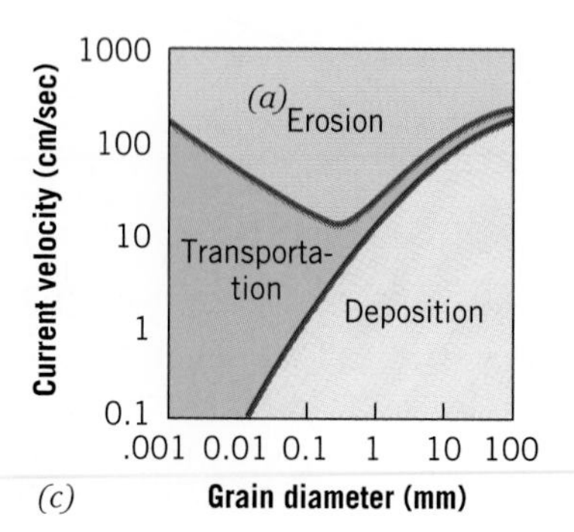

Figure 5.7

Plot of current velocity vs. sediment size for deposition, transportation, and erosion of sediments. (a) For any point in the tan region, the current velocity is insufficient to keep that sediment type in suspension. (b) For any point in the violet region, the current velocity is sufficiently swift to erode and carry away an existing sediment deposit of that size. (c) For any point in the area marked "transportation," that sediment size will be transported by the current velocity without settling out, but the current velocity is insufficient to erode away already existing deposits of that grain size. (F. Hjulstrom, 1939, Recent Marine Sediments, American Association of Petroleum Geologists.)

the water carries them to their final resting place. Actually, they go through the erosion–transportation–deposition cycle over and over again as they make their way along the various stream channels. Depending on current speed, turbulence, and sediment size, the sediments may roll or bump along the bottom, or may be carried long distances in suspension. As you might guess, finer grains are transported more easily than coarser ones. This is illustrated in Figure 5.7a, which shows the minimum speeds of water currents necessary to carry sediments of various grain sizes.

Once a sediment is deposited in a stream bed, a slightly swifter current is needed to dislodge it than is needed to keep it in suspension and carry it on its way. This is especially true for clays. These tiny flakes have a large surface area in comparison to their volume. They tend to stick together, making it difficult for water to either flow through these deposits or dislodge individual grains. Consequently, clay deposits require relatively large current speeds to dislodge individual grains and erode the deposit. In fact, the erosion of clay deposits requires much greater current speeds than does that of coarser sands, as is illustrated in Figure 5.7b.

(a)

(b)

Figure 5.8

Sediments of high and low energy environments. (a) Coarse sediments are found on beaches exposed to heavy wave activity. (b) In more protected waters, we find deposits of fine muds.

SORTING

As the sediments are carried toward the continental margins, a great deal of sorting occurs en route. Boulders may be left at the base of mountain streams, and gravels dropped off in slowly flowing rivers. As the water motion decreases, finer sediments fall out of suspension. When the streams finally reach the oceans, perhaps only sands and finer sediments remain. Some regions of the seashore are exposed to heavy wave activity, so we might find only coarse sands and gravel, because finer sediments stay in suspension in these high-energy environments. On more protected beaches, fine sands might settle out, and in very calm protected embayments, we might find primarily clays and silts (Figure 5.8).

There are some exceptions to this pattern, of course. In some coastal regions, sea cliffs crumble directly onto the beach, so there is no presorting of sediments by streams and rivers. In higher latitudes, glaciers indiscriminately carry sediments of all sizes and deposit all of them on the continental margins. In these cases, the winnowing and sorting are done completely by the ocean waves. Nevertheless, the end result is still the same: Coarser sediments end up in higher-energy environments, and finer sediments are held in suspension until they encounter lower-energy environments.

As sediments are moved about and sorted, they bump and grind against each other, knocking off sharp corners and edges. Consequently, sediments that have undergone more sorting also tend to have more rounded grains. We sometimes refer to the **maturity** of a sediment deposit, which is based on the degree of sorting and roundness of the grains (Figure 5.9).

DEPOSITION PATTERNS

Sediments accumulate roughly 100 times faster on the continental margins than the deep sea floor. This simply reflects the rapid rates of weathering and erosion of exposed land masses (Figure 5.5). Typical rates of sedimentation are 20 centimeters per 1000

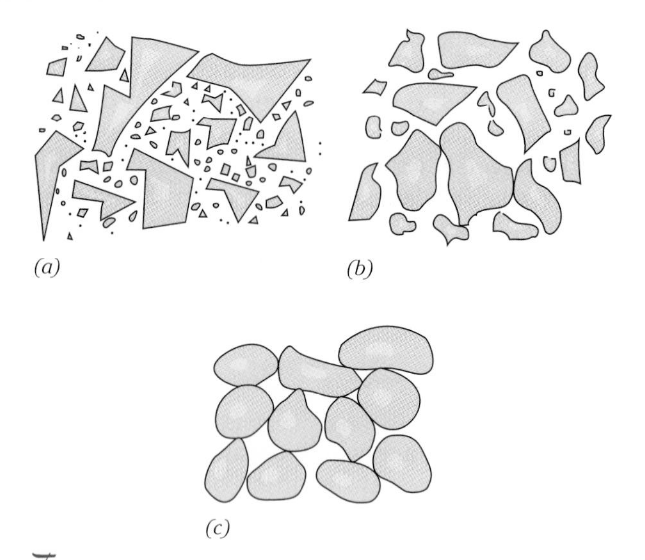

Figure 5.9

Diagram illustrating sediment deposits of increasing maturities. (a) Very immature: unsorted and very angular. (b) More mature: better sorting and less angular. (c) Very mature: the grains are all about the same size and quite rounded.

Table 5.3 Times Required for Fine Sediments to Settle Through Water Columns of Depths Characteristic of the Continental Shelf (100 m) and of the Deep Ocean Basins (5 km). Diameters Expressed in Micrometers.

Sediment diameter (μm)	Settling time through 100 m	Settling time through 5 km
10	4.6 days	230. days
1	1.3 years	63.0 years

Source: After M. Grant Gross, *Oceanography, A View of the Earth*, 4th ed., Prentice-Hall, Englewood Cliffs, NJ, p. 86.

years on the continental margins, and only 0.2 centimeters per 1000 years on the deep ocean floor, with large variations in both these numbers. Although lithogenous sediments comprise about 75% of all sediments by volume, they dominate only about 20% of the ocean area, being deposited primarily on the continental margins. Most lithogenous sediments that do reach the deep ocean floor are found in the deposits resulting from turbidity currents (i.e., *turbidite deposits*) of abyssal plains.

At this particular period in history, most lithogenous sediments are being deposited in protected coastal embayments and seas (Chapter 10), because most of the major rivers in the world discharge into these areas. Important exceptions are the Amazon and Congo Rivers, which empty directly into the Atlantic Ocean, and the Ganges and Brahmaputra Rivers, which empty directly into the Indian Ocean.

Of the lithogenous sediments that have been deposited on the continental margins, you might expect to find finer sediments as you go away from the shore, because of the longer suspension times required to reach these greater distances. Finer sediments take longer to settle out (Table 5.3). The pattern is not this simple, however, due to the many processes that have redistributed shelf sediments. Wave motion reaches down to the shelf, and so do wind-driven surface currents. Large areas of the shelves have been exposed during ice ages and subjected to erosion by wind, rain, streams, and moving ice sheets.

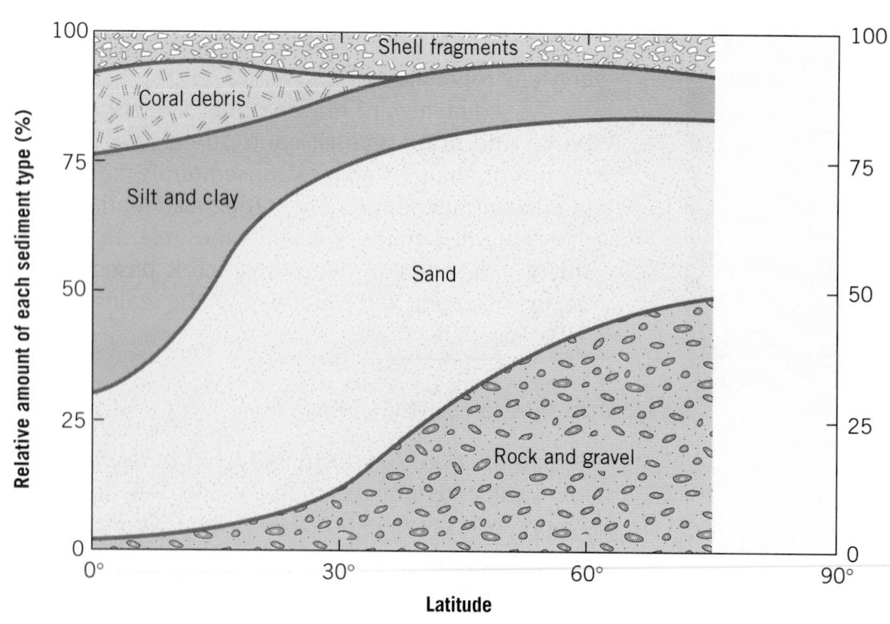

Figure 5.10

Relative abundances of various sediment types on the shelf, as a function of the latitude. (M. O. Hayes, 1967, *Marine Geology* **5**:2.)

There is a tendency for shelf sediments to be coarser at higher latitudes (Figure 5.10), which is due to both glacial activity and severe winter storms. Glaciers tend to carry large and small sediments alike, and winter storm waves tend to attack the shoreline directly. Both these processes create an abundance of rocks and coarse sediments on the continental margins. These have not been sorted by rivers and streams, as has been the case for most shelf sediments found at lower latitudes.

5.3 BIOGENOUS SEDIMENTS

THE CONTINENTAL MARGINS

Although the bulk of the sediments on the continental margins are lithogenous, a small amount are provided by organisms. Actually, biological activity is quite vigorous over most continental shelves—much more than over deep ocean basins (Chapter 11). However, the very heavy dilution of organic debris by lithogenous sediments washing off the adjacent land mass means that the biogenous component is a relatively small fraction of the total in most areas. If you closely examine a handful of sand the next time you are at the beach, you will find some tiny fragments of shells, seaweeds, and perhaps other scraps of organic debris there. These items will probably make up only a small fraction of the total, however.

There are a few shallow water coastal environments where the biogenous component dominates. These are normally associated with reefs of some sort. Plants that build reefs include mangrove trees (Figure 5.11) and some kinds of mat-forming algae. Animals that build reefs tend to use the cemented skeletons of previous generations as the platform on which new generations live. Examples include oysters, and some kinds of tube worms and corals. Like the reefs formed by algae, those formed by oysters and tube worms tend to be small. Dimensions of a few meters are typical. Coral reefs can be extremely large, however (Figure 4.22). Where these structures are found, coralline debris tends to dominate the surrounding shelf sediments and often forms a large component of the beach sand on the nearby land mass.

THE DEEP OCEAN FLOOR

Although the continental margins are dominated by the lithogenous sediments that wash off the neighboring land mass, most deep ocean areas are far removed from this prolific source. Consequently, lithogenous sediments are often scarce, and other types of sediments may dominate. The two main sources of deep ocean sediments are the tiny external skeletons of microorganisms, called **exoskeletons** or **tests**, and the very fine clays of mixed origins, called **abyssal clays**. We examine biogenous sediments in this section and abyssal clays in the next.

The accumulation of biogenous sediments on the sea floor depends on three factors: (1) production, (2) dilution, and (3) dissolution. That is, it depends on the rate of production of tests by the microorganisms (mostly in the surface waters), the

Figure 5.11

The unique roots of the mangroves allow them to grow in shallow waters, trapping sediments and forming mangrove reefs.

rate of dilution of these tests by sediments from other sources, and the rate of dissolution of these tests into the sea water. We have seen that biogenous sediments seldom dominate the continental margins because of heavy dilution by lithogenous sediments. This is also often true on the deep ocean floor bordering the continents. In addition, there are large regions of the deep ocean floor where the mineral exoskeletons of microscopic organisms dissolve as fast as they accumulate. Due to the lack of biogenous sediments in these regions, abyssal clays may dominate, in spite of their extremely slow rate of accumulation.

When the biogenous component makes up over 30% of the sediment mixture, the deposit is called an **ooze**. With the exception of regions where deep nutrient-rich waters come to the surface, there is relatively little biological activity in the surface waters of most deep ocean regions (Chapter 11). Nevertheless, sedimentation from other sources is so small that biogenous oozes still dominate altogether about 62% of the deep ocean areas.

The tiny skeletons of most marine microorganisms (Figure 5.12) are made up of either silica (SiO_2) or calcium carbonate ($CaCO_3$). Common silica skeletons include those of diatoms (plants) and radiolarians (animals). Common calcium carbonate skeletons include those of coccolithophores (plants) and foraminifera (animals). Oozes derived from microscopic skeletal debris are commonly referred to as **siliceous** or **calcareous** oozes, depending on the predominant skeletal material. Further specification can be given to identify the particular type of organism whose tests dominate the sediment, such as a *diatomaceous* or *foraminiferous* ooze.

Figure 5.12
Color enhanced scanning electron micrograph of the tiny external siliceous tests of various types of radiolaria. (Magnified about 100 times.)

DISTRIBUTION OF SILICEOUS AND CALCAREOUS OOZES

Silica slowly dissolves in sea water. Therefore, only in regions of relatively high productivity, where siliceous tests accumulate faster than they dissolve, would we find siliceous oozes among the bottom sediments. Such regions include areas near the equator, where deep nutrient-rich waters come to the surface, and regions between 50° and 65° south latitude, where surface waters are rich in nutrients and support vigorous biological activity. Siliceous oozes are not found so much in corresponding northern waters, because of heavy dilution by terrigenous sediments and because the ice cover inhibits biological productivity. Altogether, siliceous oozes dominate about 14% of all deep ocean areas.

Calcium carbonate dissolves in acids. Cooler waters tend to contain slightly more dissolved carbon dioxide, which makes them more acid.[3] Cold water masses form at high latitudes, sink to the bottom of the ocean, and flow along the bottom even passing through the temperate and tropical latitudes. Therefore, throughout most of the oceans, the deep waters are cold and slightly more acid than the waters at the surface. As a result, calcareous tests tend to dissolve in the cold deep waters of temperate and tropical latitudes, but not in the warmer waters near the surface. The polar waters are cold at all depths, so calcium carbonate dissolves at all depths in these regions.

The depth below which calcareous skeletons dissolve as fast as they accumulate is called the **calcium carbonate compensation depth**, or sometimes the *calcite compensation depth*, or *CCD* for short. It is also sometimes referred to as the **snow line**, because the off-white calcium carbonate deposits form above this line, and not below it, much like the "snow line" on continental mountains (Figure 5.13).

The snow line is usually about 4 kilometers deep in temperate and tropical latitudes, depending on the exact nature of the cold deep water masses and productivity of the surface waters. Consequently, calcareous oozes will be found only at depths shallower than about 4 kilometers, and only in temperate or tropical latitudes. Such regions include most of the oceanic ridge and plateaus. Altogether, calcareous oozes dominate about 48% of all deep ocean sediments.

[3]The "fizz" in soft drinks is due to the dissolved carbon dioxide. You may be aware that colder drinks hold this fizz better, because the carbon dioxide dissolves better in colder water. The dissolved carbon dioxide also makes soft drinks slightly acid, and they slowly dissolve the enamel of your teeth.

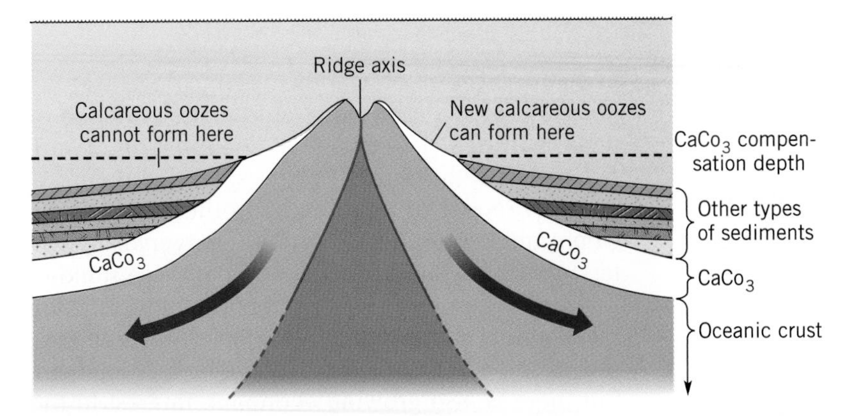

Figure 5.13

Calcareous oozes can form above the calcium carbonate compensation depth, such as on the oceanic ridge. Then it is carried away from the ridge by the motion of the underlying plate. As other sediments collect on top of this layer, they protect it from dissolution, even after it has been carried below the calcium carbonate compensation depth. The sediment cover thickens going away from the ridge. The most recent sediments are on top, and the oldest on the bottom. Near the ridge, only the most recent sediments are present. Going away from the ridge, the older underlying plate supports increasingly older layers of sediment.

5.4 ABYSSAL CLAYS

Just as fine dust in the air settles much more slowly than large stones, extremely fine clay-sized particles take much longer to settle out of the ocean water than coarser sediments (Table 5.3). Both slow rates of settling and water motions that tend to keep them in suspension mean that fine clay particles may take years, decades, or centuries to settle out of the water. During this time, ocean currents may carry them far away from the continents and into the deep ocean areas.

The combination of tiny size and long suspension time causes tiny clay particles to undergo considerable alterations during their travels. Some portions may dissolve, and other portions may be chemically altered or enlarged through interactions with the sea water solution. Consequently, the chemical composition of these deposits is often highly altered from that of the original materials. In fact, some clays have no terrigenous origins at all, forming directly from precipitation from minerals in the sea water. Because of these complex origins, abyssal clays are therefore considered neither lithogenous nor hydrogenous, but rather are usually thought of as being in a category of their own.

Not all fine clays are carried out to sea by water; some are carried by the wind. Fine airborne sediments blown from the Sahara Desert are a sizable component of the deep ocean sediments in the eastern Atlantic. In some layers, we find traces of fine dust from past volcanic eruptions. It is estimated that about 1 cubic kilometer of dust from Mt. St. Helen's (1980) was blown into the upper atmosphere, 16 cubic kilometers from Krakatoa (1883), and 80 cubic kilometers from Tambora (1815). Some of this dust was so fine that it remained in the atmosphere for more than a year and caused noticeable changes in the world climate. Nonetheless, much of these airborne clay particles eventually end up on the ocean floor, chemically altered due to their interaction with the water.

5.5 HYDROGENOUS SEDIMENTS

Compared to the lithogenous sediments that dominate the continental margins and the abyssal clays and biogenous oozes that dominate many deep ocean areas, hydrogenous sediments are much less abundant, but nonetheless important and interesting. They form in many different ways, not all of which are yet well understood. Several have commercial value. Common table salt has been an economically important hydrogenous sediment throughout recorded history and so has **limestone**, which is made of calcium carbonate and may have either hydrogenous or biogenous origins.

MANGANESE NODULES

A relatively recent discovery has been the existence of solid deposits enriched in manganese, iron, copper, nickel, and cobalt. These often form as nodules that are typically the size of golf or tennis balls

(a)

(b)

Figure 5.14

(a) Manganese nodules on the floor of the northeast Atlantic Ocean. (b) Manganese nodules that were brought up from the ocean floor with the white dredge that is partially visible behind the pile of nodules.

(Figure 5.14), but sometimes form as coatings on rocks. The appearance of these **manganese nodules** is particularly puzzling, because their rate of growth is usually very slow compared to the rate of accumulation of the granular sediments around them. They should be buried by these other sediments long before attaining any significant size. One possible explanation is that small bottom-dwelling organisms may use these nodules for protection, living under them and keeping the nodules on the surface.

Although we don't yet clearly understand how these nodules form or what prevents their burial, we do know that they are attracting commercial interest for their heavy metal content. As we deplete our easily accessible continental sources, it becomes increasingly attractive to mine these nodules.

METAL SULFIDES

In many places on or near the oceanic ridge axis, deposits are being found that are greatly enriched in metal **sulfides** (i.e., chemical combination of various metals with sulfur). This is especially true in regions of recent or ongoing hydrothermal activity, where the extruded hot **brines** (i.e., very salty waters) contain a large concentration of these sulfides. Right now there is a lot of scientific interest in the mechanisms for the formation of these brines and the precipitation of the metal sulfide deposits from them. But there is also growing economic interest in the possible commercial exploitation of these deposits for their metal content.

EVAPORITES

Some hydrogenous deposits form when evaporation removes an appreciable amount of the fresh water from the solution. As fresh water evaporates, the concentrations of the remaining salts increase, and some begin to precipitate out (Figure 5.15a). These deposits are called **evaporites**, and they are especially likely to form in shallow seas cut off from circulation with the ocean, such as the Dead Sea or Great Salt Lake. Such shallow enclosed seas and semienclosed seas were much more common in ancient times than they are today. But many of these ancient evaporite deposits are quite familiar to us.

As the evaporation continues and the salinity of the water increases, the first salts to precipitate out are the carbonates (Figure 5.15b)—first calcium carbonate (calcite) and then a mixture of calcium and magnesium carbonates (**dolomite**). The carbonates are a minor portion of the total salt content, but water does not hold them well, and that is why they precipitate first.

If evaporation continues and the salinity increases still further, sulfates begin precipitating, especially calcium sulfate (**gypsum**). After 90% of the water has been evaporated, common rock salt (NaCl) will start precipitating. This is by far the most abundant of the dissolved salts, but it is also extremely soluble. So in spite of its abundance, it doesn't start to precipitate until a very large amount of evaporation has occurred.

One particularly interesting evaporite deposit is salt beds beneath the Mediterranean Sea, which are 300 to 500 meters thick and about 5 to 6 million years old. This deposit indicates that the Mediterranean Sea has been cut off from circulation with the Atlantic Ocean, and heavy evaporation has occurred. To give some feeling for the amount of evaporation needed, if the Mediterranean were sealed off today, it

(a)

Figure 5.15

(a) Evaporite deposits, such as these at Mono Lake, California, form when water evaporates in an enclosed, or semi enclosed basin. (b) A redwall limestone deposit that has been raised above sea level, and then exposed by erosion. Limestone is made of calcium carbonate and may form either as an evaporite or from the calcareous skeletons of microorganisms.

(b)

would take about 1000 years for all the water to evaporate, and the resulting salt layer would be about 70 meters thick. Therefore, for the observed beds to have formed, the Mediterranean must have been sealed off, at least partially evaporated, and reopened many times.

CARBONATE CHEMISTRY

Because the ocean is everywhere nearly saturated with calcium carbonate, some rather minor disturbances can cause it to precipitate. Examples include the heating of surface water exposed to sunshine, or the warming of deep water as it passes over some shallow strait or bank. The chemistry of carbonate precipitation is rather important. It is the eventual resting place of most of the carbon dioxide that would otherwise be in our atmosphere. The removal of this carbon dioxide is what makes our atmosphere so different from that of Venus or Mars (Chapter 7).

As we have learned, as carbon dioxide dissolves in sea water, it makes the sea water more acid, which enables it to dissolve calcium carbonate. The reverse of this process is that as carbon dioxide is removed from the water, it becomes less acid, so excess calcium carbonate precipitates out of the solution. Carbon dioxide is removed from the solution by minor agitation or heating (as it is with a soft drink). So minor agitation or heating of sea water may cause calcium carbonate to precipitate.

The chemistry of the above paragraph goes as follows: Warmer water holds less carbon dioxide in solution. Dissolved carbon dioxide tends to release free H^+ ions in solution through the process

$$H_2O + CO_2 \longrightarrow H^+ + HCO_3^-$$

More H^+ ions means the solution is more acid. If we reverse this process, the removal of dissolved carbon dioxide reduces the number of free H^+ ions. As these H^+ ions disappear, others come off the bicarbonate (HCO_3^-) ions to replace them:

$$HCO_3^- \longrightarrow H^+ + CO_3^{2-}$$

The free carbonate ions (CO_3^{2-}) then readily combine with the metal ions such as calcium (Ca^{2+}) that are near saturation and precipitate out of the solution.

5.6 GENERAL PATTERNS IN SEDIMENTATION

We now combine our knowledge of tectonics and sedimentation processes, to summarize and review the overall distribution of sediments in the oceans. Because the oceanic crust is born at the ridge and spreads outward from there, the sea floor near the ridge axis is "brand new" and hasn't yet had sufficient time to accumulate much sediment cover. Going away from the ridge axis, however, the age of the sea floor increases. Because older crust has had more time to accumulate sediments, the thickness of the sea floor sediment cover increases by 100 to 200 meters of sediments for every thousand kilometers from the ridge axis (Figure 5.13). Throughout most of the deep ocean, sediments accumulate only very slowly, so that very thin layers correspond to thousands of years of accumulation.

Near the continental margins, the sediments can be extremely thick, not only because these regions are far from the ridge and therefore older, but also because they receive heavy sedimentation from continental sources. Abyssal plains and other extensions of the continental rise are examples of thick sediments from the continents that have reached the deep ocean floor.

Table 5.4 summarizes where the various kinds of sediments come from, and where in the ocean they are found. The distribution of lithogenous sediments, abyssal clays, calcareous oozes, and siliceous oozes is further illustrated in Figure 5.16. Lithogenous sediments dominate the continental margins. They also dominate near Antarctica and above the Arctic circle, where ice cover restricts biological productivity and the ice carries terrigenous sediments out to sea. Although clays contribute to the deep sea sediments everywhere, they accumulate only extremely slowly and so are heavily diluted by biogenous sediments in many areas. Siliceous oozes dominate in highly productive waters, where siliceous tests are produced faster than they dissolve. Calcareous oozes are found in temperate and tropical latitudes and only at depths shallower than 4 or 5 kilometers, because they dissolve in the colder deeper waters. Abyssal clays dominate these deeper waters, due to the dissolution of biogenous sediments there.

Table 5.4 Source and Distribution of Sediments

Type	Source	Where Dominant	Approximate Percent of Bottom Area Covered
Lithogenous	Continental erosion	Continental margins, abyssal plains, and very high latitudes	20
Biogenous	Skeletons of organisms	Much of deep ocean bottom, especially at high temperate latitudes, and at depths less than 5 km in temperate and tropical latitudes	50
Clays	Miscellaneous. Fine inorganic particles carried by winds or currents, underwater volcanic eruptions, etc.	Deep ocean bottom, deeper than 5 km in temperate and tropical latitudes	30
Hydrogenous	Dissolved minerals	Virtually nowhere	<1
Cosmogenous	Outer space	Nowhere	0

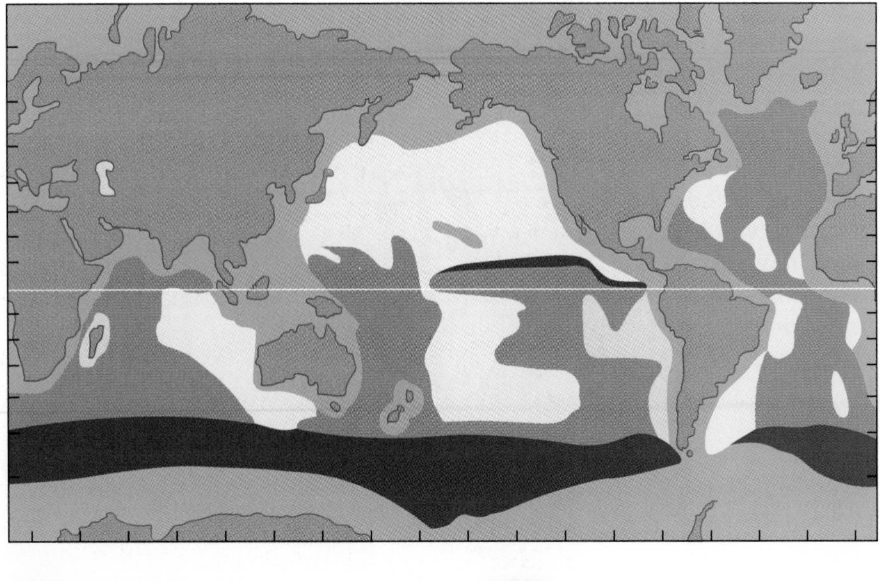

Figure 5.16

The distribution of various sediment types on the ocean bottom. Lithogenous sediments dominate the continental margins and the very high latitudes where they are carried out to sea by ice. Siliceous oozes are dominant where surface waters are cooler. Calcareous oozes are found in temperate and tropical latitudes, but not in the deepest, coldest waters. The abyssal clays cover much of the deepest parts of the ocean basins in temperate and tropical latitudes.

Altogether, about 14% of all deep ocean areas have siliceous oozes, 48% have calcareous oozes, and the remaining 38% have abyssal clays and terrigenous sediments. The Pacific Ocean is slightly deeper on the average than the Atlantic or Indian Oceans, because the oceanic ridge is a smaller fraction of the total floor area. Therefore, the Pacific Ocean has more than its share of abyssal clays and less calcareous oozes compared to the other two oceans.

5.7 COLLECTING SAMPLES

Most sediment samples are retrieved from aboard surface ships, which are separated from the ocean bottom by several kilometers of darkness and water. This extra obstacle requires extra precautions. Scientists want a sampler that is not likely to malfunction, because they do not wish to waste hours of ship cruise time with a sampler that comes up empty. For this reason, simple mechanical devices are often preferred over more complicated gadgets. To ensure that the samples are informative, we would like our collector to distort them as little as possible, and to protect them from being washed or otherwise adulterated by the ocean during their long journeys back to the ship.

SURFACE SAMPLERS AND CORERS

Sampling of surface sediments is most frequently done with a **dredge** or **grab sampler** (Figure 5.17). Dredges are especially useful for coarse sediments, such as nodules or rocks. Grab samplers have jaws that close when they reach the bottom, and they are more useful for finer sediments.

A deeper section of sediments can be sampled using **corers**. These come in a variety of forms, but often resemble very large darts hanging from cables (Figure 5.18). The "needle" of a dart is a long hollow tube, called the *barrel*, that is designed to minimize distortion as it plunges into the sediments. The barrel

Figure 5.17

Grab samplers are used for retrieving samples of bottom sediments.

usually has a liner that is removed along with the sample it contains. The distortion is greatest on the core's outer skin, so it is often scraped off before a study is made. Several hundred kilograms of weights may be added to help push the barrel into the bottom. *Piston corers* are used to get much longer cores, sometimes up to 20 meters long. Drilling allows the retrieval of even deeper sediments. It is a very difficult feat to drill into the deep ocean floor from a surface ship, but such projects have been done. The most ambitious of these drilling projects is described next.

OCEAN DRILLING

In 1968, the United States undertook a very ambitious **Deep Sea Drilling Project** to retrieve and analyze deep sea sediments, using a sophisticated drilling ship, the *Glomar Challenger* (Figure 1.15). This ship could operate in water up to 6 kilometers deep and had the ability to stabilize for weeks at a time above a bore hole. After bringing up worn drill bits for replacement, the new bit could relocate and reenter the tiny bore hole after traveling down several kilometers through the water column. Being able to accomplish this on the high seas was a marvelous feat of technology.

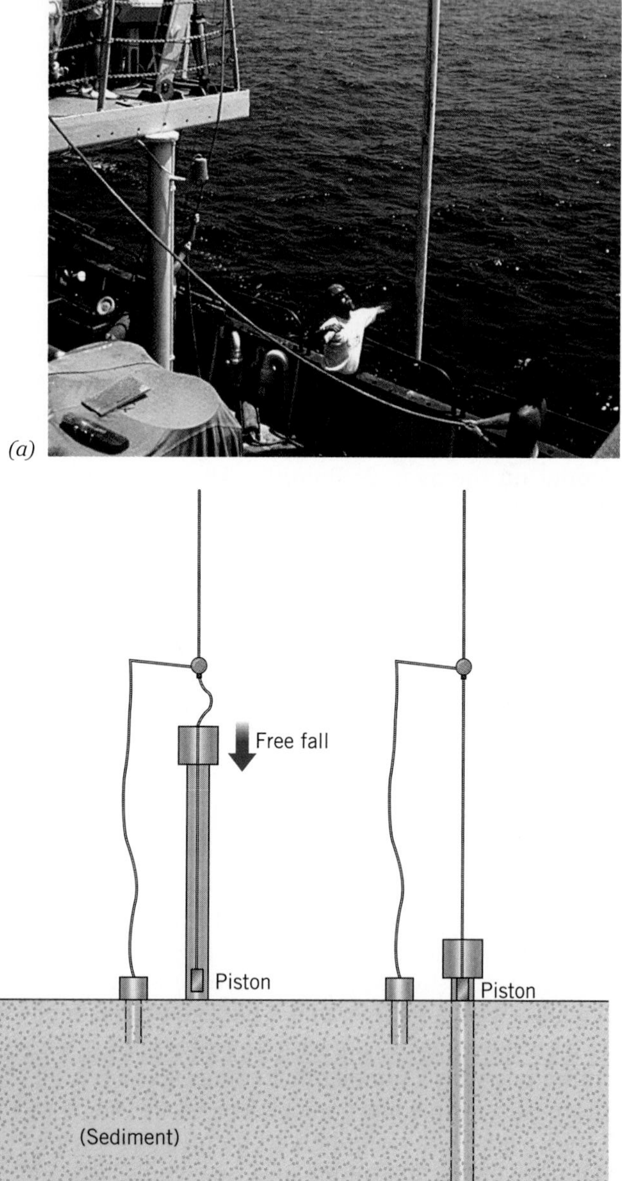

(a)

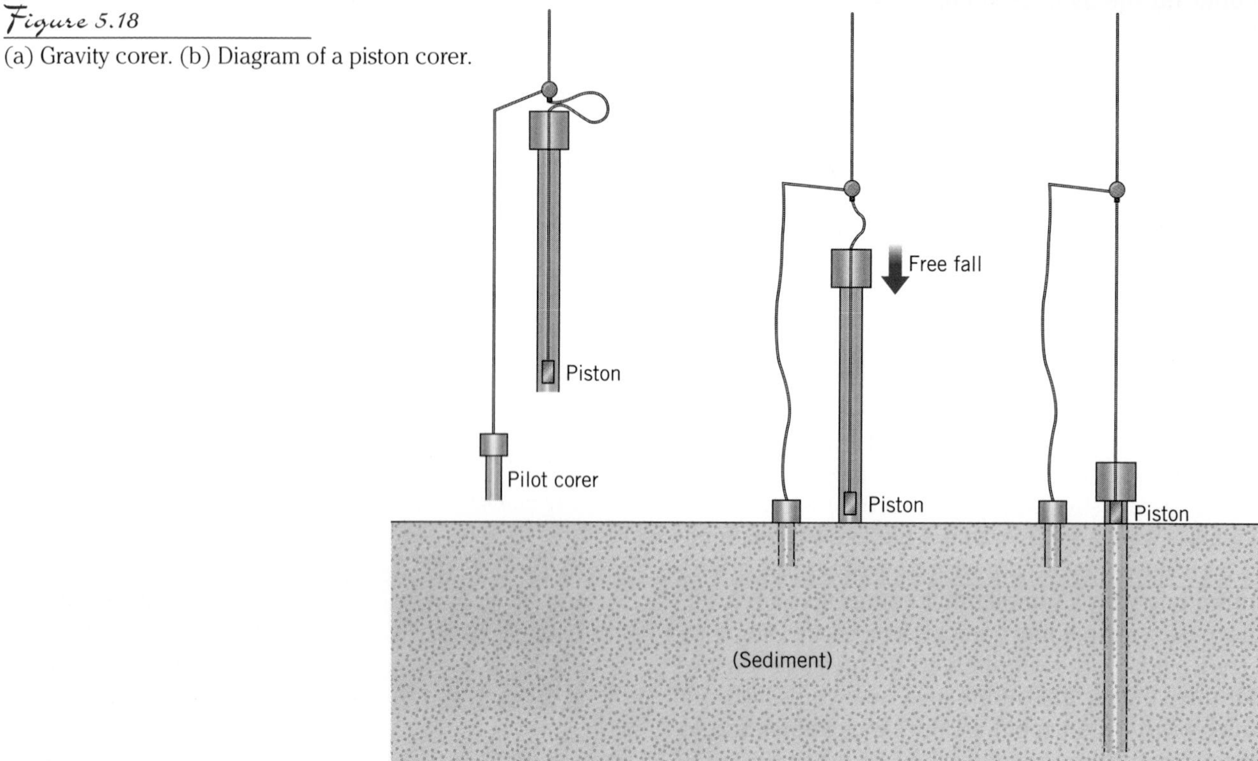

(b)

Figure 5.18
(a) Gravity corer. (b) Diagram of a piston corer.

The *Glomar Challenger* operated in all the major oceans, except for the Arctic Ocean, and drilled a total of 1092 holes at 624 sites, at locations extending from 76°N to 77°S. Its deepest hole went 1.7 kilometers into the sea floor. Analysis of the sediments retrieved from these sites has resulted in an increase in our understanding of our sea floor, and in the recent history of our planet. The ship was retired in 1983, and a larger, more modern drilling ship, the *JOIDES Resolution* (Figure 1.15), was built to replace it, making its maiden voyage in 1985. The scope of the drilling program was expanded to include sediments of the continental margins, the directorship expanded to include a larger consortium of U. S. universities, and the program name changed to *Ocean Drilling Project*.

The major advantage of drilling compared to coring is that drills can go much deeper into the sediments and older hardened sedimentary strata, therefore exploring farther back into the Earth's history. The disadvantage is the mixing of neighboring sediment layers in the drilling process, which tends to blur what otherwise might have been rather sudden changes in the sediment composition.

Summary

Sediment Materials

Sediments blanket the Earth and display a variety of sizes, shapes, and colors. Sizes vary from fine clays to large boulders. Porosity and permeability are also important sediment characteristics. Sediments may be composed of a large number of materials that are common on the Earth's surface. These usually involve chemical combinations of electron givers, such as silicon, aluminum, iron, calcium, sodium, potassium, or magnesium, together with elements or radicals that accept electrons, such as chlorine, oxygen, silicate, carbonate, sulfate, or phosphate. The material may be amorphous or crystalline, depending on whether there is some regular order in the arrangement of the atomic groups within it.

Lithogenous Sediments

Although the continental crust is igneous, much of the continental surface is covered by sedimentary rock. Consequently, most sediments are derived from the weathering of exposed sedimentary rocks, and many sediments go repeatedly through the sedimentary cycle. Sedimentary rocks form in deep and submerged layers of sediments, when neighboring grains are cemented together by the slow precipitation of minerals from the water trapped among the sediment grains. When they are exposed at the surface, fresh water dissolves these cementing minerals and, along with other agents, speeds the weathering of these rocks. Igneous rocks also weather at the surface when exposed to destabilizing and harsh environmental conditions.

The erosion of sediments from their original locations is accomplished in many different ways, but they eventually find their way into rivers and streams. There, the erosion–transportation–deposition cycle is repeated over and over until they finally reach the continental margins. By this time, there has been a great deal of sorting. Today, most deposition is taking place in coastal embayments. There is a tendency for coarser sediments to be found at higher latitudes.

Biogenous Sediments

Although there is generally a lot of biological activity on the shelf, the biogenous sediments seldom dominate, except for reef environments. Throughout most shelf regions, biogenous sediments are heavily diluted by lithogenous sediments from the neighboring land mass.

In the deep ocean, biogenous oozes dominate many large areas. Their formation depends on production in the surface waters, dilution by sediments from other sources, and dissolution by the sea water. Silica dissolves everywhere, so siliceous oozes are found only where the surface waters are so productive that the tiny skeletons accumulate faster than they dissolve, and where dilution by other sediments is small. Calcareous oozes dissolve in cooler waters, so they are found only in temperate and tropical latitudes and above the calcium carbonate compensation depth.

Abyssal Clays

Tiny clay-sized sediments from the continents may remain in suspension many years before settling out on the ocean floor. In this time, ocean currents may carry them to remote regions of the oceans, and interactions with the sea water may cause considerable change in their chemical composition. Clays may also be carried far out to sea by the winds, or may form from direct precipitation from the sea water. Abyssal clays accumulate very slowly and therefore dominate only those regions of the deep ocean floor where sediments from other sources are even scarcer.

Hydrogenous Sediments

Hydrogenous sediments form from the precipitation of materials from the sea water solution. Manganese nodules and metal sulfide deposits are attracting interest for their commercial value. As water evaporates in enclosed seas, the remaining water gets saltier, and materials begin precipitating out, forming *evaporite* deposits. Salt and gypsum are familiar and important examples. Rather minor heating or agitation of sea water tends to drive out some of the dissolved carbon dioxide, which makes the water less acid and may cause the precipitation of calcium carbonate.

General Patterns in Sedimentation

Sediments are thinnest on the oceanic ridge and get thicker with increasing distance from the ridge, due to increasing age of the underlying crust. Lithogenous sediments dominate the continental margins and neighboring deep ocean regions. Calcareous oozes dominate deep ocean sediments in tropical and temperate latitudes at depths less than about 4 kilometers. Siliceous oozes dominate those deep ocean areas where the surface waters are especially active biologically. Abyssal clays dominate large areas of the deep ocean floor, but only by default of the other kinds of sediments.

Collecting Samples

Sediment samples are collected with a variety of tools, including dredges, grab samplers, various types of corers, and drilling rigs. It is of special concern that these tools be as reliable as possible, and the sample is not unduly modified during the long trip back to the ship.

Key Terms

seismic echo profiling
lithogenous
terrigenous
biogenous
precipitation
hydrogenous
authigenic
cosmogenous
micrometeors
porosity
permeability
crystal structure
amorphous
crystalline
mineral
weathering
erosion
transportation
saturated
sedimentary cycle
maturity
exoskeletons
tests
abyssal clays
ooze
siliceous
calcareous
calcium carbonate compensation depth
snowline
limestone
manganese nodules
sulfides
brines
evaporites
dolomite
gypsum
dredge
grab sampler
corer
Deep Sea Drilling Project

Study Questions

1. What are the four basic kinds of ocean sediments? Which is least prominent?
2. Give an example of something that is porous, but not permeable. If it is full of holes, why can't water pass through it?
3. Using the lists of common electron givers and electron acceptors in Chapter 2, invent some molecules you think could make materials that are relatively common in the Earth's crust and sediments. (Be sure that the number of electrons donated equals the number accepted for each.)
4. Why are carbon and water important materials in sediments, even if they are not particularly prominent within the crust?
5. Are liquids amorphous or crystalline? Why? What about ice? Defend your answer.
6. Why are there far more sediments derived from the weathering of continental crust than from oceanic crust?
7. If the crustal materials are predominantly igneous, why are sedimentary rocks predominant on the Earth's surface?
8. How are sedimentary rocks formed? Why do they slowly crumble when exposed on the continental surface?
9. Describe how the erosion of stream bed sediments of various sizes is related to current speed. Do the same for transportation and deposition. Why are swifter currents needed to erode clay deposits than sands?
10. Would you expect beach sands to be more mature or less mature than the sediments at the base of mountains? Why?
11. Why is there a tendency for the beach sediments at higher latitudes to be coarser than

those at lower latitudes? Can you give more than one reason?

12. Why are oozes rare on the continental shelf?
13. What three conditions are necessary for oozes to form? If biological productivity is so low over deep ocean waters, why do oozes compose over half the deep sea sediments?
14. In what oceanic regions are there siliceous oozes, and why? In what oceanic regions are there calcareous oozes, and why? In what oceanic regions are there abyssal clays, and why?
15. Why are abyssal clays not considered lithogenous, even when the original clay particles come from continental rocks?
16. Where are evaporite deposits likely to form, and why? What are some of the materials found in evaporite deposits? Why does calcium carbonate precipitate out first, even though there is far less of it in sea water than many other materials? Why was this form of sedimentation more common in ancient times than it is today?
17. How would you expect the thickness of sediments to vary, going away from the ridge crest? Why? How might their composition change, and why?
18. How are samples of deep sea sediments collected? What is the main advantage and main disadvantage of drilling as opposed to coring?

Critical Thinking Questions

1. Pick out three of the things around you (e.g., the walls, windows, and floors of your room, the clothes you are wearing, or the pages of this book) and explain how the production of each depended on sediments.
2. Table 5.1 indicates that the average thickness of sediments on the continental slopes and rises is around 8 to 9 kilometers. What does this tell you about what must have happened to the underlying crust as these sediments accumulated on the continental margins? (That is, how can the sediments be below sea level if they are thicker than the ocean is deep?)
3. The book states that although about 75% of all sediments are lithogenous, only about 20% of the ocean floor is covered by them. Explain why these two numbers don't match.
4. Put a few handfuls of dirt in a tall jar, add water, shake it up, and then let it settle. Describe what you see. (Then you might let the jar sit in the Sun for a few days and see what happens next.)
5. Suggest some processes that might cause sea level to rise relative to land.
6. Materials of the continental crust, such as granite, are speckled, yet beach sands usually aren't speckled. Can you explain why the beach sands are more uniform than the materials they are derived from?

Suggestions for Further Reading

BROADUS, J. M. 1987. Seabed Materials. *Science* 235:4791, 853.

GROSS, THOMAS F. 1988. A Deep-Sea Sediment Transport Storm. *Nature* 331:6156, 518.

MAXWELL, ARTHUR E. 1993. An Abridged History of Deep Ocean Drilling. *Oceanus* 36:4, 8.

MCCAVE, N. 1988. Oceanography: Stirrings in the Abyss. *Nature* 331:6156, 484.

REA, D. K. 1987. Windfalls of Dust. *Natural History* 96:2, 28.

RONA, P. A. 1986. Mineral Deposits from Sea-Floor Hot Springs. *Scientific American* 254:1, 84.

SHEPARD, F. P. 1973. *Submarine Geology*. Harper and Row, New York.

SHEPARD, F. P. 1977. *Geological Oceanography*. Crane and Russak.

STANLEY, DANIEL JEAN. 1990. Mediterranean Desert Theory Is Drying Up. *Oceanus* 33:1, 14.

1990. Reading the Ocean's Diary. *Nature* 346:6280, 111.

If the water molecule weren't so "sticky," it would not be a liquid here at Earth temperatures, and we wouldn't have any oceans.

Water is a unique material. It is responsible for life and makes the Earth quite different from its planetary neighbors.

Six

SEA WATER

The preceding three chapters focused on the broad area of geological oceanography. We were concerned with the characteristics of the ocean floor and the processes that created and modified the oceans over long time periods. Now we turn our attention to the water itself.

It is the abundance of surface water that distinguishes the Earth from its planetary neighbors, and the exceptional properties of the water make this distinction an attractive one. Water drastically modifies the surface of this planet, provides a very mild climate, and makes life possible. Life depends on not only the water itself, but also its remarkable ability to dissolve other materials and hold them in solution. Most of these dissolved materials are also involved in the physical and chemical processes that are continually remodeling the Earth's surface.

To begin this chapter, we take a microscopic look at sea water, to see what gives it these exceptional properties, and how these properties influence the various physical, chemical, and biological processes that are so critical to our existence.

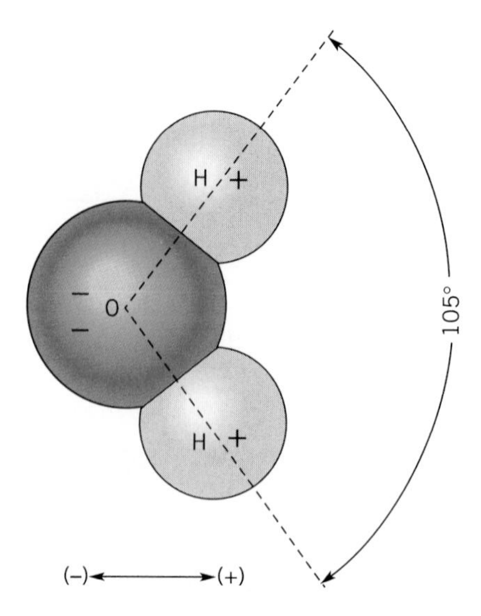

Figure 6.1

In the water molecule, the shared electrons are not shared equally. They spend more time in the vicinity of the oxygen atom and less time near the hydrogen atoms. This makes the molecule electrically polarized, with the oxygen side carrying a negative charge and the hydrogen side a positive charge.

6.1 THE WATER MOLECULE

Water is exceptional because of its molecular structure. In Chapter 2, we learned that each water molecule is made of one oxygen atom and two hydrogen atoms, which bond together by sharing electrons. But the shared electrons are not shared equally. Because oxygen is two electrons *shy* of the preferred arrangement of 10 electrons in two shells, it is a bit greedy for shared electrons. Hydrogen is more generous. Consequently, the shared electrons spend more of their time around the oxygen atom than in the neighborhood of the hydrogen atoms. This shift in the distributions of the negatively charged electrons gives negative charge to the region near the oxygen atom and positive charge to the region near the hydrogen atoms. As is illustrated in Figure 6.1, both hydrogen atoms are located on the same side of the molecule. This means that one side of the molecule has a positive charge, and the other side a negative charge. This separation of charges is referred to as electrical **polarization**.

Because opposite electrical charges attract (Chapter 2), the electrically polarized water molecule is very "sticky." The charged ends of the water molecule are attracted to oppositely charged portions of whatever other molecules it encounters, whether they be other water molecules or the molecules of other materials that the water touches. The attraction of the positively charged hydrogen end of a water molecule to a negatively charged part of another molecule is referred to as a **hydrogen bond**.

6.2 SPECIAL PROPERTIES OF WATER

The "stickiness" of the water molecule gives water some truly exceptional properties. Among these are its remarkable ability to store thermal energy, its amazingly high melting and boiling points, the porous hexagonal microscopic structure of ice crystals, and its exceptional ability to dissolve and transport substances.

TEMPERATURE AND HEAT

First let us examine some of the exceptional thermal properties of water. As we learned in Chapter 2, thermal energy is stored in the motion of the individual molecules. In solids, the individual atoms and molecules vibrate, and in liquids and gases, they can rotate and translate (i.e., move around) as well. Temperature measures the average energy per molecule, or the "violence" of the motion on a molecular scale.

As we heat something up, the molecular motion increases, and the temperature rises.[1] As we cool it down, the molecular motion decreases, and the temperature falls. If you touch your finger to the burner of a hot stove, the molecules in your finger shake so violently that some are damaged or destroyed.

The total thermal energy stored in a substance depends on both the number of molecules and the average energy of each. For example, if a cup of water has the same temperature as the Atlantic Ocean, then the average energy per molecule is the same for both. However, the Atlantic Ocean has far more thermal energy because it contains many more molecules.

For example, if a cup of water has the same temperature as the Atlantic Ocean, then the average energy per molecule is the same for both. However, the Atlantic Ocean has far more thermal energy because it contains many more molecules.

A familiar temperature scale is the Celsius scale (Chapter 2), on which the boiling point of water is 100 degrees (100°C), and the freezing point is zero (0°C). Molecular motion doesn't stop at 0°C, however. The molecules in ice crystals are actually still vibrating quite violently. In fact, the absolute minimum temperature, beyond which absolutely no more thermal energy may be removed from a substance, is not reached until 273 degrees below zero Celsius (–273°C). This point is called **absolute zero**. Scientists often prefer to use a temperature scale on which absolute zero is really zero. This is called the **absolute** or **Kelvin scale** and degrees are indicated by the symbol K. It differs from the Celsius scale by 273 degrees, so that absolute zero is 0 K, water freezes at 273 K, and water boils at 373 K. We use this scale in some of the diagrams of this section for comparing the thermal properties of water with other substances.

SPECIFIC HEAT

The amount of heat required to heat up 1 gram of a substance by 1°C is called its **specific heat**. In Chapter 2, we learned that a *calorie* is the amount of heat required to raise the temperature of 1 gram of water by 1°C, so the specific heat of water is 1 calorie per gram per degree Celsius.

Because the water molecules stick to all the other water molecules around them, it takes a great deal more energy to increase the molecular motion for water than it does for other materials. A large amount of added energy must be used to break the bonds between neighboring molecules, in order that the motions of individual molecules might increase. Consequently, the specific heat of water is much higher than that of all other common materials (Table 6.1).

MELTING AND BOILING POINTS

Heavier molecules are more sluggish, moving more slowly at any given temperature. Therefore, materials with heavier molecules generally melt and boil at higher temperatures. This tendency is illustrated in Figure 6.2. To compare the weights of various molecules, we use the **molecular mass number**, which is the number of protons and neutrons in the nuclei of the atoms comprising a molecule. (Protons and neutrons contain nearly all the mass of an atom.)

Water is an important and impressive exception to this pattern. Although the water molecule is extremely light and lively, the melting and boiling points of water are exceptionally high. In Table 6.1 and Figure 6.2, you can see that the methane molecule and water molecule have nearly the same mass, but methane melts and boils at much lower temperatures. Even molecules considerably heavier than water have lower melting and boiling points. For example, oxygen (O_2) and nitrogen (N_2) have nearly twice the molecular mass, and carbon dioxide (CO_2) has nearly three times the mass. However, all of these have much lower melting and boiling points. The

Table 6.1 Specific Heats for Various Common Substances, in Terms of How Many Calories of Heat are Required to Heat up 1 Gram by 1°C

Substance	Specific Heat (cal/g/°C)
Alcohol	0.58
Aluminum	0.21
Copper	0.09
Gold	0.03
Leather	0.36
Marble	0.21
Salt	0.21
Sugar	0.27
Synthetic rubber	0.45
Water	**1.00**
Wood	0.42

[1]When the substance is undergoing a phase change, such as melting or boiling, something else happens. We will discuss this situation later.

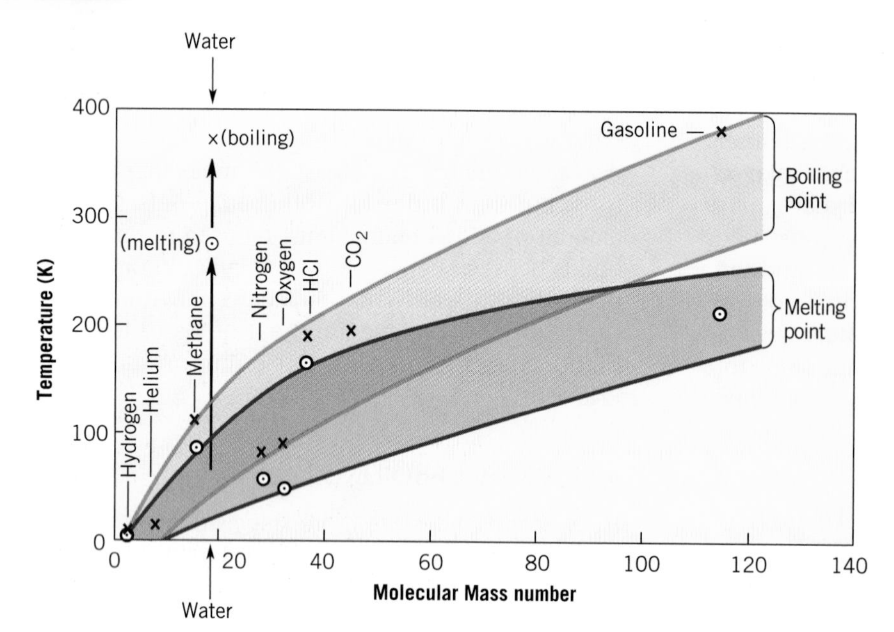

Figure 6.2
Melting and boiling points of various common substances, plotted as a function of their molecular weight. In general, heavier, slower-moving molecules require higher temperatures for both melting and boiling. But water is a clear exception. Why? (Note that temperatures are on the absolute scale, where absolute zero is 0 K, and water freezes at 273 K.)

very light and lively water molecule is found mostly in the liquid and solid states on Earth, whereas the heavier oxygen, nitrogen, and carbon dioxide molecules are found as gases.

The reason for the elevated melting and boiling points of water is that the highly polarized water molecules tend to stick together. Consequently, when ice melts, it takes a good deal more energy to shake the molecules apart from each other than it would for other substances of similar molecular masses. Similarly, when water is boiling, it takes more energy to shake water molecules loose from the liquid surface and into the vapor (i.e., gaseous) phase than it would if the water didn't stick together so strongly. It is clear that if the water molecule weren't so "sticky," it would not be a liquid here at Earth temperatures, and we wouldn't have any oceans.

It is clear that if the water molecule weren't so "sticky," it would not be a liquid here at Earth temperatures, and we wouldn't have any oceans.

LATENT HEAT

For a similar reason, water has a very high **latent heat of fusion**, which is the amount of energy needed to melt 1 gram of a solid. Once you bring ice to the melting point, you must still add a lot of heat just to melt it. All this energy goes into shaking the molecules loose from their frozen, rigid, stuck-together crystalline structures without raising the temperature at all. That is, the added heat changes ice at 0°C to liquid water at 0°C.

Water also has an exceptionally high **latent heat of evaporation**, which is the amount of energy needed to evaporate 1 gram of a liquid. A large amount of energy is needed to shake the water molecules loose from the liquid surface and into the gaseous state. This is why it takes so long for a pot of water on the stove to boil dry. You must add a lot of heat to evaporate even small amounts of water. It is also why you feel cool when you get out of a shower or pool. As the water evaporates from your skin, it removes a great deal of thermal energy from your body.

The latent heat of fusion for water is 80 calories per gram. The latent heat of evaporation depends on the water's temperature, being 540 calories per gram at the boiling point (100°C), and somewhat more at temperatures characteristic of the ocean surface (e.g., 585 calories per gram at 20°C). These numbers are both much higher than for any other common substance, simply because the water molecules are so sticky that it is difficult to shake them apart.

Whatever energy is required to break the bonds between neighboring molecules when a material is melted or evaporated, this same energy is later released when these bonds reform. For example, the same amount of energy that was needed to melt ice is released again when the water freezes. And the same amount of energy that was needed to evaporate water is released again when this water vapor condenses.

On a molecular scale, the attraction between neighboring molecules might be something like the attraction between a stone and the Earth. As you throw a stone upward, it slows down as it goes higher. But as it falls back toward Earth, it speeds up again. Similarly, in a material where neighboring molecules attract each other, their separation consumes energy, but that energy is returned when the molecules come back together.

As you throw a stone upward, it slows down as it goes higher. But as it falls back toward Earth, it speeds up again. Similarly, in a material where neighboring molecules attract each other, their separation consumes energy, but that energy is returned when the molecules come back together.

ICE

Another interesting property of water is that the colder solid phase is *less* dense than the liquid phase (Figure 4.6). Ice floats. Water freezes *over*. This is opposite the behavior of most other materials, for which the cooler, denser portions sink, and the frozen portions collect on the bottom rather than at the top. The reason for this strange behavior can be traced back to the shape of the water molecule. As the water cools, the thermal motion of the molecules lessens, and they begin sticking together in vibrating, rigid structures, with the positive part of one molecule attracted to the negative part of the next. The angular shape of the individual molecules makes for an angular hexagonal crystal structure which is porous, resembling chicken wire in three dimensions (Figure 6.3). Because of all the tiny holes in this crystal struc-

= Oxygen

= Hydrogen

= Water molecule

= Water molecule tipped so that one hydrogen is interacting with the plane above or below

(a)

(b)

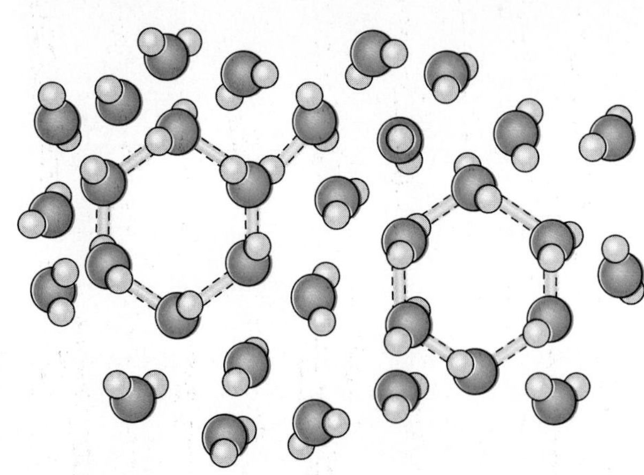

(c) Ice melt

Figure 6.3

(a) Molecular view of one layer in an ice crystal. The porous, hexagonal structure is clearly evident. Because of the pores, ice is less dense than water and floats. (b) The hexagonal structure of the crystal lattice is reflected in the shapes of ice crystals. (c) As ice melts, the hexagonal lattice begins to break apart. Some of the hexagon fragments remain intact, however, even after the ice has melted. Continued heating continues to break up these hexagon fragments, which allows the molecules to pack more closely. Consequently, the density of cold fresh water actually decreases with increasing temperature until 4°C.

ture, ice is less dense than liquid water, and it floats on top.

If you add heat to ice at 0°C, it starts to melt. The added energy causes the molecules to shake more violently. The hexagonal structures start breaking apart, and the pores disappear. Actually, even after the ice has melted, there is still some tendency for neighboring molecules to stick together in fragments of the porous crystalline structure (Figure 6.3c), and consequently, pure water doesn't reach its highest density until 4°C. Dissolved salts change the picture somewhat, because the salt ions tend to help separate the water molecules and fill in the pores in the crystalline fragments when the ice melts. Consequently, for water saltier than 2.5%, the water reaches its maximum density at the freezing point, rather than a few degrees above it.

SOLVENT PROPERTIES

Another important property of water that is attributed to its polarized molecule is that it is an extremely good solvent. That is, it dissolves things well (Figure 6.4). It is one of nature's best, and definitely the best one that comes in appreciable quantities. It is this property of water, in particular, that makes life possible.

Water molecules tend to stick to the molecules of other materials and tear them loose. This is how the material is dissolved. The only materials that are not soluble in water are those whose molecules have extremely even charge distributions, so that neither end of the polarized water molecule will find something to pull on. The water molecules are sort of like miniature hungry piranha, and they can find some way of getting their teeth into most other materials. Organic molecules, such as those that living organisms are made of, have no loose charges on them. Otherwise, they would dissolve in water and we would be in big trouble. Because organisms must take advantage of the special solvent properties of water without themselves being dissolved, these organisms must be made of these special organic molecules.

The most soluble materials tend to be the **ionic salts**, such as NaCl, that separate into charged ions when dissolved (Table 6.2). Their high solubility is

Table 6.2 Dissociation of the Most Common Ions in Sea Water

Ion	Percent as a Free Ion
Positive	
Sodium (Na^+)	99
Magnesium (Mg^{2+})	87
Calcium (Ca^{2+})	91
Potassium (K^+)	99
Negative	
Chloride (Cl^-)	100
Sulfate (SO_4^{2-})	50
Bicarbonate (HCO_3^-)	67
Bromide (Br^-)	100

Figure 6.4
Water is an excellent solvent. The dissolution of minerals by underground water can carve caverns such as this.

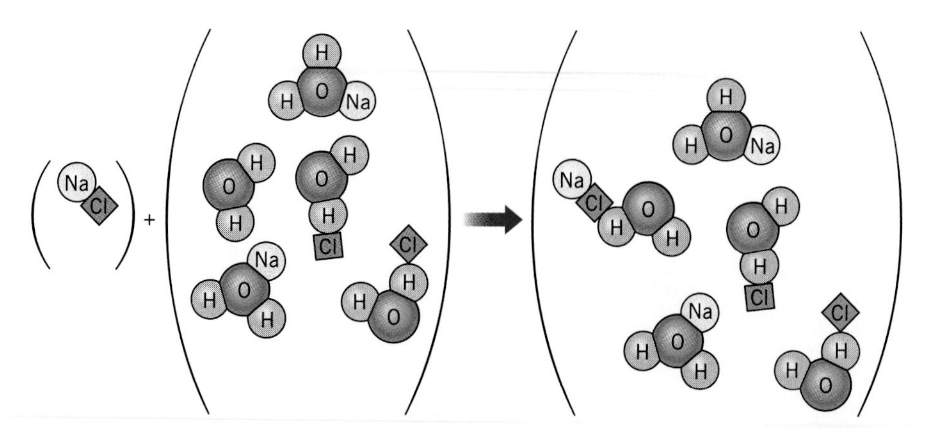

Figure 6.5

As more and more salt molecules dissolve in the water, there are fewer and fewer unattached water molecules left. Eventually, attraction by the water molecules is reduced to a point where incoming salt ions are attracted as strongly by other salt ions as they are by the water, and so they begin to crystallize out of the solution. When this happens, we say that the solution is saturated.

attributed to the separation of charges among the bonded atoms, which makes it easier for the polarized water molecules to stick to them. Most sodium chloride in sea water is found as separate Na^+ and Cl^- ions. Relatively little is found as NaCl molecules or microscopic crystals.

When the water is holding as much of a certain material as it can hold, it is said to be "saturated" with that material, be it a salt or gas. The amount of a salt that water can dissolve before it becomes saturated depends on the balance between how strongly an ion or molecule of that salt is attracted by the water molecules and how strongly it is attracted by other ions or molecules of its own kind. The balance of these two opposing forces varies greatly from one salt to another, so some salts are much more soluble than others.

As salt crystals are first immersed in water, the salt begins dissolving quickly. But as the solution becomes saltier, more and more water molecules become tied up with the dissolved salt ions, and fewer and fewer water molecules are available for new salt ions (Figure 6.5). Consequently, as the solution becomes saltier, the water's affinity for additional salt molecules becomes less and less. Eventually, a point is reached where a salt molecule is attracted equally strongly by the remaining salt crystals as by the water. At this point, as many salt molecules will crystallize out of the solution as dissolve into the solution. The solution gets no saltier, and the salt crystals become no smaller. The solution is saturated.

EFFECTS OF THE SALTS

Since opposite charges attract, dissolved salts tend to have their positive parts attached to the negative end of the water molecule, and negative parts attached to the positive end of the water molecule, neutralizing the charges at both ends. In this way, dissolved salts slightly mollify the exceptional properties of water by reducing the effect of the charge separation. For example, the salt in sea water lowers its freezing point to −1.9°C, and its specific heat is reduced by a few percent.

When water evaporates, the dissolved salts are left behind (Figure 5.15). Similarly, when water freezes, the ice crystals attract more strongly molecules of their own kind, and the dissolved salts tend to be rejected by the ice. In fact, only about one-third of the original salt is incorporated into the ice (depending on how fast it freezes). Sea ice, then, is only about 1% salt and would be fit drink for thirsty sailors, if necessary, even though sea water is not.

6.3 COSMIC ORIGINS OF THE OCEAN

The exceptional properties of water, as outlined in the previous section, will be used throughout this book in the explanation of important processes on this planet—particularly in its oceans. It is no surprise that these properties make this water-covered planet a very special place. But we have yet to understand why this blessing was granted to Earth, and not its neighbors.

Table 6.3 Molecular Mass Numbers of Some Lighter Molecules That Might Be Found in Planetary Atmospheres, Using the Most Common Isotope for Each Element

A common mineral in ordinary crustal rocks is included for comparison.

Molecule	Constituents	Mass Number of Molecule
H_2	$2(H^1)$	2
He	$1(He^4)$	4
CH_4	$1(C^{12}) + 4(H^1)$	16
NH_3	$1(N^{14}) + 3(H^1)$	17
H_2O	$1(O^{16}) + 2(H^1)$	18
N_2	$2(N^{14})$	28
O_2	$2(O^{16})$	32
CO_2	$1(C^{12}) + 2(O^{16})$	44
$FeMgSiO_4$	$1(Fe^{56}) + 1(Mg^{24}) + 1(Si^{28}) + 4(O^{16})$	172

PLANETARY ATMOSPHERES

Earth has 1.4 billion cubic kilometers of liquid water on its surface, which is about 0.25 cubic kilometers for every person. Venus and Mars have little surface water and none as a liquid.[2] Mercury and our Moon have no surface water at all. Jupiter and the other **outer planets** (i.e., Jupiter, Saturn, Uranus, and Neptune) have huge atmospheres (i.e., gaseous outer coverings) of hydrogen and helium gases. Both these gases are completely missing on Earth. Since all these planets condensed from the same primordial cloud of interstellar gas and dust, all their constituent materials should be the same. But judging from surface appearances, they aren't. So what makes us so special? Where did our atmosphere and water come from, and why are we so different?

We will answer these questions by first examining some of the more common light materials that might be good candidates for planetary atmospheres and oceans (Table 6.3). Then we will learn how Earth differed from the other planets in the way it acquired and kept these materials.

Hydrogen and helium, the lightest and simplest of the elements, are by far the most abundant materials in the Universe (Table 2.1). For this reason, you might think they should be the predominant gases in the atmospheres of the planets. However, the **inner planets** (Mercury, Venus, Earth, and Mars) and our Moon are too close to the Sun and, therefore, too warm to be able to hold onto these two gases. At these warmer temperatures, the light hydrogen and helium molecules are moving very fast. At any instant, some are moving faster than the velocity needed to escape from the planet, so they do. In order to retain these very light molecules, a planet must either be cold, so the molecules move slowly, or it must have a very strong gravity.

Mercury is the smallest and closest to the Sun of these planets. It is so hot, and its gravity is so weak, that it cannot hold onto any gases at all. By contrast, Earth is larger and farther from the Sun. It is sufficiently cool and has strong enough gravity that it can hold onto all gases except for the very lightest, hydrogen and helium. Any hydrogen we might have on Earth is bound to heavier atoms, such as oxygen (H_2O). But except for some small amounts that may still be trapped in the Earth's interior, we have no pure hydrogen or helium in the gaseous form.

The main reason that the outer planets are so much larger than Earth is that their greater distance from the Sun, and consequent cooler temperatures, have enabled them to retain these very abundant light gases (Figure 3.3b). Earth and the other warm inner planets couldn't. They represent the small "residue" that is left after the hydrogen and helium are gone (Figure 1.2).

PRIMORDIAL ATMOSPHERES

In fact, the inner planets began with no atmospheres at all. The collision and compression of the accumulating materials and their closeness to the young Sun made them far too hot. Such high temperatures drove off not only hydrogen and helium, but the other gases as well. The atmospheres of these inner planets were acquired later, coming from their own interior regions, after the planets had cooled to the point that they could retain the gases that reached their surfaces.

The process by which gases and other light mate-

[2]Both Mars and Venus have water vapor in their atmospheres, and Mars has ice on its surface as well.

rials (including water vapor) reach the surface is called **outgassing**. Under the high pressures of the interiors of planets, gases are dissolved in the interior materials. When magmas reach the surface (e.g., volcanism), they release these gases. The process of outgassing is verified by looking at the materials coming out of volcanoes today and also examining the bubbles in ancient lava flows. Both methods reveal that both the rates of outgassing and the materials being outgassed are indeed appropriate for explaining the present atmosphere and oceans on Earth. There are some interesting anomalies in the abundances of carbon dioxide and oxygen, however, that we will examine later in this chapter.

Today, the atmospheres of the inner planets are very small and light compared to the heavier minerals in their interiors, even though this is the reverse of the abundances of these materials in the universe as a whole. The reason, of course, is that the planets originally could hold no atmospheres at all. So even those gases that might stay on the planet today were gone from the very beginning.

In summary, the overall picture is that initially the inner planets were very hot and had no atmospheres. Subsequent outgassing from their interiors accounts for their present atmospheres. They are still too warm and too small to retain the very light gases of hydrogen and helium, however.

EVOLUTION OF ATMOSPHERES

Since the inner planets are not able to hold onto hydrogen and helium, the question becomes, "Aside from hydrogen and helium, what other light gases might be found in planetary atmospheres?" In Chapter 2, we learned that oxygen, carbon, and nitrogen follow hydrogen and helium in order of relative abundances. In forming molecules, any one of these would most likely combine with hydrogen, simply because the abundance of hydrogen makes it most likely to be encountered. Therefore, the next group of moderately light molecules in the universe are those composed of carbon, nitrogen, or oxygen combined with hydrogen (Figure 2.5). These combine to form methane (CH_4), ammonia (NH_3), water (H_2O), and fragments of these, such as CH, NH, and so on. Since carbon and oxygen are considerably more common than nitrogen, we might expect water and methane to be more common than ammonia. They are.

Still heavier and less common would be molecules involving two or more of these moderately light atoms, such as O_2, CO_2, and N_2. From Table 2.1, you can see that slightly less common than nitrogen are several metals, such as silicon, magnesium, iron, and aluminum. Consequently, we might expect the next echelon of heavier and slightly less common molecules in the universe to be things like silicates and the oxides of various metals. These make up the minerals of the solid Earth and neighboring planets. But these would not be in planetary atmospheres. The molecular weights of some common light molecules that might be found in planetary atmospheres are listed in Table 6.3.

From this information, we might expect that water (H_2O), ammonia (NH_3), and methane (CH_4) would be the most common materials in the atmospheres of the inner planets. However, several processes work to tear the hydrogens off these molecules. These processes include the temperature and pressures these gases encounter when still in planetary interiors (i.e., before being outgassed), and the exposure to the very energetic ultraviolet rays from the Sun when in planetary atmospheres (i.e., after being outgassed). The hydrogen removed by these processes tends to form H_2 molecules and escape from the planets. The remaining carbon, nitrogen, and oxygen tend to combine to form carbon dioxide (CO_2) and nitrogen (N_2), in addition to various oxides and carbonates that remain in the mineral planetary interiors. Planetary atmospheres, then, should contain large amounts of CO_2 and N_2. From the atomic abundances listed in Table 2.1, we might expect carbon dioxide to be more abundant than nitrogen, in general. It is.

Water is also fairly abundant in planetary atmospheres, because the water molecule is both more common and more stable than methane or ammonia. But water is involved in a variety of physical and chemical processes that make its abundance vary. On Earth, water is primarily a liquid; relatively little is in our atmosphere. On Mars, it is found mostly as ice—mostly *permafrost* beneath the surface. On Venus, it is largely missing. Because of Venus' hot temperature, all water on the surface would immediately evaporate and join the atmosphere as gaseous water vapor. But in Venus' atmosphere, the water molecules are exposed to intense heating and ultraviolet rays from the Sun, which have torn apart the molecules, releasing the hydrogen.

As we have seen, larger planets have more internal heat sources and also greater difficulty in cooling. Consequently, larger planets should undergo more geological activity and outgas heavier atmospheres. Indeed, this is true. Of the four inner planets, Mercury is the smallest, is geologically inactive, and has no atmosphere (Figure 1.2). Mars is the next smallest, shows small amounts of geological activity

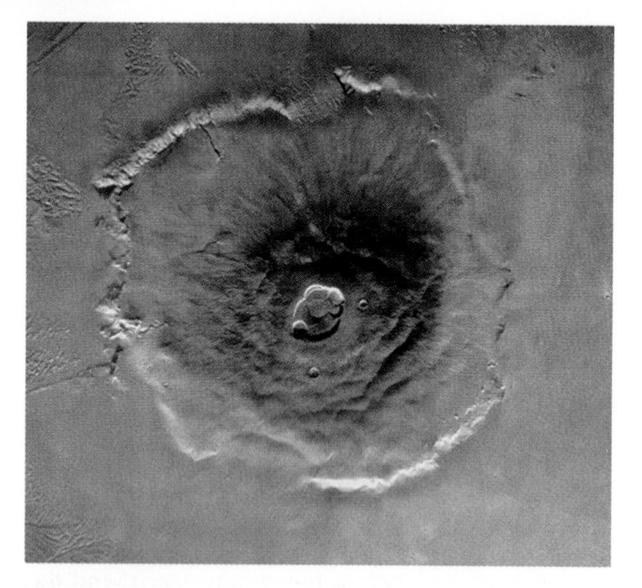

Figure 6.6

Olympus Mons is one of the 12 very large ancient volcanoes on the surface of Mars, which are no longer active. These volcanoes tell us of past volcanic outgassing there. Olympus Mons is much larger than any volcano on Earth, being more than 500 km across at its base, and about 30 km high.

(Figure 6.6), and has the next smallest atmosphere. Venus is yet larger, shows greater geological activity, and has by far the heaviest atmosphere of the three. Both Mars' and Venus' atmospheres are primarily carbon dioxide, with nitrogen the second most abundant component, just as expected (Table 6.4).

ODDBALL EARTH

This brings us to the Earth. The Earth is slightly larger than Venus, and therefore should be slightly more active geologically, and have a slightly heavier atmosphere. Indeed, it is more active geologically, but its atmosphere is not nearly as heavy as that on Venus.[3] Although Earth does have a great deal of nitrogen in its atmosphere (78%), it is lacking the expected larger abundance of carbon dioxide. Carbon dioxide constitutes about 96% of the atmospheres of Venus and Mars, but only 0.03% of Earth's atmosphere (Table 6.4). Furthermore, Earth has the huge amounts of surface water, mostly in the oceans, that the other inner planets do not. How did the Earth manage to keep its water and lose its carbon dioxide?

Table 6.4 The atmospheres of Venus and Mars, compared to Earth. Listed are the surface pressures (relative to sea-level pressure on Earth), and the relative abundances of three major gases in each atmosphere. Notice that the Earth has somehow lost its carbon dioxide and gained oxygen.

Planet	Pressure	Carbon Dioxide (%)	Nitrogen (%)	Oxygen (%)
Venus	90	96	4	0
Mars	0.007	95	3	0
Earth	1	0.03	78	21

The answers to these two questions are interrelated. Ultraviolet radiation from the Sun is absorbed by atmospheres, so only gases that can reach the outer portions of an atmosphere can be dissociated (i.e., torn apart) by it. Most of Earth's water is liquid, and even the minor amounts of water vapor in the air cannot reach very high elevations before cooling and condensing into clouds and droplets that fall back to Earth. Consequently, almost all the water is kept as a liquid on the Earth's surface. Very little water vapor reaches the outer portions of our atmosphere where it could be torn apart by solar ultraviolet. In contrast, water on Venus stays in the vapor phase because of the high temperatures there. It reaches the outer regions of the atmosphere through atmospheric turbulence, where it is torn apart by solar ultraviolet. This explains why we have so much surface water and mostly in the liquid form, whereas Venus has so little and all in the vapor form.

For carbon dioxide, the situation is reversed. Venus has an abundance (96% of its atmosphere), and we have very little. The reason we have so little is that carbon dioxide dissolves readily in our oceans. If you put a balloon over a soft drink bottle and then heat it up or shake it to drive out this gas, you will find that the volume of carbon dioxide dissolved in the soft drink is much larger than the bottle itself! After the carbon dioxide dissolves in our oceans, it undergoes various chemical reactions, most of which lead to the formation of carbonate precipitates, such as limestone. Marine plants also consume some. Through these processes, our oceans have removed most of the carbon dioxide from our atmosphere. Today, there is more than 60 times as much carbon dioxide in the oceans as in our atmosphere, and there is thousands of times more than that tied up in limestone and other marine sediments.

If our oceans had not removed the carbon dioxide from our atmosphere, then our atmosphere would

[3]Actually, if Earth were as hot as Venus, its oceans would evaporate, giving Earth an atmosphere three times as heavy as Venus'. In addition, all the carbon dioxide that has dissolved in our oceans over the ages would have remained in our atmosphere, making it much heavier.

indeed be dominated by carbon dioxide and be much heavier than Venus', just as expected. The removal of the carbon dioxide has made our atmosphere much lighter than Venus' and left nitrogen as the dominant gas here.

If you put a balloon over a soft drink bottle and then heat it up or shake it to drive out this gas, you will find that the volume of carbon dioxide dissolved in the soft drink is much larger than the bottle itself!

The other unique feature of the Earth's atmosphere is the oxygen in it. The element oxygen is extremely reactive chemically, and it binds very strongly with other elements such as carbon, silicon, aluminum, iron, magnesium, and so on. So although oxygen is very abundant in the materials making up all planets, it is almost always chemically combined in the minerals of the planets' interiors. As a free gas in the atmosphere, it is unique to Earth. The reason for this is very special: We have plants that can release oxygen from carbon dioxide, and the other planets don't.

6.4 SOURCES OF THE SALTS

Since the time of the *Challenger* expedition, scientists have been systematically collecting and analyzing water samples from all parts of the oceans. Because water is such a good solvent, it is no surprise that just about everything imaginable can be dissolved in it. Of the 92 naturally occurring elements, more than 80 have been found in sea water (Table 6.5). There is every reason to believe that the others are also present, but in trace amounts that have escaped detec-

Table 6.5 Abundances of the Various Elements in Sea Water

Element		Concentration (ppb)	Element		Concentration (ppb)
Oxygen	O	857,000,000	Nickel	Ni	2
Hydrogen	H	108,000,000	Vanadium	V	2
Chlorine	Cl	19,000,000	Manganese	Mn	2
Sodium	Na	10,500,000	Titanium	Ti	1
Magnesium	Mg	1,350,000	Tin	Sn	0.8
Sulfur	S	890,000	Cesium	Cs	0.5
Calcium	Ca	400,000	Antimony	Sb	0.5
Potassium	K	380,000	Selenium	Se	0.4
Bromine	Br	65,000	Yttrium	Y	0.3
Carbon	C	28,000	Cadmium	Cd	0.1
Strontium	Sr	8,000	Tungsten	W	0.1
Boron	B	4,600	Cobalt	Co	0.1
Silicon	Si	3,000	Germanium	Ge	0.06
Fluorine	F	1,300	Chromium	Cr	0.05
Argon	A	600	Thorium	Th	0.05
Nitrogen[a]	N	500	Silver	Ag	0.04
Lithium	Li	170	Scandium	Sc	0.04
Rubidium	Rb	120	Lead	Pb	0.03
Phosphorus	P	70	Mercury	Hg	0.03
Iodine	I	60	Gallium	Ga	0.03
Barium	Ba	30	Bismuth	Bi	0.02
Indium	In	20	Niobium	Nb	0.01
Zinc	Zn	10	Lanthanum	La	0.01
Iron	Fe	10	Thallium	Tl	<0.01
Aluminum	Al	10	Gold	Au	0.004
Molybdenum	Mo	10	Cerium	Ce	0.005
Copper	Cu	3	Rare earths		0.003-0.0005
Arsenic	As	3	Protoactinium	Pa	2×10^{-6}
Uranium	U	3	Radium	Ra	1×10^{-7}

[a]Nutrient nitrogen only; the dissolved gas is not included.

Table 6.6 Maximum Amounts of Various Excess Volatiles That Could Have Come from Crustal Rocks

Material	Fraction That Could Have Come from the Crust (%)
Water (H_2O)	1
Carbon (C)	1
Chlorine (Cl)	2
Nitrogen (N)	1
Sulfur (S)	20

tion so far. These observations rouse our curiosity. We want to know where these various materials came from, how they got into the water, why they are present in the observed abundances, and what roles they play in the various oceanic and biological processes.

THOSE TAKEN FROM THE SOIL

At one time, people thought that the ocean was salty because rains and subsequent water drainage dissolve the salts from the soil and carry them out to sea. There are two ways of checking this theory. One is to look at the salts being carried to sea by rivers and streams, and the other is to look at the amounts of salts that have been depleted from weathered crustal rocks. Both approaches involve many uncertainties. For example, when analyzing the salts carried in river water, we must estimate how many are entering for the first time, and how many are just returning after a sojourn on land, such as those dissolved from sedimentary deposits. But by carrying out these studies, scientists have found that there are many salts whose abundances are quite compatible with the available supply in the crust. This group includes the most common dissolved metals, such as sodium, calcium, magnesium, and potassium.

THE EXCESS VOLATILES

However, a very important group of materials, including chlorine, bromine, sulfur, nitrogen, carbon, and the water itself, are far too abundant to have come from the crustal rock (Table 6.6). They are called the **excess volatiles**: "excess" because of their excessive abundance, and "volatiles" because they tend to be lighter than most materials in the Earth. The explanation of where these materials could have come from eluded scientists for some time.

Once we understood the geological history of the Earth, with its plate motions and volcanic outgassing, we also understood where the water came from. It was then logical to suspect that the other excess volatiles were likewise a product of outgassing from the Earth's mantle. Indeed, a check of the bubbles trapped in solidified igneous rock shows they contain the excess volatiles in about the correct ratios. In another approach, scientists analyze fallen meteorites[4] to get an estimate of the abundances of these materials deep within the Earth. The conclusion is that the mantle has a large supply of excess volatiles. Only 10% of the mantle's water supply needed to be outgassed to account for all the present oceans, for example, and this is compatible with present-day rates of volcanic outgassing from the interior.

CHANGES OVER TIME

Because the water and other excess volatiles were outgassed together, the oceans have always been about as salty as they are now. They are not getting saltier with time as was once suspected. Most salts were outgassed with the water. Others were dissolved from the crustal materials, but even these entered the ocean rather quickly as rain and run-off dissolved them and carried them to the ocean. Thus, the dissolved salts have come along with the water, rather than the water coming first, and the salts later.

Of course, there have been slight changes in the ocean's chemistry over time. We know, for example, that the accumulation of free oxygen in our ocean and atmosphere has been relatively recent.[5] Biological activity also affects other aspects of the ocean's chemistry, and we will study some of these later in this book. But overall, these changes have been relatively minor, and the ocean's salinity has always been pretty much as it is today.

HYDROTHERMAL ACTIVITY AND THE SALT BALANCE

Because the ocean's salt content remains so remarkably constant over time, we know that the rate at which each salt enters the ocean must be exactly canceled by the rate at which it is removed. Most salts enter the ocean from rivers and streams. Other sources include the dissolution of atmospheric car-

[4]Remember that the Earth's crust is light and not representative of Earth's internal composition. To get an idea of what might be deep down within the Earth, we look at the compositions of other objects, such as meteorites, that formed along with the Earth in the early Solar System.

[5]Therefore, animals have evolved only recently.

bon dioxide to form carbonates and the slow dissolution of various materials from sediments. The main way that most salts are removed from the solution is through the formation of sediments. Other losses include the removal of materials by organisms and the evaporation of sea spray.

For most salts, the equations balance; the rate of entry balances the rate of removal, at least to within the errors of our measurements. But until the late 1970s, there were some glaring exceptions. Then large pools of hot brine were found on the Red Sea floor, and soon thereafter, hydrothermal vents were found in many places along the East Pacific Rise. These discoveries provoked a great deal of interest among oceanographers. Samples of vent water were taken, and laboratory experiments were initiated in which sea water was circulated through cracks in hot rocks like those of the oceanic crust. The results of these tests indicate that the vents do indeed account for the salts whose equations previously didn't balance.

As a result of these experiments and measurements, we've learned a great deal about these hydrothermal vents and hot brines. As we learned in Chapter 4, water flows down through cracks in the hot crustal rocks on the oceanic ridge, where it becomes heated by the hot rocks and interacts chemically with them. The superheated water then rises back to the surface and reenters the ocean through vents (Figures 4.14 and 6.7). The interactions with the hot rock surfaces remove magnesium from the water. Oxygen is removed from the sulfate ion (i.e., SO_4^{2-}) to form sulfide (i.e., S^{2-}). Calcium, manganese, lithium, barium, and other metals are added to the water, and many of these metals combine with the sulfur to form sulfides. The sulfides of many heavier metals, such as copper, nickel, cadmium, selenium, uranium, and chromium, are fairly insoluble and precipitate out of the water upon exiting the vent and cooling. These deposits create the metal-rich sulfide ores that are often found in these vent areas.

These are the general patterns. The details are sensitive to temperatures, interaction times, flow rates, and dilution by cooler waters. Consequently, the detailed chemistry of vent water varies considerably from one vent to another. Altogether, the total flow of water through these vents is very small compared to the water entering the ocean from rivers, yet their influence on the ocean's chemistry is comparable for many salts. Particularly significant is the removal of sulfates and magnesium, and the addition of sulfides and many metals other than magnesium.

Figure 6.7

Although most hydrothermal activity is found on or near the ridge axis, it is also exposed a few places on the continental surfaces, such as this.

6.5 MATERIALS IN OCEAN WATER

Now that we know where the materials of sea water come from, we turn our attention from their origins to their abundances and behaviors. Because sea water contains so many different materials, it is convenient to group them into several general categories and examine one category at a time.

SUSPENDED MATERIALS

Sea water contains a great deal of suspended material, in sizes ranging from grains of sand to tiny particles only a few molecules across. Water motion helps keep these materials in suspension. If a jar of sea water were placed on a shelf and let sit motionless for a long period of time, the suspended materials would eventually settle out. This is the distinction between materials that are suspended and those that are dissolved. Dissolved materials would remain in the water no matter how long you wait. Dissolved materials are in the form of molecules and molecular fragments, which the individual water molecules hold and do not let go.

The concentrations of suspended materials vary greatly from stormy turbid coastal waters to the quiet clear cool waters of the deep oceans. Some of this suspended matter slowly dissolves, and some settles out, adding to the permanent sediment on the ocean bottom. Particles of organic matter have other possible fates as well. They may be consumed directly by organisms, stick to sediment grains, or decompose through bacterial action and return to solution as nutrients, to be used by plants in food production (Chapter 11).

The suspended particles have a noticeable effect on the seawater chemistry. Some exchange their molecular components with those dissolved in the water. They also provide surfaces onto which dissolved materials may precipitate out of solution, and from which other materials may be dissolved into solution. Because of their small sizes, they provide a remarkably large surface area for interaction with the water. A wheelbarrow load of fine clay sediments, for example, has about 10 square kilometers of surface area.

DISSOLVED MATERIALS

Dissolved materials are individual atoms, ions, molecules, or molecular fragments that are bonded in some way to water molecules. The dissolved substances are conveniently grouped into the following four broad categories, listed in order of decreasing abundance:

1. Major constituents
2. Dissolved gases
3. Nutrients
4. Trace elements

Sometimes, there is a fifth category, called *minor constituents* that is slipped in between *major constituents* and *dissolved gasses* on the above list. These are salts whose abundances are much smaller than those of the leading major constituents, much larger than the *trace elements*. In this book, however, we consider these salts simply as lesser members of the major constituent category.

The **major constituents** (sometimes called *major salts*) comprise about 99.7% of all dissolved materials, so the remaining 0.3% belong to the last three categories (Figure 6.8). This does not mean that the last three categories are unimportant, however. Indeed, life as we know it could not have developed and survived on this planet were it not for them. A great deal of effort is directed toward studying these materials, far in excess of what you might think from their small relative abundances.

The concentrations of the dissolved gases, nutrients, and trace elements are greatly affected by organisms, the climate, and some geochemical processes. This means that it is impossible to say exactly what the relative proportions of these various materials are. As a very rough approximation, the dissolved gases are about 100 or 200 times more abundant than the nutrients, and the nutrients are thousands of times more abundant than trace elements (Table 6.7).

The salinity (Chapter 1) of sea water is measured in terms of how many grams of total dissolved material there are per kilogram of sea water, or equivalently, *parts per thousand*. The symbol for parts per

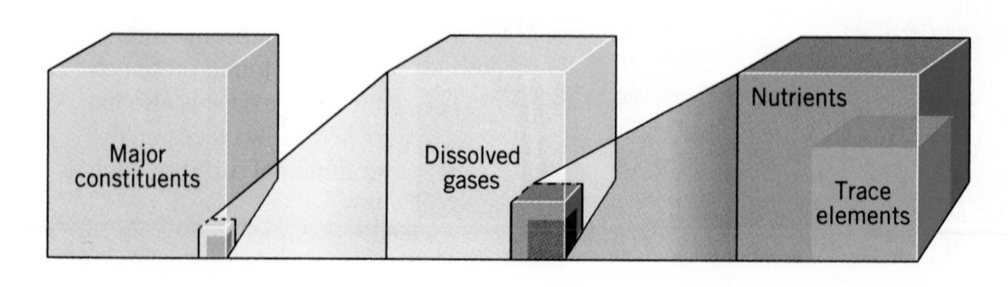

Figure 6.8
Sketch illustrating the relative proportions of the major constituents, dissolved gases, nutrients, and trace elements among the materials dissolved in sea water.

Table 6.7 Some Typical Values for the Concentrations of Some of the Important Members of Each of the Four Categories of Sea Water Constituents, by Weight

Each is expressed in parts per thousand, parts per million, and parts per billion. Clearly, parts per thousand is the most appropriate unit for the major constituents, parts per million is most appropriate for dissolved gases and nutrients, and parts per billion is most appropriate for trace elements.

	Parts per Thousand	Parts per Million	Parts per Billion
Major constituents			
Cl^-	19.3	19,300	19,300,000
Na^+	10.7	10,700	10,700,000
SO_4^{2-}	2.7	2,700	2,700,000
Mg^+	1.3	1,300	1,300,000
Dissolved gases			
CO_2	0.09	90	90,000
N_2	0.014	14	14,000
O_2	0.005	5	5,000
Nutrients			
Si	0.003	3	3,000
N	0.0005	0.5	500
P	0.00007	0.07	70
Trace elements			
I	0.000060	0.060	60
Fe	0.000010	0.010	10
Mn	0.000002	0.002	2
Pb	0.00000003	0.00003	0.03
Hg	0.00000003	0.00003	0.03

thousand is ‰, which is an easily recognized extension of the more familiar symbol for parts per hundred, %. Most sea water is found to have salinity in the range of about 34 to 36‰.

Just as a baker would use different units of measure for the salt than sugar or flour in a recipe, an oceanographer finds it appropriate to use different units for describing the concentrations of the different types of materials. The parts per thousand units used for the major salts are much larger than units appropriate for describing the dissolved gases or nutrients, and these, in turn, are much larger than those appropriate for trace elements (Table 6.7).

Major Constituents As we have seen, 99.7% of all dissolved materials are the major constituents, so variations in the concentrations of the other dissolved materials have very little effect on the overall salinity of a sample. The relative concentrations of the most abundant dissolved salts are listed in Table 6.8. By analyzing seawater samples gathered from all parts of all major oceans, we find that the major constituents occur everywhere in the same relative proportions (see "Rule of Constant Proportions," Chapter 1). This uniformity tells us that the oceans are very thoroughly mixed, just like the ingredients in a baking batter that you have stirred hundreds of times.

This uniformity tells us that the oceans are very thoroughly mixed, just like the ingredients in a baking batter that you have stirred hundreds of times.

Table 6.8 Amounts of the Principal Salts and Ions in Sea Water of Salinity 34.32‰

Material	Grams per Kilogram of Sea Water	Percent of Total Salt by Weight
Chloride (Cl^-)	18.980	55.04
Sodium (Na^+)	10.556	30.61
Sulfate (SO_4^{2-})	2.649	7.68
Magnesium (Mg^{2+})	1.272	3.69
Calcium (Ca^{2+})	0.400	1.16
Potassium (K^+)	0.380	1.10
Bicarbonate (HCO_3^-)	0.140	0.41
Bromide (Br^-)	0.065	0.19
Boric acid (H_3BO_3)	0.026	0.07
Strontium (Sr^{2+})	0.013	0.04
Fluoride (F^-)	0.001	0.003

Source: From Sverdrup, Johnson and Fleming, copyright 1942, renewed 1970, *The Oceans*, Prentice-Hall, Englewood Cliffs, NJ.

The only major component of sea water that varies is the freshwater content. In areas of high precipitation or much freshwater discharge, the salinity is low. In sunny warm regions, evaporation leaves the surface waters saltier than normal. Whatever the salinity, when chemists analyze the dissolved materials alone, they find that 55.04% by weight is the chlorine ion, 30.61% is sodium, 7.68% is sulfate, 3.69% is magnesium, and so on. That is, regardless of the salinity of the sample, the ratios of the various salts among themselves do not change.

This *rule of constant proportions* is subject to two qualifications. First, there are some coastal regions where discharge from rivers and streams is sufficient to have a noticeable effect on the ratios. For these minor coastal areas, the rule is violated. Second, the rule applies to the major constituents only. We know that the dissolved gases, nutrients, and trace elements show variations in their relative abundances due to various processes. But these processes have little overall effect on salinity, because these things altogether amount to only about 0.3% of the total dissolved material.

Dissolved Gases The **dissolved gases** are usually measured in terms of the number of milliliters of the gas that are dissolved in 1 liter of water. This unit, *milliliters per liter*, is written symbolically as **ml/l**. Sometimes, other units are used, such as *parts per million* (**ppm**).[6] Having different units is not really as confusing as it might appear. What is usually of interest is how the concentrations of the various gases compare with each other, and how they vary with position in the ocean and with time. What unit we use to make these comparisons is unimportant, as long as we are consistent (Table 6.9).

Many processes affect the concentrations of the gases. Biological processes can have particularly pronounced effects on oxygen concentrations, due to its short supply. Oxygen concentrations vary from less than 1 ml/l up to nearly 10 ml/l (or roughly 1–14 ppm), and carbon dioxide is usually present in abundances of around 45 to 55 ml/l (or about 90–110 ppm). Dissolved nitrogen is about twice as abundant as dissolved oxygen, although not as interesting, because it is not used by organisms. All other gases are present in much smaller concentrations.

An analysis of our atmosphere shows that dry air is about 78% nitrogen, 21% oxygen, and nearly 1% argon. Adding up these three, you can see that all the other gases must amount to very little. Indeed, the next most abundant atmospheric gas, carbon dioxide, constitutes only about 0.03% of dry air. In the ocean, the relative abundances of these gases are remarkably different, as can be seen in Table 6.10. Even though the ocean is not nearly saturated with carbon dioxide, it contains 60 times more of this gas than the atmosphere. In contrast, the ocean's content of oxygen and nitrogen is only about 1% of that in the atmosphere. Obviously, carbon dioxide is extremely soluble in water.

Table 6.9 The Three Common Units for Measuring the Concentrations of Dissolved Gases

Gas	Milliliters per Liter	Parts per Million	Millimoles per Liter
Conversion from 1 ml/l			
CO_2	1	1.96	0.045
N_2	1	1.25	0.045
O_2	1	1.43	0.045
Typical concentrations			
CO_2	40	78	1.79
N_2	10	12.5	0.45
O_2	5	7.2	0.23

The concentration of carbon dioxide tends to increase with depth in the ocean. The deepest waters formed initially near the surface and at high latitudes, where the cold temperatures facilitated the dissolution of CO_2. (Your experience with carbonated beverages should tell you that colder water holds carbon dioxide better.) These cold water masses took this carbon dioxide with them as they sank and flowed along the bottom to their present deep ocean locations. The intermediate and surface water masses were formed in warmer climates, and consequently, they dissolved less carbon dioxide during their formation. The concentrations of dissolved nitrogen show a similar small increase with depth and for the same reason.

Biological organisms also have a noticeable influence on the concentrations of dissolved gases, especially near the surface where most marine life is found (Chapter 11). Respiration depletes the oxygen and produces carbon dioxide, whereas photosynthesis (i.e., the production of food by plants) replenishes oxygen, but depletes carbon dioxide. So the concentrations of these two gases vary, depending on which of these two processes is dominant. In the sunlit surface waters where photosynthesis dominates, carbon dioxide is depleted and oxygen plentiful. Just below this surface layer reside many animals who can hide from predators in these darker waters, but who still feed on the food produced by the plants at the sur-

[6]You may think that a unit of ml/l should be parts per thousand. However, a milliliter of a gas has much less material than a milliliter of a liquid or solid. A milliliter of gas per liter of liquid is typically one or two parts per million by weight (Table 6.9).

Table 6.10 Comparison of the Abundances of Various Gases in Sea Water with Their Abundances in the Atmosphere

Gas	Abundance in Dry Air (%)	Abundance in Sea Water (ppm)	Ratio of Total Amount in the Oceans to Total Amount in the Atmosphere
N_2	78.0	12.0	0.004
O_2	21.0	7.0	0.01
CO_2	0.03	90.0	62.0

face. Due to the respiration of these animals, oxygen is depleted and carbon dioxide enriched. This region is called the **oxygen minimum layer** (Figure 6.9).

Biological activity has very little effect on dissolved nitrogen gas, because very few organisms can use nitrogen in this form. Typical values for nitrogen concentrations are around 10 ml/l, slightly below saturation. Argon is present only in very small concentrations, and being chemically inert, it is uninteresting anyhow.

Nutrients **Nutrients** are those materials needed by plants that are in short supply relative to their need. The nutrients are equally important, but much less abundant than the dissolved gases. Because of their short supply, the addition of relatively small amounts of nutrients to a plant community often results in a tremendous increase in plant activity. The nutrients of greatest general interest are those containing nitrogen and phosphorus.

The nutrient nitrogen comes in the form of the **nitrate** ion (NO_3^-), the nitrite ion (NO_2^-), or in some cases the ammonium ion (NH_4^+). A typical value for the concentration of nutrient nitrogen in the ocean is 0.5 parts per million by weight, with large fluctuation due to biological activity. Dissolved nitrogen gas (N_2) is *not* usable except by a few specialized nitrogen-fixing plants, and so it is not considered a nutrient. Plants usually obtain their phosphorus through phosphate (PO_4^{3-}). The concentration of phosphorus in sea water is typically 0.07 parts per million, again with wide variations. Other substances can be nutrients for certain organisms, such as the silica used in some skeletons.

The large variation in nutrient concentration is primarily the result of biological activity. Although nutrient nitrogen is in greater supply than phosphorus, it is also in greater demand for the fabrication of organic tissues. Consequently, the shortage of nutrient nitrogen usually is more constraining to plants than the shortage of phosphorus.

Like carbon dioxide, nutrients are also used by plants in the synthesis of new organic matter. Therefore, they are removed from the water during photosynthesis. They are returned to the water when these organic tissues are oxidized during respiration, which is performed either by the plants themselves, animals that have fed on these plants, or bacteria that are decomposing the organic materials.

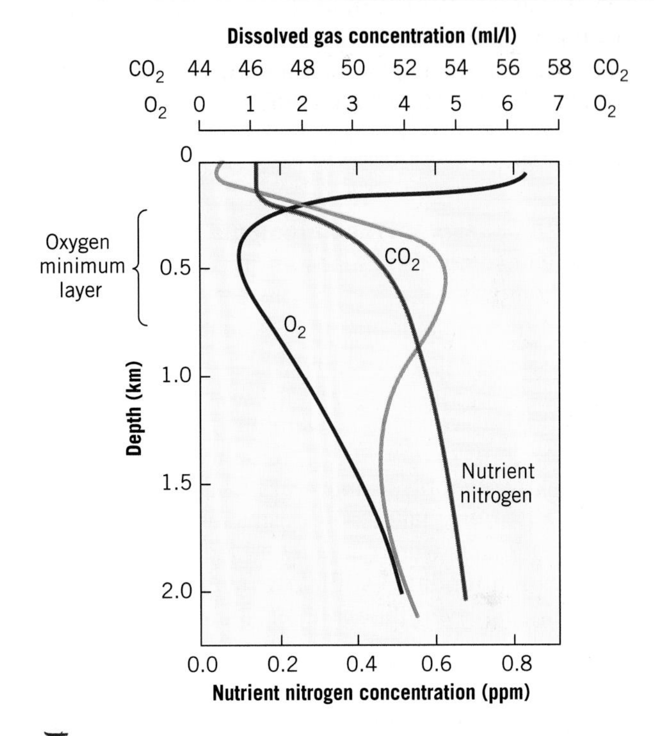

Figure 6.9

Plot showing a typical variation in the concentrations of dissolved oxygen, carbon dioxide, and nutrient nitrogen with depth. In the sunlit surface waters, plants consume carbon dioxide and nutrients while releasing oxygen. Immediately beneath this layer, respiration by animals releases carbon dioxide and nutrients, but consumes oxygen, creating the oxygen minimum layer. Deep water masses form at high latitudes and generally contain ample dissolved carbon dioxide and nutrients.

Since nutrients are consumed during photosynthesis and released during respiration, the variation in nutrient concentration parallels that of carbon dioxide (Figure 6.9). In the sunlit surface waters, photosynthesis removes the nutrients and carbon dioxide, by incorporating them into the organic matter produced. Just below this, animal respiration and metabolism release the carbon dioxide and nutrients, while removing the oxygen. At greater depth, there is less animal activity, less oxygen consumed, and fewer nutrients released.

Bacterial decomposition of organic detritus removes oxygen and releases carbon dioxide and nutrients, just as animal respiration does. Although this occurs at all depths, the influence of the bacteria is particularly prominent within the bottom sediments, because organic detritus that has escaped consumption near the surface tends to collect on the ocean floor.

Trace Elements **Trace elements** are those elements whose concentrations are measured in parts per billion or less. Concentrations are so small that it is sometimes a major accomplishment to detect certain trace elements at all, not to mention making a quantitative determination of their abundances. (That is why they are called trace elements.) The biological importance of certain trace elements was realized when early attempts to make artificial sea water for use in aquariums failed. Organisms died even when the various salts were carefully added to reproduce all known abundances. Clearly, some vital things were missing, even though their abundances in sea water were so small that they escaped detection.

The biological impact of trace elements is especially crucial for the more complex organisms with more complex and specialized systems, such as ourselves. Sometimes, this "crucial" effect is positive, in that an organism needs the material to survive. Sometimes, it is negative, in that an excess may cause harm or death (Figure 6.10). The negative effects are becoming a major problem in areas where human processes are polluting the marine environment with excesses of certain trace materials.

Because of their importance, organisms have become quite proficient at hoarding and concentrating certain trace elements in their own tissues. This ability to hoard clearly has been important to the development and livelihood of the organisms needing these trace elements. But now, with the increased pollution of our environment, it works to our detriment, as organisms are concentrating some poisonous pollutants in their tissues as well. These tend to get increasingly concentrated at each step of the food ladder, and by the time the organic material reaches secondary carnivores such as ourselves, it may be unfit to eat.

Figure 6.10
Many trace elements that are needed by organisms in small amounts become toxic when their concentrations in live tissues become excessive. The inherited ability of organisms to hoard certain needed trace elements was advantageous when these materials were scarce, but becomes detrimental when these same materials are encountered in more concentrated form, such as in certain pollutants. The concentrations of these materials increases at each step of the food ladder.

6.6 SALINITY VARIATIONS

Virtually all significant variations in salinity are caused by **physical processes**, such as freezing, thawing, precipitation, or evaporation (Figure 6.11). These processes simply add or subtract fresh water from the seawater solution, but have no effect on the salts. This is the reason why the various major constituents are always present in the same ratios. Since freezing, thawing, precipitation, and evaporation occur right on the surface, the surface waters show the greatest variations in salinity. Deeper waters remain fairly uniform.

For example, the surface salinity near land is usually low, due to freshwater runoff. Also, surface salinity is reduced at high latitudes, due to precipitation, snow and ice melt, and reduced evaporation. By contrast, the water tends to be salty near the tropics where the abundant sunshine evaporates lots of wa-

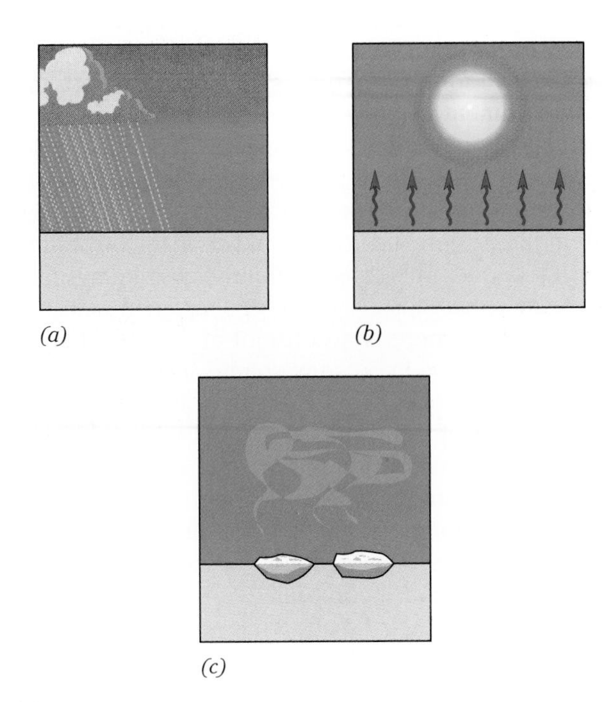

Figure 6.11

Illustration of some important physical processes that cause changes in the salinity of sea water. (a) Precipitation adds fresh water to the mixture. (b) Evaporation removes fresh water from the mixture. (c) The formation of sea ice removes fresh water from the mixture.

ter from the surface (Figure 6.12a). The salinity is not quite so high along the equator, because the daily afternoon showers return some of the fresh water to the ocean surface there. In temperate latitudes, there are seasonal variations in surface salinity, due to the seasonal changes in climate. Most deep waters were originally formed in polar regions, where the waters were cold but not exceptionally salty. Therefore, deep waters tend to be cold, but often not as salty as the surface in temperate and tropical regions (Figure 6.12b).

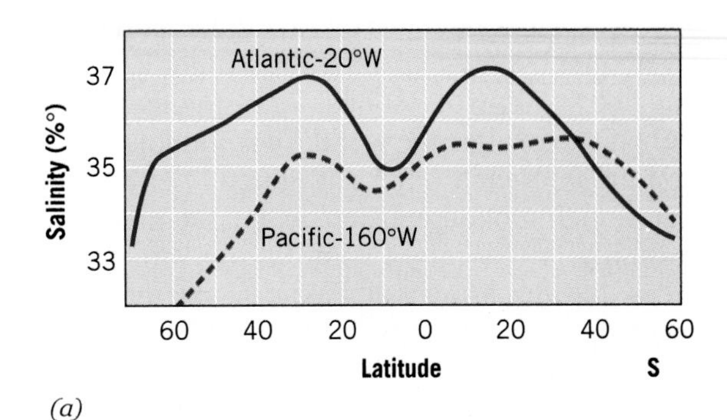

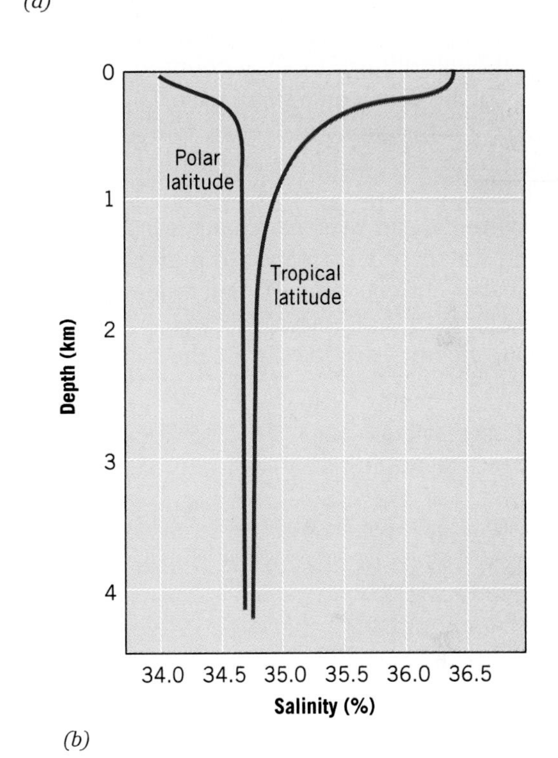

Figure 6.12

(a) Variation of surface salinity with latitude for the Atlantic and Pacific Oceans during summer in the Northern Hemisphere. Can you explain the dip in the curves near the equator and at high latitudes? (From Peter K. Weyl, 1970, *Oceanography*, John Wiley and Sons, New York.) (b) Typical plot of the variation of salinity with depth. In some regions, subsurface currents bring waters formed at different latitudes and in different seasons, causing corresponding wiggles in the lines. Can you explain the variations near the surface?

6.7 MIXING

As we have seen, the fact that the major salts are in the same ratios throughout the oceans indicates that the oceans are well mixed. There are two types of processes at work to accomplish this mixing: *Diffusion* works on a microscopic scale, and *convection* works on a larger scale.

Diffusion is the spreading of a material due to the thermal motion of the molecules. This motion is surprisingly fast on a molecular scale. A typical dissolved ion moves at several hundred meters per second! On a macroscopic scale, however, diffusion progresses very slowly, being typically a few millimeters per minute. The reason for this apparent contradiction is that these fast-moving salts keep bumping into water molecules. In fact, a typical salt ion would undergo over 10 billion collisions per second! After each collision, it bounces off in a different direction and frequently ends up where it started. It is something like an army of fast blind ants starting out from an ant hill to explore the world. They stumble around, bumping into each other and going in circles, and most of their motion ends up getting them nowhere. Gradually, a few might stumble out into the distances, but the army will spread at relatively slow speeds due to all the wasted motion. Similarly, diffusion of salts homogenizes a solution on a small scale, but it works very slowly at appreciable distances (Figure 6.13).

Although diffusion can account for the homogeneity of the ocean on a small scale, large-scale motion (i.e., convection) is needed to account for its mixing from one area to another. Near the surface, this mixing proceeds quickly, where the waves, surface currents, and seasonal changes insure that the surface waters of all oceans are thoroughly mixed. The deeper currents proceed much more slowly, requiring thousands of years to mix well.

One method used to estimate the mixing time for deep water employs **carbon-14** (Figure 6.14). This radioactive form of carbon is produced through the bombardment of our atmosphere by very energetic particles from outer space, called *cosmic rays*. It decays with a half-life of 5600 years, which means that at the end of any 5600-year period, the remaining abundance is one-half of what it was at the beginning of that period. The constant flow of cosmic rays into our atmosphere insures that it gets replaced as fast as it decays, so there is a constant amount of this radioactive carbon in our atmosphere at all times.

Once removed from the atmosphere, however, the carbon-14 continues decaying without being replaced. By measuring the amount of depletion of carbon-14 relative to the normal carbon-12, we can deduce how long it has been since the carbon was in the atmosphere. This is done, for example, to date fossils, which had incorporated carbon from the atmosphere into their tissues while the organism was alive. This method is also used to determine how long it has been since any particular deep water mass was at the surface and in contact with the atmosphere. The ability to date the ages of waters at

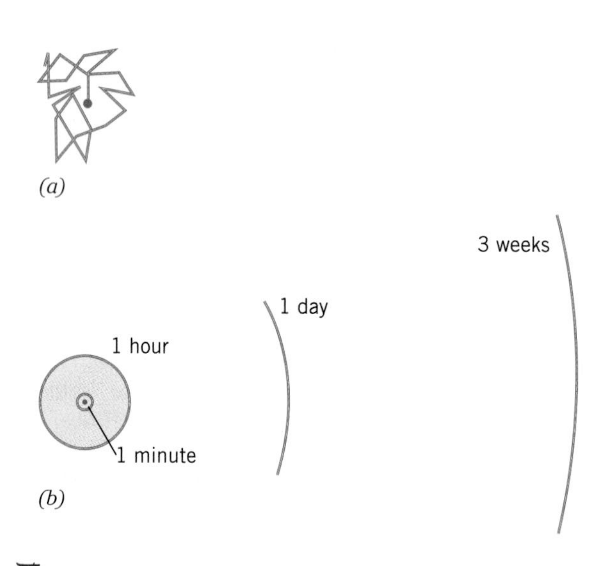

Figure 6.13

Diffusion. (a) On a microscopic scale, the individual molecules move at very high speeds, typically several hundred meters per second. However, their direction is random, changing directions after each collision. The result is that their net progress is small. (b) Full-scale representation of the progress of salt ions in water from a point source due to diffusion only (i.e., no water motion). The circles indicate the average distance from the source for the ions after 1 minute, 1 hour, 1 day, and 3 weeks, respectively. After 1 year, the average salt ions would be only 60 centimeters from the source.

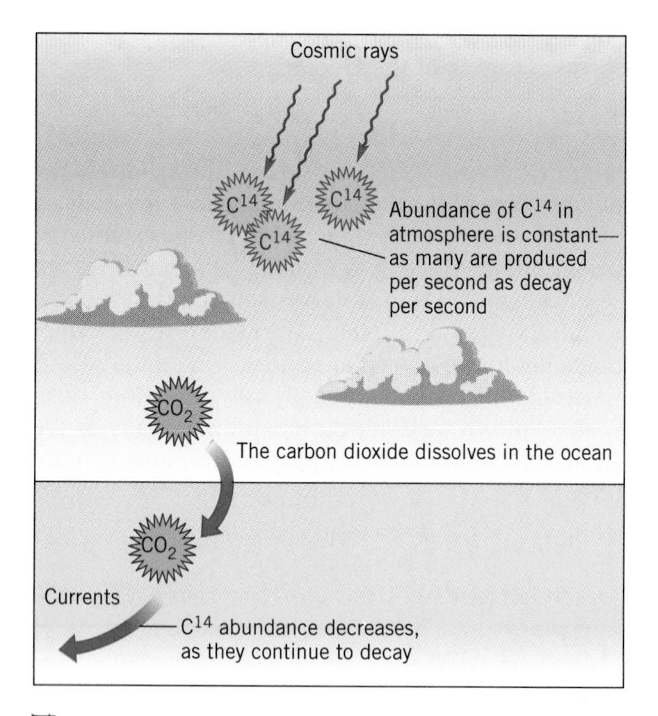

Figure 6.14

Illustration of how we can tell how long it has been since a particular water mass has been at the surface by the depletion of carbon-14 in the dissolved CO_2. When at the surface, the water absorbs carbon dioxide from the air. After it sinks beneath the surface, the carbon-14 decays without being replaced. The longer the water has been away from the surface, the less carbon-14 is left.

various places allows us to identify the evolution of the various deep water masses, without having to follow them for the thousands of years that they are underway.

Summary

The Water Molecule and Special Properties of Water

In the water molecule (H_2O), the shared electrons spend most of the time in the vicinity of the oxygen atom. This gives the hydrogen side of the molecule a net positive charge and the oxygen side a net negative charge. This unsymmetrical charge distribution makes the water molecule sticky and is responsible for a wide variety of unique properties, including very high melting and boiling points, high specific heat, high latent heats of fusion and vaporization, and a porous ice crystal structure. It also makes water an excellent solvent.

Cosmic Origins of the Ocean

Originally, the inner planets of our Solar System had no atmospheres at all, because they were too hot. Their present atmospheres were acquired by subsequent outgassing of gases trapped in their interiors. Larger planets should have larger abundances of these gases, due to greater geological activity. The very light and abundant gases hydrogen and helium cannot be retained in the atmospheres of the inner planets. These planets are the residue of denser materials left over after these two light gases are gone. Among the remaining light volatile materials, we would expect methane, ammonia, and water to be most prevalent. However, they are easily broken apart, the hydrogen escapes, and the remaining materials form carbon dioxide and nitrogen gases. Consequently, we would expect these two gases to dominate the atmospheres of the inner planets.

Indeed, this is true for all the inner planets except for Earth. Earth has managed to keep its water and lose its carbon dioxide. Most of Earth's water remains a liquid and on the surface, protected from the ultraviolet rays that cannot penetrate far through our atmosphere. Most of Earth's carbon dioxide dissolved in the water and then underwent chemical reactions that removed it from the water. Therefore, Earth's outer environment differs from that of the neighboring planets in that it has lost most of its carbon dioxide, but has *not* lost its water. Earth's atmosphere also differs from that of its neighbors in that it contains free oxygen, which has been put there by plants.

Sources of the Salts

Some of the salts in ocean water were dissolved from weathered crustal rocks. This is the source of most of the common metals. But many of the most prominent ingredients of ocean water, including the water itself, could not have come from this source. These *excess volatiles* were outgassed from the Earth's interior. The salts have always joined the oceans along with the water, so the oceans have always been about as salty as they are today.

Along the oceanic ridge, sea water flows down through cracks in the young hot rocks, becomes heated, and rises back up through hydrothermal vents. Its interaction with the hot rocks tends to deplete magnesium and sulfate from this water, and add calcium and other metals. Some heavy metals form insoluble sulfides, which deposit on the neighboring ocean floor as the vent water cools.

Materials in Ocean Water

Sea water contains particles in suspension, which provide a surprising amount of surface area and have a significant impact on both the chemistry of sea water and its biological activity. Dissolved materials are about 34 to 36‰ of sea water by weight. Of these, 99.7% are the *major constituents*. The remaining 0.3% is mostly dissolved gases, with nutrients and trace elements being present in even much smaller amounts.

The major salts are everywhere present in the same relative proportions. Only the amount of water in the mixture varies. By contrast, there are large variations in the abundances of the dissolved gases and nutrients. Organisms have pronounced influences on these concentrations. Deep water tends to be relatively rich in dissolved carbon dioxide and nutrients. Trace elements are present only in very tiny amounts, but are essential to life, especially the more complex forms. Organisms have the ability to hoard these necessary but rare items, which sometimes works to their detriment.

Salinity Variations and Mixing

Variations in salinity are due to physical processes only, such as freezing, thawing, evaporation, and precipitation. Since these occur right on the surface, surface waters show the widest variations in salinity. The fact that the major salts are everywhere in the same relative proportions indicates that ocean waters are very well mixed. On a microscopic scale, this mixing occurs through diffusion, and on a larger scale, convection does it. Large-scale mixing of deeper waters can be measured by using radioactive carbon-14.

Key Terms

polarization
hydrogen bond
absolute zero
absolute scale
Kelvin scale
specific heat
molecular mass number
latent heat of fusion
latent heat of evaporation
ionic salts
outer planets
inner planets
outgassing
excess volatiles
major constituents
‰
dissolved gases
ml/l
ppm
oxygen minimum layer
nutrients
nitrate
nitrite
trace elements
physical processes
diffusion
carbon-14

Study Questions

1. What property of a molecule does temperature measure? Which has more thermal energy: a cup of hot coffee, or an iceberg? Defend your answer.
2. Why is the specific heat of water so much higher than that of other materials whose molecules are of similar weight? Explain why water's latent heat of fusion and latent heat of evaporation are so high.
3. Why does water freeze over rather than under?
4. Why are many of the salts dissolved in the oceans in the form of charged ions rather than neutral atoms or molecules? How do the salt molecules get pulled apart? Why don't organic materials dissolve very easily in water?
5. As a salt dissolves in water, eventually the water becomes saturated. Explain why. Why can water hold more of some salts than others?
6. Why are Jupiter and Saturn so much larger than the inner planets?
7. Why does Earth have so much more surface water than Venus? Why does Earth have so much less carbon dioxide in its atmosphere than Venus?
8. What are *excess volatiles*? Where did they come from? How do we know?
9. What are the four general categories of substances dissolved in the oceans? Give at least two examples of each. What units are used for describing the concentrations of each? What fraction of all dissolved materials belongs to the category, *major constituents*?
10. Most organisms actually consume and release larger quantities of major salts than nutrients. Why, then, do biota have such a profound influence on the concentrations of the nutrients and negligible influence on the concentrations of the major salts?
11. Over the ages, some organisms have developed the ability to hoard some important trace elements within their tissues. Why might this be advantageous to them? How might this ability work to the organism's detriment in today's world?
12. If the speeds of dissolved salts at any instant are hundreds of meters per second, why would it take an hour for an average salt ion to get just 1 centimeter from where it started?
13. Explain how we can tell how long it has been since a deep water mass has been at the surface.

Critical Thinking Questions

1. How did Venus and Mars get their atmospheres? Why are they made primarily of carbon dioxide (CO_2) and nitrogen (N_2), rather than methane, ammonia, and water? Referring to Table 2.1, about how many times more abundant in the Universe is carbon than nitrogen? Oxygen than nitrogen? Why do you suppose carbon dioxide is more abundant in Venus' and Mars' atmospheres than nitrogen?
2. Give two reasons why a bucket of silt dumped into the ocean would have a greater influence on the ocean's chemistry than a bucket of gravel.
3. Referring to Figure 6.9, explain the variations in oxygen concentration with depth in the ocean. Do the same for carbon dioxide and for the nutrients.

4. Would land organisms or marine organisms be in greater danger of suffering from lack of one or more trace elements? Why?
5. Referring to Figure 6.12b, explain the variation of salinity with depth in tropical latitudes. Do you suppose the profile for the polar latitude was taken in late spring or late fall? Why?

Suggestions for Further Reading

CAMPBELL, A. C. ET AL. 1988. Chemistry of Hot Springs on the Mid Atlantic Ridge. *Nature* 335:6190, 514.

CAMPBELL, P. 1987. Solar System: Growing Planetary Atmospheres. *Nature* 327:6123, 554.

HOLLAND, H. D. 1984. *The Chemical Evolution of the Atmosphere and Oceans*. Princeton University Press, Princeton, New Jersey.

JENKYNS, HUGH C. 1993. Early History of the Oceans. *Oceanus* 36:4, 49.

KERR, R. A. 1988. Ocean Crust's Role in Making Seawater. *Science* 239:4837, 260.

KNAUTH, PAUL. 1993. Ancient Sea Water. *Nature* 362:6418, 290.

MACINTYRE, F. 1970. Why the Sea Is Salt. *Scientific American* 223:5, 104.

Nature. 1986. The Origins of the Earth's Oceans. 322:6082, 779.

SCHUHOF, JULIA. 1993. Old Salt Sea Tells Tales of the Past. *Sea Frontiers* 39:5, 11.

Sky and Telescope. 1984. Active Volcanoes on Venus? 67:2, 129.

TOGGWEILER, J. R. 1990. Bombs and Ocean Carbon Cycles. *Nature* 347:6289, 122.

A rainfall of 1 centimeter releases more heat than one entire day's sunshine.

Hurricane Gladys, about 240 km southwest of Tampa, Florida.

Seven

THE OCEAN AND OUR CLIMATE

Because of the remarkable properties of ocean water, the climate everywhere on Earth is extremely mild compared to that of our planetary neighbors. In this chapter, we study ocean–atmosphere interactions for the purpose of understanding the Earth's climate and the ocean's role in controlling it.

The moderating influence of the ocean is illustrated in Table 7.1 and Figure 7.1. Table 7.1 compares seasonal temperature variations for three Canadian cities at the same latitude. For all three, the winds tend to come from the west, so the West Coast city, Victoria, is more strongly influenced by the adjacent ocean than the East Coast city, St. Johns. Winnipeg is on the Canadian Great Plains and far removed from either coast. The moderating influence of the ocean is clear.

At 40° N latitude, the surface temperature of the Pacific Ocean changes by around 6°C or 8°C during the year. At the other extreme, the seasonal average temperature fluctuation on the interior of the Asian continent at the same latitude is about 38°C (from –4 to +34°C). But all such variations on Earth are small compared to the 120°C average seasonal and daily fluctuations on Mars, where there are no oceans at all (Figure 7.1).

7.1 OCEAN–ATMOSPHERE INTERACTIONS

Both the ocean and atmosphere are several thousand times wider than they are thick, which makes their relative dimensions like those of a thin sheet of paper. Like two neighboring pages in a closed book, these two fluid layers are intimately connected and their behaviors are closely related. For example, heat and moisture from the oceans drive the winds in the atmosphere. The winds, in turn, generate waves, drive surface currents, and influence the deep ocean currents as well.

Both the ocean and atmosphere are several thousand times wider than they are thick, which makes their relative dimensions like those of a thin sheet of paper. Like two neighboring pages in a closed book, these two fluid layers are intimately connected and their behaviors are closely related.

Table 7.1 Temperatures (°C) for Various Canadian Cities of the Same Latitude Showing Role of Oceans in Moderating the Climate

	Victoria	Winnipeg	St. Johns
Mean January Minimum	2	–22	–7
Mean July Maximum	20	27	21

Our program for this chapter will be to first study the important features of both the atmosphere and ocean surface waters. We then investigate the transfer of heat between the two in enough detail that we can understand the Earth's heat budget; that is, how the heat gets to the Earth, what happens to it while it is here, and how it leaves. Finally, we study wind and weather patterns. Wind and weather are driven by heat received mostly from the ocean, and as we will learn in future chapters, they return some of this energy to the ocean in the form of waves and surface currents.

7.2 THE ATMOSPHERE

STRUCTURE

In Chapter 2, we learned that pressure increases with depth in any fluid. Unlike the water of our oceans, the gases of our atmosphere are compressible. As the pressure increases, the gases are compressed, and as the pressure decreases, gases expand. One consequence of this is that there is no clean outer edge of the atmosphere as there is with the ocean. The atmosphere just gets thinner and thinner. It is impossible to tell where it stops and "outer space" begins (Figure 7.2). As a rule of thumb, both the pressure and density decrease by a factor of $\frac{1}{2}$ for every 6 kilometers of elevation.

We will learn in this chapter that the Earth's atmosphere is fairly transparent to the incoming sunlight. Therefore, incoming sunlight tends to heat the Earth's surface, but not the atmosphere itself. The warmed Earth then heats the atmosphere via interactions we will study in this chapter. That is, the heating of the atmosphere comes primarily from the Earth below, not the Sun above, and the oceans play the primary role in this heat transfer.

This heating from below drives the atmosphere's motions, so it is no surprise that the atmospheric layer closest to the Earth's surface has the most inter-

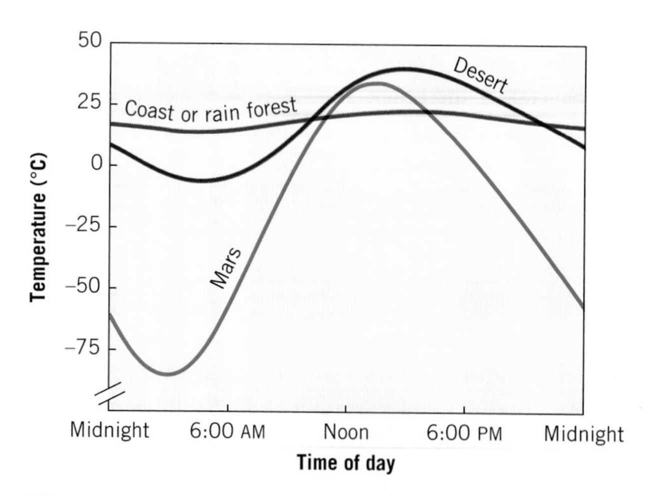

Figure 7.1
Comparison of typical daily temperature fluctuations in various environments. The remarkable thermal properties of water moderate the day–night temperature fluctuations near the coast and in rain forests. Contrast these temperatures with those on a desert far from a coast, and on Mars, where there is no liquid surface water at all. (The length of a day on Mars is $24\frac{1}{2}$ hours, which makes comparisons easy.)

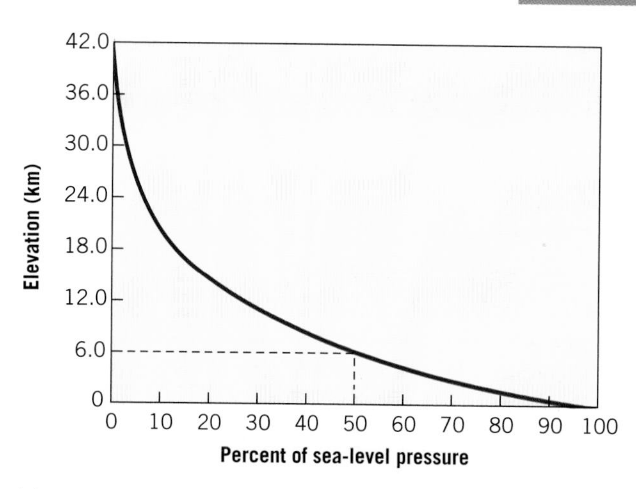

Figure 7.2
Plot of air pressure (or density) as a function of elevation. Notice that the pressure reduces by roughly a factor of $\frac{1}{2}$ for every 6-kilometer increase in elevation. (Atmospheric pressure at sea level is about 10^5 Newton/m^2, or 14.7 lb/in.^{2}.)

esting and dynamic activity, such as strong winds, storms, cloud formation, and so on. This lower layer is called the **troposphere**. It extends to an altitude of about 12 kilometers, and contains 75% of all air. One consequence of this activity is that the troposphere is well mixed in comparison to the more static layered structure of the thinner upper atmosphere.

TEMPERATURE PATTERNS

Another consequence of mixing is that the temperature of the troposphere decreases with increasing altitude. This temperature variation is caused primarily by **adiabatic processes**, which are those processes involving expansion or compression, but not the addition or removal of heat. Rising air encounters lower pressure, so it expands, and the expansion causes it to cool. Conversely, falling air encounters higher pressure, so it is compressed and heated. In this manner, the vertical motions of the air masses cause the temperature to decrease with increasing altitude, reaching about −60°C at the troposphere's upper boundary.

You are familiar with some of the consequences. For example, you undoubtedly know that air tends to be cooler in the mountains than at sea level, and snow tends to be found on the tops of mountains, and not at their bottoms. In fact, we can define the troposphere as the lower region of the atmosphere where the temperature falls with increasing elevation. Where this pattern stops marks the top of the troposphere, because that is where the vertical mixing stops.

There are some interesting variations in the atmosphere's temperature at elevations well above the troposphere. But these are not as significant as they might seem. At these high elevations, the atmosphere is very thin. Very minor amounts of heating can cause large temperature changes, because there is only a small amount of material for this energy to heat. The bombardment of the upper atmosphere by ultraviolet rays creates layers of certain types of materials that absorb only a very tiny fraction of the incoming solar energy. But the air is so thin there that even this tiny amount of heating causes a striking increase in temperature.

COMPOSITION

As we will see in the next section, the atmosphere's moisture content is small and variable. However, the relative abundances of the other gases are the same the world over. From Table 7.2, you can see that dry air is mostly nitrogen and oxygen, with only very small amounts of the other gases. The uniform composition demonstrates that the atmosphere is well mixed.

There are some important variations in the abundances of trace gases, especially in urban and industrialized areas, near active volcanism, and in the outer reaches of the atmosphere, where bombardment by ultraviolet rays and particles from outer space creates some interesting molecules and ions. Typical concentrations for such trace gases are parts

Table 7.2 Composition of Dry Air

Gas	Percent of Total Molecules in Air
Nitrogen	78
Oxygen	21
Argon	1
Carbon dioxide	0.03
Neon	0.002
Helium	0.0005
Krypton	0.0001
Hydrogen	0.00001
Xenon	0.00001
Others	Still Less

per million or less. So they cannot be considered major components of the atmosphere, even though their effects on life may be quite important.

MOISTURE CONTENT

The only major gas in the atmosphere whose abundance varies considerably from place to place and from time to time is water vapor. The reason for this variation is that the time scales for evaporation and precipitation are short compared to the time required for mixing to even out the varying concentrations. It takes weeks or months for the lower atmosphere to undergo significant mixing, as you can appreciate by watching the day-to-day motions of various air masses outlined in the evening weather reports. By contrast, through evaporation and precipitation, the moisture content of air at any one location can change significantly over the period of just a few hours. Therefore, mixing is too slow to even out variations in the air's moisture content over large distances.

The atmosphere's water content is extremely sensitive to temperature, for reasons that involve molecular structures and motion. We have seen that water molecules are very "sticky." They tend to stick together to form tiny water droplets (clouds, fog, dew, etc.), which grow as additional water molecules join them. In this way, they tend to remove themselves from the air in the form of precipitation. At higher temperatures, molecules move faster. The more forceful collisions make it less likely that two water molecules will stick together when they collide, and those that do may be broken apart again as other fast molecules collide with them. For this reason, at higher temperatures it becomes increasingly difficult for water molecules to stick together to form droplets, and so more moisture can remain in the air. In a very hot, steamy environment, such as a tropical forest right after a heavy rain, the atmosphere's water vapor content can be as high as 4%, but that is exceptional. Normally, it is below the 1% level and has a world average of less than $\frac{1}{2}$% of the gases in the troposphere. It is extremely rare above the troposphere, because the cool temperatures at the top of the troposphere act as a "trap" that forces the water vapor to condense by that point and not rise any further.

Table 7.3 Water Vapor Content of Saturated Air as a Function of the Temperature

	Temperature (°C)						
	0	5	10	15	20	25	30
Water vapor content when saturated g/m^3	7.7	11	15	21	28	37	50

When air is holding all the moisture it can, we say that it is "saturated," a word we have encountered before with reference to sea water holding as much of a particular salt as it can.[1] The maximum amount of moisture that air can hold increases roughly by a factor of 2 for every 10°C increase in the temperature. For example, from Table 7.3 you can see that if air at 20°C is holding 28 grams of water vapor per cubic meter of air (about 2.3% of the air's total composition), then it is saturated. If you were to cool down this air to 10°C, then it would be holding too much moisture, and about half the water would condense and precipitate out as fog, dew, rain, and so on.

7.3 THE OCEAN SURFACE WATERS

OVERALL CHARACTERISTICS

Interactions between the ocean surface waters and atmosphere dominate the Earth's climate. Because of turbulence and mixing that are caused by wind, waves, density variations, and surface currents, these interactive surface waters extend to depths of several hundred meters. As is illustrated in Figure 7.3, the surface waters in the tropics tend to be warm and salty, due to heating and evaporation by the intense sunlight in these latitudes. In temperate latitudes, the surface waters vary with the seasons, but are always warmer and lighter than the deep waters below. In

[1]When the air is holding all the moisture it can, we also say that the relative humidity is 100%.

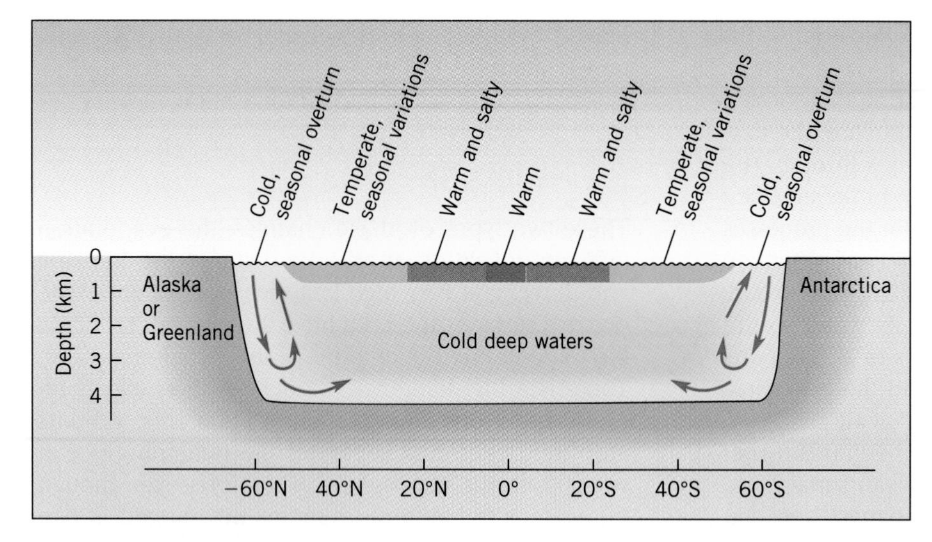

Figure 7.3

Cross section of an ocean, showing variation of surface waters with latitude.

the polar regions, the surface waters are cold. During the winter, they cool still further, becoming sufficiently dense to sink and mix with the deeper waters.

The polar water's influence on the world's climate is much smaller than that of the temperate and tropical waters for several reasons. First, the tropical and temperate regions encompass a much greater fraction of the ocean surface, since the world is much fatter at the equator than it is at higher latitudes. Second, large regions of polar waters are covered by ice, which insulates them from the atmosphere and prevents their influencing the climate. Third, the processes that exchange heat between ocean and atmosphere operate much more slowly when the water is cold.

THERMAL INERTIA

The ocean's dominant moderating influence on our climate is due to its **thermal inertia**. That is, the ocean's temperature tends to remain steady; the ocean does not undergo the large day–night or seasonal temperature changes that other substances would display. The reasons for this can be traced to the following properties of water:

1. Specific heats
2. Light penetration
3. Mixing
4. Phase changes

The specific heat[2] of water is much greater than that of the Earth's crustal materials. For example, if you put equal amounts of heat into equal masses of soil and water, the soil will heat up five times faster. The soil will also cool off five times faster when equal amounts of heat energy leave. Both ocean and land experience the same Sun in the day, the same darkness at night, and the same seasonal variations in sunlight. But water has a much higher specific heat than land, so the oceans experience correspondingly much smaller temperature fluctuations.

On land, heating and cooling occur only at the surface. You may have experienced the difficulty in walking barefoot on a beach on a hot afternoon. Burning feet can be cooled by pushing them just a few centimeters beneath the surface, where the sand is refreshingly cool.

On land, heating and cooling occur only at the surface. You may have experienced the difficulty in walking barefoot on a beach on a hot afternoon. Burning feet can be cooled by pushing them just a few centimeters beneath the surface, where the sand is refreshingly cool. Virtually all the temperature fluctuations take place right at the surface of the sand. By contrast, sunlight penetrates several meters into the ocean. Compared to land where it is concentrated at the surface, the Sun's heat is distributed throughout a much thicker surface layer in the oceans. The actual thickness of this directly heated layer depends on how far the sunlight can penetrate, ranging from less than a meter in very turbid coastal waters to several tens of meters in very clear oceanic waters.

[2]As explained in Chapter 6, *specific heat* is the amount of heat energy needed to raise the temperature of 1 gram of a substance by 1°C.

Mixing by waves and turbulence also helps distribute the Sun's heat through the surface waters. This is particularly crucial in tempering the cooling of surface waters at night. Although much of the daytime incoming solar energy penetrates through the water and would be distributed even in the absence of mixing, the same cannot be said for the processes that remove heat (evaporation and radiation), which operate at the very surface only. If there were no ongoing mixing processes bringing new water to the surface, the very thin surface layer of the oceans would get much colder at night, and the water just beneath the surface wouldn't cool off at all.

Another extremely influential mechanism that moderates the temperature of the ocean surface water is **phase change**. This includes changes between liquid and vapor phases at all latitudes and those between liquid and solid phases at very high latitudes. Phase changes are the single most important mechanism for the transfer of energy between the ocean and atmosphere. Of the solar energy absorbed by the world's oceans, more than half goes into evaporating water, and less than half into increasing the water's temperature. That is, more than half turns into latent heat and less than half into **sensible heat** (i.e., temperature changes you can "feel"). This is particularly true in the tropics, where both warmer air and more intense sunshine cause greater evaporation.

In the daytime, the removal of latent heat by evaporated water tends to keep the surface cool, in spite of intense sunshine. At night, the reverse process occurs. As the night air cools, moisture condenses. On land, it often condenses as dew on weeds or blades of grass, whereas on the oceans, it condenses directly back onto the ocean's surface. This condensation releases the latent heat back to the ocean, keeping the water surface warm at night. In short, the evaporation of water from the ocean's surface keeps the surface cool during the day, when it should be warming, and the condensation of moisture onto the surface keeps it warm at night, when it should be cooling. Thus, evaporation and condensation tend to reduce temperature fluctuations. Seasonal temperature variations are moderated in the same way.

At very high latitudes, it is primarily the freezing and melting of ice that moderates the ocean's surface temperature. In winter, ice forms on the cooling surface. This releases latent heat, which keeps the remaining water from getting colder. During the summer, the reverse happens. As the ice melts, it absorbs heat from the water, keeping it cool. Ice is used to keep drinks cool for the same reason. Notice how this process parallels that at lower latitudes. At higher latitudes, latent heat of fusion is absorbed and released by thawing and freezing, and at lower latitudes, latent heat of evaporation is absorbed and released by evaporation and condensation.

ICE INSULATION

These two types of phase changes, the evaporation and condensation at all latitudes and the freezing and thawing at very high latitudes, occur right at the ocean's surface, and both have the effect of reducing the temperature fluctuations of the remaining water. There is a big difference, though, in their effects on moderating our climate. When ice forms, it floats atop the ocean and separates the remaining ocean water from the atmosphere. Therefore, even though the water's temperature remains unchanged, it can no longer influence the climate. Consequently, interior areas on the Arctic ice sheet that are far from open water experience some extremely harsh temperature variations.

7.4 ENERGY TRANSFER BY RADIATION

Having reviewed the important properties of the atmosphere and the ocean's surface waters, we can now study the interactions between the two. We begin with the transfer of heat, which is carried out via the processes of conduction, radiation, and phase change. **Conduction** is the transfer of heat by contact, such as happens when you put a cold pot on a hot stove burner. When hot, fast-moving molecules collide with cold, slow-moving ones, energy is transferred to the colder ones.

> Air is a very poor thermal conductor. For this reason, materials that provide thermal insulation, such as blankets, sweaters, jackets, and building insulation, are full of tiny air pockets.

Conduction is by far the least important of the processes that transfer heat between ocean and atmosphere, because air is a very poor thermal conductor. For this reason, materials that provide thermal insulation, such as blankets, sweaters, jackets, and building insulation, are full of tiny air pockets. Measurements of heat transfer from ocean to atmos-

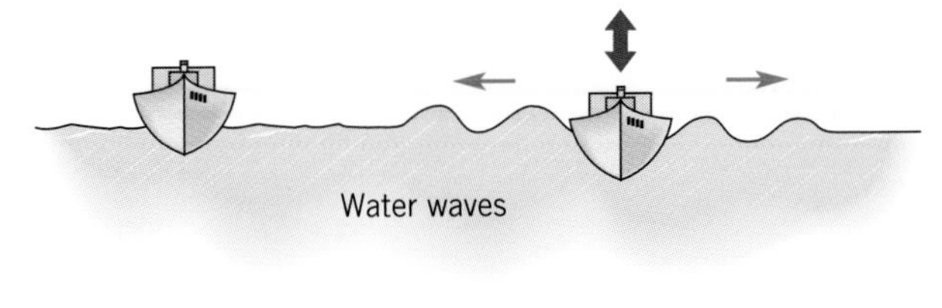

Figure 7.4

When you bob one toy boat, it sends out waves. Other toy boats then also bob as these waves pass by. In a similar fashion, electromagnetic waves are generated and detected by wiggling electrical charges. Whereas water waves travel relatively slowly across the surface of water, electromagnetic waves travel extremely quickly through space in all directions.

phere indicate that on the average over the world, conduction accounts for only 6% of the energy transfer, whereas radiation accounts for 41%, and evaporation/condensation for 53%.

We now examine **radiation**, a process that can transfer thermal energy between widely separated objects. Not only is it one of the two dominant processes for the transfer of heat between ocean and atmosphere, but it is also the process by which the ocean receives its energy from the Sun in the first place, and the mechanism by which Earth exhausts its excess heat back into outer space. Energy transfer by radiation is also crucial in some of the more local and time-varying climatic behaviors, and many of the environmental problems we are facing.

ELECTROMAGNETIC WAVES

All things radiate energy in the form of **electromagnetic waves**. These waves are generated by accelerating electrical charges, such as the wiggling atoms in the filament of a hot lightbulb, or the oscillating electrical currents in a radio broadcast tower. These waves travel through space in the form of oscillating electric and magnetic fields. They can be absorbed by other electrical charges they encounter. In other words, if you jiggle an electrical charge somewhere, it sends out electromagnetic waves, which then cause other electrical charges to jiggle as these waves pass by.

Electromagnetic waves are in some ways similar to the water waves generated when you jiggle a toy boat in a bathtub (Figure 7.4). If you move the toy boat up and down slowly, you generate long, low-energy waves. If you jiggle the boat up and down quickly, you create short, choppy, high-energy waves. As these waves travel across the tub, they may encounter other toy boats, making them bob up and down as they pass under. Shorter, higher-energy waves make these boats bob up and down more quickly than do the longer, lower-energy waves.

In a similar fashion, rapidly jiggling electrical charges, such as those found in the atoms and molecules of very hot objects, create electromagnetic waves of short wavelengths and high energies. The slower-moving atoms and molecules of cooler objects generate longer, lower-energy electromagnetic waves. Once created, these waves travel through space. If they encounter other charged particles, they make them jiggle as they pass by. This is why you feel warm when standing in sunshine, and it is how your radio antenna picks up signals sent out from a broadcast tower.

Unlike water waves, electromagnetic waves of all wavelengths travel at the same speed through space: 300,000 kilometers per second. This is extremely fast. In fact, everything in the atomic world seems fast relative to human time scales. For example, radio waves typically have **frequencies** (i.e., rates of oscillation) of millions of cycles per second, and light waves from the Sun have frequencies nearly a million billion (10^{15}) cycles per second. Consequently, the "toy boat in a bathtub" analogy for electromagnetic waves must be thought of as being run in very slow motion and very highly magnified.

THE SPECTRUM

Electromagnetic waves are placed into several broad categories according to wavelength, with **radio waves** being the longest, and *high-energy x-rays* or **gamma rays** being the shortest. In between these two extremes are **microwaves**, **infrared waves**, **visible light**, **ultraviolet rays**, and **x-rays**, as indicated below:

Wave	Typical Wavelength
Radio waves	meters or kilometers
Microwaves	oven size or smaller
Infrared waves	hair width size
Visible light	a hundredth of a hair width
Ultraviolet rays	a few molecules across
x-rays	smaller than an atom
Gamma rays	smaller than an atomic nucleus

A more accurate definition of categories is provided in Figure 7.5.

Shorter wavelengths correspond to faster oscillations (i.e., higher frequencies) and more energy. For example, radio waves and microwaves are long, low-

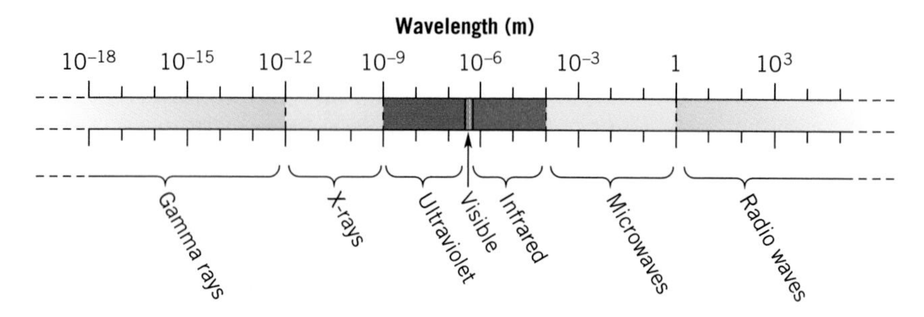

Figure 7.5

Electromagnetic waves come in all wavelengths, from the very long radio waves to the very short wavelength gamma rays. Visible light encompasses only a very small part of the total spectrum. For an expanded view of the visible part of the spectrum, see Figure 7.7. (The notation used is powers of 10; it indicates where the decimal belongs: $10^3 = 1000$, $10^{-3} = 0.001$, etc.)

energy waves and have no known biological consequences. But infrared waves carry enough energy to trigger your sensory nerves (i.e., you can feel the heat), and visible light can trigger your optic nerves as well. Ultraviolet radiation has even more energy and can give you a sunburn. X-rays and gamma rays can burn you all the way through, because these more energetic forms of electromagnetic waves carry enough energy to actually damage or destroy the rather delicate organic molecules within you.

We are particularly interested in visible light, both because it is the range of wavelengths that we see, and it is the region in which most of the incoming solar energy is found. The visible band is very narrow, including wavelengths only in the range from about 0.4 to 0.7 μm.[3] Our eyes perceive different wavelengths within this visible band as different colors, ranging from violet, which has the shortest wavelength and highest energy, to red, which has the longest wavelength and lowest energy. This makes it obvious why the bands bordering the visible range on the two sides are called *ultraviolet* and *infrared*, respectively.

THERMAL MOTIONS

The thermal motion of atoms and molecules in materials is very chaotic. At any instant, some are moving or vibrating rapidly, and others more slowly. So electromagnetic waves of all wavelengths are generated (Figure 7.6a). But the general tendency is for hotter objects to generate shorter, higher-energy waves, due to the more rapid thermal motions.

The surface of the Sun, for example, is at a temperature of about 5600 K, and it generates waves mostly in the visible range, with somewhat less infrared and ultraviolet, considerably less microwaves and x-rays, and so on. In contrast, the surface of the Earth is much cooler, being around 290 K on the average, so the Earth's radiation is concentrated in the infrared range (Figure 7.6b).

TRANSPARENCY AND ABSORPTION

We have seen that the energy of electromagnetic waves can be absorbed by charged particles they encounter. Frequently, charged particles are bound up in atoms and molecules. The manner in which they are bound may allow them to absorb certain energies, but not others. Therefore, these charged particles may absorb electromagnetic waves of certain frequencies, and not others.

For example, the electrical charges in window glass cannot oscillate with the frequencies of visible light. Therefore, they cannot absorb these wavelengths, and visible light passes through without obstruction. We say that window glass is *transparent* to visible light. These same electrical charges in window glass can absorb energy at other wavelengths, however. For example, high-energy ultraviolet rays cannot pass through. We say the glass is *opaque* to this ultraviolet. One consequence of this is that although you can see through window glass just fine, you cannot get a sunburn through it. It transmits visible light, but not the more energetic ultraviolet.

TRANSMISSION THROUGH WATER

Sea water is a fairly good electrical conductor, and the electrical charges within it have a great deal of freedom. As a result, most electromagnetic radiation incident on water is quickly absorbed. Solar infrared radiation is absorbed within the first few millimeters of the ocean's surface, and it would cause significant warming of this thin surface layer were it not for

[3]The symbol μm stands for *micrometer*, which is equal to one-millionth of a meter, or equivalently, one-thousandth of a millimeter, or roughly one-twentieth of a hair width.

waves and other turbulent mechanisms that mix this heat downward quickly.

Microwaves are also quickly absorbed by water. In microwave ovens the microwave energy is absorbed by the water in the food that is being cooked. If you put soup or other material with a high water content in the oven, you will notice that the outer few millimeters heat up quickly, demonstrating that microwaves are mostly absorbed within the first few millimeters of water, just like infrared.

The absorption of radio waves in the oceans makes communications with submarines difficult. For research submersibles, the communication can be transmitted from surface ships, via sound or through connecting transmission cables. But for naval submarines, the presence of the surface ship would be a dead giveaway to the submarine's position. In order to remain hidden, military submarines must remain submerged, and this ensures that they cannot communicate with the outside world. Communications for command and control of these submarines is an obvious problem, which could become quite dangerous in crisis situations. So far, no one has been able to solve this problem.

Microwaves are also quickly absorbed by water. In microwave ovens the microwave energy is absorbed by the water in the food that is being cooked.

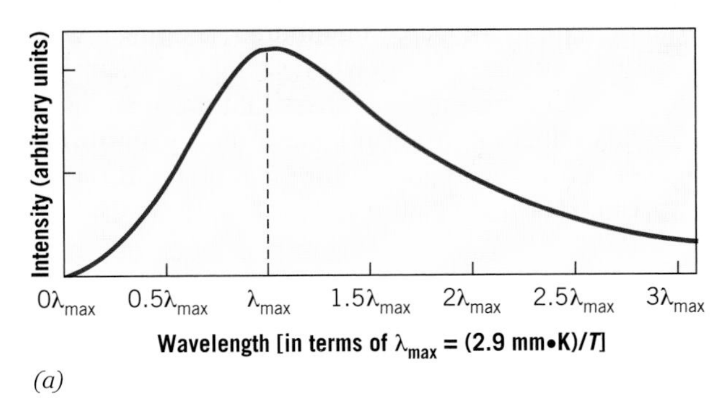

(a)

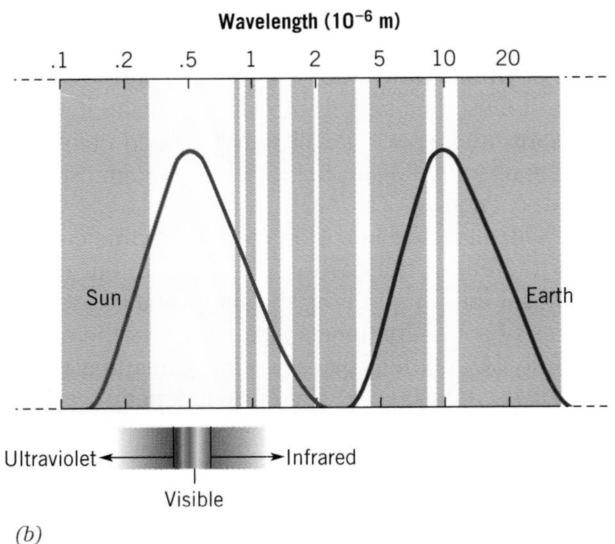

(b)

Figure 7.6

(a) Plot of intensity of radiated energy vs. wavelength. Any object radiates at all wavelengths, but peaks at a wavelength (λ) determined by λ= (2.9 mm)/T, where temperature T is measured in degrees Kelvin. Thus, higher temperatures mean shorter peak wavelengths. (b) Plot of intensity vs. wavelength for radiation received from the Sun and emitted from the Earth. Shaded areas are wavelengths for which the Earth's atmosphere is opaque. Opacity in the ultraviolet is largely due to ozone (O_3), and opacity in the infrared is largely due to carbon dioxide (CO_2) and water (H_2O) in the atmosphere.

Although water is opaque to most incident electromagnetic waves, it is fairly transparent to a small band of wavelengths centered in the visible part of the spectrum. The penetration length for light in water depends both on how murky the water is and on the color of the light. For clear ocean water, blue-green penetrates the best, going nearly 40 m before its intensity is reduced by 50%. The extremes of the visible spectrum fare less well (Figure 7.7). Red and violet go only about 4 m with 50% reduction in inten-

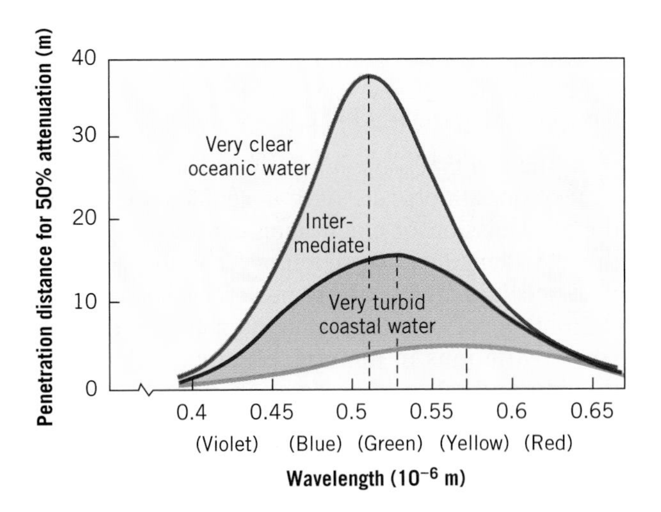

Figure 7.7

Plot of the distance that light can travel through sea water before being diminished by 50% for the various wavelengths. The better a color travels through the water, the more of that color we see when we look into the water. Notice that very clear water tends to be blue-green and very turbid water yellow-green.

sity. This explains why thin layers of water look clear, but thicker layers look blue-green. All colors make it through thin layers, but only the blue-green part of the spectrum makes it through thicker layers.

TRANSMISSION THROUGH AIR

The major components of our atmosphere—nitrogen, oxygen, and argon—are transparent to a large portion of the electromagnetic spectrum, extending from radio waves through visible light. However, there are some rather minor components of the atmosphere, such as water vapor and carbon dioxide, that absorb waves in large portions of the ultraviolet, infrared, and microwave parts of the spectrum. Due to the complexities of these absorbing molecules, however, there are some bands of wavelengths to which the atmosphere is transparent. These transparent bands are appropriately called windows. It is through "windows" in the infrared, for example, that infrared photography, and night vision telescopes are possible.

Nonetheless, most of the ultraviolet, infrared, and microwave wavelengths are absorbed by these minor constituents of the atmosphere. If you tried to "look" through our atmosphere in these wavelengths, you wouldn't be able to see very far. One consequence of this is that Earth-based telescopes use either radio waves or visible light, because the other wavelengths don't make it down through our atmosphere. If we want to look into outer space using microwaves, infrared, or x- rays, for example, we have to put these telescopes in Earth-orbiting satellites, so their "vision" is not obscured by our atmosphere.

THE GREENHOUSE EFFECT

The fact that some relatively minor constituents of our atmosphere have a significant effect on the transmission of electromagnetic waves means that the alteration of these minor components by human activities might have some serious environmental consequences. In particular, human activities that cause variations in the concentrations of carbon dioxide and other trace gases in our atmosphere have a significant impact on the atmosphere's ability to transmit infrared radiation. This gives rise to a concern commonly referred to as the *greenhouse effect*, or *global warming problem*. In addition, the release of certain chemicals used as propellants and refrigerants has an effect on the concentrations of *ozone*, which is responsible for protecting us from incoming solar ultraviolet radiation. We will study each of these problems in a little more detail now.

Infrared radiation is absorbed by water and carbon dioxide molecules in the lower atmosphere (Figure 7.6b).[4] This doesn't have much effect on the solar radiation coming in, because most of the solar radiation is visible light. But it does tend to trap this heat on the Earth's surface, because the water and carbon dioxide block the Earth's infrared radiation from getting back out. In this way, water and carbon dioxide help make the atmosphere act like a one-way blanket, allowing energy to come in but obstructing its departure. This keeps the Earth's surface and lower atmosphere warm.

This is called the **greenhouse effect**, because greenhouses work essentially the same way. The glass is transparent to the incoming solar radiation, but opaque to the outgoing infrared, keeping it warm inside. Your car heats up when sitting in the sunshine for the same reason. The gases that are responsible for this effect are appropriately called **greenhouse gases**. Carbon dioxide is the main greenhouse gas, but others listed in Figure 7.8 also contribute.

All the heat that comes in goes back out, but it hangs around for awhile before leaving. Any energy radiated outward from the surface stands a good chance of being absorbed by a water or carbon dioxide molecule in the atmosphere before making it through. When this molecule reemits this energy, it could go in any direction, so there is a 50-50 chance that it will go downward toward the Earth. If it does this, it must then start its trip all over again. If it goes toward outer space, it still stands a good chance of being absorbed again higher up in the atmosphere. And so it goes.

Our heavy reliance on fossil fuels this century has increased the carbon dioxide content of our atmosphere. In addition, the deforestation of large areas of some continents also increases the atmospheric CO_2 content by removing plants that would otherwise consume it. The current general warming trend in our Earth climate appears to be the result of this increase in carbon dioxide and the other greenhouse gasses (Figure 7.8), although there is still disagreement regarding other factors that may also be contributing. In Figure 7.9, you can see that human processes are just a very small part of the Earth's overall CO_2 budget. Nonetheless, this small part may be tipping an otherwise very delicate balance. We don't know how much we can play with these factors and still have an inhabitable planet, and probably we don't want to find out. But you can see why there is some concern.

[4]These are not the only molecules in our atmosphere that absorb infrared, but they are the main ones.

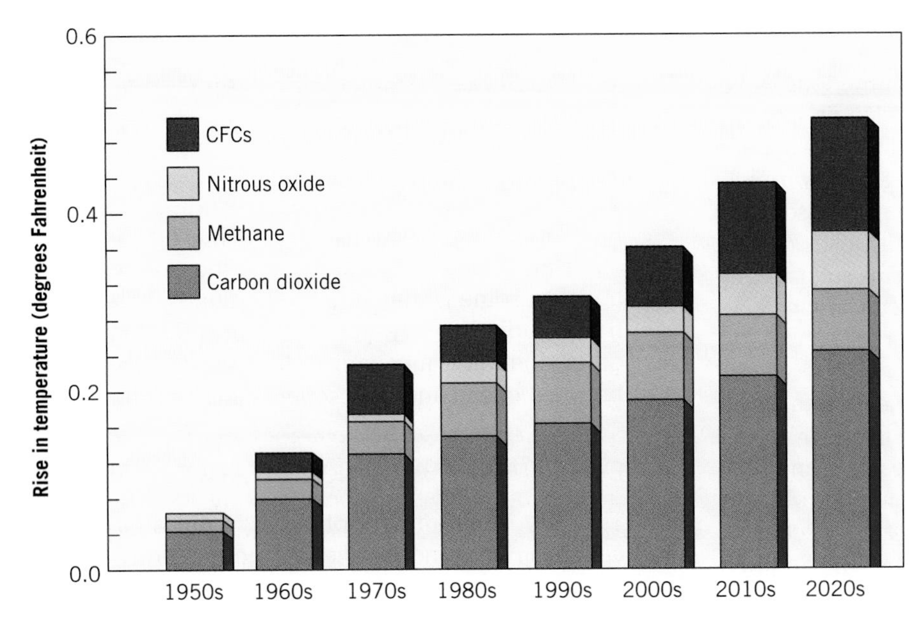

Figure 7.8

Estimates of global rise in temperature due to various gases from human processes. Although carbon dioxide will remain the main culprit, other gases will become increasingly important. CFCs stands for *chlorofluorocarbons*, which are used as propellants and refrigerants. (After Revkin, 1988, *Discover* **9**:10, 55. Data from the Electric Power Research Institute.)

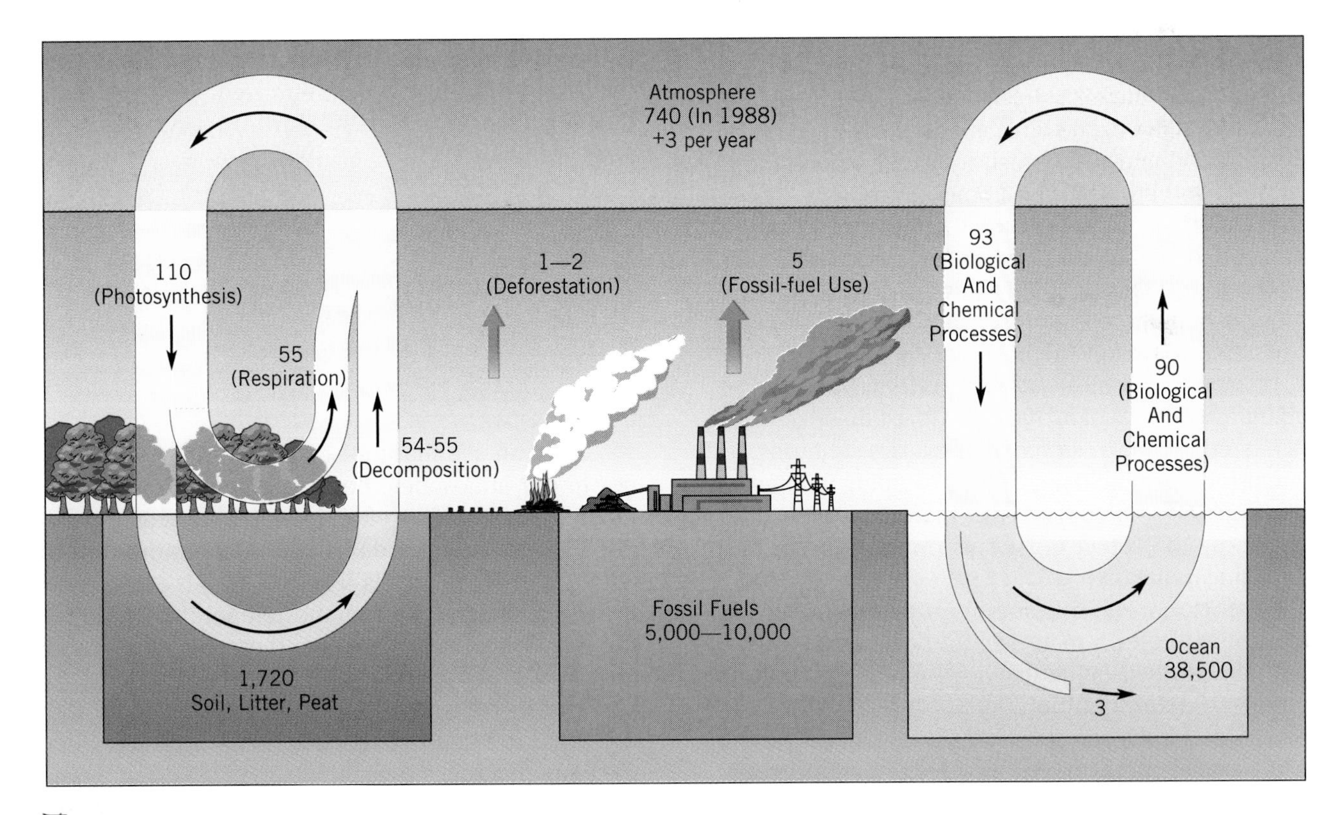

Figure 7.9

Natural and human processes that affect the abundance of carbon dioxide in the atmosphere. Numbers are in billions of metric tons of carbon, for both the reservoir contents and the annual fluxes into and out of the atmosphere. Natural processes remove about as much per year as they add. Although the input from human processes is small in comparison, it upsets the otherwise delicate balance, with the result that the net amount of carbon dioxide in the atmosphere increases by about 3 billion metric tons per year. (After Stephen H. Schneider, 1989, *Scientific American* **261:3**, 73. Data from Bert Bolin.)

THE OZONE PROBLEM

We now turn our attention from the long-wavelength infrared to the short-wavelength ultraviolet. Whereas the absorption of infrared by certain trace gasses causes concern over possible global warming, the absorption of ultraviolet by ozone has essentially no impact on our climate. Rather, it poses certain threats to the health of living organisms such as ourselves and those that live near the ocean surface.

High in our upper atmosphere, some oxygen (O_2) molecules are split into individual oxygen atoms by incoming high-energy ultraviolet rays from the Sun. These oxygen atoms are very reactive and tend to combine chemically with a large number of materials. Some combine with other oxygen molecules to form molecules of O_3, called **ozone**. The requirements of a sufficient supply of both oxygen and high-energy ultraviolet rays mean that the ozone is found mostly in a rather narrow band in our upper atmosphere. Above this band, the oxygen is too sparse, and below this band, the ultraviolet is too sparse, having been absorbed by the oxygen above. Any ozone that is produced in the lower atmosphere doesn't last long, because the air is thick. Molecules collide frequently, and the ozone molecules quickly find something appropriate onto which they can unload their extra oxygen atom. In the upper atmosphere, however, the air is thin. Molecules collide more infrequently, so the ozone lasts longer. Because of the continual irradiation by high-energy ultraviolet rays from the Sun, there is always a very small but significant amount of ozone in the upper atmosphere.

Ozone (O_3) itself is extremely effective in absorbing ultraviolet rays of all types—not just those of very high energy. Ultraviolet radiation causes sunburns. It can also do genetic damage. By absorbing incoming ultraviolet rays, the ozone layer performs the service of helping to protect organisms on Earth from this damaging radiation.

Unfortunately, our society is releasing trace amounts of certain materials into our atmosphere that have a devastating effect on the ozone. These materials include chlorofluorocarbons, which are used as propellants in spray cans and as refrigerants. In the upper atmosphere, they serve as **catalysts** (i.e., chemical helpers) to remove the extra oxygen from O_3 molecules, turning them back into O_2. The effectiveness of these catalysts is enhanced by microscopic ice crystals, so the depletion of the ozone from the upper atmosphere is most pronounced high above Antarctica, where it is referred to as the *hole in the ozone layer*. The depletion of ozone increases our exposure to the damaging ultraviolet radiation, which is the reason for our concern.

7.5 PHASE CHANGES AND OUR CLIMATE

In the preceding two sections, we studied heat transfer by radiation. We learned that visible light from the Sun warms the ocean surface, and the ocean surface transfers some of this heat to the atmosphere by infrared radiation. We now turn our attention to the other dominant mechanism for the transfer of heat between ocean and atmosphere: phase change.

EVAPORATION AND PRECIPITATION

The very small amount of water in our atmosphere carries a huge amount of energy in the form of latent heat (Chapter 6)—the heat that was required to evaporate it. When moisture condenses, the energy originally used for evaporation is released. This is one major reason why foggy nights are so much warmer than clear nights. A rainfall of 1 centimeter releases more heat to the Earth's surface and atmosphere than one entire day's sunshine. In fact, there is enough latent heat in the Earth's atmosphere right now to equal that received by the entire Earth in 4 days of sunshine!

A rainfall of 1 centimeter releases more heat to the Earth's surface and atmosphere than one entire day's sunshine.

You can see that the small amount of water vapor in our atmosphere has a huge effect on our climate. It is the major reason why arid deserts experience so much wider day–night temperature fluctuations than damp forests (Figure 7.1). It is also one of the main reasons why inland temperature fluctuations are so much larger than those on or near the oceans. The evaporation of water on hot days absorbs heat, reducing the temperature rise, and the condensation in cooler times releases heat, reducing the temperature fall.

To better understand evaporation and condensation, consider the motions of the water molecules. Due to collisions, these motions are quite chaotic, and at any instant some move faster than others. Some travel fast enough to burst free of the water surface and join the atmosphere as water vapor. This is how water evaporates. At higher temperatures, more molecules are moving fast enough to evaporate, and so the water evaporates faster. Since the fastest-mov-

Figure 7.10
Air pressure decreases with altitude. Rising air expands, cools, and moisture condenses. Conversely, falling air is compressed, heated, and dry. Therefore as air passes over an island, land mass, or mountain range, the rising air on the windward side produces precipitation, whereas the falling air on the leeward side produces arid conditions.

ing molecules are the ones that evaporate, those left behind are slower-moving, on the average, which means that the temperature of the remaining water is lower. This is why evaporation is a *cooling process*, and it is why you feel cold when you step out of a shower. It is also one main reason why the ocean's temperature remains cool even after an entire day's or summer's sunshine. The fastest-moving molecules join the atmosphere, taking their energy with them as latent heat. They leave the slower-moving ones behind so that the remaining water is colder. Averaged over the world, 53% of the heat transfer from ocean to atmosphere is accomplished by evaporation.

PRECIPITATION ON LAND

Over the oceans, there is more evaporation than precipitation, and over the land, the reverse is true. Averaged over the whole world, about 0.97 meters of water are evaporated from the ocean per year, about 0.88 meters are returned as direct precipitation, and 0.09 meters are returned as runoff from land.

One reason that precipitation exceeds evaporation on land is that the land has less exposed water surface. The less water exposed to the atmosphere, the less will be evaporated. Land masses also facilitate precipitation by their daily and seasonal temperature fluctuations. Typically, temperatures vary by 15—25°C, which is considerably larger than the temperature fluctuations at sea. When the temperature drops, moisture condenses.

Land also encourages precipitation because of its altitude. Air coming off the oceans must rise to get over the continent. As it rises, the air expands and cools. The cooler air cannot hold so much moisture, so some condenses and falls on the continents. Good examples are provided by the Olympic Mountains in the state of Washington and the Sierra Nevada in California. In both these regions, winds come predominantly from the west. As air rises to pass over these mountain ranges, it deposits about 10 times the annual rainfall on the windward side as does the predominantly falling air on the leeward side (Figure 7.10). This effect is also prominent on tropical islands. The windward sides of the islands are lush, and the leeward sides arid.

FORMATION OF SEA ICE

In higher, colder latitudes, the freezing and thawing of ice play a similar role in climate moderation, as do the evaporation and condensation of moisture elsewhere. Heat is absorbed when ice melts, and heat is released when ice freezes. That is, heat is absorbed in warm weather and released in cold weather. Either way, the absorption or release of latent heat opposes the temperature change.

Sea ice does not form easily. Sea water gets denser as it gets colder, all the way down to the freezing point.[5] As surface water cools, it sinks. This brings new water to the surface, which also cools and sinks, and so goes the cycle (Figure 7.11a). The cooling of the surface causes this continual turnover, and the water all the way to the bottom would normally have to be cooled to the freezing point for ice to begin forming. This would require more cooling than is possible in one winter, so ice would not form under these conditions.

If the surface layer could be prevented from sinking, however, then it could freeze. This would happen in shallow near-shore waters, or where denser, saltier water lies immediately beneath the surface (Figure 7.11b). For example, if a thin layer of fresh water from a river or ice melt would spread out

[5]This is in contrast to fresh water, which is densest at 4°C above the freezing point.

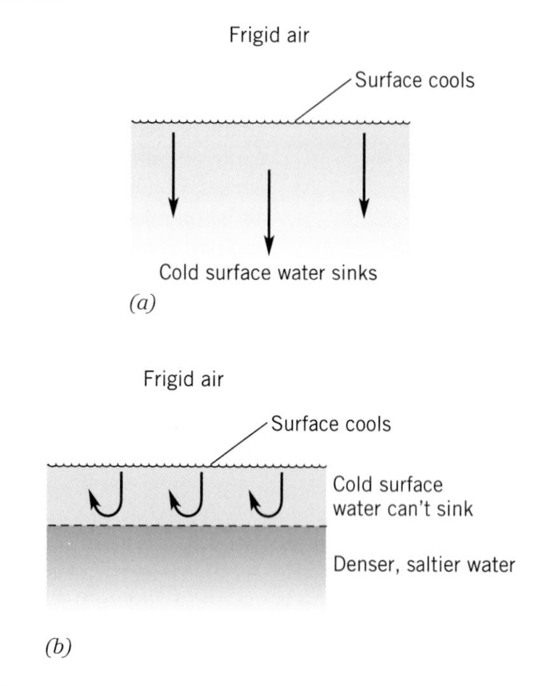

Figure 7.11

(a) Sea water gets denser as it cools, so it sinks. This makes it impossible for the surface to freeze, unless (b) it is prevented from sinking by a denser saltier layer beneath it. (c) In general, sea ice forms only when the surface is calm, because heavy waves would cause mixing, eliminating the needed layering.

across the salt water, then this thin surface layer could easily freeze. The surface layer does not have to be fresh; it only need be less salty than the water beneath. Thicker surface layers take more time to freeze, because more water must be cooled to the freezing point. For example, arctic sea ice extends twice as far southward in the northern Pacific Ocean (to about 60°N latitude) as it does in the Atlantic (to about 75°N latitude), because the surface layer in the northern Atlantic is thicker and takes more time to freeze.

Large waves tend to mix the surface layer with the saltier water below, destroying the thin surface layer that is needed for ice to form. Therefore, sea ice usually forms where the surface is quiet, such as protected waters near land or other floating ice (Figure 7.11c).

Ice is a good insulator. Consequently, where the ocean is covered with ice, it loses contact with the atmosphere. Cut off from the moderating influence of the ocean, air temperatures above the ice can get extremely cold. For the same reason, the water beneath the ice is no longer affected by the cold air above, so no more water freezes. Consequently, sea ice never gets very thick.

TYPES OF SEA ICE

Sea ice is further described as being fast ice, pack ice, or polar ice. **Fast ice** refers to ice sheets that extend out from land. These sheets often form quickly in cold weather, because the nearby land might offer protection from wind, waves, and mixing. Also, fresh water flowing into the ocean from land might create a layer of fairly fresh water on the surface, which freezes relatively easily.

Pack ice forms at sea, usually beginning as surface slush and small platelets that look like icy pancakes (collisions between them round off the corners). These freeze together and thicken to make larger sheets, but ocean surface currents and thermal stresses tend to keep pack ice in the form of large plates with seams and sometimes open water between. Both fast ice and pack ice are seasonal and normally reach a maximum thickness of only 1 or 2 meters. Pack ice may be a little thicker where collisions and compression between plates might cause overthrusting and pressure ridges. Icebreakers can generally navigate through both fast and pack ice.

The **polar ice** is more permanent, lasting typically 2 to 4 years. Although the polar cap itself re-

Figure 7.12
Huge tabular bergs such as this can be as much as several hundred kilometers across. They are more common in the southern oceans, where they break off from Antarctic ice shelves.

mains forever, the ice does not. Because it flows, new polar ice is forming in some places while older ice is being released at others. Polar ice can get up to 3 or 4 meters thick, and thicker where there is overthrusting or pressure ridges.

ICEBERGS

Knowing that sea ice is thin, you might wonder where the large icebergs come from, such as those that sometimes threaten North Atlantic shipping lanes. These come from land, primarily Greenland. Large glaciers slowly slide seaward from the coastal mountains and break off in bergs of various sizes. The largest bergs may float out to sea several hundreds or thousands of meters before breaking away from the parent glacier. These large flat **tabular bergs** often have several square kilometers of surface area and may be several hundred meters thick (Figure 7.12). Smaller irregular chips are called **splinter bergs** or *castle bergs* (Figure 4.6 background). One of these sank the Titanic.

In general, splinter bergs are more common in the North Atlantic; most of them are created as huge chunks of ice **calve** (i.e., break off) from glaciers flowing to sea from Greenland's coastal mountains. The larger tabular bergs are more common near Antarctica, because of the huge thick glacial ice shelves that flow out into the ocean from this continent. Some of these ice shelves are several thousand kilometers wide.

7.6 THE EARTH'S HEAT BUDGET

We now look at the various paths followed by energy on Earth. That is, we study how much of the incoming solar energy reaches the Earth's surface, and what becomes of that which does. Our account of all the incoming and outgoing heat is called a *heat budget*. Like any budget, it must balance. The total of everything coming in must equal the total of everything going back out. Between the "coming" and the "going," that energy is involved in some very interesting processes, which will be studied throughout the remainder of this book.

OVERALL

On the average, roughly half the solar energy incident on our outer atmosphere actually reaches the Earth surface. The rest is reflected by clouds or absorbed and reemitted from the upper atmosphere. Averaged over the whole Earth, day and night, and all seasons, the half that reaches the Earth's surface arrives at a rate of 175 Watts per square meter, of which 90% (157 W/m^2) is absorbed. (A Watt is a rate of a joule per second.)

On the average, the Earth must radiate back out into space as much energy as it receives from the Sun. Otherwise, the Earth would be getting hotter all the time. As we have seen, the incoming solar energy

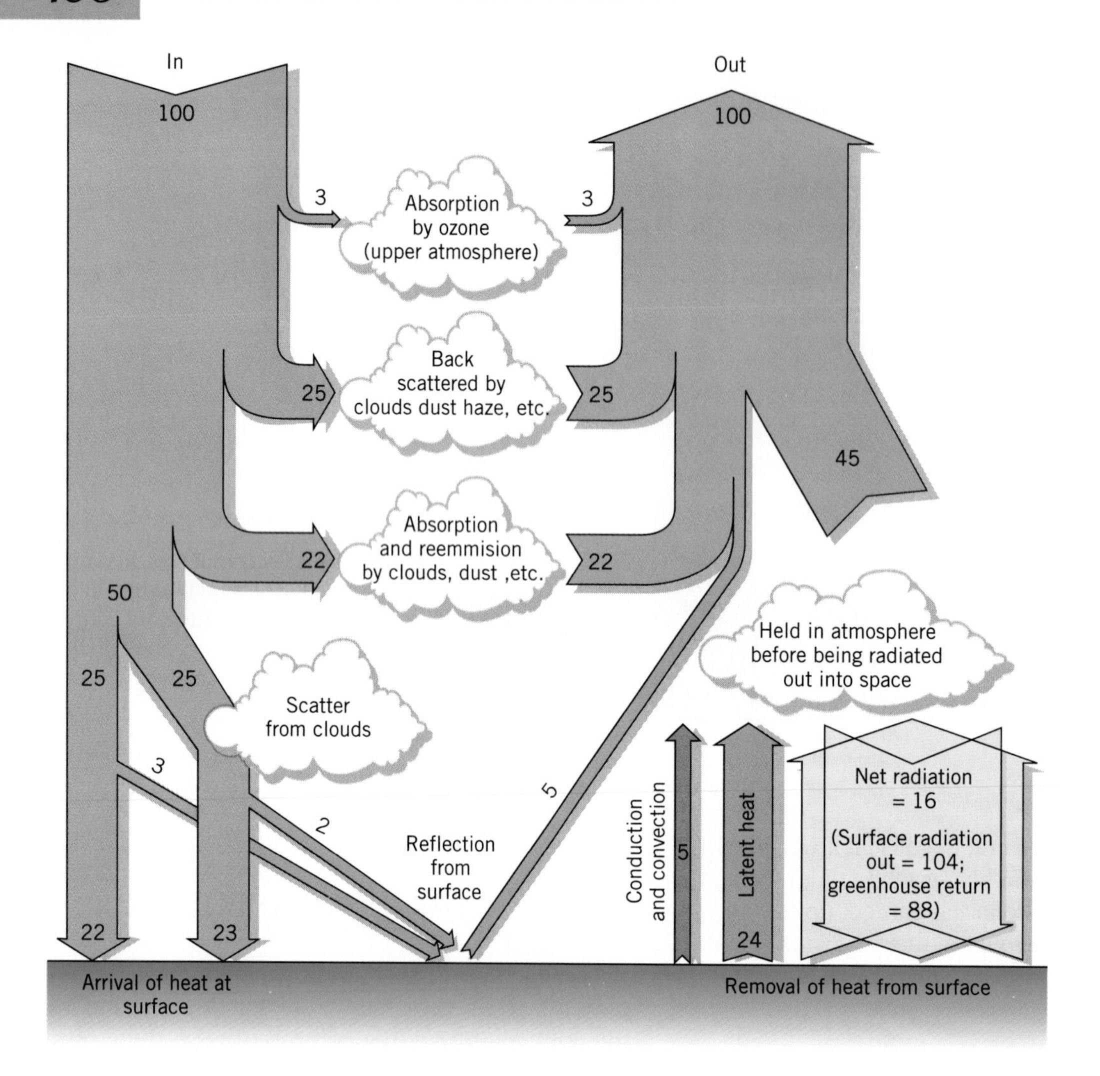

Figure 7.13

Illustration of the Earth's heat balance, showing the various avenues followed by incoming and outgoing heat. Numbers indicate the percent of the total heat flow. Notice that only 50% of the incoming heat actually reaches the surface, with 5% being reflected and the remaining 45% absorbed. Notice also the huge influence of the greenhouse trapping of the radiated heat leaving the surface.

is primarily at visible wavelengths to which our atmosphere is transparent. That which reaches the Earth's surface takes only a very tiny fraction of a second to pass through our atmosphere.

The outward energy flow, however, experiences a great deal more obstruction. The Earth is much cooler than the Sun, so it radiates its energy in the longer, infrared wavelengths, to which our atmosphere is not transparent. Therefore, other mechanisms become important for transfer through our lower atmosphere. The various paths taken by the incoming and outgoing energy are summarized schematically in Figure 7.13.

HEAT LEAVING THE EARTH'S SURFACE

The main obstruction to heat leaving the Earth's surface is the lower 12 kilometers of the atmosphere, where the air is thick and contains water vapor. Above that, the air is thin, has virtually no moisture, and presents little obstruction to the radiation of heat into outer space. So the final fate of the outward bound energy is to be radiated out into space from our upper atmosphere as infrared radiation. The only complicated part of the process is how the heat gets through the lower atmosphere.

Three mechanisms transport heat upward from the Earth's surface through this lower layer. The main one is the transport of latent heat. As we have seen, evaporation removes heat from the surface. As this water vapor reaches the cooler upper atmosphere, it condenses and releases the stored heat energy up there. Storms are important agents in this process (Section 7.8). Altogether over the entire Earth, 53% of

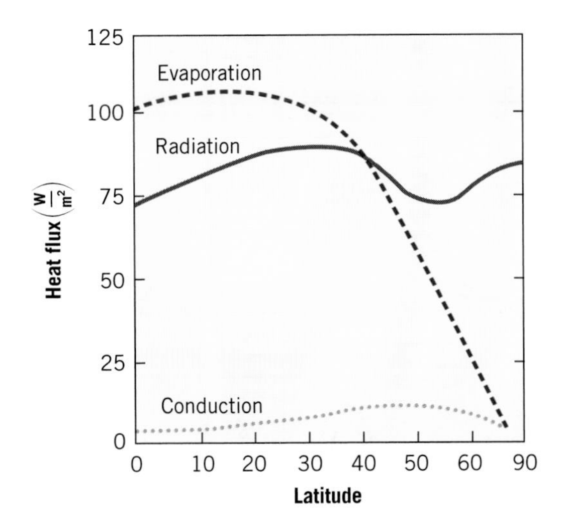

Figure 7.14

Heat transfer from ocean to atmosphere by each of the three processes as a function of latitude, measured in terms of joules per second (i.e., Watts) of heat released by each square meter of ocean surface area (yearly average). The horizontal scale is made so that equal distances represent equal amounts of the Earth's area. At higher latitudes, the Earth's area is smaller and the scale correspondingly compressed. (After A. Defant, 1961, *Physical Oceanography*, Vol. 1, Pergamon Press.)

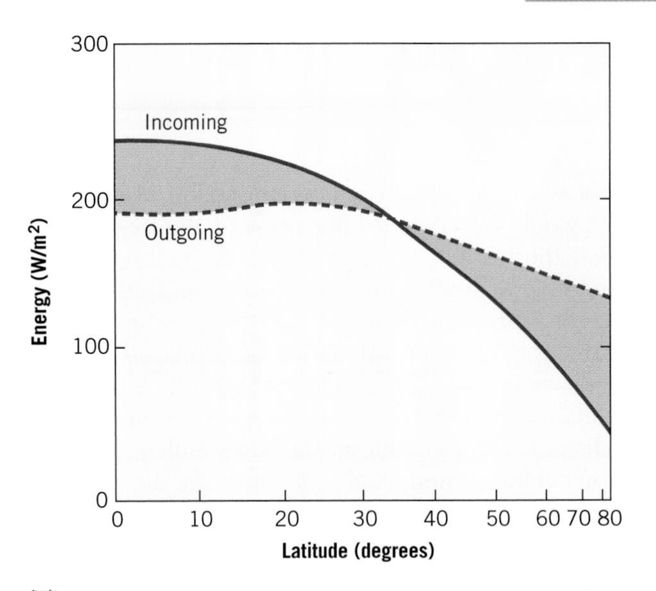

Figure 7.15

Plot of incoming and outgoing heat as a function of latitude, based on satellite measurements. Notice that the tropics receive more heat than they release into space, and the reverse is true in the polar regions. Clearly, some heat received in the tropics is being transported to higher latitudes before being released.

the exiting heat[6] gets through the Earth's lower atmosphere this way.

Another 37% of the exiting heat gets through via infrared radiation, which we have seen is a rather arduous process, due to repeated absorption and reemission of the infrared as it goes. The remaining 10% of the exiting heat reaches the upper atmosphere via turbulence and overturn in the lower atmosphere, as warm air rises and cooler air falls. Although some energy is transferred from ocean to atmosphere via conduction, none reaches the upper atmosphere that way, because air is such a poor conductor of heat. Turbulence and convection is needed to carry this heat upward.

There are some interesting regional variations in this pattern (Figure 7.14). In the tropics, the warmer air holds more moisture, which is particularly effective at blocking the infrared radiation. So in this region, the dominant mechanism for the removal of heat from the Earth's surface is the evaporation of water from the ocean or land surface. At higher latitudes, temperatures are cooler. There is less evaporation of water and less moisture in the air. The primary mechanism for the release of heat from the Earth's surface is direct radiation. Although the cooler ocean surface emits less infrared than it does in the tropics, the reduced moisture in the air means that this radiation gets through with less obstruction.

[6]To compare this to the numbers of Figure 7.13, notice that 53% of 45 is 24, the number given in that figure.

POLEWARD HEAT TRANSPORT

Although we know that the total heat into the Earth has to equal the total heat out, the actual measurement of heat going in both directions yields a surprise. Near the equator, the Earth receives more heat than it gives off, and near the poles it returns more than it gets (Figure 7.15). About 20% of the heat that is received in the tropics is transported poleward before being reemitted. How is this accomplished?

Our swift ocean surface currents, such as the Gulf Stream, carry only about 10 to 20% of the necessary amount of heat. Surprisingly, the major portion is carried as latent heat. Some of the warm moist air from the tropics is carried to higher latitudes before the moisture is released. (See the section on storms at the end of this chapter.) From weather forecasts, you might have noticed the heavy rainfall associated with tropical air masses that move into temperate latitudes. Latent heat is removed from the tropics as the water evaporates and released at higher latitudes when this moisture condenses.

EVAPORATION, PRECIPITATION, AND SURFACE SALINITY

Based on the results of the previous section, we now consider the difference between the rates of evaporation E and precipitation P as a function of latitude. We would expect this difference (E-P) to be positive near the tropics, where evaporation is larger, and negative at high latitudes, where precipitation is larger. Indeed, that is the way things are (Figure 7.16). The dip in the curve near the equator is a result of heavy rainfall there. The intense sunshine not only evaporates water, but also heats the air so much that it rises. Rising moist air means large rainfall, creating tropical forests and jungles on land. The dip, then, is due to the high amount of precipitation near the equator.

When sea water evaporates, it leaves salt behind. Similarly, the return of fresh water to the surface reduces the salinity in regions where precipitation is larger. As a result, the surface salinity of the oceans at various latitudes mimics the plot of the difference between evaporation and precipitation, as is seen in Figure 7.16.

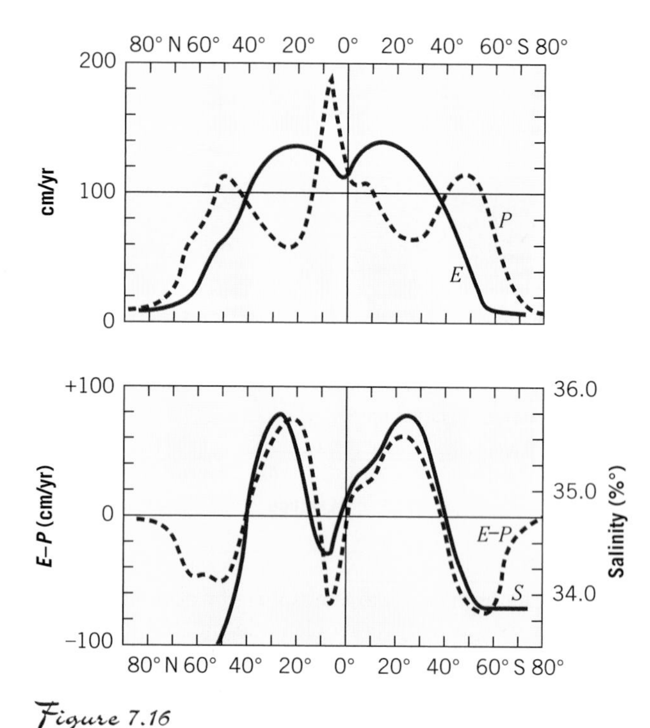

Figure 7.16

(above) Plot of evaporation E and precipitation P as a function of latitude. (below) Plot of the difference (E-P) and surface salinity S as a function of latitude. Note that the meteorological equator is about 5° north of the geographical equator. (After G. Wust, 1936, *Oberflachenselzgehalt, Verdunstung und Niederschlag auf dem Weltmeere*, Festschrift Norbert Krebs, Stuttgart pp. 347-359.)

7.7 GLOBAL WIND PATTERNS

The heat-transfer mechanisms explained in the preceding sections are the ways that our atmosphere receives heat from the Earth beneath it. Although we say that the atmospheric circulation is driven by *solar heating*, we should more properly say that atmospheric circulation is driven by *Earth heating*. The incoming sunlight passes through the atmosphere and heats the Earth's surface, and the Earth, in turn, heats the atmosphere. This heating drives the atmospheric motion and is responsible for the winds and weather that we experience.

RISING AND FALLING AIR

As we learned in Section 7.2 of this chapter, as air rises through the atmosphere, it expands and cools adiabatically. After rising a few kilometers, it will be quite cool, even though it may have been hot when on the Earth's surface. If the air contains significant moisture, the colder temperature will make it condense, and clouds will form (Figure 7.17a). Conversely, as air at high altitudes sinks downward toward the Earth's surface, it is compressed and heated adiabatically. The warmer air can hold more moisture, so any clouds will evaporate, and the skies will be clear (Figure 7.17b). These general patterns for rising and falling air are important in understanding our weather patterns.

Air can rise or fall for several reasons. It could be forced up to pass over a mountain range, or descend into a valley on the other side. It could be forced up by a mass of denser air that wedges beneath it or forced down by one that is less dense and rides above it. But these are local and temporary effects. In this section, we will be interested in more global and persistent patterns in the circulation of the atmosphere and the primary driving force for these is the tendency for hot air to rise, and cold air to fall. The oceans are instrumental in the heat transfer that is needed to drive these motions.

Air rises in equatorial regions for two reasons. The obvious and most important one is that warmer air is less dense and therefore lighter. The other reason is more subtle. The water molecule is actually lighter than the nitrogen or oxygen molecule, so moist air is lighter than dry air. Because of heavy evaporation from the oceans (and also transpiration from rain forests on the continents), the air's moisture content is greater in equatorial regions than else-

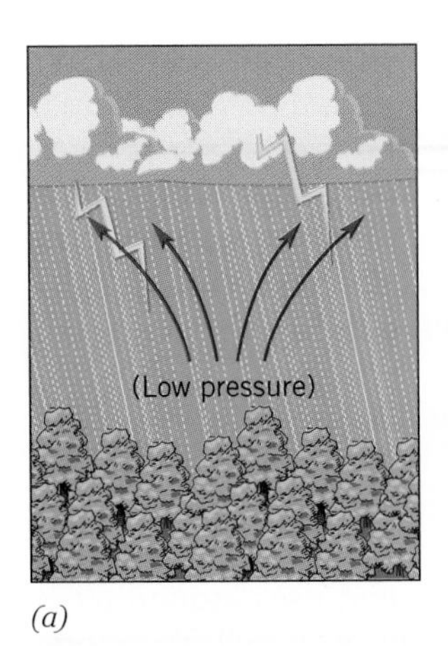

(a)

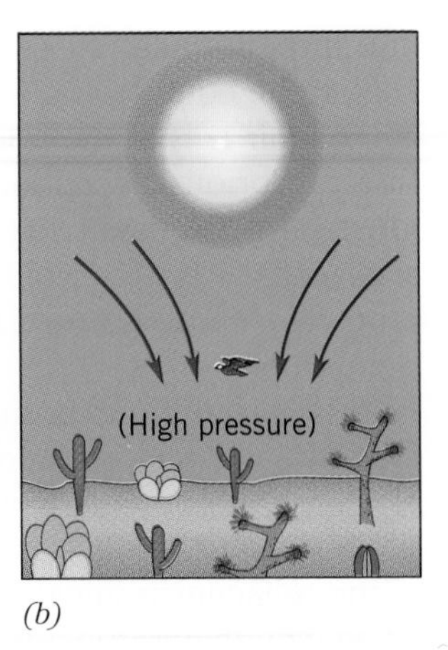

(b)

Figure 7.17

(a) As air rises, it expands and cools. (Lower pressure at higher altitudes allows expansion.) The cooling causes the moisture to condense. Where air is rising, there is reduced atmospheric pressure on the Earth's surface. (b) Falling air masses are compressed and heated, resulting in clear sunny skies. Falling air is denser and compresses the air beneath, causing high pressures. For these reasons, low pressures are usually associated with storms, and high pressures with clear skies.

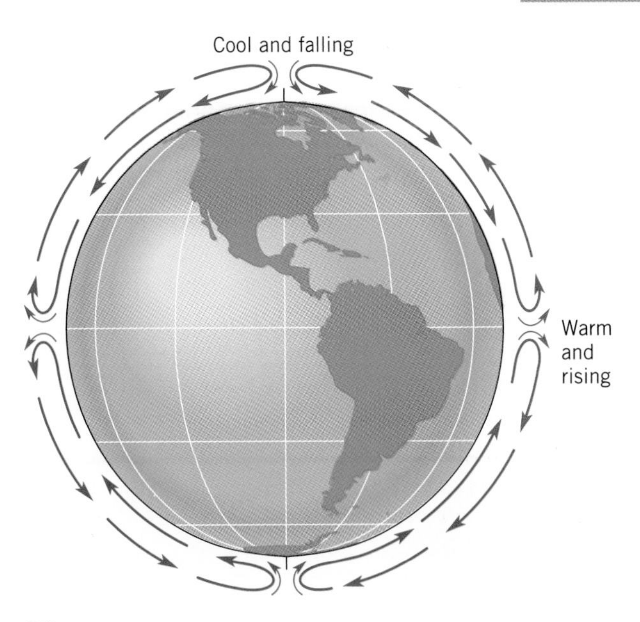

Figure 7.18

If the Earth were motionless but still had the Sun overhead and heating the equator, then the warmer, lighter air of the equator would rise and the cooler polar air would fall. The simple cellular pattern would be that of air rising at the equator, moving poleward aloft, falling at the poles, and returning to the equator along the surface.

where. Equatorial air rises, then, primarily because it is hot, but also because it is moist. As we noted in the preceding paragraph, as the warm moist air rises, it expands, cools, and the moisture condenses. Consequently, there are predictable afternoon rainstorms in these equatorial regions.

Near the poles, the high-altitude air is very cold and therefore dense. So it falls. Falling air means clear skies, so the climate is clear and cold in these polar regions.

If the Earth were not spinning, the circulation of our atmosphere would be the simple pattern shown in Figure 7.18. The air would rise near the warm equator, move poleward aloft, fall in the cooler polar regions, and then return to the equator along the ground. The surface winds would always blow from the poles toward the equator, being **northerlies** (i.e., coming from the north) in the Northern Hemisphere and *southerlies* in the Southern Hemisphere.

CORIOLIS DEFLECTION AND SURFACE WINDS

However, the real Earth spins, and the simple picture of Figure 7.18 is modified by the Coriolis deflection of the moving air (Chapter 2). Winds initially going north or south end up going east or west instead. In the Northern Hemisphere, for example, the air rising at the equator and heading poleward curves toward the right and soon is heading east. It cools at these higher altitudes and higher latitudes, so that it falls around 30°N. That is, it never gets to the North Pole as it would if the Earth weren't spinning. In fact, it only gets about one-third of the way there.

After falling, it then returns south along the surface toward the equator. But as it begins to move southward, it is deflected to the right by the Coriolis effect and heads west instead (Figure 7.19). So surface winds within the tropics blow toward the west. Interestingly, it is the custom for meteorologists to identify winds by the direction they are blowing "from," rather than "toward." Winds blowing toward the west, then, are called *easterlies*. These easterly winds in the tropics are called **trade winds**, a name originating in colonial times. Merchants learned that their travel times from Europe to the colonies could be greatly reduced by sailing southward to the Canary Islands before beginning their westward course across the Atlantic Ocean. They could count on these winds for a speedy westward voyage.

Near the North Pole, where the cool air falls and begins its southward journey, it is also deflected to the right by the Coriolis effect. As a result, the prevailing surface winds in Arctic regions also blow toward the west. They are called the **polar easterlies**. Not only does this air fail to reach the equator, but in fact it seldom works its way farther southward than about 60°N before the warmer temperatures of these lower latitudes cause this polar air to start rising again.

Thus, the Earth's spin severely modifies the circulation pattern depicted in Figure 7.18. The air rising at the equator does not make it to the pole, and the air falling at the pole does not make it to the equator. Instead, the circulation in each hemisphere is broken into three circulation cells. In the tropical cell, air rises at the equator and falls around 30°N. In the polar cell, air falls at the pole and rises at about 60°N. Due to Coriolis deflection, the surface winds blow mostly in the east–west directions, rather than north–south, as they would if the Earth weren't spinning. The same pattern is found in the Southern Hemisphere.

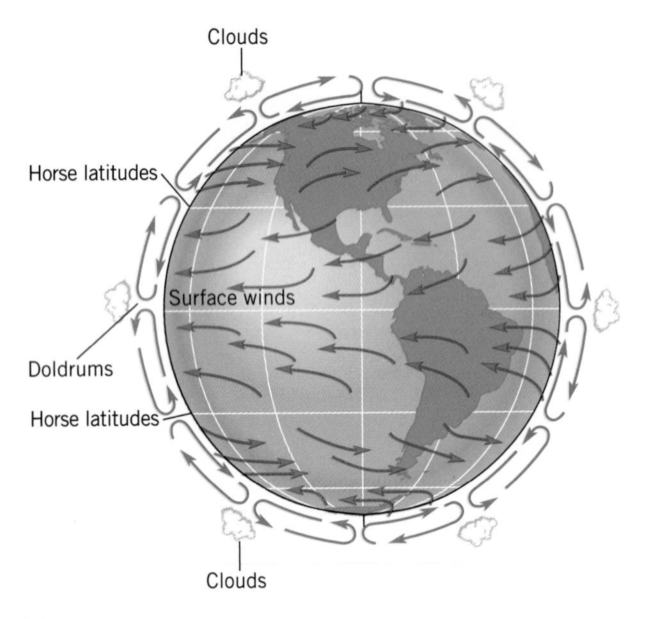

Figure 7.19
For the real spinning Earth, the atmospheric circulation pattern has three separate cells in each hemisphere. Air initially moving toward the equator is deflected westward by to the Coriolis deflection, and air initially moving toward the pole is deflected eastward. The directions of the prevailing surface winds in each zone are indicated.

The air motions in the tropical and polar cells drive the cell sandwiched between them. This *temperate cell* is particularly interesting because it is driven backward. Air falls around 30°N along with that of the adjacent tropical cell and rises around 60°N along with that of the adjacent polar cell. Therefore, air in this cell is falling where it is warmer and rising where it is cooler, which is backward from its natural tendency, and also backward to the direction of the other two cells.

This causes a great deal of unpredictability in the wind and weather in this intermediate cell. Ironically, most of the world's population lives within these intermediate latitudes where the weather is more unpredictable. As is illustrated in Figure 7.19, the Coriolis deflection of these surface winds leads to the **prevailing westerlies**, (i.e., blowing from the west), but due to the unpredictability of this cell, it is far from "prevailing." In the tropics, the trade winds are as predictable as sunrise tomorrow, but this is far from true for the westerlies in temperate latitudes.

Figure 7.20
Sea-level pressures, averaged over a year, indicating the regions of high and low pressures and general pattern of the surface winds. (From Alyn C. Duxbury, 1971, *The Earth and Its Oceans*, Addison-Wesley, Reading, MA.)

SOME CONSEQUENCES

As we will learn in the next chapter, the ocean's dominant surface currents are driven by the trade winds, which drive currents toward the west within the tropics, and the prevailing westerlies, which return them eastward in temperate latitudes. The routes followed by these surface currents have enormous influence on the Earth's regional climates.

The equator and regions around 30°N and 30°S have rather light surface winds and had bad reputations among earlier sailors for their poor sailing conditions. Air motions are more vertical than horizontal, which was of little help to sailing ships. In the equatorial **doldrums**, the rising air causes predictable afternoon showers, making the continents lush in these equatorial regions (Figure 7.17a). In the **horse latitudes**, around 30°N and 30°S, air is sinking. The skies are clear, and the continents are arid (Figure 7.17b). At sea, these regions are sometimes referred to as *maritime deserts*. To a sailor who is stranded by the light winds under sunny skies with no chance of fresh water from rain, they might be just as deadly as being stranded on a large continental desert.

Where warm light air is rising, both its lower density and its removal from the region results in low pressures. Similarly, where denser air is falling, both its greater density and the compression of the air beneath results in high pressures. Consequently, there is a tendency for low barometric pressures and storms in the doldrums, and high barometric pressures and clear skies in the horse latitudes. Likewise, there tends to be low pressures and storms where air is rising near 50–60° north and south latitudes, and high pressures and clear skies where air is falling near the poles. Figure 7.16 shows this variation of precipitation with latitude and its effect on the ocean's surface salinity. Figure 7.20 shows the barometric pressures and predominant surface winds averaged over a year.

The Earth's surface winds, as illustrated in Figures 7.19 and 7.20, are an important application of something we encountered in Chapter 2. If the Earth were not spinning, the air would flow from high pressures toward low pressures (*pressure gradient force*). However, the Coriolis deflection causes the winds to flow sideways to the pressure gradient. This behavior is somewhat modified near the ground where friction from trees, buildings, and other obstacles slow the air down. At slower speeds, the Coriolis deflection is reduced, so the air tends to flow somewhat more in the direction favored by the pressure gradient. But the overall pattern, including that of the very thin layer near the ground, is for surface winds to flow in east–west directions on Earth, whereas the pressure gradients are in the north–south directions. This curious result of the Coriolis deflection, where fluids flow sideways to the direction of the pressure gradients, is called **geostrophic flow**. We will encounter this again when studying ocean currents.

7.8 LOCAL VARIATIONS

The circulation patterns presented so far work best for a world with a uniform oceanic surface and whose spin axis is not tilted. In our world, however, both continents and seasons provide small and predictable modifications to these patterns. Fortunately, most of the important modifications are simple to understand.

LAND-SEA VARIATIONS

Most of the world's population lives in temperate latitudes where seasonal patterns are important. The land masses become cold in winter and hot in the summer, whereas the ocean surface temperatures remain moderate. Because the atmosphere receives its heat from the Earth beneath, where the Earth is cold, the air is cold. Cold air tends to fall, and falling air creates high atmospheric pressure. Conversely, where the Earth is warm, the air above it tends to be warm. Warm air tends to rise, and rising air leaves behind low pressures.

Therefore, in winter, there is a tendency for falling air and higher pressures to be over the colder land, and rising air and lower pressures over the warmer ocean (Figure 7.21b). Winds tend to blow from land toward the ocean (high pressure toward low pressure). In the summer, the reverse is true (Figure 7.21a). Land masses are warmer, so that is where the low pressures will be found. Winds blow from the ocean toward the land. In southern and southeastern Asia, these seasonal wind patterns are called *monsoons*. In the Southern Hemisphere, there is relatively little land mass, so the effect is stunted there.

A similar pattern is noticed in coastal areas, but on a daily basis. In the early mornings, when the land is cool compared to the ocean, air rises over the ocean and falls over the neighboring coast, resulting in a **land breeze** blowing from the land toward the ocean (Figure 7.22). But in the afternoon, after the land has heated up, the pattern reverses. Air rises over the warmer land and falls over the cooler oceans, causing a **sea breeze** to blow landward from the ocean.

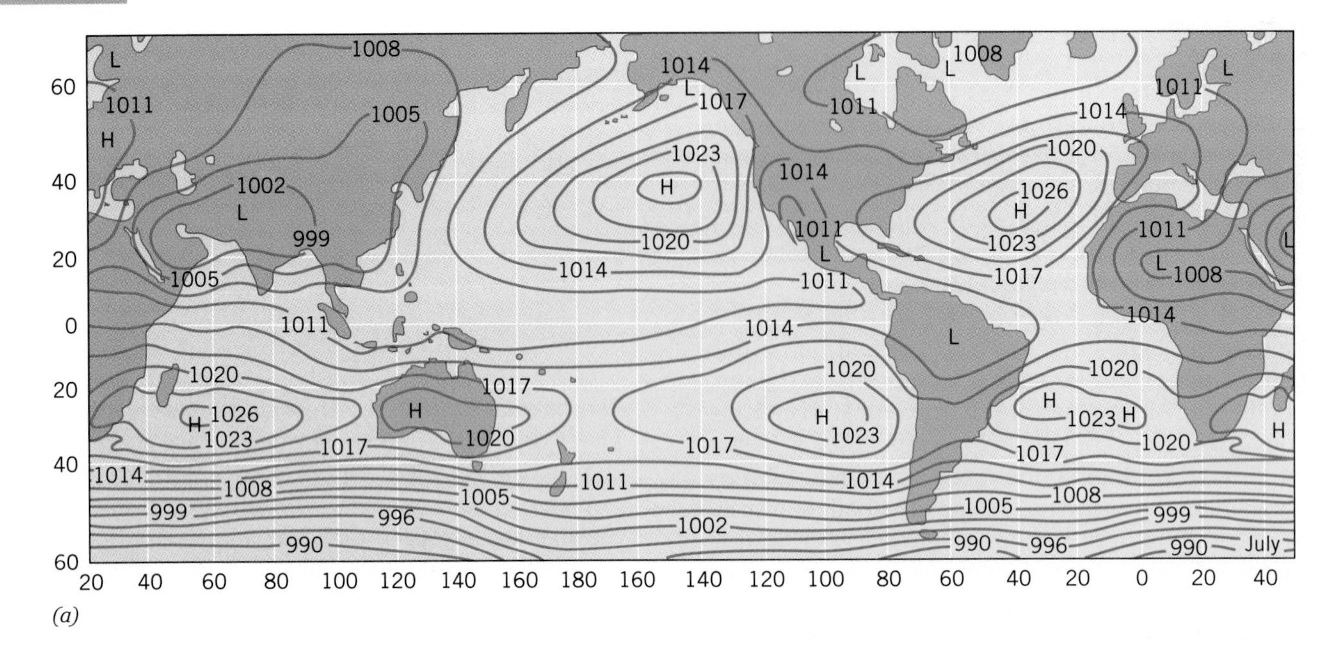

(a)

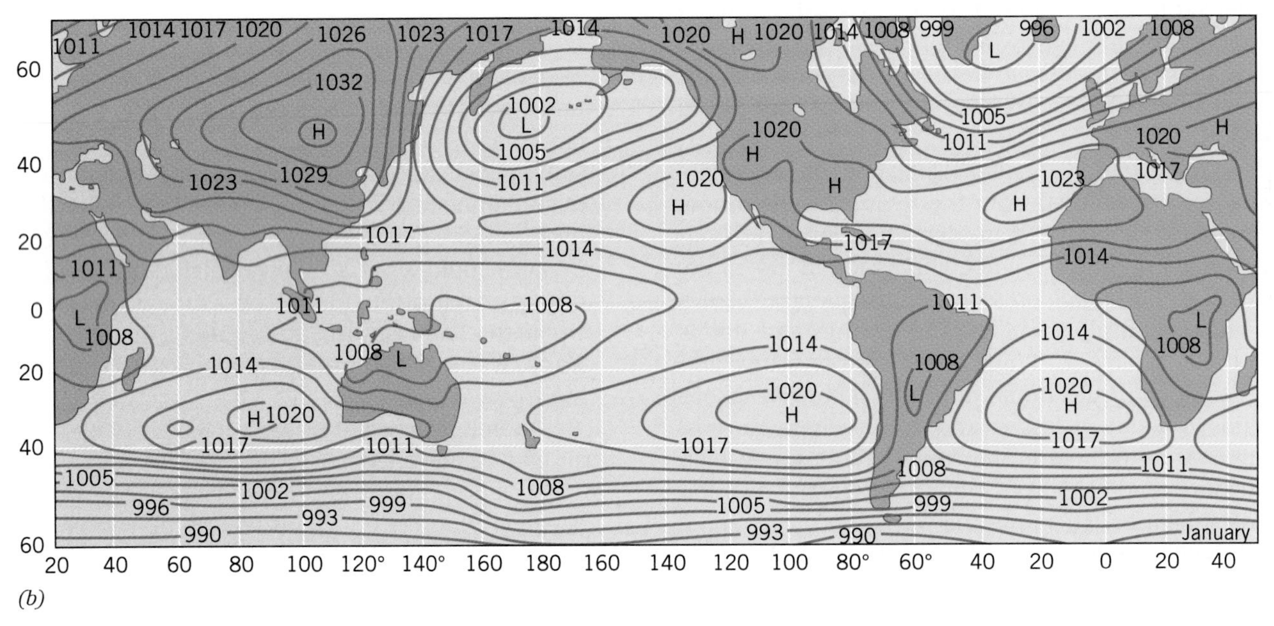

(b)

Figure 7.21

Mean sea-level pressures for the months of July (a) and January (b), in units of millibars. (After Y. Mintz and G. Dean, 1952, Geophysics Research Paper 17, Geophysics Research Directorate, Air Force Cambridge Research Center.)

STORMS

Storms are particularly effective in transporting latent heat from the Earth's surface to higher altitudes. Whenever other mechanisms are not getting heat back up through the atmosphere fast enough, nature employs storms to accomplish that task.

Although storms can be generated by a variety of mechanisms, an important component in fueling all major storms is the release of latent heat. As warm moist air rises, it expands, cools, the moisture condenses, and the rain falls. The released latent heat further warms the air around it and causes continued rising. Rising air leaves low pressure in the region beneath it. Neighboring air rushes toward this low pressure like it would toward a giant vacuum cleaner, spiraling inward due to the Coriolis deflection (Figure 7.23). If this neighboring air is also warm and moist, it will also rise through the storm center, causing further condensation, additional release of latent heat, and continued fueling of the storm.

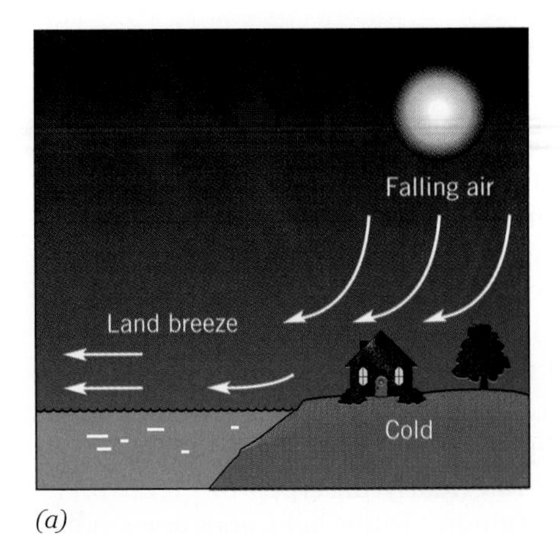

(a)

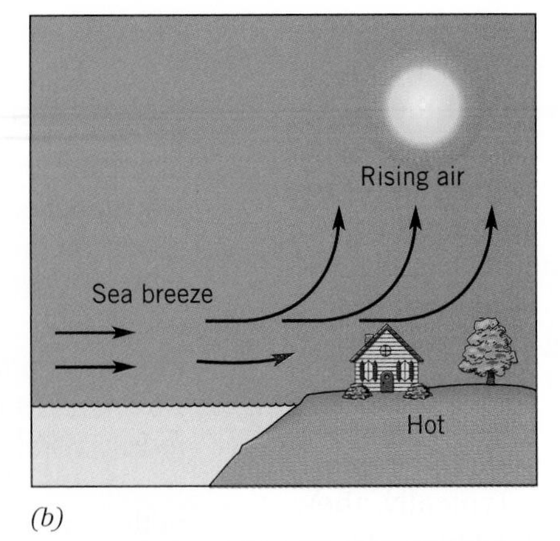

(b)

Figure 7.22
(a) In the early morning, the land is cooler than the ocean. So the cooler air over the land falls and blows out to sea. (b) In the afternoon, the land is warmer. The warmer air over the land rises, and air from the oceans blows landward to replace it.

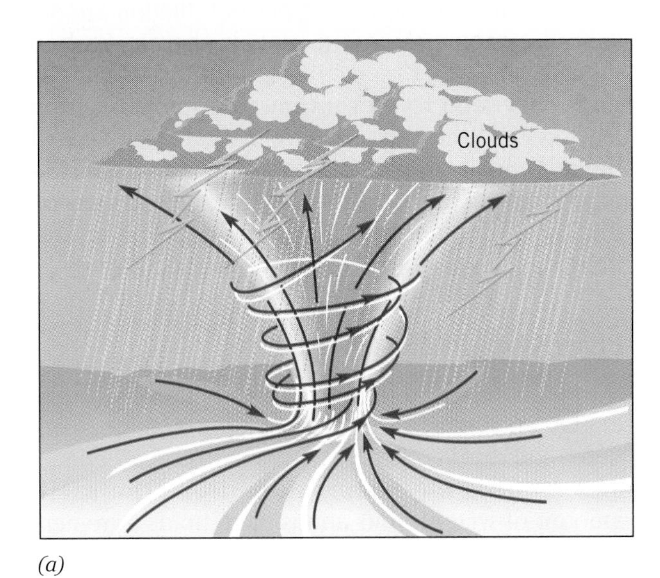

(a)

Figure 7.23
(a) Illustration of the fueling mechanism for violent storms. As warm, moist air rises, condensation releases latent heat, which keeps the air warm and rising and leaves a low-pressure center below. Incoming air masses experience Coriolis deflection and spiral into the low-pressure center before they begin to rise and continue the cycle. (b) Computer enhanced and vertically exaggerated satellite image of Hurricane Andrew (August 25, 1992). Heavy clouds form where warm moist air is rising.

(b)

The strength and duration of a storm depend on both the potency and amount of the fuel. That is, it depends on both how warm and humid the air is, and how much of it there is. In the spring and summer, intense solar heating often creates expansive regions of hot humid air over the tropical ocean. As this air starts to rise, it may generate particularly violent and long-lived storms, called *hurricanes* in the Atlantic Ocean and *typhoons* in the Pacific and Indian Oceans. The requirements of warm surface waters and the Coriolis deflection of incoming winds dictate that these storms tend to form in the tropical oceans, but not near the equator.[7] Typically, they form around 5 to 20° north and south latitudes. Once started, these storms travel away from the region where they started and continue until the incoming air is dry. Without the condensing moisture to fuel the storm, it fizzles. Unfortunately, dry air is often not encountered until the storm runs into land and wreaks havoc along the coast.

Summary

Ocean–Atmosphere Interactions

Earth has an extremely mild climate, due to the moderating influence of the oceans. The ocean and atmosphere are very thin in comparison to their breadth, and they are closely interconnected.

The Atmosphere

The lowest 12 kilometers of the atmosphere contain most of the air, and are called the troposphere. The atmosphere's pressure and density decrease with altitude, as does the temperature in the troposphere. Dry air is mostly nitrogen and oxygen. Air cannot hold much water because the water molecules tend to stick together and form droplets. Warmer air holds more water.

Ocean Surface Waters

The ocean's interactive surface waters extend to several hundred meters' depth due to mixing by waves, surface currents, and turbulence. The oceans' thermal inertia is due to the high specific heat of water, the penetration of light through water, mixing of surface waters, and phase changes of water at the surface.

[7] There's no Coriolis deflection on the equator.

Energy Transfer by Radiation

Electromagnetic waves all travel the same speed, are generated by accelerating electric charges, and are absorbed by other charges as they pass by. Shorter wavelengths carry more energy. The electromagnetic spectrum includes radio waves, microwaves, infrared waves, visible light, ultraviolet rays, x-rays, and gamma rays. The hotter the object, the more violently the atoms and molecules shake, and the shorter the wavelengths of the radiation it emits. The Sun emits mostly visible light, and the Earth emits infrared.

Water absorbs nearly all types of electromagnetic waves, although visible light goes reasonably far through water, with blue-green being transmitted the best. Air transmits radio waves and visible light well, but most of the other types of radiation are absorbed by some trace materials in the air. Carbon dioxide and water molecules absorb infrared radiation. They tend to prevent infrared radiation from leaving Earth, creating the *greenhouse effect*. Several human endeavors are affecting the carbon dioxide content of the atmosphere. Some chemicals are reducing the ozone in the upper atmosphere, and therefore increasing our exposure to ultraviolet.

Phase Changes and Our Climate

Evaporation during the daytime or summer absorbs energy and reduces the temperature increase. Condensation during the night or winter releases energy and reduces the temperature drop. The small amount of water in our atmosphere holds a tremendous amount of heat energy. At high latitudes, the freezing and melting of ice moderate the water temperature in the same way. Ocean processes at higher latitudes have lesser effect on our climate because the oceans are smaller and the ice tends to prevent the air from interacting with the ocean. Sea ice comes in various forms and requires a thin layer of lighter surface water in order to freeze over. It never freezes very thick. Large icebergs come from continental glaciers, not the oceans.

The Earth's Heat Budget

Of all the solar energy incident on the outer atmosphere, about half reaches the Earth's surface, and most of this half is absorbed. The warmed Earth heats the atmosphere mostly through latent heat and radiation and, to a very small extent, conduction. As heat is leaving the Earth, the lower atmosphere presents a major obstruction. More than half is carried upward as latent heat in water vapor. Another major portion eventually makes it through as radiation, and a lesser amount of heat is carried upward by turbulence.

In polar regions, most of the Earth's heat leaves

the ocean surface as radiated energy. But in the tropical latitudes, the atmosphere's moisture content is high and most heat is carried from the ocean surface as latent heat. Some of this latent heat is transported to higher cooler latitudes, where the moisture condenses. The ocean's surface salinity reflects the difference between the rates of evaporation and precipitation.

Global Wind Patterns and Local Variations

Solar heating and evaporation cause air to rise at the equator, and air tends to fall in the cold polar regions. Were it not for the Coriolis effect, surface winds would blow from the poles toward the equator. The Coriolis deflection of these moving air masses causes the circulation to break up into three cells in each hemisphere. In polar and tropical regions, surface winds blow toward the west, and at temperate latitudes, they blow east. The middle of the three cells is the most unstable and unpredictable. Due to Coriolis deflection, fluids on the Earth's surface tend to flow sideways to the pressure gradients, an effect called *geostrophic flow*.

Near the equator and 50–60° north and south are rising air, low pressures, and storms. Near the poles and 30° north and south are falling air, high pressures, and clear skies. There is low pressure in the equatorial doldrums and high pressure at the horse latitudes. Seasonal changes in the barometric pressure patterns in temperate latitudes are caused by the greater seasonal heating and cooling of land masses. A similar effect occurs daily in coastal areas. Storms are fueled by condensation in rising warm, moist air masses.

Key Terms

troposphere
adiabatic processes
thermal inertia
phase change
sensible heat
conduction
radiation
electromagnetic waves
frequency
radio waves
gamma rays
microwaves
infrared waves
visible light
ultraviolet rays
x-rays
windows
greenhouse effect
greenhouse gases
ozone
catalyst
fast ice
pack ice
polar ice
tabular bergs
splinter bergs
calve
northerlies
trade winds
polar easterlies
prevailing westerlies
doldrums
horse latitudes
geostrophic flow
land breeze
sea breeze

Study Questions

1. Why does the temperature decrease with altitude in the troposphere?
2. Give four reasons why there are greater temperature fluctuations over land than sea. Explain each.
3. How are electromagnetic waves generated, and how are they detected? Why do hotter objects tend to radiate waves of shorter wavelength?
4. Why is the ocean usually blue-green?
5. Explain the greenhouse effect.
6. From a molecular point of view, why can warm air hold more water vapor than cool air?
7. Suppose two identical pans were sitting beside each other, one containing warm water and the other cool water. From which would the water evaporate faster, and why? Why does the evaporation of some of the water make the remaining water cooler?
8. Explain why there tends to be more precipitation over land than sea. Give two reasons.
9. What special condition is needed for sea ice to form, and why is this condition necessary? Why doesn't sea ice ever get very thick?
10. Of the heat that is *radiated* directly into outer space from the Earth's surface, more goes out in polar regions than in equatorial regions, even though the equator is hotter. Why?
11. What is the primary mechanism by which heat is removed from the Earth's surface in equatorial regions? How does this result in a net transport of heat away from the equator and toward the poles?
12. Explain the surface salinity variations illustrated in Figure 7.16.
13. Of the three atmospheric circulation cells in the Northern Hemisphere, why is the middle one most unstable and unpredictable? Why do the winds come predominantly from the west in these temperate latitudes?
14. Why does rising air expand? Why is it associated with rainfall? Why are there clear skies where air falls?

15. Why are there usually low atmospheric pressures near the equator and high atmospheric pressures around 30°N and 30°S? Explain seasonal changes in barometric pressure patterns in northern temperate latitudes.
16. Explain morning and afternoon land–sea breeze patterns in coastal areas.
17. How do hurricanes (or typhoons) start, and what keeps them going? Why do they fizzle when they run into land?

Critical Thinking Questions

1. Pressure increases by about one atmosphere for every 10 meters of depth in the ocean. This tells us that the weight of the atmosphere above us is about the same as the weight of 10 meters of water. Using this information and that in Tables 6.4 and 3.1, answer the following questions. If all the water were to evaporate from the oceans, atmospheric pressure would increase by about how many times? Would this make our atmosphere heavier than Venus'?
2. It has been suggested that as we burn more fossil fuels, we might eventually reach a point where we trigger a *runaway greenhouse effect*, where temperatures would keep rising even if we stopped burning fossil fuels. How could this happen? (*Hint*: Think about water vapor as a greenhouse gas, and what happens to the atmosphere's water vapor content as its temperature rises.)
3. If the Earth were not spinning, what direction would the surface winds blow where you live? Why? What direction does it blow on the real Earth at 15° south latitude, and why?
4. The text implies that clear skies are normally associated with cooler weather. Although true, this statement often surprises people. But think of the four cases: winter days, winter nights, summer days, summer nights. In only one of these four cases are clear skies sometimes warmer. Which one is it? Why not the others?

Suggestions for Further Reading

BLUEFORD, JOYCE. 1988. El Niño: Ocean Temperature Affects Weather and Life Cycles. *Instructor* 98:1, 74.

1989. *Earth Science*. Ocean/Atmosphere Interaction Wreaks Havoc. 42:4, 9.

HARDY, JOHN T. 1991. Where the Ocean Meets the Sky: The Ocean's Skin Is the Richest, Most Extensive Habitat of All. *Natural History*, May, 58.

HORGAN, J. 1989. Greenhouse America: A Global Warming May Destroy U. S. Forests and Wetlands. *Scientific American* 260:1, 20.

HOUGHTON, R. A. AND G. M. WOODWELL. 1989. Global Climate Change. *Scientific American* 260:4, 36.

KASTING, J. F. ET AL. 1988. How Climate Evolved on the Terrestrial Planets. *Scientific American* 258:2, 90.

MACINTYURE, F. 1974. The Top Millimeter of the Ocean. *Scientific American* 230:5, 62.

MASON, B. J. 1993. Role of the Oceans in Global Climate. *Contemporary Physics* 34:1, 19

MAUL, GEORGE A. 1992. Global Temperature and Sea Level Change. *Physics and Society* 21:4, 6.

RAMANATHAN, V., B. R. BARKSTROM, AND E. F. HARRISON. 1989. Climate and the Earth's Radiation Budget. *Physics Today* (May), 22.

REVKIN, C. 1988. Endless Summer: Living With the Greenhouse Effect. *Discover* 9:10, 50.

RYAN, P. R., ed. 1986. Changing Climate and the Oceans. *Oceanus* 29:4, 1.

SHACKLETON, N. J. 1990. Estimating Atmospheric CO_2. *Nature* 347:6292, 427.

WAGNER, A. J. 1989. Persistent Circulation Patterns. *Weatherwise* 42:1, 18.

WALLACE, JOHN M. 1992. Effect of Deep Convection on the Regulation of Tropical Sea Surface Temperature. *Nature* 357:6375, 230.

ZEBIAK, S. E. 1989. Ill Winds: How Events in the Tropics Throw the World's Weather out of Whack. *The Sciences* 29:2, 26.

When an equatorial current runs into a continent, it "squirts" out the sides, just like water from a garden hose that strikes the side of a building.

Variations in the height of the ocean surface due to surface currents. The highest levels are indicated by white (e.g., in the western Pacific), and the lowest by the purple (e.g., near Antarctica). Total vertical relief from lowest to highest is about 2 meters.

Eight

OCEAN CURRENTS

Scotland, Scandinavia, southern Greenland, Alaska, and the coast of Antarctica are all at similar latitudes–about 60° north or south of the equator. Although all receive about the same sunshine during the year, their climates are quite different. Some have supported forests, grasslands, and thriving human societies for ages. Others are frozen, bleak, and virtually lifeless. The difference is that some are bathed by warm ocean currents, and others are not. In this chapter, we study these and other ocean currents, drawing from our knowledge of atmospheric circulation learned in the preceding chapter.

One consequence of the intimate connection between ocean and atmosphere is that each drives the other. Wind and weather are driven by the heat that the atmosphere receives from the Earth (mostly the ocean) beneath it. Some of this energy is returned to the ocean as the winds drive currents along the ocean surface. The wind and weather also produce the denser water that sinks and flows through the dark deep regions of the ocean. Because the ocean is so massive, however, its motions are much steadier and more predictable than those of the atmosphere.

8.1 SURFACE AND DEEP WATERS

It is convenient to divide ocean currents into two categories: those that flow along the surface, and those that do not. It is often convenient to further subdivide these two main categories—especially the deeper waters—into various **water masses**. Each water mass is distinguished by properties that include the water's temperature, salinity, density, and trace chemicals. Most deep waters originally formed near the surface, but at different places and under different conditions that gave them different properties. Once these deep waters have formed and sunk beneath the surface, they are insulated from those surface processes that might change their characteristics. Therefore, they retain these properties for long periods of time and over long distances.

The surface currents are subject to the forceful but sometimes capricious influences of the wind and weather, whereas the deeper currents are not. The surface currents tend to be swifter, but the deeper currents are more voluminous. Surface water is less dense than the water beneath it. Within the deeper water, the denser water sinks or slides downward and buoys the less dense water upward as it goes. Therefore, an understanding of the factors that influence the density of sea water is essential for understanding both the difference between the surface and deep waters, and the motions of the various deeper water masses as well.

WATER DENSITY

The three factors determining the density of sea water are its pressure, temperature, and salinity. The density increases slightly with increased pressure or salinity, but decreases with increased temperature.

For purposes of studying deep currents, we wish to know the densities of water masses at the same pressure. For example, if two water masses encountered each other at the same depth, which would sink and which would rise? Because we are interested in the *relative* densities of water masses when under similar pressures, we measure all densities at the same pressure. It is most convenient to do this aboard ship or in our laboratories, so the standard pressure used for these measurements is atmospheric pressure. We can simply retrieve the sample from depth and measure its density when it arrives at the surface.

Since all densities are measured at atmospheric pressure, they depend on only the temperature and salinity of the sample. Densities are generally in the range of 1.024 to 1.028 grams per cubic centimeter. For comparison, the density of fresh water is 1.000 grams per cubic centimeter.[1] Because the first two digits are always the same, it is customary to ignore the 1 and move the decimal point over three places. The resulting number is called the σ_t or **sigma-tee**. For example, if the density is 1.026 grams per cubic centimeter, then the σ_t of the sample is 26.

VARIATIONS IN TEMPERATURE AND SALINITY

Water is nearly incompressible, so that large changes in pressure cause only very slight changes in its density. Even at the tremendous pressure of the deep sea floor, which is around 4000 metric tons per square meter, the water's density is only about 2% greater than that at the surface. Therefore, as a water sample is brought up from the deep ocean floor, it expands only very slightly. This slight expansion causes slight cooling, amounting to only 0.1 or 0.2°C. Nonetheless, the temperature of the sample when it is brought aboard ship, called the **potential temperature**, is slightly cooler than its temperature at depth, called its **in situ** temperature.

[1] This is by definition of a gram, which is the mass on 1 cubic centimeter of fresh water.

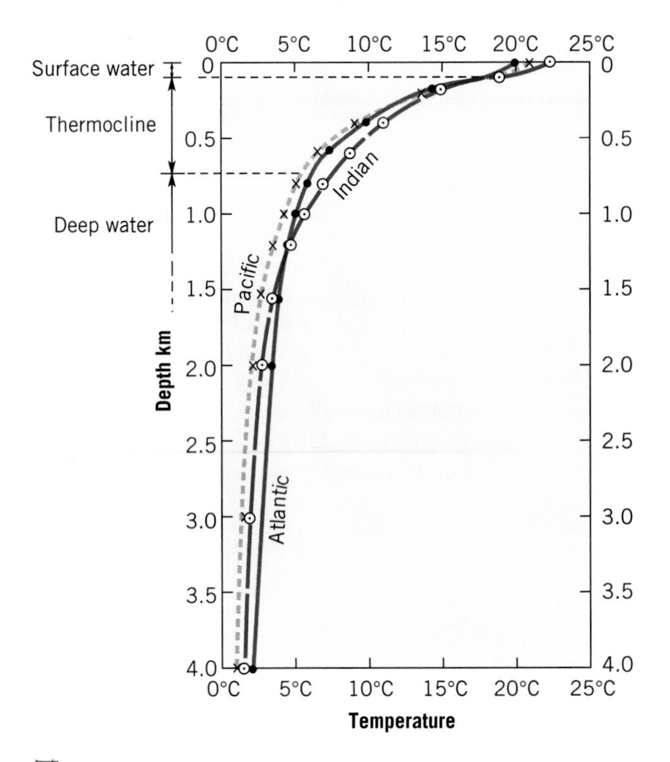

Figure 8.1

Vertical temperature profiles for all three oceans, averaged over the tropical and subtropical regions of each. (Data from: A. Defant, 1961, *Physical Oceanography*, **Vol. III**, Pergamon Press.)

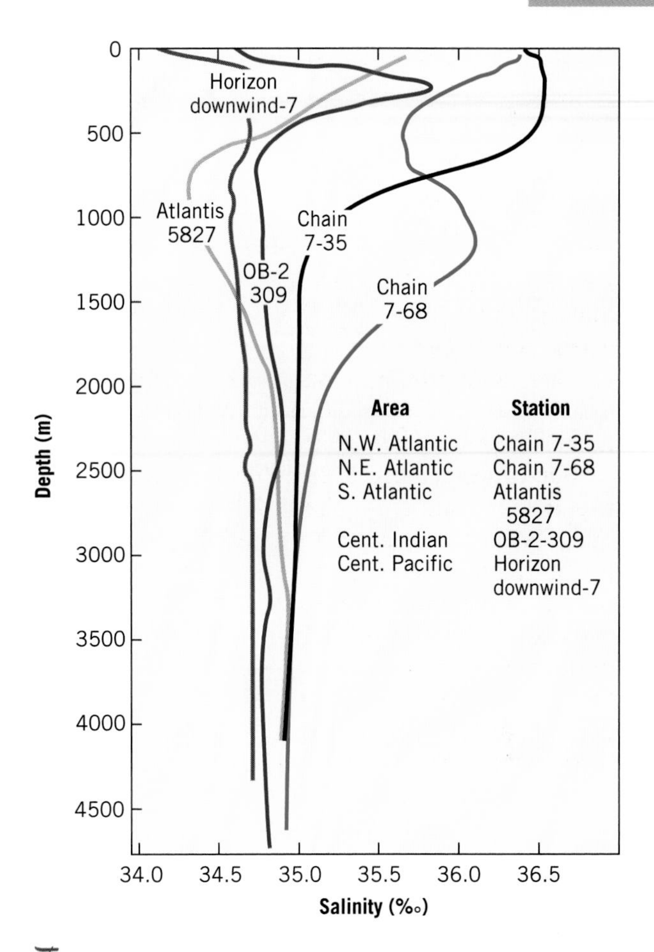

Figure 8.2

Some examples of how salinity varies with depth at five different places: three in the Atlantic Ocean, one in the Indian Ocean, and one in the Pacific Ocean. Notice that the two stations with low surface salinity are near the equator where precipitation is high. The others are in the horse latitudes where evaporation is high. The bulge at 1000 to 1500 meters in the Chain 7-68 profile is due to warm salty water flowing out of the Mediterranean Sea. (From Hugh J. McLellan, 1965, *Elements of Oceanography*, Pergamon Press.)

The largest variations of both temperature and salinity are found near the surface (Figures 8.1 and 8.2). The reason for this is that changes in temperature and salinity are almost entirely due to solar heating, evaporation, precipitation, and the freezing or melting of ice (Figure 5.21), and all these processes occur at the surface.[2] Most subsurface water masses originally formed when they were at the surface, so their various temperatures and salinities reflect the different regions and conditions where they first formed.

SURFACE WATERS

For the surface waters, the temperature, salinity, and density vary with latitude in a pattern illustrated in Figure 8.3. It is what you would expect. In the tropics the surface water is warmer, and this warmer water is less dense. The salinity is high in the tropics due to heavy evaporation. There is a dip at the equator due to the heavy afternoon rainfalls there. At high latitudes, the lower rate of evaporation means lower salinity. Notice in particular that the far southern waters are denser than their far northern counterparts, even though their temperatures are the same. The difference is that the southern waters are saltier.

From this diagram, it should be clear that the water's density is generally more sensitive to variations in temperature than variations in salinity. Also, it should be clear to you why the deep water masses are formed at high latitudes, and why those formed in the south are generally denser than those formed in the north.

The influence of surface processes like heating, cooling, evaporation, and precipitation is mixed

[2]The water exiting hydrothermal vents affects the salinities of local waters near these sources, but it does not much affect the salinities of the major water masses.

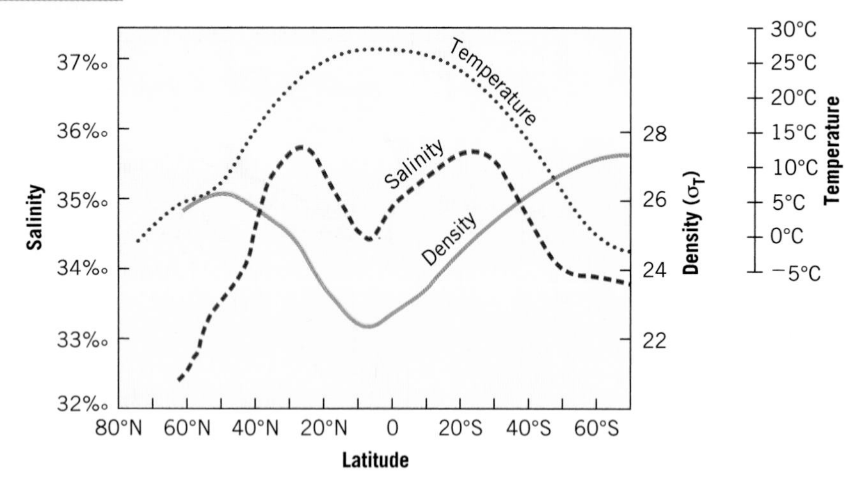

Figure 8.3

Density of surface waters as a function of latitude, averaged over all oceans. The average temperature and salinity are also plotted. Can you explain their overall patterns, such as the peak in salinity near the tropics and that in temperature near the equator? Can you explain the density variations in terms of these two factors? (From G. L. Pickard, 1964, *Descriptive Physical Oceanography*, Pergamon Press.)

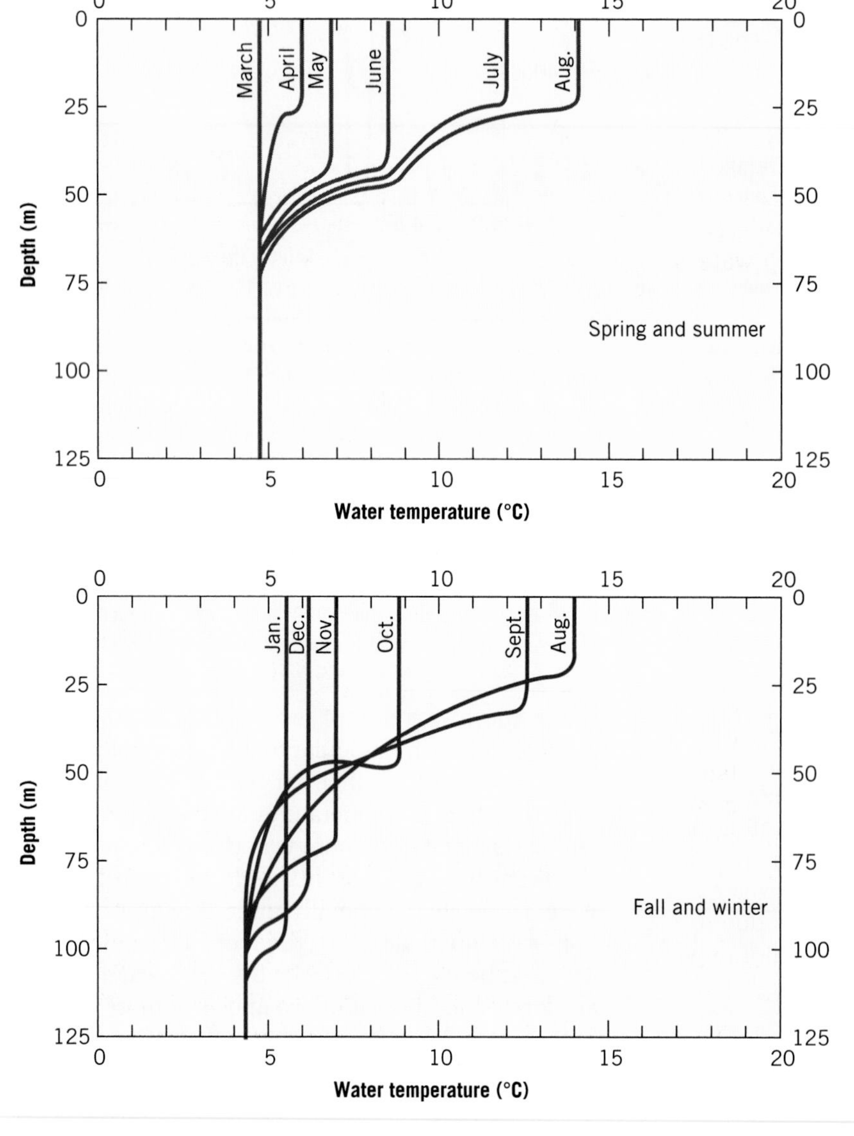

Figure 8.4

Typical seasonal changes in the temperature of near-surface water for subarctic latitudes. Turbulence from waves keeps the water near the surface uniform and well mixed. Notice that the smaller waves of summer do not mix the surface heat as deeply as the larger waves of the stormy winter months. (After A. J. Dodimead, F. Favorite, and T. Hirano, 1963, Review of the Oceanography of the Subarctic Pacific Region, *International North Pacific Fisheries Commission, Bulletin 13*.)

downward by waves and turbulence. The exact patterns depend on the latitude, but the pattern of Figure 8.4 is typical at high latitudes. In general, summer heating does not mix as deeply as winter cooling. One reason for this is that the summer waves are smaller. Another is that the warmer summer surface waters are lighter and, therefore, have a greater tendency to stay close to the surface.

Because of mixing, the boundary between surface and deeper water is somewhat blurred, typically beginning at 100 to 400 meters' depth and extending several hundred meters downward from there. Such a boundary region, over which the density of water changes quickly, is called a **pycnocline** and is due to a rapid change in temperature (**thermocline**), or salinity (**halocline**), or both (Figure 8.1).

Surface waters remain above and distinct from deeper waters at tropical and temperate latitudes. At high latitudes, however, there is seasonal **vertical mixing** (*overturn*). The lighter surface waters appear in the summer, as the warming sun and melting ice raise the water's temperature and lower the salinity. As winter comes, however, these lighter surface waters disappear again. As ice freezes, it tends to eject the salt, making the winter surface water cold, salty, and dense. This water is often so dense that it sinks, and deeper water comes up to replace it. This is the mechanism by which some important deep water masses are formed.

DEEPER WATERS

Although the surface water in temperate and tropical regions is warm, the bulk of the ocean's water is not near the surface, and most of this subsurface water was formed in cold polar regions. The temperature profiles of the three world oceans, averaged over the tropical and subtropical regions, are shown in Figure 8.1 You can see that only the water near the surface is warm. The majority of the water in our oceans is between 0 and 3°C, as is seen in Figures 8.1 and 8.5a.

Because the subsurface waters are much more voluminous than those at the surface, they show up as much thicker bulges and dents in the plots of temperature or salinity vs. depth, such as those of Figure 8.2. As you can see in this figure, variations are most pronounced near the surface, and the deepest water masses are fairly uniform. The deepest water masses include most of the water in the ocean and have salinities in the range between 34.5 and 35.0‰ (Figures 8.2 and 8.5b).

Deep water masses form at or near the surface and then sink, carrying with them the characteristics acquired while they were forming. At depth, they are insulated from the capricious surface environment, so they remain pretty much the same as when they were formed. There may be some slight changes due to slow mixing with neighboring water masses, respiration of organisms, hydrothermal activity, or heat received from the ocean bottom. However, the deep waters are so voluminous that these processes have very little effect overall.

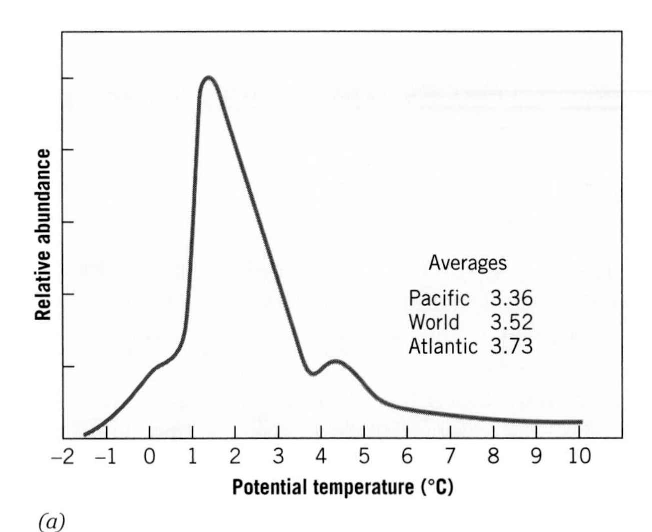

(a)

(b)

Figure 8.5

Plot of the relative abundances of ocean waters as a function of (a) potential temperature, and (b) salinity. The deep waters are by far the most voluminous and abundant. So the fact that most of the ocean water is between 0 and 3°C and has salinities in the range 34.5 to 35‰ is due to the deep waters (Figures 8.1 and 8.2). (After R. B. Montgomery, 1958, *Deep Sea Research* **5**.)

8.2 FORCES THAT DRIVE OCEAN CURRENTS

THE PRIMARY DRIVERS

The primary forces that drive the ocean currents are easy to understand. The surface currents are driven by the winds. The deeper currents are driven by gravity; denser waters flow downhill. They sink until they encounter either the ocean floor, or another water mass that is denser than themselves, and then they flow downhill along this interface. For both the surface and deeper currents, however, the motion is severely modified by ocean boundaries, the Coriolis effect, and pressure gradients.

THE CORIOLIS EFFECT

Because the major ocean currents travel large distances over long periods of time, the effect of the Earth's spin is very pronounced. As deep water flows downhill along the deep ocean floor, it tends to curve to the right (Northern Hemisphere) and bank up against the right-hand margin of the basin as it goes (Figure 8.6). On the surface, as water enters and leaves a coastal embayment with the rising and falling of the tides, the water banks up against the side on its right as it enters and again banks up against the side on its right (i.e., the other side) as it leaves.

A similar thing happens to wind-driven surface currents, which we now examine in a little more detail. First, consider the idealized situation where the water is deep and subject to no constraints other than the force of the wind. As might be expected, when currents are driven by the wind, the water on the very surface moves the fastest. Deeper water must be dragged along by the water above. In the Northern Hemisphere, the surface water tends to flow to the right of the wind direction, with the Coriolis effect tending to deflect it further to the right and the wind trying to straighten it back up in line with itself. An equilibrium is reached, which for perfect conditions results in the surface water flowing at 45° to the right of the wind.

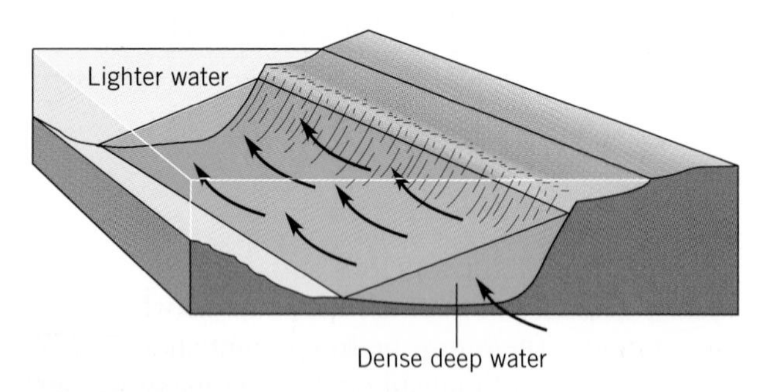

Figure 8.6
Deep currents flow extremely slowly and therefore are heavily influenced by the Coriolis effect. In the Northern Hemisphere, they bank up against the right-hand margin of the ocean basins as they flow.

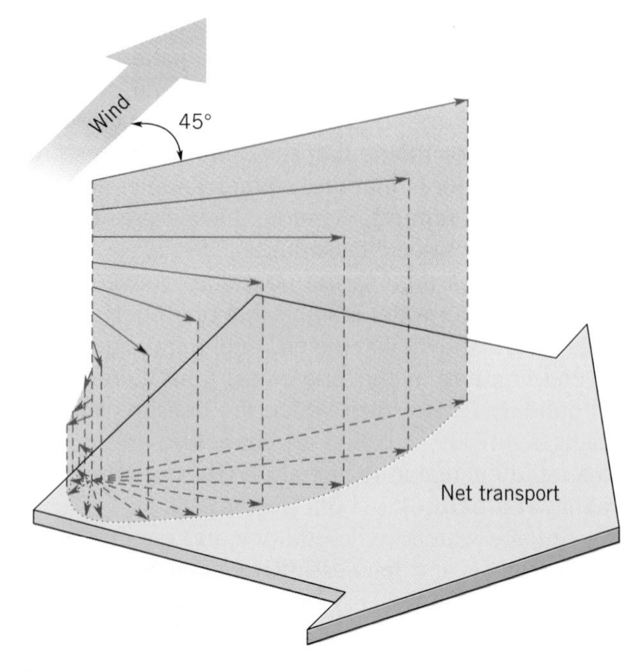

Figure 8.7
The Ekman spiral for the Northern Hemisphere. Due to the Coriolis deflection, the water flows to the right of the wind direction. The surface current flows 45° to the right of the wind, and successively deeper layers flow more slowly and further to the right. Under the ideal conditions of water being subjected to no other forces or constraints, the net water transport would be 90° to the right of the wind direction.

Below the surface, each layer of water is dragged along by the layer above it. Because of the Coriolis effect, each layer tends to go to the right of the direction it is being dragged, so each layer moves to the right of the layer above it. The resulting flow pattern is called the **Ekman spiral**, and illustrated in Figure 8.7. At each level in the Ekman spiral, the current is flowing slower and farther to the right, compared to the layer above. That is, the spiral describes the direction and intensity of current flow at various levels below the surface. Because the Coriolis deflection increases at higher latitudes, the spiral is "wound" more tightly.[3] At 35° north latitude, for example, the

[3]It is not wound at all at the equator, where there is no Coriolis effect.

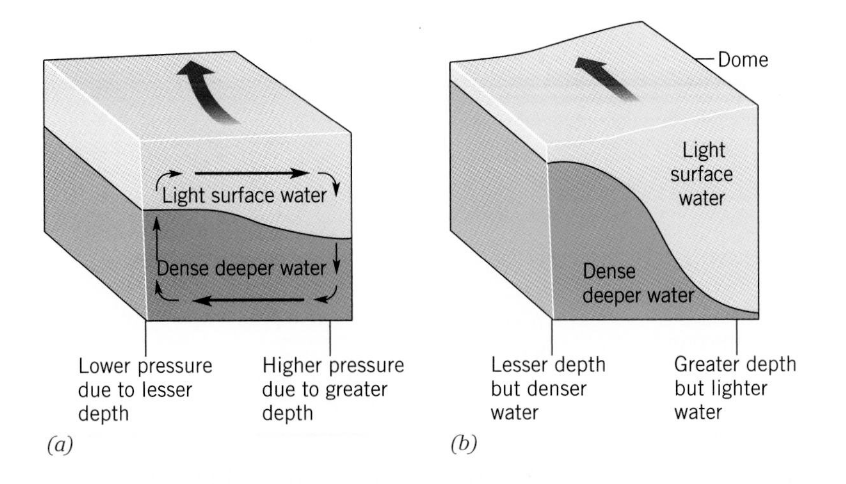

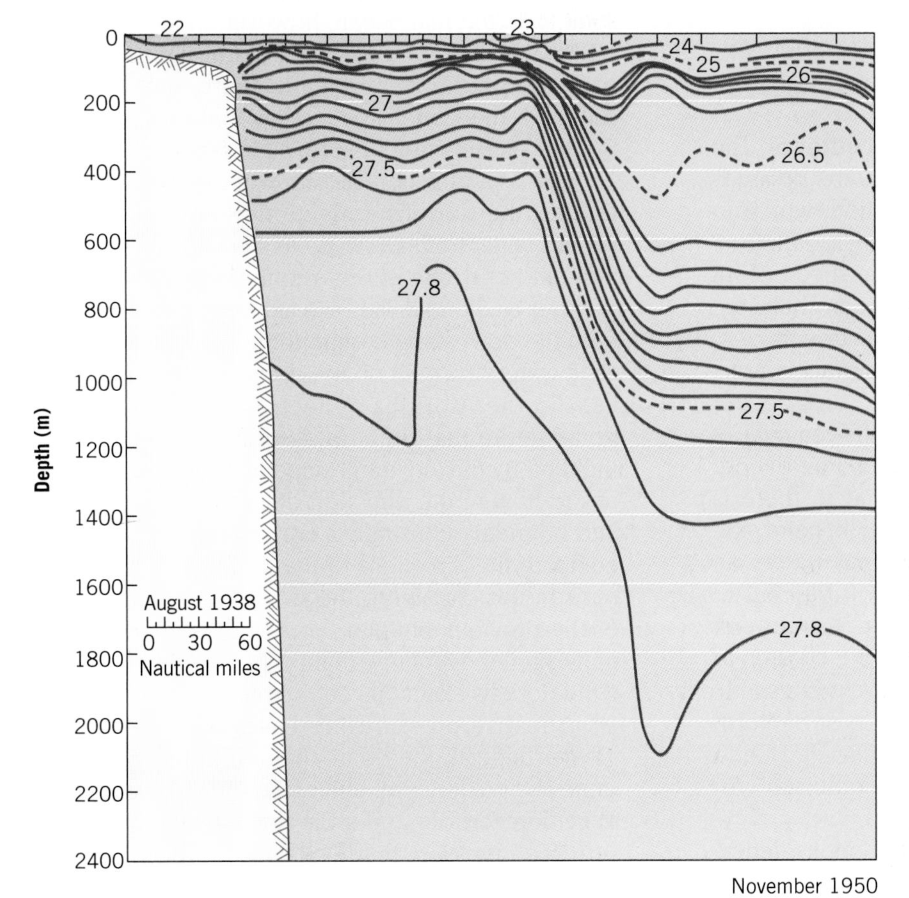

Figure 8.8

Coriolis deflection of currents causes surface water to accumulate on the right (Northern Hemisphere). On the left, deeper water rises to replace the departing surface water. This results in a slope to the density contours across the current. (a) Beginning stages. (b) After equilibrium is reached. (c) Cross section of the Gulf Stream between Cape Cod (left) and Bermuda (right), showing contours of constant density. (The numbers should be read as follows: 27 means 1.027 g/cm^3, 27.5 means 1.0275 g/cm^3, etc.) Denser water is on the north (left) side and lighter water is on the south (right) side. Vertical exaggeration is about 500 times. (F. C. Fuglister, Woods Hole Oceanographic Institution.)

depth at which the current flows at 180° to the direction of the surface current (and 225° to the wind direction) is about 100 meters. At this point, the current velocity is only 4% as great as that at the surface.

In deep water, the average of all these current components gives a *net Ekman transport* that is 90° to the right of the wind direction. In shallower water, the bottom of the Ekman spiral is missing, and the net water transport is not so far to the right.

DENSITY CONTOURS

In the Northern Hemisphere, the deflection of surface waters to the right results in the accumulation of light surface waters on the right-hand side of major surface currents (Figure 8.8). As the surface waters slide from left to right, deeper water rises on the left to replace them. This is illustrated in Figure 8.8c, which shows the water density at various depths in a cross section across a surface current. The lines connect points of

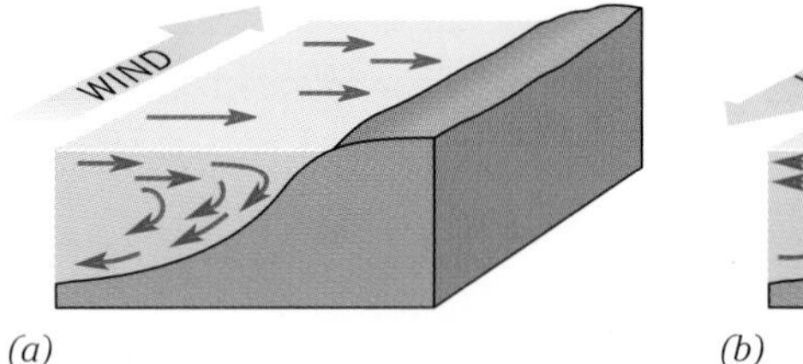

(a)

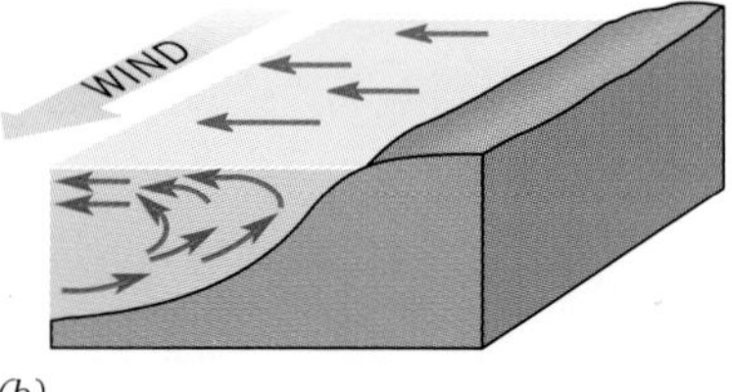

(b)

Figure 8.9

Coastal upwelling and downwelling in the Northern Hemisphere. (a) When the coast is to the right of the current, the Coriolis deflection causes downwelling along the coast. (b) When the coast is to the left of the current, the Coriolis deflection creates upwelling along the coast.

equal density and are called **density contours**. The amount of slope to these density contours depends on the strength of the Coriolis deflection, which in turn depends on the latitude and current speed. Consequently, one way of determining the speed of an ocean current from a ship at sea is to measure the slope of the density contours across the current.

Notice that the accumulation of surface waters on the right (Northern Hemisphere) forces deeper underlying waters down. Conversely, the removal of surface waters from the left side means that deeper waters must rise to replace them. That is, there is **downwelling** of surface water on the right-hand side of the current and **upwelling** of deeper waters on the left. For coastal currents, there will be coastal downwelling if the coast is to the right of the current, and there will be coastal upwelling if the coast is to the left (Figure 8.9). The California current is a southward-flowing current along the West Coast of North America, meaning that the coast is on its left. Due to the upwelling of cooler water, San Francisco can boast of being an "air conditioned city."

In the Southern Hemisphere, the Coriolis deflection is to the left. Therefore, the downwelling is on the left side of the current and the upwelling on the right.

DOMES AND PRESSURE GRADIENTS

From the discussion of the previous section, you might conclude that wind-driven surface currents should always flow sideways to the wind direction—to the right of the wind in the Northern Hemisphere, and to the left in the Southern Hemisphere. However, that is not the way things are. Many of the major surface currents flow parallel to the wind, and not sideways to them as you might expect. We now investigate why that is.

The Ekman spiral results from the Coriolis effect and describes the flow of surface currents in the idealized situation where the water is subject to the force of the winds only, and no other forces or constraints. In the real ocean, however, there indeed are other very important forces and constraints, such as pressure gradient forces and the ocean margins. Although the Coriolis deflection is always present, it is often severely modified by other factors.

Just as a plow pushes snow or dirt to the sides as it goes, the Coriolis deflection of surface currents pushes water to the the side as they flow, resulting in a mound or **dome** of water to the side of the current (Figures 8.8a and 8.8b). The accumulation continues until the downslope component of gravity tending to slide water downhill off the dome is sufficient to prevent further Coriolis deflection of water into the dome. From that time on, the Coriolis deflection is nullified by this downhill slope and the water continues to flow in the direction determined by the wind and boundary constraints. For this reason, when you compare the directions of the dominant surface currents in this chapter to those of the dominant winds of the previous chapter, you will see that they are quite similar over large regions of the open ocean, as if the Coriolis effect played no role.

Because pressure increases with depth, it is greater beneath the dome than off to the side at the same level. Therefore, below the surface, water tends to get forced from under the dome by the extra pressure (i.e., pressure gradient forces). Thus, once the dome has developed, the pressure gradient force beneath the surface counteracts the Coriolis deflection of more water, just as the downslope component of gravity counteracts the Coriolis deflection of water on the very surface. That is, once the dome has built up, the water both on and below the surface tends to flow sideways along the dome in the direction of the wind. The Coriolis deflection is nearly nullified by the downslope component of gravity at the surface and pressure gradient forces beneath the surface (Figure 8.10). The only evidence of Coriolis deflection is the presence of the dome and the slope to the density contours beneath the surface.

Typical surface currents are 50 to 1000 kilometers in width, and the dome's upward slope is only 1 or 2 meters across its entire breadth. Therefore, the dome is not noticeable, except with special satellite imaging techniques, and our illustrations of this are exaggerated considerably.

GEOSTROPHIC FLOW

Suppose that the wind stops. In this case, gravity would cause water on the very surface to start flowing downhill off the dome, and the pressure gradient would force out subsurface water from under the dome. But these waters flowing away from the dome would experience Coriolis deflection and end up flowing sideways along the edge of the dome again, just as they were when the wind was blowing. We encountered this type of flow, called *geostrophic flow*, in the previous chapter, where we learned that winds tend to flow sideways to the pressure gradients. The same thing happens in the oceans, and it has the same name.

Most major ocean surface currents are a combination of *wind-driven* and *geostrophic*. The winds are the basic driving force behind these currents, but geostrophic effects ensure that the currents continue to flow at a fairly constant rate even during periods when the wind is stopped. The ocean surface currents, then, are much more steady than the winds that drive them.

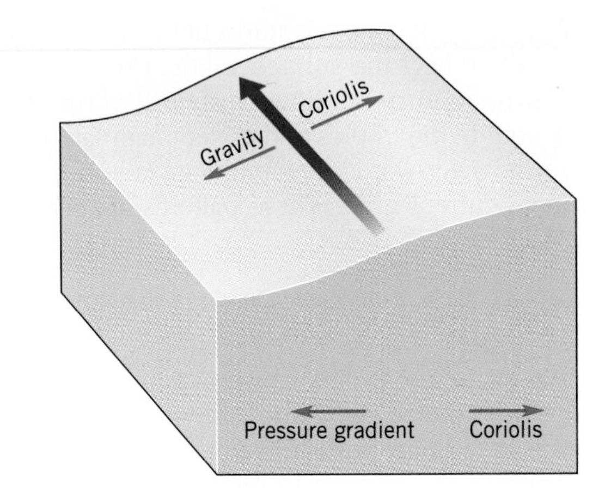

Figure 8.10

When equilibrium is reached, the surface water tends to flow in the direction of the wind and sideways to the dome. Further Coriolis deflection of water into the dome is opposed by the downslope component of gravity for water on the very surface, and the pressure gradient force for water beneath the surface.

8.3 LARGE-SCALE SURFACE CURRENTS

GENERAL FEATURES

Figure 8.11 illustrates an idealized ocean, with the surface waters being driven by the wind patterns as seen in Figures 7.19 and 7.20. As we learned in the previous section, the Coriolis deflection causes slight

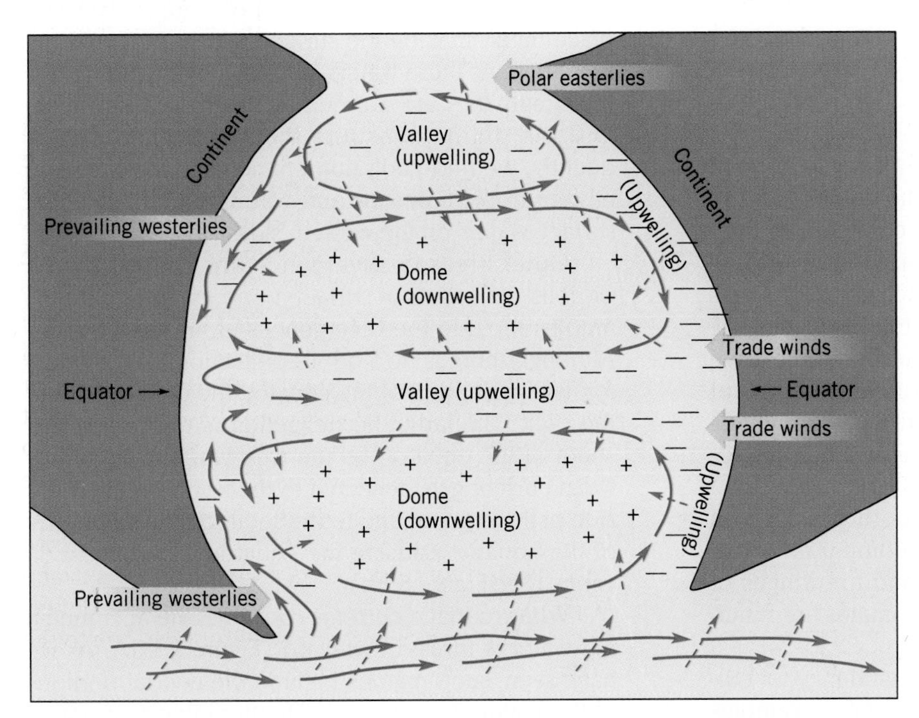

Figure 8.11

Circulation of wind-driven surface currents in an idealized ocean. Direction of the dominant surface winds (trade winds, westerlies, and polar easterlies) is indicated with the yellow arrows. Flow of surface currents is indicated with solid arrows. The direction of the Coriolis deflection of surface waters is indicated with dashed arrows. Because of this Coriolis deflection, surface water accumulates in some regions, forming domes and causing downwelling. These regions are indicated with + signs. Surface waters are deflected away from other regions, forming valleys and causing upwelling. These regions are indicated with – signs.

variations in the surface elevations, measuring 1 or 2 meters over distances of several hundred kilometers. In the Northern Hemisphere, there is a slight dome and downwelling on the right side of the current, and a slight valley and upwelling on the left. In the North Atlantic, for example, you can see that there is downwelling in the central subtropical region, and upwelling along the equator and the continental margins. In the Southern Hemisphere where the Coriolis deflection is to the left, the domes and downwelling are on the left of the current direction, and the valleys and upwelling on the right.

The surface current patterns of the idealized oceans of Figure 8.11 are clearly seen in the real oceans, as illustrated in Figure 8.12. Within the tropics, the trade winds drive the **equatorial currents** westward. In each ocean, this westward flow is eventually blocked by a continent, causing the equatorial current to split, some flowing northward and some southward along the ocean's western margins.

At temperate latitudes, the prevailing westerlies drive back the surface currents eastward. The most pronounced of these eastward temperate currents is the **Antarctic Circumpolar Current** or **West Wind Drift** of the Southern Hemisphere. You can see that it flows entirely around the globe, because there are no continents to block its passage as there are for its Northern Hemisphere counterparts.

In this fashion, the trade winds and westerlies combine to drive the tropical and temperate surface waters around the oceans in huge **current gyres**. This circular motion is clockwise in the Northern Hemisphere and counterclockwise in the Southern Hemisphere. Therefore, the flow is *anticyclonic* (i.e., opposite to the direction of winds around low-pressure centers; see Chapter 2). These large current gyres are the dominant features in the tropical and temperate oceans in both hemispheres.

One important consequence of these current gyres is to carry heat away from the equator and to higher latitudes. Tropical waters get heated in their long westward trek across the oceans under the guidance of the trade winds. Upon reaching the ocean's western boundary, they then flow toward higher latitudes, where the prevailing westerlies then drive them back toward the eastern ocean margins at higher latitudes. This has an appreciable warming effect on the climates of both Alaska and Europe, for example, that are at latitudes quite a bit higher than their climates would indicate. Both Alaska and Northern Europe are at latitudes comparable with parts of Greenland in the north, and the fringes of Antarctica in the south, but their climates are much warmer.

At very high latitudes, there is a tendency for current gyres to flow in the opposite direction. Examples shown in Figure 8.12 include the gyre east of Greenland, the one west of Alaska, and those in the Southern Hemisphere near Antarctica. Again, these reflect the pattern of surface winds in these regions. The polar easterlies drive the return flow of water carried east by the prevailing westerlies. These high-latitude gyres are much smaller than those at lower latitudes, both because of interference by continents, and because the Earth is not as wide at these high latitudes. (Flat projections of the spherical Earth, such as that of Figure 8.12, tend to stretch and exaggerate these high-latitude regions.)

In summary, the main features of the surface currents simply reflect the surface winds. Due to these winds, surface currents flow westward across the oceans both in the tropics and at very high latitudes, and they flow eastward at intermediate latitudes, creating large current *gyres*. These patterns are seen in Figure 8.12.

CONVERGENCE AND DIVERGENCE

The flow of these currents causes surface waters to diverge in some regions and converge in others (Figure 8.13). Where surface waters *diverge*, deeper waters rise to the surface to replace them (Figure 8.14a). Such upwelling generally results in cooler surface temperatures and flourishing biological communities, because the upwelled deep waters are both cold and rich in nutrients.

Among the important regions of divergence are the eastern sides of all oceans in the tropics, where trade winds blow the surface waters away from the continents (Figure 8.14b). Deeper water rises to replace them, making the surface waters cool and nutrient-rich. As these waters are blown westward across the ocean, they are warmed by the tropical sunshine, and the plants consume their nutrients. Consequently, we find cool, nutrient-rich surface waters in eastern tropical oceans and warm, nutrient-deficient surface waters on the western sides.

Other areas of divergence are caused by the Coriolis deflection of surface currents. Important examples include the divergences along the equator. Although there is no Coriolis deflection on the equator itself, just north of the equator, the deflection is to the right. Similarly, the deflection is to the left, just south of the equator. So the westward-flowing equatorial current gets deflected to the north on the north side of the equator and to the south on the south side of the equator, causing the equatorial current to diverge. (See Figure 8.11.)

Where surface currents *converge*, the accumulating water is thrust downward (Figure 8.14b). As we have seen, such regions include the central portions of the major current gyres. Another important region

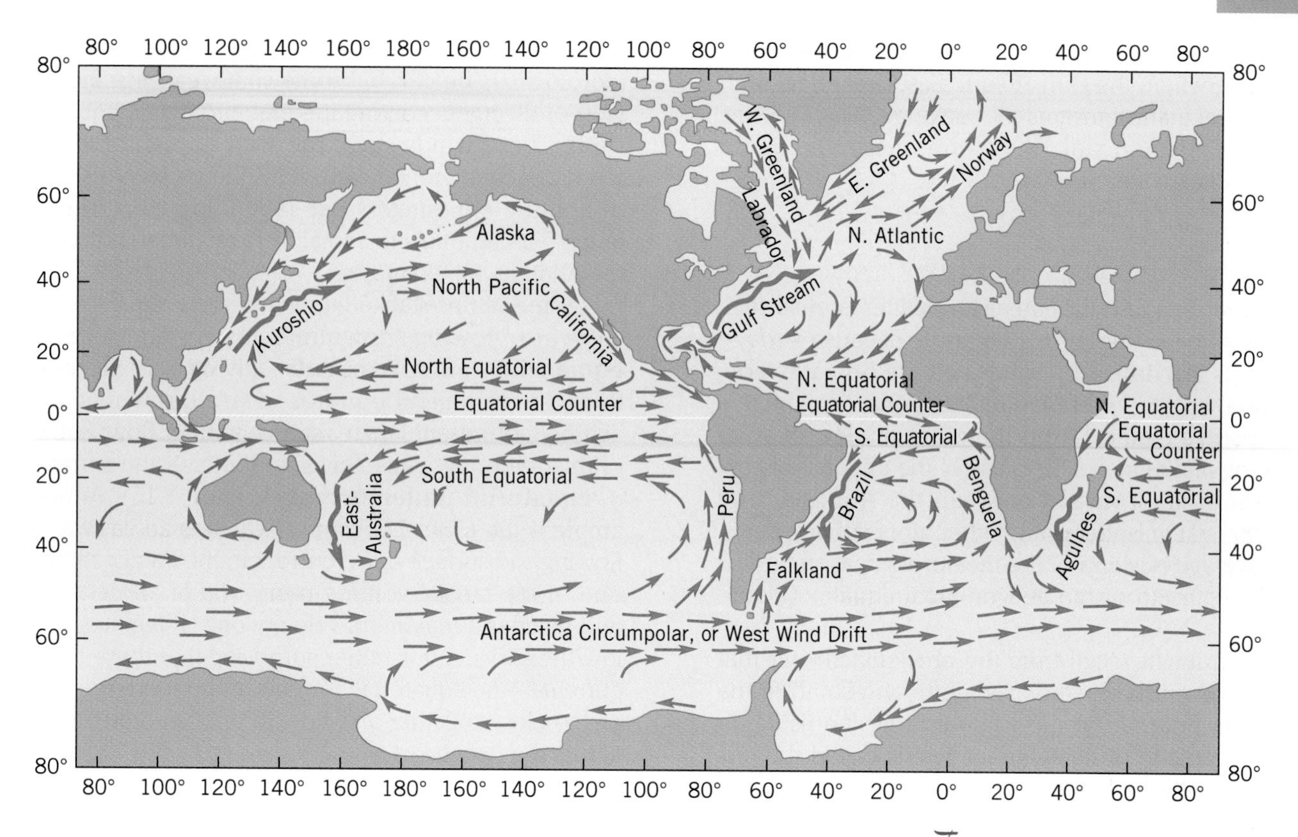

Figure 8.12

Circulation of the oceans' surface waters. The names of the major currents are indicated.

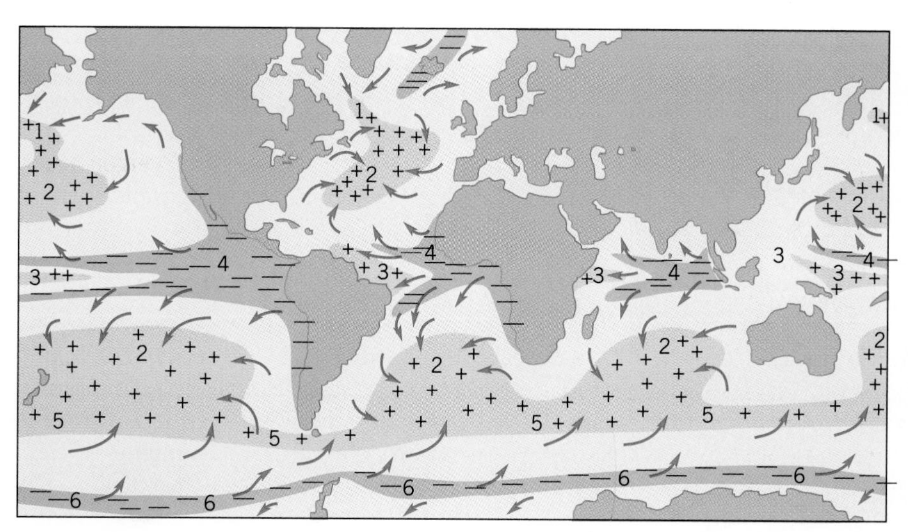

Figure 8.13

General areas of convergence (+) and divergence (–) of surface currents in the oceans. They are named as follows: (1) Arctic convergence, (2) subtropical convergence, (3) equatorial convergence, (4) equatorial divergence, (5) Antarctic convergence, (6) Antarctic divergence.

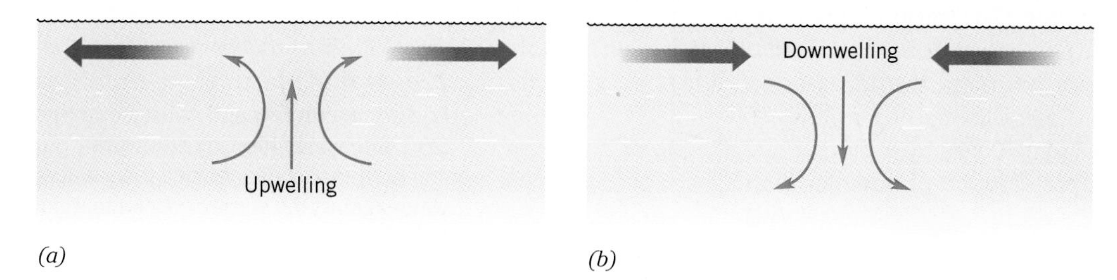

Figure 8.14

(a) Upwelling. Where surface currents diverge, deep water must surface to replace it. (b) Downwelling. Where surface currents converge, the water is forced downward.

of convergence is south of Greenland in the North Atlantic, where the Gulf Stream, East Greenland, and West Greenland currents all converge (Figure 8.12). As they mix and sink, they form one of the major deep ocean water masses.

EQUATORIAL COUNTERCURRENTS

We now examine the interesting patterns of surface currents near the Earth's equator. Because surface currents are driven by winds, we are primarily concerned with the **meteorological equator**, which is the Earth's equator from the point of view of solar heating and climates. Because of the unequal distributions of land masses between the Northern and Southern Hemispheres and other more subtle effects, the average position of the meteorological equator is about 5° north of the geographical equator (Figure 7.16).

You might recall from the previous chapter that the trade winds are strongest north and south of the equator, but that on the equator itself, the rising air creates a zone of light surface winds called the *doldrums*. Therefore, the westward-flowing equatorial currents are strongest just north and south of the meteorological equator, but somewhat weaker in the region of divergence along the meteorological equator itself. These winds stack the water up against the western ocean margin, with a resulting upward slope of about 4 centimeters per 1000 kilometers of distance. This amounts to a total of 15 centimeters across the Atlantic, for example.

Some of this stacked-up water then returns back eastward, flowing "downhill" in swift and narrow **equatorial countercurrents**, staying very close to the meteorological equator where the winds are lighter and present less resistance to their flow. Some of this returning water flows underneath the surface in **equatorial undercurrents** (Figure 8.15). An example is the *Cromwell current*, which is an eastward-flowing, subsurface undercurrent in the Pacific. It has one of the largest volume transports of all currents and reaches a maximum velocity only 100 meters below the surface. It is rather surprising that these large currents—the equatorial currents and undercurrent—are so close together and flow in opposite directions across the ocean.

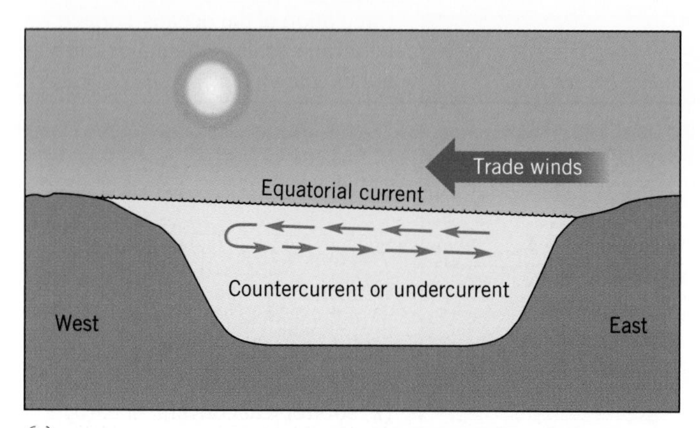

(a)

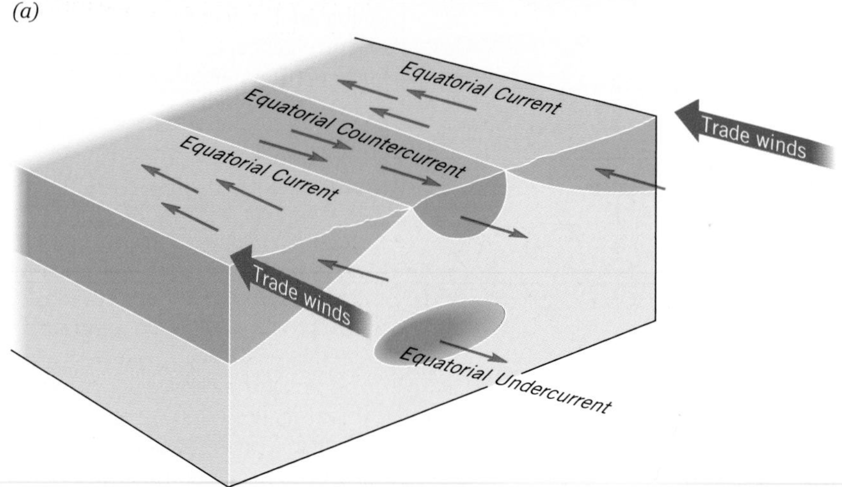

(b)

Figure 8.15

Equatorial currents and countercurrents (a) Trade winds blowing westward across tropical oceans stack the surface water against the western ocean margin. Countercurrents flow back downhill eastward across the oceans, sometimes beside and sometimes beneath the westward surface currents. (b) Perspective drawing illustrating equatorial currents, countercurrents, and undercurrents.

EL NIÑO AND THE SOUTHERN OSCILLATION

Normally, there is considerable upwelling along eastern tropical ocean margins, because the trade winds blow the surface water westward away from the adjacent continental margins. Deep water surfaces to replace it. Each year in October, however, the trade winds slacken and the upwelling is reduced. The warm tropical surface water that was blown up against the western margin begins to flow back downhill eastward across the ocean. Arriving back at the ocean's eastern boundary, it further suppresses the normal upwelling along the coast.

In the Pacific Ocean, where the effects are most pronounced, this occurrence is referred to as **El Niño** (meaning "the Christ Child" in Spanish), because this warm surface water arrives at the South American coast each year around Christmas time. Normally, the trade winds pick up again in early spring, the surface waters are again blown westward across the ocean, and everything is back to normal.

Sometimes, however, the trade winds fail to strengthen in the spring, and the warm surface waters remain off the South and Central American coasts for an entire year or longer. Upwelling continues to be suppressed or reduced by the overriding warm surface water and the failure of the trade winds to blow it away. This greatly reduces the nutrients at the surface, so plant productivities decline as well as the animal life that it normally supports. This is disastrous to the fishing industry, which normally flourishes in the upwelled waters. Although a small El Niño occurs every year, it is this occasional extended event that is commonly referred to as the **El Niño condition**.

The El Niño condition also influences climates on the two sides of the tropical Pacific Ocean. Normally, the surface water on the eastern side is cool, having risen from depths to replace the surface waters blown away by the trade winds. On the western side, the surface water is warm, because it has been receiving solar heating during its entire westward trip across the tropical ocean. Since warm air rises, we normally have rising warm moist air, low pressures, and storms on the warm western side of the ocean. (Remember that rising air masses leave low pressures

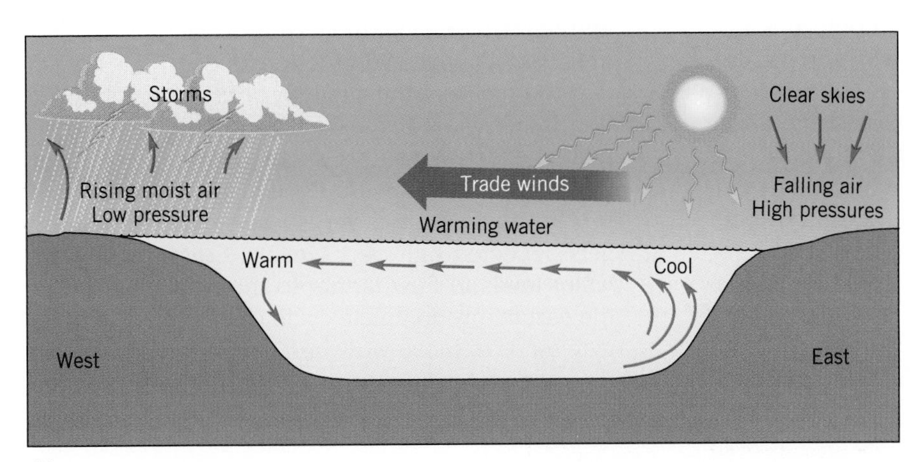

(a)

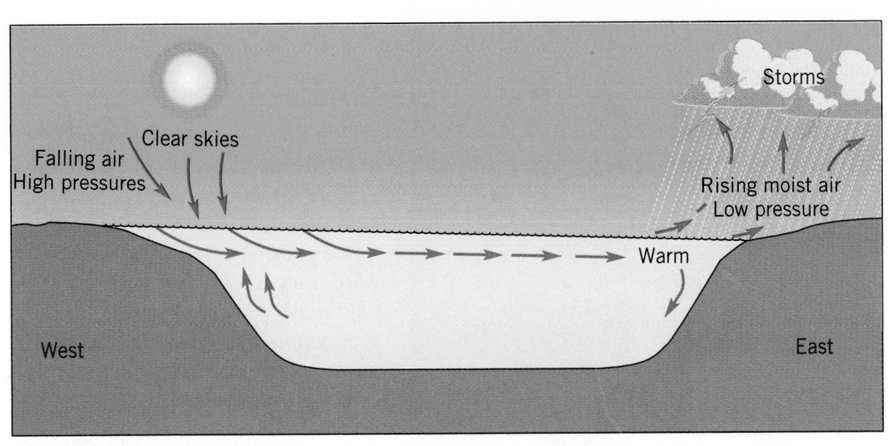

(b)

Figure 8.16

Southern oscillation. (a) Normally, surface waters are driven westward by the trade winds, and they get warmed by the tropical sunshine as they go. Plant life consumes the nutrients, so they arrive on the western sides relatively warm and lifeless. The warm surface waters create warm moist air above them, which rises and yields low pressures and storms on the western sides of the tropical oceans. On the eastern sides, however, are cooler surface waters and drier, falling air masses, high pressures, and clear skies. (b) During an extended El Niño event, trade winds are weak and the warm surface waters slosh back eastward across the tropical ocean. The warm, moist, rising air masses, and consequent low pressures and storms are found on the eastern side, rather than the west.

behind. As they rise, they expand and cool, causing moisture to precipitate, with resulting storms.) Similarly, we normally find falling dry air on the cooler eastern side, along with high pressures and clear skies (Figure 8.16a).

But during an extended El Niño condition, the pattern is reversed (Figure 8.16b). The warm water is on the eastern side of the ocean rather than the west. So the rising warm moist air, low pressures, and storms are found on the eastern side rather than the west. This exchange of pressure systems and weather patterns between the east and west sides of the tropical ocean during an El Niño condition is sometimes referred to as the **southern oscillation**.

WESTERN BOUNDARY CURRENTS

After studying the circulation of our atmosphere, with the *trade winds* blowing westward across the oceans in tropical latitudes and *prevailing westerlies* blowing eastward at higher latitudes, we can understand why the current gyres should form the dominant surface current pattern in the low and midlatitudes. What isn't so obvious, however, is the reason why these currents should be so swift and narrow along the oceans' western boundaries. What could possibly cause this **western intensification** of surface currents?

The Gulf Stream is just one of these swift narrow western **boundary currents**. A look at Figure 8.12 shows that there are similar intensified boundary currents along the western edges of all oceans, in both hemispheres. From considerations of the winds alone, you should think that each current could be half an ocean wide. But they are not, so we are faced with the problems of explaining why they are so swift and narrow, and why they occur on the western ocean margins only.

There are three related processes that contribute to the creation of strong narrow western boundary currents, all of which are products of the Earth's rotation and atmospheric circulation. The first cause is that when an equatorial surface current runs into the continent on the ocean's western margin, it "squirts" out the sides, just like water in a stream that strikes a rock, or water from a garden hose that strikes the side of a building.

When an equatorial surface current runs into the continent on the ocean's western margin, it "squirts" out the sides, just like water in a stream that strikes a rock, or water from a garden hose that strikes the side of a building.

Second, the Coriolis deflection is stronger in the portion of the gyres at higher latitudes, where these eastward-flowing waters are deflected toward the equator (Figure 8.17). This pinches the equatorial currents and tends to prevent them from leaving the equator until they reach the very western end.

The fact that the strength of the Coriolis deflection increases at higher latitudes produces another related effect. When this water is flowing east, it gets deflected quickly toward the equator, whereas when it is flowing west, it is very close to the equator and gets deflected only very weakly. Consequently, the water tends to flow farther to the west than toward the east in any complete cycle (Figure 8.17), and the gyre tends to move westward across the ocean each time the water flows around it. This westward tendency forces the gyre up against the western margins

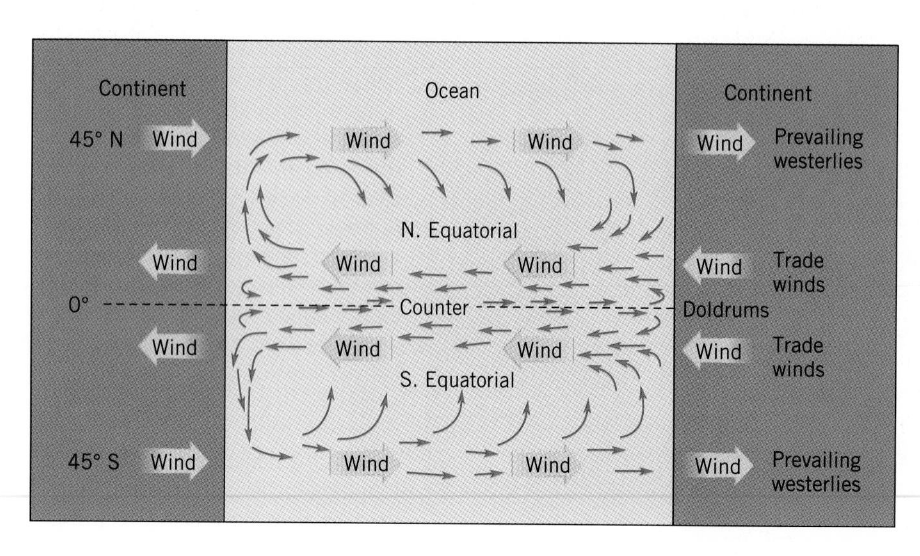

Figure 8.17

Wind and ocean circulation patterns in an idealized ocean. Because the Coriolis effect is stronger at higher latitudes, these eastward flowing waters are strongly deflected back toward the equator. The surface waters tend to flow toward the equator everywhere except along the very western edge of the ocean, and the gyres migrate westward, squashing up against the continents along the western margins of the oceans.

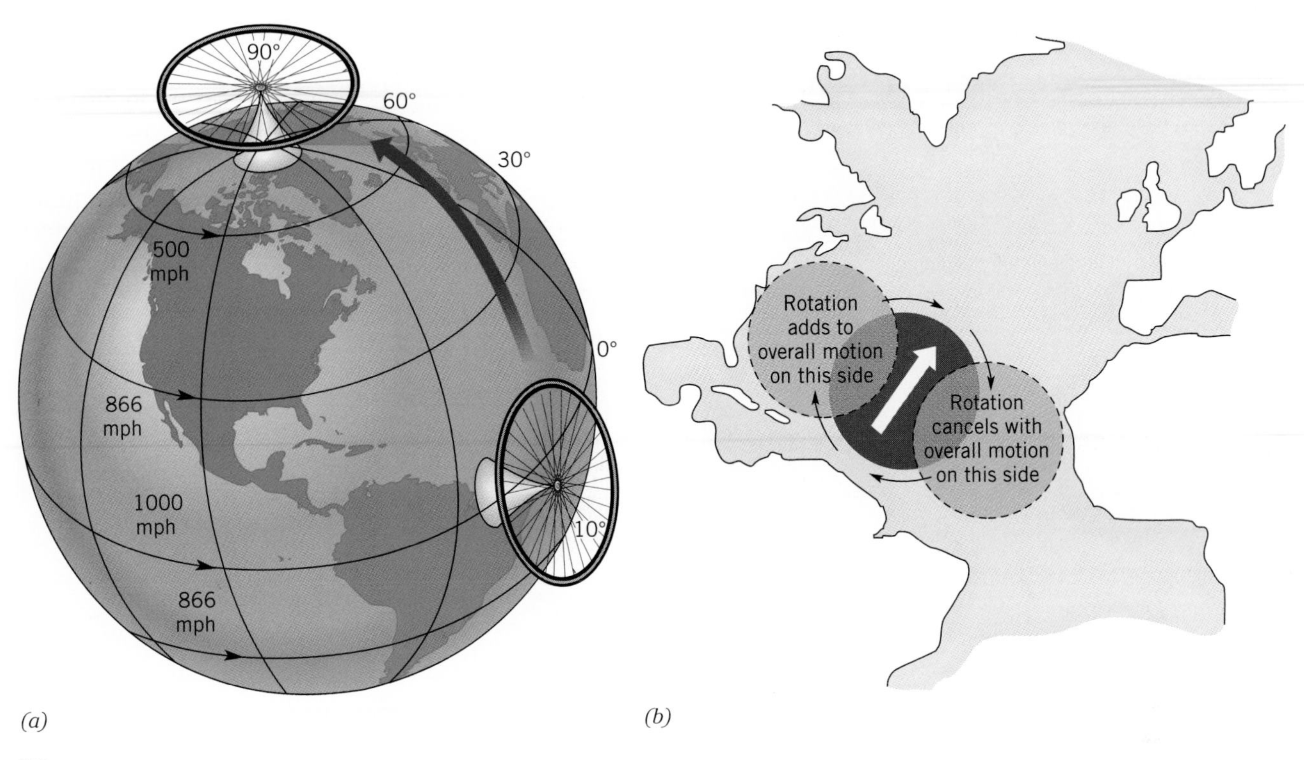

Figure 8.18

(a) If a nonspinning bicycle wheel on the equator is transported very carefully by its base to the North Pole, it still won't spin. But the Earth spins beneath it, making it seem to an earthbound observer that the wheel has acquired a clockwise spin, making one revolution every 24 hours. (b) As a mass of water moves northward, it appears to pick up a clockwise spin relative to the Earth. (left) On the western margin of the mass, both the velocity due to the spin and that due to the motion of the mass are in the same direction, so they add together to make a very large total velocity. (right) On the eastern margin, the two velocities are in opposite directions, so they counteract each other, leaving little net velocity.

where the currents are correspondingly compressed and intensified.

The third cause of the intensified western boundary currents is also related to the Earth's rotation, through the apparent change in the rotational state of objects moving north or south along the Earth's surface. To illustrate this effect, imagine a large horizontal bicycle wheel on frictionless bearings sitting motionless on the equator (Figure 8.18a). Suppose that we very carefully transport this wheel to the North Pole, touching only the stand and being very careful not to touch the wheel itself.

Under these conditions, the wheel will still have no spin at all when we set it down at the North Pole. It will still be stationary relative to the Sun, Moon, and stars. However, the Earth will be spinning beneath it once every 24 hours. Since we are earthbound observers, it will appear from our point of view that the Earth is still and the wheel is spinning, rather than vice versa. From our point of view, the wheel has somehow miraculously acquired a spin in the clockwise direction that gives it one complete revolution every 24 hours.

The same thing appears to happen to air or water masses as they move north or south along the Earth's surface. As they go, their rotational motion appears to change. A mass of water starting on the equator with no spin at all appears to acquire a spin as it goes. The farther poleward it goes, the faster it appears to spin. At intermediate latitudes, this amount of spin is less than one complete revolution per day. But if the water mass is wide, it doesn't have to spin very rapidly for the outer regions to have high speeds. For example, a water mass 1000 kilometers wide and spinning only half a revolution per day would have to move at 65 kilometers per hour on its outer boundary!

Because of the dominant wind-induced current gyres, surface waters in the western portions of all oceans are traveling away from the equator. As these water masses go toward the pole, they acquire the appropriate spin (clockwise in the Northern Hemis-

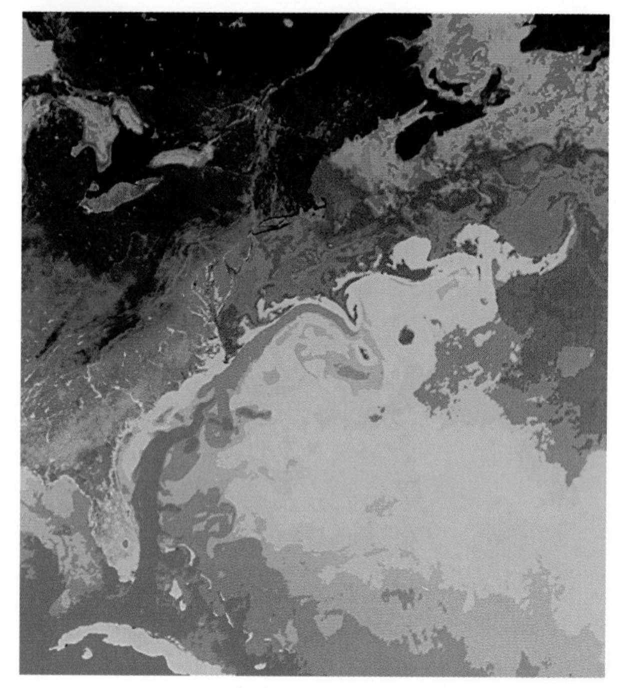

(a)

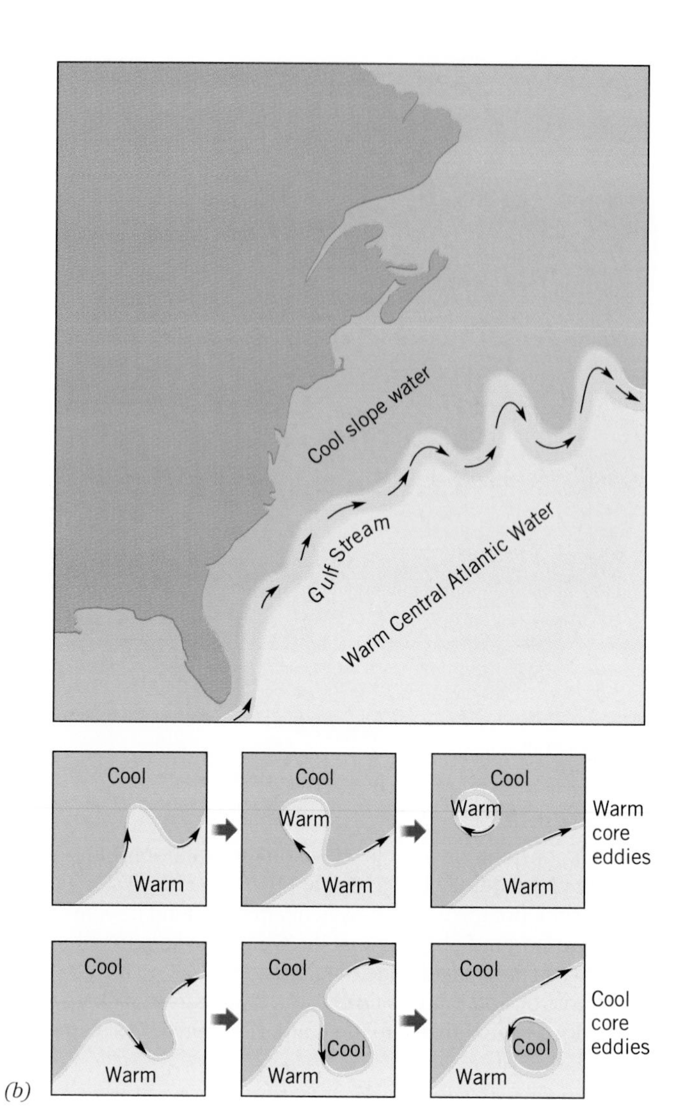

(b)

Figure 8.19

(a) Satellite infrared photo of the Gulf Stream. The warmest waters are red, and the coolest are violet. (b) As the Gulf Stream passes the U.S. eastern seaboard, it begins to meander. Sometimes, the meanders become so pronounced that they break off and form separate eddies. Eddies on the west carry pockets of warmer water from the east and eddies on the east carry pockets of cooler water from the west.

phere and counterclockwise in the Southern Hemisphere) to make the current on the very western edge extremely swift, as is illustrated in Figure 8.18b.

THE GULF STREAM

One of the best known of these swift, narrow western boundary currents is the Gulf Stream, which is located just off the eastern continental margin of North America (Figure 8.19). Having a width of about 50 to 75 kilometers and depth of 1.5 to 2 kilometers, it flows northeastward with a velocity of 3 to 10 kilometers per hour. Daily tidal oscillations move the entire Gulf Stream back and forth some 6 to 8 kilometers per day as the tides come and go. Sometimes, there are countercurrents flowing in the opposite direction down its eastern side, and sometimes meandering currents and spinning whirlpools, called *eddies*, move down its western side, bringing cold northern water down the eastern seaboard of North America.

There is a sharp distinction between the western edge of the Gulf Stream and the neighboring **slope water**, that lies between it and the North American coast. It is usually even possible to distinguish this edge visually. Much of the water in the Gulf Stream has come from the South and Equatorial Atlantic, and most of it has spent a long time at the surface. Consequently, most of its nutrients have been consumed. Being deficient in nutrients and suspended organic matter, there is little biological activity, and it is a rich deep blue color. This is in stark contrast to the green color of the neighboring slope water, which is cool, nutrient-rich, and contains an abundance of plankton.

The warm Gulf Stream heats the air above it, causing it to rise and clouds to form. Consequently, you might be able to tell when you are approaching the Gulf Stream by looking at the sky. The edge of the Gulf Stream can also be distinguished from the neighboring slope water by the waves. Just as a rider in a convertible notices a different wind velocity than a stationary pedestrian, the wind-driven waves on the swiftly flowing Gulf Stream waters differ from those on the neighboring slope water. This visual distinction is more difficult on the southeastern side of the Gulf Stream, however, because there is not the distinct edge and abrupt change in current speeds, as there is on the northwestern side.

Just as a rider in a convertible notices a different wind velocity than a stationary pedestrian, the wind-driven waves on the swiftly flowing Gulf Stream waters differ from those on the neighboring slope water.

As the Gulf Stream reaches higher latitudes off the U. S. East Coast, it begins to meander. The loops sometimes become so pronounced that they break off as independent eddies, as illustrated in Figure 8.19b. The surface of the central Atlantic to the east of the Gulf Stream is warm, and the slope water to its west tends to be cool. Therefore, as these eddies break off, those going into the cool slope water region have cores of warm water from the central Atlantic, and those that go into the warm central Atlantic side have cores of cool slope water. These independent eddies last typically 1 to 2 years, leaving circular pockets of warm water from the central Atlantic within the cool slope waters, and pockets of cool slope water within the warm central Atlantic.

COASTAL WATERS

In temperate latitudes, the coastal waters are usually distinctly different from those further out to sea. The primary cause of this is the Coriolis deflection[4] of the ocean currents in the major ocean current gyres. This deflection is always toward the center of the gyre and away from the coast, as illustrated in Figure 8.11. The departing surface waters must be replaced, and these replacement waters are of quite a different nature.

Along the east coasts of the continents, there is a tendency for the departing surface waters to be replaced by cold water from high latitudes creeping down along the continental margins (Figure 8.11). In the North Atlantic, this cold water comes down as far as North Carolina from the area around Labrador and Greenland. In the Pacific, cold water from the Bering Sea creeps down along the coast as far as Japan.

Along the continental west coasts, such as those of North and South America, Europe, and Africa, continental shelves are narrower. The departing coastal surface water is normally replaced by cool deep water upwelled from the nearby basins. In California, Mexico, Chile and Peru, for example, this keeps the coasts cool and "air conditioned," and frequently provides chilling surprises for inexperienced bathers who expect the water temperature to match the climate.

During the El Niño condition, the warm surface water that has sloshed back across the Pacific and up against the Peruvian Coast tends to then creep up along the coasts to both the north and south. These El Niño waters fill in behind the current gyres along the coast and suppress or reduce the upwelling. Although they provide pleasant warm surprises for coastal bathers, they are not so welcome to fishermen, who rely on the cool upwelled nutrient-rich waters for the health of their industry.

In summary, on the continental east coasts, the departing surface waters are generally replaced by cold waters from higher latitudes that flow down along the continental margins. Along the west coasts, the departing surface waters are normally replaced by cooler upwelled waters. In the Pacific, warm El Niño waters occasionally creep up from low latitudes along the American West Coast.

8.4 DEEP CURRENTS

GENERAL FEATURES

As surface currents are driven by wind, deeper currents are driven by gravity. Denser waters sink. Those that are densest sink to the bottom and then flow along the bottom to the deepest parts of the ocean floor. Others sink until they encounter denser water beneath them and then they flow outward along the top of these denser water masses. In this way, the deep water is layered, with each layer being very thin in comparison to its breadth. If you will remember that the relative dimensions of the ocean itself are those of a sheet of paper, then you will realize how proportionally thin these horizontal layers must be. Obviously, we have to use a great deal of vertical exaggeration in our sketches of the ocean layers.

Deep currents are much slower and more voluminous than surface currents. Typical flow speeds

[4]Usually, the Coriolis deflection is severely reduced, but not quite completely cancelled by the presence of domes.

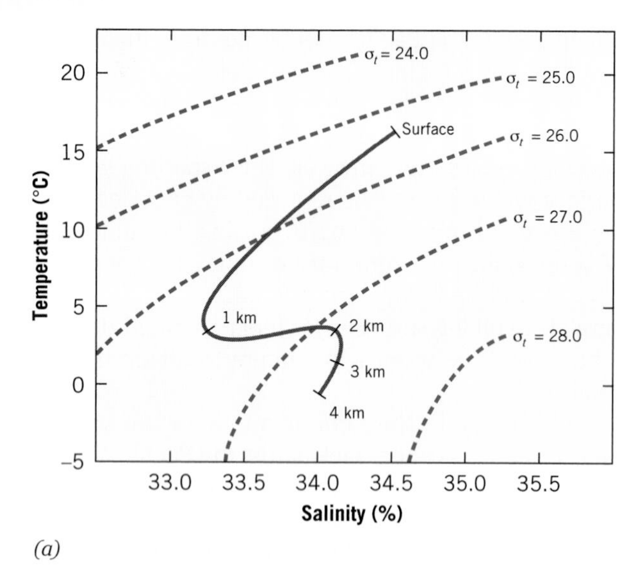

(a)

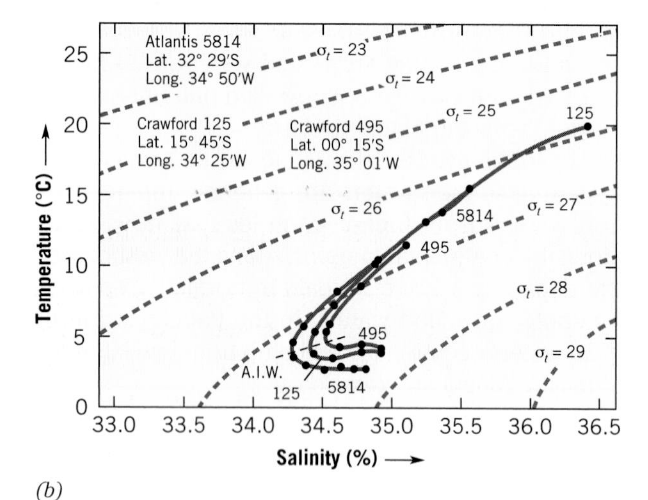

(b)

Figure 8.20

(a) An example of how a T-S diagram might appear at some oceanographic station. Deeper water needn't always be colder, nor always more saline, but it is always denser. (b) T-S curves for three different stations in the South Atlantic, showing the layer of Antarctic Intermediate Water sandwiched between layers with higher salinities. (From Hugh J. McLellan, 1965, *Elements of Oceanography*, Pergamon Press.).

are only a few tens of kilometers per year! These slow speeds result from both the fact that the densities of the various water masses differ only very slightly from each other, and that the slopes along which they flow are inclined only very slightly. By analogy, a balloon rolls down a very gentle incline much more slowly than a bowling ball rolls down a steep slope.

Because they flow so slowly, deep currents are heavily influenced by the Coriolis effect. Most deep waters form at the surface in cold high latitudes and

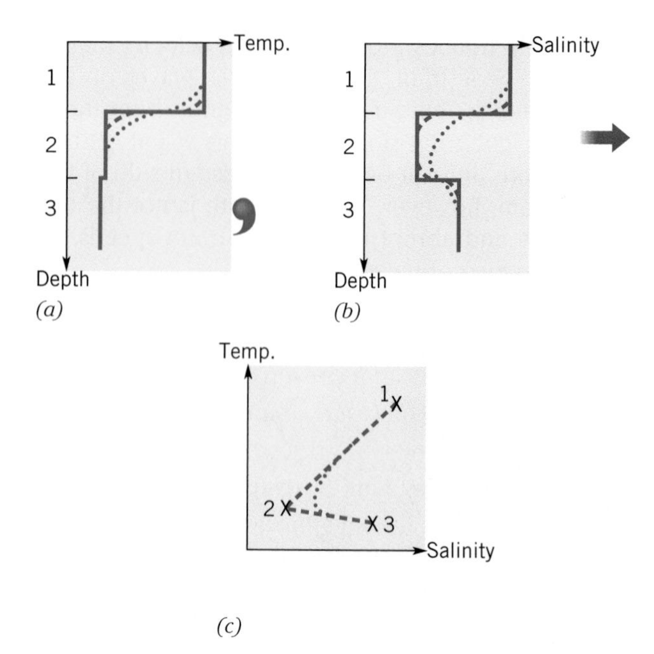

(a) (b) (c)

Figure 8.21

Suppose three distinct water masses (1, 2, and 3) are layered vertically and have temperatures and salinities as indicated by the solid lines in (a) and (b). The positions of these three batches on the T-S diagram on the right (c) are indicated by the X's. As they mix across the boundaries, the resulting temperature and salinity pattern would be given by the dashed and then the dotted lines in these figures.

then sink and flow along the ocean bottom toward the equator. As they go, they are strongly deflected toward the right in the Northern Hemisphere (Figure 8.6) and to the left in the Southern Hemisphere. That puts them up against the western ocean boundaries in both hemispheres. Therefore, deep currents tend to be intensified along the western ocean boundaries like surface currents, but flowing in the opposite direction.

T-S DIAGRAMS

The densities of the various important water masses differ only slightly, so direct measurements of density differences are difficult. Fortunately, there is an easier way of doing it. The density of sea water is a known function of temperature and salinity, both of which are easy to measure. Once these two factors are known, the density can be easily calculated from them. In fact, it is usually calculated and recorded automatically by the device that makes the measurements.

Electronic methods are generally used in making temperature and salinity measurements. The electrical conductivity of many materials depends on their temperatures. Other devices generate electrical volt-

Figure 8.22
Antarctic Bottom Water, the densest, deepest water mass, is formed beneath Antarctic ice during the southern winter. (This photo of Antarctica was obviously taken during the summer.)

ages in an amount dependent on their temperatures. So if we immerse either of these kinds of devices into the water, then the electrical conductivity of the first kind of device, or the voltage produced by the second, will tell us the temperature of the water.

Electronic devices can also be used to determine salinity. The saltier the water, the better it conducts electricity. So the salinity of a sample can be measured by immersing two wires and measuring the ease with which electrical current flows between them. Since both the temperature and salinity can be determined electronically, these measurements can be performed *in situ*, at the end of a long cable extending deep into the ocean.

The dependence of density on temperature and salinity can be displayed visually with a chart called a **T-S diagram**. From the T-S diagram of Figure 8.20a, you can see that the temperature is measured along the vertical axis and salinity along the horizontal one. Notice how the density (dashed lines) increases as you go down (lower temperature) and to the right (higher salinity).

At any one location or "station" at sea, the temperature and salinity of samples from various depths are taken and marked on the same T-S diagram. These points are then connected by smooth lines, so that the properties of water between the sampled points can be estimated. Examples of some real T-S diagrams, illustrating changes in density with depth, are given in Figure 8.20b. Can you guess which samples come from tropical regions and which from high latitudes, by the temperature and salinity near the surface?

T-S diagrams help oceanographers understand both the nature of the deeper water masses and the amount of mixing across the boundaries that separate them. For example, Figure 8.21 shows three distinct waters arranged vertically as they should be, with the lightest on top and the densest on bottom. Before mixing, each would be represented by the appropriate X on the T-S diagram, but as they mix across the interface between layers, they go to the dashed and then the dotted lines of that figure. If any T-S diagram made from measurements taken at sea (e.g., Figure 8.20b) looks like this one, we can conclude that there were these three distinct deep water masses, with some mixing across the boundaries.

DEEP WATER MASSES

Although the ocean's warm surface waters are most familiar to us and most directly influence our commerce, fisheries, and climate, the deep waters are far more stable and voluminous. The ocean basins are filled nearly to the surface with cold dense water masses, which initially formed at high latitudes.

The very densest major deep water mass forms during the southern winter beneath the Antarctic ice shelf, primarily in the Weddell Sea in the South Atlantic (Figure 8.22). This very densest of the major deep water masses is called the **Antarctic Bottom Water** (**ABW**). It sinks to the bottom and then flows outward along the ocean floor, beneath the other deep water masses. The shape of the ocean bottom confines it primarily to the Atlantic Ocean, where its slow deliberate flow carries it northward beyond the equator.

Of course, the same thing happens beneath the northern ice in the northern winter. However, the "Arctic Bottom Water" produced in this way does not appear in our major oceans, because the Arctic Ocean is surrounded by continents and sills. The deep water can't get out.

Another major water mass formed near Antarctica is the **Antarctic Circumpolar Water (ACW)**. Driven by the prevailing westerlies, the Antarctic Circumpolar Current flows eastward around the globe. Continents do not obstruct its flow, although it does get pinched a bit in getting past Cape Horn. The water in this current continuously encircles the globe near the Antarctic Circle. This water mass extends from the surface to the sea floor, except in the South Atlantic where the Antarctic Bottom Water flows beneath it.

The most prolific producer of deep water in all oceans is a region south of Greenland in the North Atlantic, where several major surface currents converge. Here, the warm salty Gulf Stream converges with the cold, not-so-salty East Greenland and West Greenland Currents. These merging surface waters sink for two reasons. First, "down" is the only place they can go. The second reason is more subtle. Because the lines of constant density on a T-S plot are curved (see Figure 8.20a), when waters of the same density but different temperatures and salinities are mixed, the resulting water mass is denser than either of the original waters. So the denser mixture would tend to sink even if it wasn't being forced downward by the convergence. This sinking of merging surface waters is sometimes referred to as **caballing**. The water mass produced by these converging currents south of Greenland is called **North Atlantic Deep Water**. It is produced so abundantly that it fills most of the Atlantic Ocean (Figure 8.23a).

The North Atlantic Deep Water flows southward beyond the equator and all the way to the Antarctic

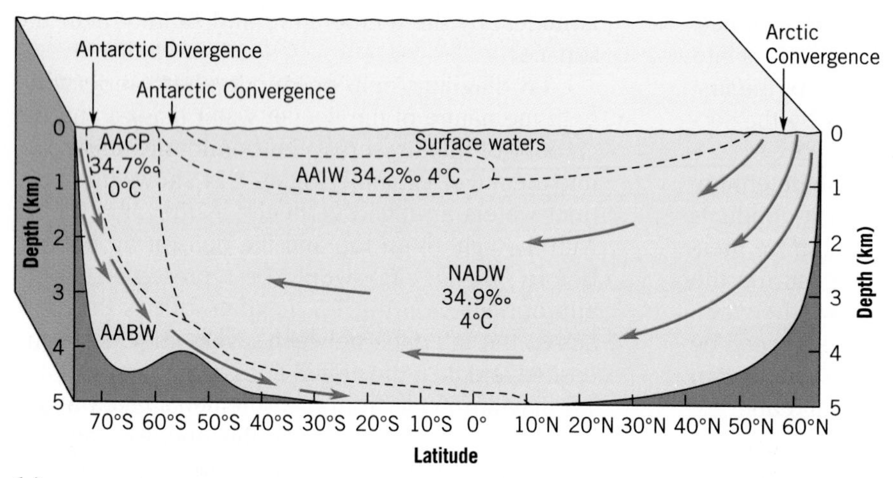

(a)

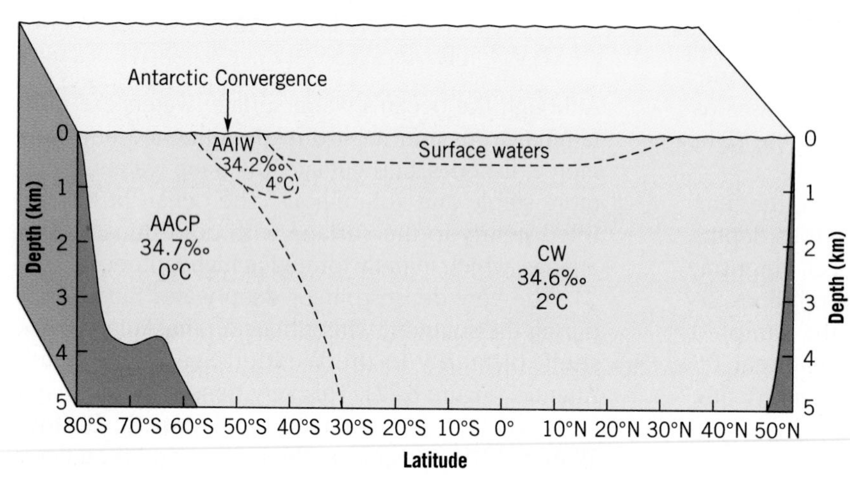

(b)

Figure 8.23

(a) Vertical cross section of the Atlantic Ocean extending from Antarctica to Greenland, showing the locations of the various water masses (AABW = Antarctic Bottom Water; AACP = Antarctic Circumpolar Water; AAIW = Antarctic Intermediate Water; NADW = North Atlantic Deep Water). (b) Vertical cross section of the eastern Pacific Ocean extending from Antarctica to the Aleutian Islands, showing the locations of the various water masses (AACP = Antarctic Circumpolar Water; AAI = Antarctic Intermediate Water; CW = Common Water).

Circumpolar Current, which mixes with it to produce another water mass, called **Common Water**. This mixture is carried by the Antarctic Circumpolar Current around the Cape of Good Hope into the Indian Ocean, and around Australia and New Zealand into the Pacific, where it fills most of these two oceans as well (Figure 8.23b).

INTERMEDIATE WATERS

These four water masses—Antarctic Bottom Water, Antarctic Circumpolar Water, North Atlantic Deep Water, and Common Water—are the deepest and most voluminous of the deep water masses. Closer to the surface are many smaller and more local water masses, generally produced by the convergence of surface currents.

One of these, which appears in all three oceans, is called *Antarctic Intermediate Water.* It is formed at the Antarctic Convergence. The Coriolis deflection of the Antarctic Circumpolar Current is to the left (Southern Hemisphere), or northward, causing it to converge with the waters along its northern boundary. The mixing of these waters in this convergence produces this water mass.

In the Atlantic, a significant subsurface water mass is the warm, salty Mediterranean Water, which flows out over the Gibraltar Sill and down to a depth of 1 to 1.5 kilometers, as is illustrated in Figure 10.26b. It is also seen in Figure 8.2 as the high salinity bulge on the Chain 7-68 station profile.

8.5 TIDAL CURRENTS

In the next chapter, we will study waves. We will learn that the water motion beneath waves is cyclical. After each complete wave passes by, the water ends up where it started from. In order to determine the overall drift motion of a water mass, our observations must span hundreds of wave cycles, so that the overall motion of the water is not obscured by the rapid but cyclical motion associated with the waves.

Most surface waves complete their cycles in a matter of seconds, and they do not affect water motion deeper than a few tens of meters. If these were the only waves on the oceans, then deep water motion could be determined almost instantaneously, and even the wavy surface waters could reveal their overall motions within a matter of tens of minutes. However, there are some types of waves that cycle much more slowly and reach all the way to the bottom. The most prominent of these is tides.

Tides are long low waves that cause the sea level to rise and fall periodically in most coastal areas. In deep ocean regions, the ocean surface rises and falls less than a meter, but the horizontal motion is typically 6 to 8 kilometers. That is, there is typically thousands of times more horizontal motion than vertical motion associated with the tides, and this horizontal motion extends all the way down to the deepest parts of the ocean.

The tides are sufficiently important and complicated that we devote a good portion of the next chapter to their study. Here we simply point out that most of the time throughout most of the ocean, the tides come and go twice a day, at roughly 12-hour intervals, which means that it takes about 6 hours to come in and 6 hours to go back out. In any 6-hour period, the Earth has spun one-quarter of a revolution, so the Coriolis deflection of the moving water is considerable. In fact, the horizontal water motion is more circular than back and forth, going clockwise in the Northern Hemisphere and counterclockwise in the Southern Hemisphere.

The speeds of these cyclical **tidal currents** are typically 1 to 2 kilometers per hour, which is much faster than the flow speeds of the major deep water masses. Therefore, when we wish to study the flow of these deep waters, our instruments must perform these measurements over periods of several months. The tidal currents are fast, but cyclical. They end up going nowhere, so that over long time periods, the slow overall motion of the water dominates.

8.6 LOCAL AND TEMPORARY CURRENT PATTERNS

In the preceding sections, we studied the dominant features of large-scale ocean currents. These have global effects on the world, including our climate, commerce, fisheries, and the locations of human populations. There are many coastal and temporary current patterns that may not have global impact, but are crucial to local coastal regions.

COASTAL CURRENTS

Coastal waters tend to have relatively large variations in temperature and salinity, both because they are comparatively shallow, and because they are heavily influenced by the freshwater run-off from the adjacent land mass. This run-off tends to vary with the seasons, remain near the surface, and hug the coast (Figure 5.5). Due to the Coriolis effect, the freshwater discharge from rivers and streams tends to curve to

Figure 8.24
Strong storm winds blow surface water shoreward, causing a rise in sea level along the coast and driving large storm waves into shore.

the right in the Northern Hemisphere upon entering the ocean, flowing southward along the coast on eastern seaboards and north along western seaboards. The winds sometimes reinforce this flow, and sometimes oppose it. Farther out, the Coriolis deflection of surface currents in the main gyres tends to drive surface waters away from the local coasts, and so it must be replaced by waters either upwelled from depth, or flowing down along the coast from other regions. All these factors make coastal currents complex. We study them in more detail along with other features of the coastal ocean in Chapter 9.

STORM SURGES

Some of the world's greatest natural disasters have been caused by local surface currents called **storm surges**, or sometimes *storm tides*. These are caused by a combination of low atmospheric pressure and strong winds that cause sea level to rise to exceptional levels along a coast.

In the preceding chapter, we learned that violent storms, such as hurricanes and typhoons, are fueled by the condensation of moisture in rising warm, moist air masses. The condensation releases heat, which keeps the air masses warm and rising. The rising air leaves low pressures beneath it, which brings in more warm moist air masses to continue the cycle. The incoming winds whirl around the storm center due to the Coriolis effect (Figure 7.23), driving strong surface currents as they go.

Just as high atmospheric pressure pushes the ocean surface down, low pressure causes the ocean surface to rise. By itself, this would cause a change in sea level of less than a meter and wouldn't be too alarming. More frightening are the surface currents driven by the strong storm winds associated with these low-pressure regions. When the storm reaches shore, these winds may drive surface water into the coast, raising the sea level several additional meters and inundating some coastal areas that are normally dry (Figure 8.24).[5] A strong storm surge may cause local sea level to rise 3 or 4 meters and may last from 3 to 5 hours. Large destructive storm waves ride atop these elevated waters. Even though many coastal structures might survive the strong currents of the flooding waters, most cannot survive the battering of these storm waves.

Some coastal areas are accustomed to large tidal variations. In these regions, storm surges are not as dangerous, unless they happened to be in progress during a period of high tide. But many coastal areas are not accustomed to large variations in sea level, creating a false sense of security and temptation to build homes and buildings too close to the water.

[5]Actually, the impact of a storm at sea may generate large long waves that travel across the ocean like the tsunamis that we study in the following chapter. As these waves reach shore, they cause the sea level to rise and fall. So the storm needn't reach shore for the coast to experience a storm surge. The largest and most devastating surges, however, arrive with a storm.

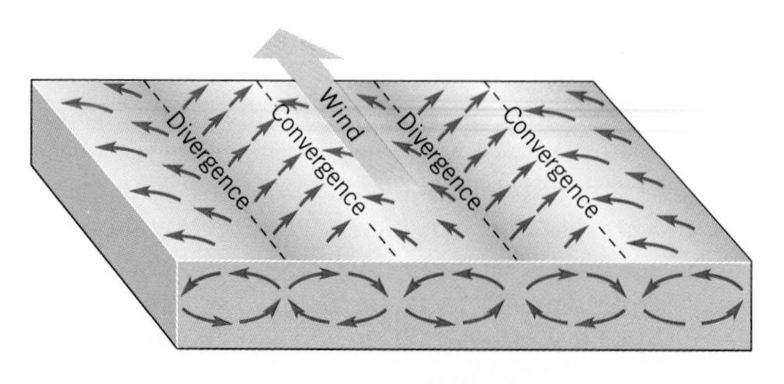

Figure 8.25
Langmuir circulation.

These areas are particularly vulnerable to devastation by storm surges. One gruesome example of this occurred in Texas in 1900, when the city of Galveston was destroyed with a loss of 6000 lives—the worst natural disaster in U.S. history. Galveston was built on a low-barrier island in the Gulf of Mexico, where a small tidal range gave a false sense of security. Even larger disasters have occurred many times along the coast of the Bay of Bengal in southern Asia. Here are many heavily populated, low-lying regions that are unaccustomed to large variations in sea level and therefore particularly vulnerable. The largest known death toll due to natural disasters was recorded in this area in 1970, when a storm surge took an estimated half-million lives.

The desire of many people to live near a coast places heavy pressure on coastal development and tends to make governments relaxed in their regulations. Because storms are infrequent, people can easily be persuaded to buy property in a development that looks stable. Satellite weather-tracking systems can normally provide sufficient warning to evacuate endangered coastal areas, but the early warning system didn't work in Bangladesh in 1991 where the endangered area was large, heavily populated, and had inadequate communication and transportation systems. An estimated 130,000 lives were lost. Irrespective of improvements in communication and evacuation, property damage will increase as these endangered areas continue to be developed.

LANGMUIR CIRCULATION

When winds begin to blow across the water at speeds of more than a few meters per second, the surface water develops an interesting circulation pattern. Rather than simply flowing directly in line with the wind, the water flows in "corkscrew" motions, in long parallel **Langmuir cells** (Figure 8.25). Such cells are each typically a few meters wide and a few meters deep, oriented lengthwise parallel to the wind direction. Neighboring cells have alternate clockwise and counterclockwise corkscrew motions, which means that the boundaries between these cells alternate between zones of converging and diverging surface waters. Debris and foam floating on the surface are carried into zones of convergence, making long visible slicks, parallel to the wind direction.

8.7 MEASURING OCEAN CURRENTS

The overall motion of ocean currents is often masked by more rapid temporary and oscillatory motions, such as those due to varying winds and waves on the surface, or tidal currents throughout the ocean. Therefore, the measurement of ocean currents normally requires observations over an extended period of time. Ships themselves drift with the surface currents, and shipboard time is normally too expensive to afford continual monitoring of current measuring devices over the periods of weeks or months that may be needed. Therefore, current measuring devices are usually made to be dropped off at one time and then retrieved sometime later. Finding and retrieving devices that have been at sea for weeks or months can be quite challenging and require some forethought in the design of the equipment.

Current measuring techniques can be put into two classes, direct and indirect, according to whether we are actually measuring the water's motion, or inferring that motion from some other effect. An example of an **indirect technique** would be using the slope in the water surface, or the slopes of the density contours to determine the speed of a surface current. For deep water masses, we might infer the current's speed from characteristic water properties or trace chemicals that we know originated at some other point in the ocean.

The **direct techniques** fall into one of two categories: the *Eulerian* techniques measure how fast the water passes a fixed point on Earth, and the *Lagrangian* techniques measure how fast the Earth passes a fixed point in the water. In the first case, we would typically have several current meters at various positions along on a line that is anchored to the ocean floor on one end and held upward by a buoy on the other. Each current meter would measure and record current speed and direction at its location above the ocean floor. Often, propellers are used to measure current speed and vanes to measure direction, although electrical or acoustic methods may also be used. Because radio waves do not travel through water, communication between these current meters and surface ship is usually done acoustically. Acoustic signals from a buoy help the ship locate its position, and an acoustic signal from the ship may detach the meter system from the anchor, so that it can float to the surface and be retrieved.

Lagrangian techniques involve putting buoys in the water and measuring how far they move from the drop-off point over time. For currents on the very surface, the buoys can be as simple as bottles or floats with notes asking the discoverer to report the time and location of discovery. More commonly, the buoys carry radios that broadcast their positions, either continuously, or after a period of time when they are ready to be retrieved. These can be monitored by observations from aboard ship, airplanes, or Earth-orbiting satellites. For deeper currents, the buoys must be carefully weighted so that they are neutrally buoyant at the desired depth, and furthermore, they must have the ability to rise to the surface and report their position when they are ready to be retrieved.

Summary

Surface and Deep Waters

Throughout most of the ocean, the thin layer of surface water is quite distinct from the cooler, denser water below. Water's density depends on temperature, pressure, and salinity. For comparative purposes, the densities of all water samples are measured at atmospheric pressure. Different deep water masses can be identified by their characteristic temperatures and salinities.

Due to their interaction with the atmosphere, the temperature and salinity of surface waters are variable, and the currents flow swiftly compared to the deeper waters below. The deeper waters retain the properties they acquired when at the surface. During the winter at high latitudes, freezing ice creates cold dense surface waters and causes vertical mixing.

Forces That Drive Ocean Currents

The driving force for deep ocean currents is gravity. Denser waters sink and flow downhill toward deeper regions. The ocean's surface currents are driven by the winds. Because of the Coriolis effect, the current direction spirals downward, with each layer flowing more slowly and farther to the right (Northern Hemisphere) compared to the layer above. In deep water and under ideal conditions, the net *Ekman* transport is 90° to the right of wind direction. The same happens in the Southern Hemisphere, but with the Coriolis deflection to the left.

In the Northern Hemisphere, the Ekman transport causes surface water to collect on the right side of the current direction, creating downwelling there, with corresponding upwelling of deeper water on the left. This gives a slope to the density contours across the current and stacks the surface water into a dome on the right. When equilibrium is reached, the downhill component of gravity sliding water back off the dome and the pressure gradient forces beneath the dome just nullify further Coriolis deflection into the dome, so the water flows sideways along the dome in the direction of the wind. Due to geostrophic flow, this pattern would continue even if the wind temporarily stopped. The corresponding pattern in the Southern Hemisphere has the dome and downwelling on the left.

Large-Scale Surface Currents

The dominant surface currents are driven westward across the oceans by the trade winds and returned eastward at higher latitudes by the prevailing westerlies. Where surface currents diverge, there is upwelling of deeper waters, and where surface currents converge, there is downwelling. The trade winds stack water up against the western sides of the oceans, and some of this returns eastward across the oceans in equatorial countercurrents and undercurrents. The El Niño condition is a result of the slackening of the trade winds for a prolonged period of time. It causes an oscillation in the weather patterns on opposite sides of the Pacific.

Surface currents are strongly intensified along the western margins of all oceans. The reasons for this include water squirting out the sides when equatorial currents run into the western ocean margins, the Coriolis deflection of currents in the gyres, and the change in rotational state of water masses moving to higher latitudes. The Gulf Stream is one well-known example of these intensified western boundary currents.

Deep Currents

Temperature and salinity are easier to measure than the density itself and can be determined electronically. T-S diagrams are convenient tools for examining the properties of various waters. The densest of all deep water masses is the Antarctic Bottom Water, formed beneath the freezing ice in the Weddell Sea. North Atlantic Deep Water is formed at the convergence of surface currents just south of Greenland. It fills most of the Atlantic Ocean and mixes with Antarctic Circumpolar Water to form Common Water.

This is then carried into the other two oceans by the Antarctic Circumpolar Current. Other interesting deep water masses include Arctic Bottom Water, which forms beneath the northern ice sheet; Antarctic Intermediate Water, which forms at the Antarctic Convergence along the northern border of the Antarctic Circumpolar Current; and Mediterranean Water, which flows into the Atlantic over the Gibraltar Sill.

Tidal Currents, and Local and Temporary Current Patterns

Tidal currents are relatively fast, but cyclical. Because of Coriolis deflection, the tides go around in circles, rather than just oscillating back and forth. Coastal waters tend to be variable, and coastal currents are quite different from the flow of surface waters farther out to sea. Storm surges are created when water is driven landward by strong storm winds. Langmuir circulation results when strong winds blow across relatively calm water.

Measuring Ocean Currents

Indirect measurement techniques involve inferring current speeds from some other property that is caused by the current. Direct techniques can involve either a fixed instrument that measures how fast the water passes a fixed point on the ocean floor, or floating instrument that measures how fast the ocean floor passes a fixed point in the water.

Key Words

water masses
sigma-tee
potential temperature
in situ temperature
pycnocline
thermocline
halocline
vertical mixing
Ekman spiral
density contours
downwelling
upwelling
dome
equatorial currents
Antarctic Circumpolar Current
West Wind Drift
current gyres
meteorological equator
equatorial countercurrents
equatorial undercurrents
El Niño
El Niño condition
southern oscillation
western intensification
boundary currents
slope water
T-S diagram
Antarctic Bottom Water
Antarctic Circumpolar Water
caballing
North Atlantic Deep Water
Common Water
tidal currents
storm surge
Langmuir cells
indirect techniques
direct techniques

Study Questions

1. Explain the Ekman spiral. Why does each layer flow farther to the right (Northern Hemisphere) than the layer above it?
2. Why is there upwelling along much of the U.S. West Coast?
3. Why is there a slope to the density contours across a current?
4. Explain how geostrophic effects enable surface currents to keep on flowing long after the driving winds have stopped.
5. Explain the reason for the equatorial currents, equatorial countercurrents, and Antarctic Circumpolar Current.
6. What is *El Niño*? Why do surface temperatures along the eastern ocean margins rise during El Niño? Why do fishing industries suffer?
7. How and why do weather patterns change during the El Niño condition?
8. Why are surface currents so markedly intensified along the western ocean boundaries?
9. What are some of the ways in which the Gulf Stream can be visually distinguished from the neighboring slope water? Why does it have these distinctive characteristics?
10. How does the density of sea water vary with changes in pressure? Temperature? Salinity? Why does the potential temperature of a sample differ from the *in situ* temperature? Which would be higher, and why?
11. Why is there seasonal overturn of surface waters at high latitudes, but not at low latitudes?
12. The temperatures and salinities of several different deep water masses are given in Figures 8.23a and 8.23b. From the temperature and salinity of any one of these, find its density using a T-S diagram (e.g., Figure 8.20).
13. Explain how any four of the major deep water masses are formed.
14. Why do tidal currents go around in circles rather than back and forth?
15. What kinds of areas are particularly vulnerable to storm surges, and why?

Critical Thinking Questions

1. According to Figure 8.3, the surface water at high northern latitudes is not as salty as those at high southern latitudes. Why do you suppose this might be?
2. Make a figure like 8.21, but do it for three water masses that are different from those of this figure.
3. In Figure 8.13, explain what causes the antarctic divergence and the subtropical convergence.
4. In general, would you expect the surface water to be colder in a region of divergence or convergence? Why?

Suggestions for Further Reading

ANDERSON, IAN. 1990. Oceanographers Make Noise to Test the Water. *New Scientist* 125:1703, 34.

CALKINS, JAY R. 1989. Modeling the Ocean in Motion. *The Science Teacher* 56:1, 28.

CLERY, DANIEL. 1992. Satellite Records Ocean's Moving Story. *New Scientist* 135:1834, 19.

The Economist. 1990. Ears Across the Water (Anatomy of Oceans). 314:7648, 80.

JENKINS, WILLIAM J. 1992. Tracers in Oceanography. *Oceanus* 35:1, 47.

KERR, RICHARD A. 1987. Another El Niño Surprise in the Pacific, but Was it Predicted? *Science* 235:4790, 744.

KERR, RICHARD A. 1988. The Weather in the Wake of El Niño. *Science* 240:4854, 883.

KERR, RICHARD A. 1989. Did the Ocean Once Run Backward? *Science* 243:4892, 740.

MACLEISH, WILLIAM H. 1989. Painting a Portrait of the Stream from Miles Above and Below. *Smithsonian* 19:12, 42.

MCCARTNEY, MICHAEL S. 1994. The Atlantic Ocean: Progress in Describing and Interpreting Its Circulation. *Oceanus* 37:1, 1.

MCCARTNEY, MICHAEL S. 1994. Towards a Model of Atlantic Ocean Circulation: The Plumbing of the Climate's Radiator. *Oceanus* 37:1, 5.

MONASTERSKY, RICHARD. 1993. Getting the Drift of Ocean Circulation. *Science News* 144:8, 117.

Open University. 1989. *Ocean Circulation*. Pergamon Press.

PEDLOSKY, JOSEPH. 1990. The Dynamics of the Oceanic Subtropical Gyres. *Science* 248:4953, 316.

SOMBARDIER, LAURENCE. 1992. Surface Current Trackers of the World's Oceans. *Oceanus* 35:2, 6.

TOLMAZIN, D. 1985. *Elements of Dynamic Oceanography*. Allen and Unwin.

WHITEHEAD, J. A. 1989. Giant Ocean Cataracts. *Scientific American* 260:2, 50.

WHITWORTH, T. 1988. The Antarctic Circumpolar Current. *Oceanus* 3:2, 53.

The light from any one object does not garble your view of others. Waves from any source proceed as if the others weren't there. This is true of all waves.

Surfer on a plunging breaker.

WAVES AND TIDES

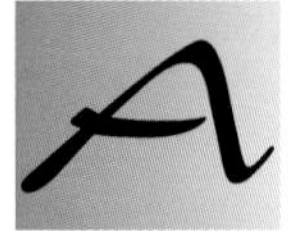

mong the most familiar and universal features of the ocean are the waves on its surface. Waves are traveling oscillations that can transfer energy over large distances, even though the water itself goes nowhere. It just oscillates as the waves pass by. Although this chapter deals specifically with ocean waves, all waves have similar properties, so much of this chapter's material applies to waves of all kinds.

As waves pass through a medium, the oscillations can be sideways to the direction the wave is traveling, as in waves on a string, or the oscillations can be parallel to the way the wave is moving, as in sound waves. These are called *transverse* and *longitudinal* waves, respectively, and are illustrated for waves on a slinky in Figure 3.17a. Waves on the ocean are simultaneously both transverse and longitudinal. You may have noticed that you are carried up and forward as the tops of waves come by, and you are drawn down and back in the depressions between them. That is, you are carried around in a circle each time a wave passes, going up, forward, down, and back. Waves whose oscillations are circular are called **orbital** waves.

Waves cover the ocean and range in size from tiny ripples to giant tides. Small waves ride on larger ones, and these, in turn, ride on waves that are still larger. When you are at an ocean beach, you may notice small ripples on the larger waves that are crashing in the surf. You might also notice that the height on the beach which the breakers reach changes slowly over periods of several minutes, indicating the presence of long undulations in the sea surface. Almost surely you would notice the gradual 12-hour rise and fall of the sea surface, due to the still longer waves called the tides.

Waves of all sizes eventually reach some shore where their energy is deposited. This energy is involved in many important coastal processes (Chapter 10), but is also receiving attention as a possible source of energy for our energy-hungry society. These concerns will be addressed more thoroughly when we look at the ocean as an energy source in Chapter 14.

9.1 WAVE PROPERTIES

CAUSES

The most familiar ocean waves are caused by the wind. These include everything from tiny ripples to long undulations that require patience to notice. Actually, waves are generated at the interface between any two fluids that are moving relative to each other. For example, the passage of one air mass over another creates waves that may be revealed by rows of long parallel clouds (Figure 9.1). If we were birds, we might be more familiar with these. When fresh water from a river mouth flows out over the denser salt water of the ocean, waves are generated between the two. These subsurface waves are called **internal waves**. If we were fish, we might be more fa-

Figure 9.1

Waves form along the interface between any two fluids, if there is relative motion between the two. In this photo, clouds indicate waves between two layers of air.

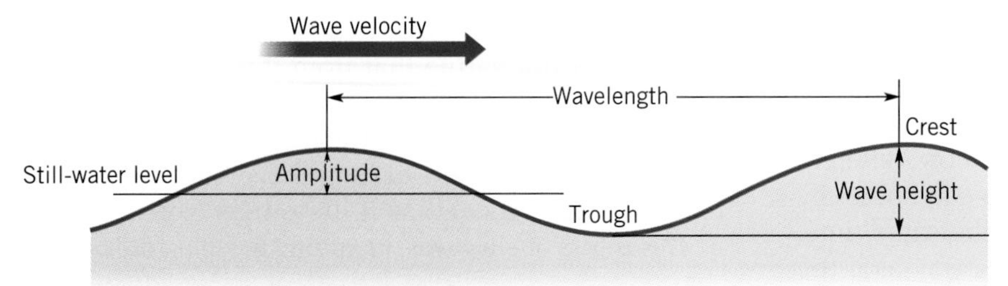

Figure 9.2
Illustration of some wave terminology.

miliar with these. However, we land animals live at the bottom edge of the atmosphere, and so we are most familiar with the waves between the air and water surfaces.

Although the wind-driven waves are the most obvious to us from a casual glance at the ocean surface, there are other important forces that generate waves. Many of these other forces generate long waves that cause long-period fluctuations in sea level. Underwater land movements, such as earthquakes, landslides, or volcanic eruptions, generate **tsunamis**, whose waves typically are from tens to hundreds of kilometers in length. Still longer are the **tidal waves**,[1] or **tides**, that are caused by the gravitational forces of the Moon and Sun.

In this chapter, we will focus mostly on the waves just listed because of their importance. But there are many other ways that waves are generated. Waves can be generated by schools of fish or whales, or the calving of icebergs. Waves caused by moving ships are a nuisance in harbors, and longer waves caused by severe winds or changes in atmospheric pressure can increase the destructive power of coastal storms.

Scientists classify waves as being **free** or **forced**, based on whether they ever get free of the force that generates them. All waves are forced to begin with, but most become free shortly after their formation. For example, if a gust of wind or splashing whale generates waves near New Zealand, these waves will continue on long after the wind or whale has stopped. In fact, as we watch these waves cross the ocean, we won't be able to say for sure what caused them, because they are traveling freely and independent of the force that generated them. By contrast, the tides are generated by the gravitational force of the Sun and Moon, which do not shut off. The Sun and Moon continue to constrain the behavior of the tide all the way across the ocean. The tide, therefore, is not "free", but rather a "forced" wave.

[1]Beware of a popular misuse of the word "tidal wave" in place of "tsunami." Please note the correct usage: Tidal waves are the waves associated with the daily tides.

DESCRIPTION

To describe waves, we use words defined below. Some are illustrated in Figure 9.2.

crest — The very top of any wave.

trough — The depressed hollow between two crests.

wave height — The vertical distance between the top of one wave crest and the bottom of the neighboring trough (symbol H).

significant wave height — The average wave height of the highest third of the waves in the region.

wavelength — The horizontal distance between any point on one wave and the corresponding point on the next wave (symbol L).

wave steepness — The ratio of wave height to wavelength (H/L).

still water level — The average level of the sea surface. The level that the surface would have in the absence of waves.

amplitude — The maximum vertical displacement of the sea surface from the still water level. Half the wave height.

period — The time it takes for one complete wavelength to pass a stationary point (symbol T).

wave speed — The velocity with which waves travel (symbol v).

wave group — A long sequence of similar waves traveling together in the same direction.

deep water waves — Waves that are in water that is deeper than half their wavelength.

shallow water waves — Waves that are in water that is shallower than one-twentieth their wavelength.

SUPERPOSITION

One interesting feature of waves is that the motion of any one wave is completely independent of the motion of any other. Even when two or more different groups of waves are superposed upon each other by passing through the same region at the same time, each wave group will continue on its way as if the others were not there. The ocean surface is a collage of many different waves of all shapes and sizes and going in many different directions. Each of these continues on its way across the ocean, unaffected by the others. The same is true for sound and light waves. When two persons are talking, or two musical instruments playing, you can hear both. The sound from one does not garble that from the other. Similarly, when you look at several different objects, you can see each quite distinctly. The light from any one does not garble your view of the others. The waves from any source proceed as if the others weren't there. This is true of all waves, including those on the ocean.

When you look at several different objects, you can see each quite distinctly. The light from any one does not garble your view of the others. The waves from any source proceed as if the others weren't there. This is true of all waves, including those on the ocean.

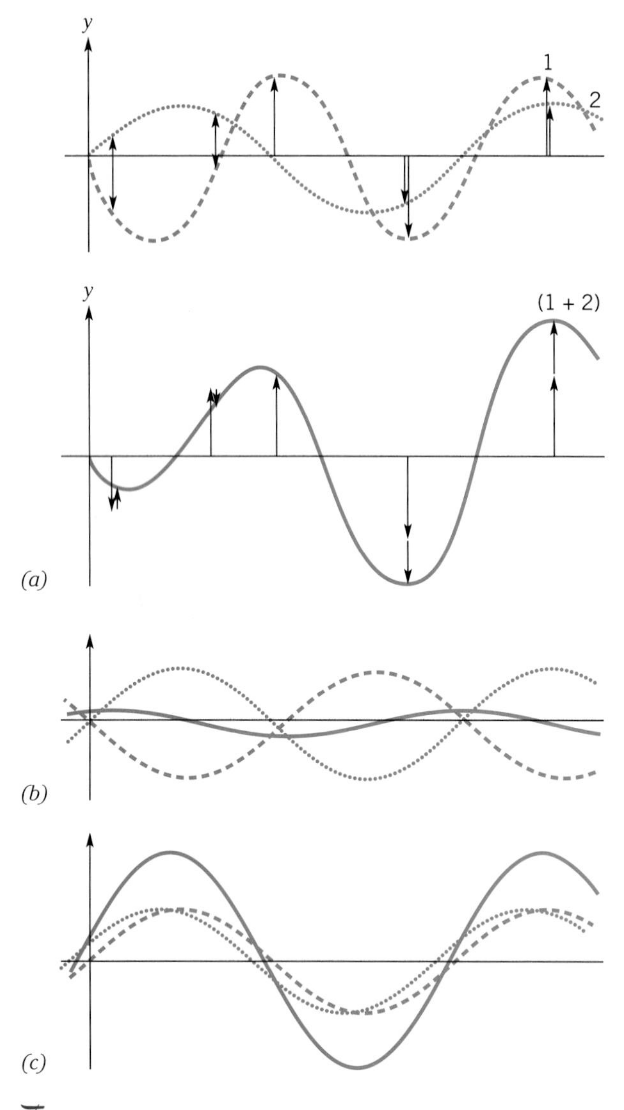

Figure 9.3

(a) In the superposition of waves, the resulting displacement is the sum of the individual displacements. The superposition of the two waves indicated by the dashed and dotted lines yields the wave indicated by the solid line. (b) Destructive interference. (c) Constructive interference.

At any instant, the net displacement of the sea surface from the still water level is simply the sum of the displacements of the individual waves there (Figure 9.3). This property is referred to as **superposition**. Sometimes and in some places, the waves from different wave groups may superpose crest on trough, so that the individual displacements cancel, and there is little or no net displacement of the sea surface. These are regions of **destructive interference** (Figure 9.3b). At other times and places, the different waves may superpose crest on crest and trough on trough, so that the net displacement of sea level is much greater than the displacement due to the individual waves (Figure 9.3c). These are regions of **constructive interference**.

When two or more wave groups of different wavelengths are in the same region and going in roughly the same direction, there will be some places where the waves interfere constructively and other places where they interfere destructively, as illustrated in Figure 9.4. Among other things, this is the

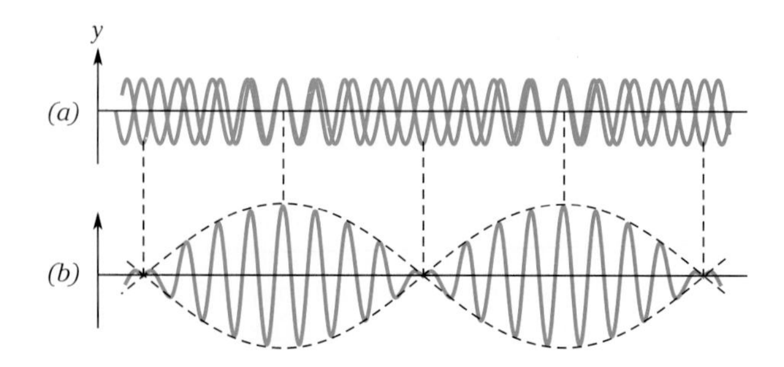

Figure 9.4

The superposition of two waves of slightly different wavelengths (a) yields alternate regions of constructive and destructive interference (b). (From Halliday and Resnick, 1981, *Fundamentals of Physics*, 2nd ed., John Wiley & Sons, New York.)

reason for the familiar phenomenon called **surf beat**. A series of relatively large waves comes into shore in succession, followed by a series of relatively small waves, followed by large waves again, and so on. Alternate regions of constructive and destructive interference between the incoming wave groups cause alternate regions of large and small waves, which you observe.

Standing waves result when two equal waves are going in opposite directions. These waves have the characteristic up and down oscillations of the water surface, without the horizontal progression of the wave form. They are quite common in coastal areas where incoming waves can reflect off of steep rigid barriers, such as seawalls, ships' hulls, or breakwaters. They are also prominent in swimming pools and harbors. The incoming and reflected waves are identical, but move in opposite directions. As is illustrated in Figure 9.5, this superposition yields some regions where the water has large amplitude oscillations, called **antinodes**, separated by points where there is no vertical motion at all, called **nodes**, or *nodal points*.

If the incoming waves approach at an angle to the boundary, the interference between the incoming and reflected components produces an interesting pattern called **edge waves**, which are "standing" in the in–out direction, but move in the direction parallel to the boundary (Figure 9.6).

SEICHES

When water is confined within a protected harbor or embayment, it may be vulnerable to a type of standing wave called a **seiche**, in which the water sloshes rhythmically back and forth across the confinement. You can observe a seiche in your home. If you would

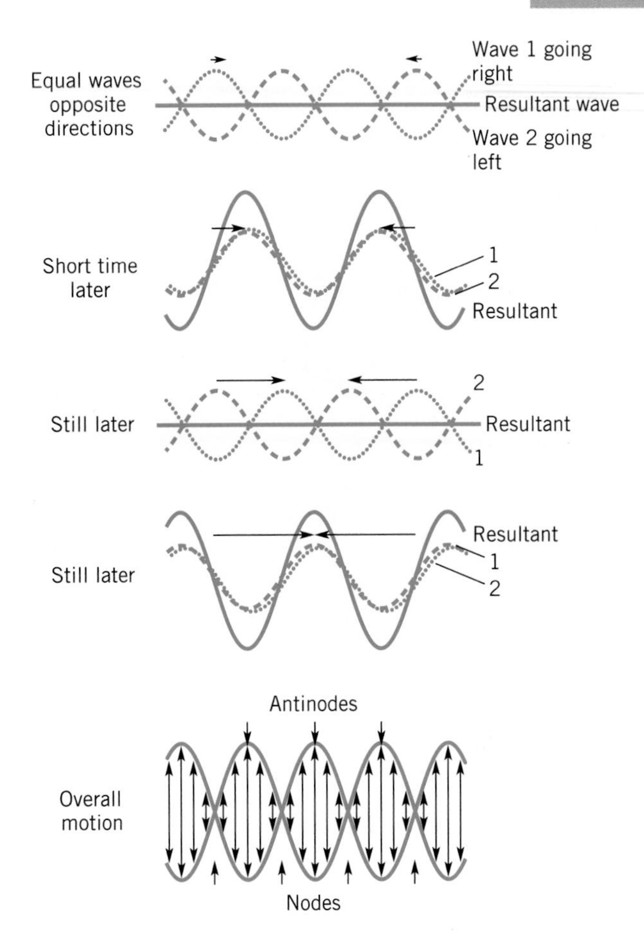

Figure 9.5

Top four sequences: Standing waves result from the superposition of two similar waves going in opposite directions. The dotted and dashed lines represent the individual waves, and the solid line the resulting net displacement of the water surface. The bottom sketch shows the overall motion of the resulting standing wave, with regions of large vertical motion (antinodes) separated by regions of no vertical motion at all (nodes).

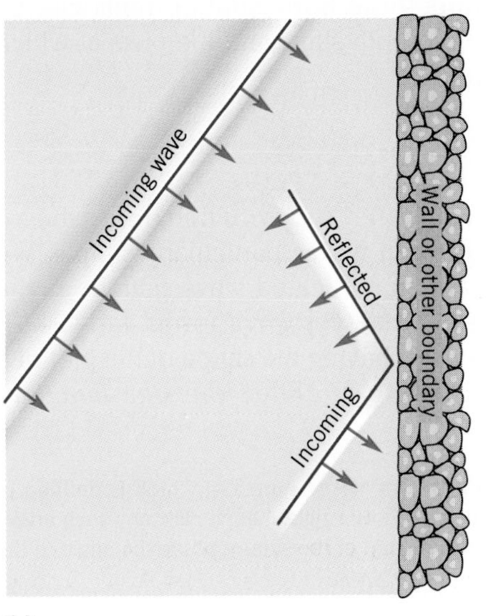

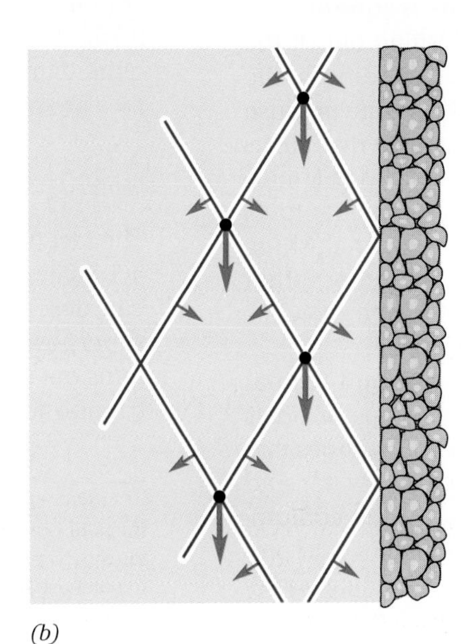

Figure 9.6

(a) When waves come in at an angle to a steep smooth wall or other boundary, they are reflected at an angle. (b) The interference between incoming and reflected waves produces a pattern of *edge waves* that move sideways parallel to the boundary.

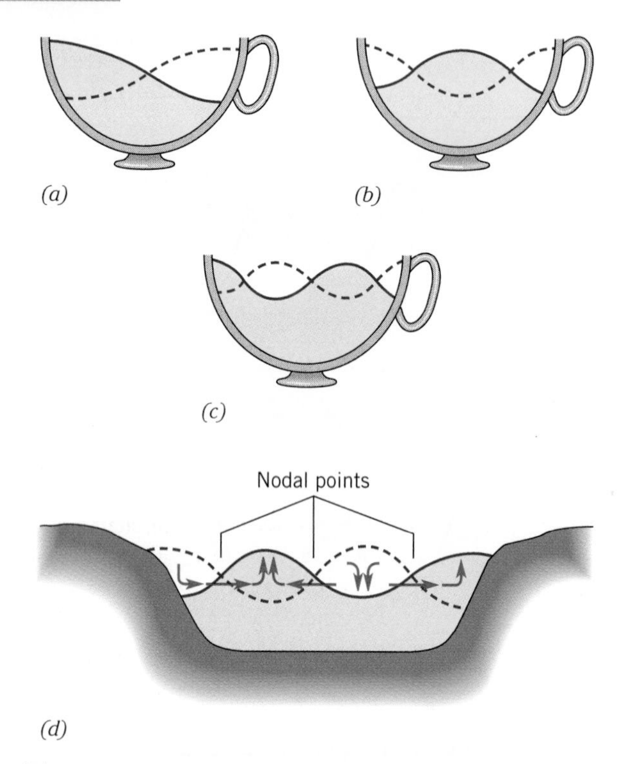

Figure 9.7

When water is confined, waves bounce back and forth across the confinement, creating standing waves. Like a child being pushed in a swing, if the frequency of the driving force just matches the natural resonant frequency for the water in the confinement, then the amplitude builds and the motion is called a *seiche*. (a–c) Seiches of water in a teacup for three different resonant frequencies (i.e., three different wavelengths). (d) Seiche in a harbor, indicating the position of the nodes, where the horizontal currents are strongest. Arrows indicate water motion as the water surface goes from dashed to solid line position.

stir your coffee or tea back and forth at the appropriate frequency, called the **resonant frequency**, or *resonance*, it will quickly begin sloshing out of the cup and onto the saucer and table. The same is true for water in a bowl, sink, or bathtub. When you push the water back and forth with just the right rhythm, its motion quickly builds up and water sloshes out, making a mess. Although the seiches most familiar to us appear to have just one mound of water sloshing from one side of the container to the other, other higher-frequency (i.e., shorter-wavelength) seiches are possible too, as illustrated in Figure 9.7.

In designing a harbor, breakwaters and natural land forms are used to try to protect ships or boats from waves. However, the same rigid boundaries that protect boats also confine the enclosed water and allow it to slosh back and forth. One of the compromises that must be reached in these designs is to provide both sufficient protection from incoming waves

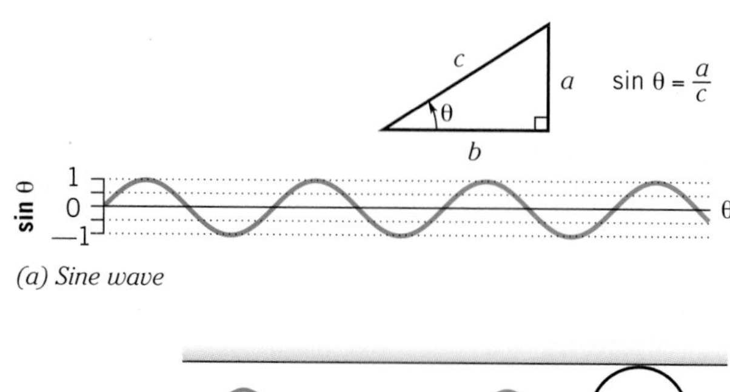

(a) Sine wave

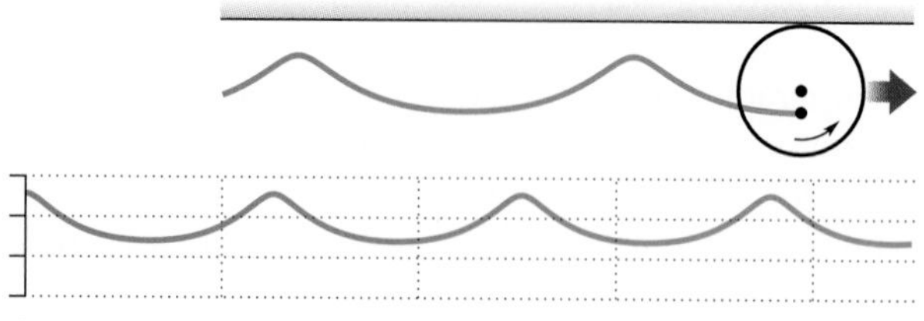

(b) Trochoidal wave

Figure 9.8

(a) Plot of the sine of an angle as a function of the angle, that is, a *sine wave*. The sine of an angle is defined with reference to a right triangle. It is the ratio of the side opposite the angle to the hypotenuse (the longest side). (b) Plot of a trochoidal wave. A trochoid is the path followed by a spot somewhere inside a wheel as the wheel rolls along the ground. In the case of the ocean waves, the trochoid is upside down (i.e., the wheel is rolling along a ceiling).

and also sufficient avenues for the water in seiches to slosh out of the harbor before further buildup.

Typical resonant periods for harbors are on the order of 5 to 10 minutes and are driven by such factors as long low incoming waves or gusty winds. In larger estuaries with longer resonant periods, the tides may provide an appropriate driving force. The Bay of Fundy is a famous example. When a seiche is in progress, there can be exceptionally strong horizontal currents, especially near the nodes (Figure 9.7d). In harbors these horizontal currents can be quite damaging to both ships and the piers at which they are moored.

SINE WAVES

The fact that waves obey superposition makes the exact analysis of ocean waves particularly simple. We can describe any complicated wave pattern as the superposition of whatever wave forms we wish. It turns out that waves having the shape of the plot of a trigonometric function, called the *sine function*[2]

[2]The *sine* is a trigonometric function, which is defined in terms of one of the angles of a right triangle. For any such angle, the sine is equal to the ratio of the side opposite the angle to the hypotenuse of the triangle.

(Figure 9.8a), are particularly simple to deal with mathematically. Therefore, we write any complicated wave pattern as the superposition of these simple **sine waves** (Figure 9.9). To understand the behavior of the complicated wave patterns we observe, we need only to understand the behavior of the individual sine wave components. For this reason, the analyses of ocean waves (in fact, analyses of all types of waves in all media) are often done for sine waves alone.

Very rarely is any ocean wave a perfect sine wave. In fact, we can show mathematically that the simplest free water wave of uniformly repeating shape and with all parts going in the same direction and at the same speed on an otherwise unperturbed water surface has a mathematical form called *trochoidal*. This is similar to a sine wave, but a little more peaked at the crest and broader in the trough (Figure 9.8b). Since trochoids, like all wave forms, can be written as superpositions of sine waves, which form to use is simply a matter of mathematical convenience.

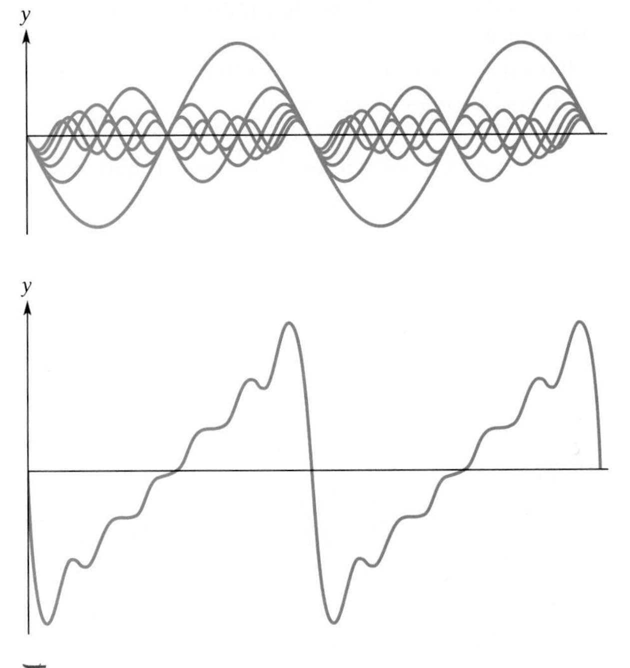

Figure 9.9
Illustration of how the superposition of several smooth sine waves (top) can yield one large jagged wave (bottom). (From Halliday and Resnick, 1981, *Fundamentals of Physics*, 2nd ed., John Wiley & Sons, New York.)

The energy carried by waves is of two forms: potential energy (energy of position) and kinetic energy (energy of motion). (These are defined in Chapter 2.) To understand the potential energy, imagine you could lift one region of the water surface and hold it motionless for a moment. If you let go, then the crest starts to fall, and the trough starts to rise. Thus, the initial energy of position is starting to move something. It is easily shown that the energy carried by sine waves is half potential and half kinetic. The same must be true for all waves, since all waves can be made of the superposition of sine waves.

RESTORING FORCE

When the sea surface is displaced from the still water level, some force must act to try to return the water to its equilibrium position. Without this **restoring force**, elevated regions would remain elevated, depressions would remain depressed, and waves would not move.

An important restoring force is gravity. The most familiar free waves are driven by gravity and called **gravity waves**. An elevated portion of the sea surface gets pulled back down. As it falls, the water gets forced out from under it, causing the neighboring sea surface to rise. This is how the wave travels (Figure 9.10).

Because of inertia, a mass in motion tends to keep on moving, unless acted on by a force. At the still water level, there is no net force on the moving

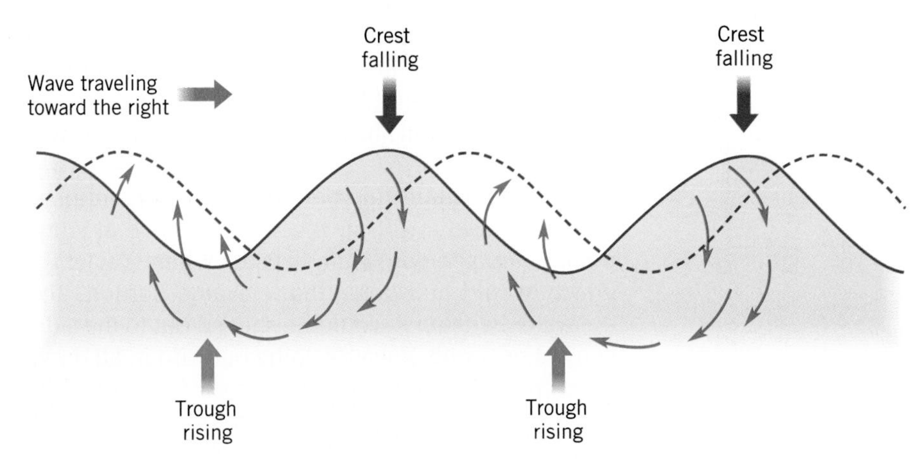

Figure 9.10
Waves travel because gravity pulls the water in the crests downward. Forced out from beneath the falling crests, the falling water pushes the former troughs upward, and the wave moves to a new position, as indicated. (Notice that the actual motion of the water itself beneath these waves is circular or *orbital*, which confirms our experience that we are carried up and forward as the wave approaches, and down and back as it passes.)

water surface, so the falling portions keep on falling, and the rising portions keep on rising. The former crest becomes a trough, and the former trough a crest. So goes the cycle. The water surface continues to oscillate as the waves pass. Although this description applies most literally to standing waves, the same principles apply to traveling waves, except that the up and down motions of closely neighboring regions are sequential, or staggered, rather than simultaneous.

This molecular attraction is what causes water to pull itself together to form droplets, and the reason why it is easier to wipe up a spill with a damp cloth than a dry one.

For very small waves, another restoring force becomes important. This is the **surface tension** of the water. The cause of surface tension is the electrical polarization of the water molecule. In Chapter 6, we learned that this electrical polarization makes them very "sticky" (Figure 6.1). Neighboring molecules are strongly attracted to each other, with the positive side of one molecule attracted to the negative side of the next, forming hydrogen bonds. Among other things, this molecular attraction is what causes water to pull itself together to form droplets, and the reason why it is easier to wipe up a spill with a damp cloth than a dry one.

The attraction between molecules on the very surface of water is particularly strong. Normally, the attraction is in all directions, but at the water's surface, it is concentrated in the horizontal direction. The resulting surface tension acts like a thin invisible stretched membrane over the water's surface. The effect of this "membrane" can be observed by carefully placing pins, needles, small metal blades, or other small metal objects on the surface of water in a glass. These objects are denser than water and should sink, but will be kept afloat on this invisible membrane. Surface tension is also the reason you can fill a water glass slightly more than full, and the reason small drops hang beneath a leaky faucet until they grow so large that their weight pulls them off.

Normally, the amount of restoring force provided by surface tension is insignificant compared with that provided by gravity. But tiny waves weigh very little, and the force of gravity on them is very small. For tiny waves shorter than about 2 centimeters, surface tension takes over as the dominant restoring force. Like a stretched membrane, the surface tension tries to flatten the water surface, pulling raised regions down and depressed regions up.

The tiny waves that are driven by surface tension are called **capillary waves**. You see capillary waves when you blow across a hot bowl of soup, or when the wind starts up over a quiet body of water. They "rough up" the water surface, thereby increasing the friction and facilitating the energy transfer between wind and water. They have the interesting property that shorter capillary waves travel faster. The opposite is true for the larger gravity waves (Figure 9.11).

The tiny waves that are driven by surface tension are called capillary waves. You see capillary waves when you blow across a hot bowl of soup, or when the wind starts up over a quiet body of water.

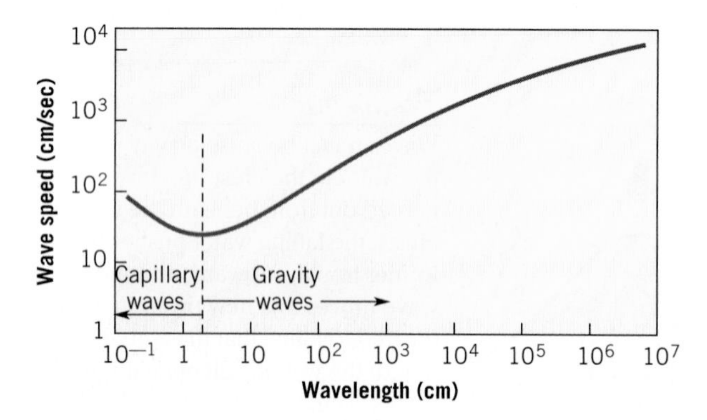

Figure 9.11

Plot of wave speed vs. wavelength for waves in deep water. The slowest waves are those with a length of 1.7 centimeters; they travel at a speed of 23 centimeters per second.

DIFFRACTION

The restoring force not only causes waves to travel, but it also causes waves to bend and spread out behind obstacles that they pass. This makes waves different from other moving things. For example, you could not hope to hit someone with a snowball if he or she were standing behind some obstacle. Things would be different, however, if you could throw a wave.

To understand this bending behavior, suppose you were able to lift up the water surface at some point in an otherwise calm flat sea. When you let go, gravity would pull down that elevated portion, and the water beneath it would be shoved out to the side, generating a ring of waves going outward in all directions from the original disturbance (Figures 9.12a and 9.12b). We can apply this idea to an extended elevated region, such as a long straight wave crest. As

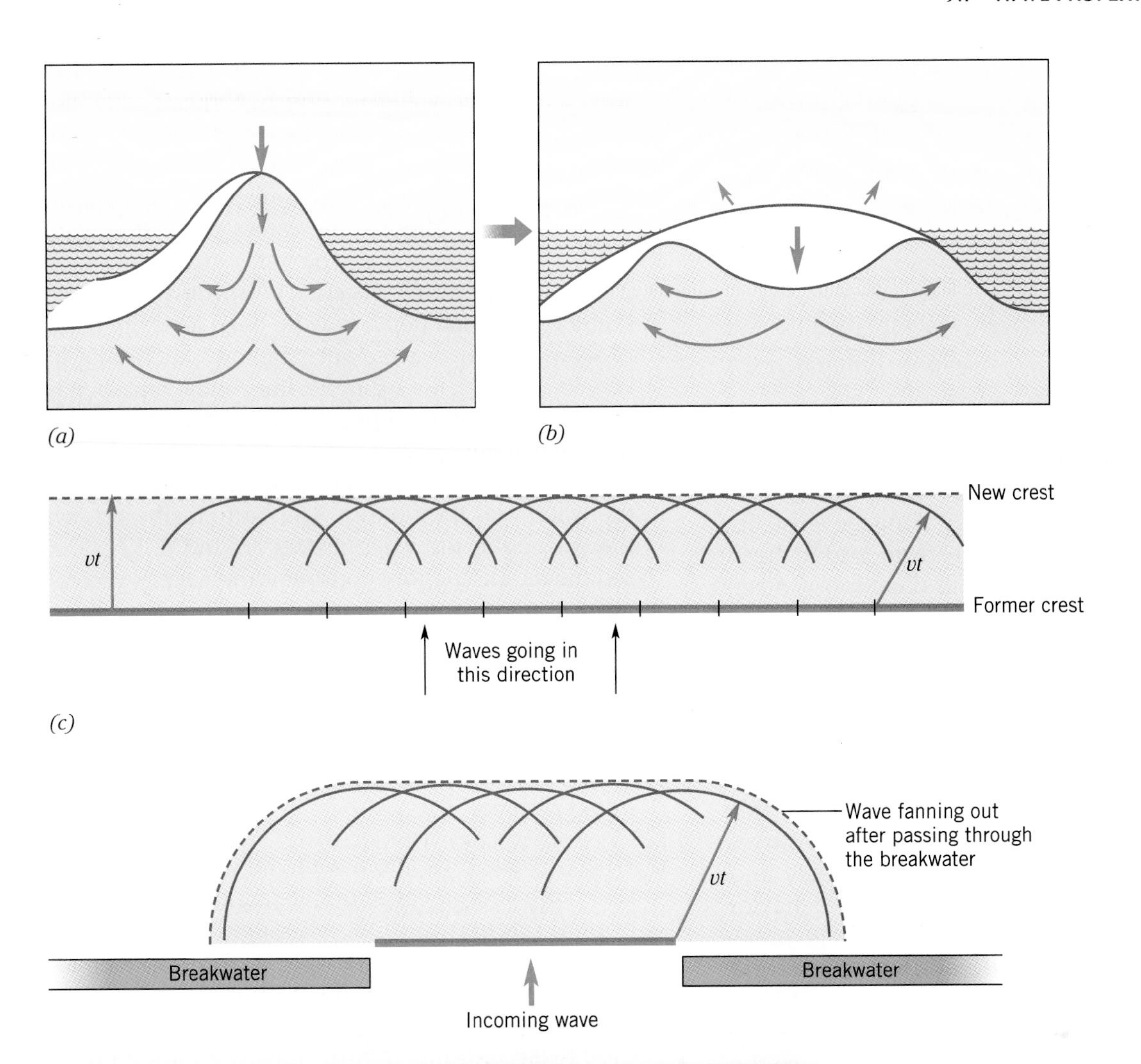

Figure 9.12

(a) Consider a bump in the water surface. As gravity pulls it down, the water beneath it is forced out to the sides. (b) This flow of water from under the bump generates a ring of waves traveling outward. (c) Viewing from above, in a long straight wave each point on the wave crest acts like a point source for a new ring of wavelets going out in all directions (Huygens' principle). Where these wavelets constructively interfere is the new position of the wave. (Distance traveled is wave speed times time, vt.) (d) Waves coming through a gap in a breakwater fan out in compliance with Huygens' principle. (e) Aerial photo of waves coming through the breakwater at the mouth of Morro Bay, California.

gravity pulls the crest back down, each point along the crest acts as a source for generating a ring of wavelets going outward in all directions (Figure 9.12c). This idea is called **Huygens' principle**, and it applies to all kinds of waves.

> You could not hope to hit someone with a snowball if he or she were standing behind some obstacle. Things would be different, however, if you could throw a wave.

As a long straight wave crest falls, the water beneath it cannot flow out sideways along the crest, because the water is flowing out from under these neighboring parts of the falling crest too. Consequently, the water beneath a falling wave crest must flow out toward the front or back, causing the old wave to continue moving forward and a new wave to come in from the rear.[3]

As waves pass an obstacle, such as a breakwater or ship, that obstacle blocks the waves, creating a *shadow zone* behind it. As the waves outside this shadow zone continue on, they generate new waves moving not only forward, but also to the side and into the shadow zone (Figure 9.12d). This behavior of waves, which causes them to fan out and go in behind obstacles, is called **diffraction**. You should now be able to explain diffraction in terms of Huygens' principle. As a wave fans out, the energy in the original wave is spread out over a larger area, so the wave is smaller than the original wave. Nonetheless, mooring a boat behind a breakwater will not protect it completely from incoming waves (Figure 9.12e).

ORBITAL MOTION

As we noted earlier, objects floating on or beneath the surface undergo circular oscillations as waves pass by. As a crest approaches, they are carried up and forward. After the crest passes, they are carried down and back again (Figure 9.13). This circular orbital motion of the water decreases with depth beneath the wave. At a depth greater than half of the wavelength, the water motion becomes negligible. Therefore, if the water is deeper than a half a wavelength, the bottom will have little effect on the wave, and the wave will have little effect on the bottom. Under these conditions, the waves are called *deep water waves*. Notice that whether or not the water is "deep" depends on the wave's perspective, not ours. Water of a certain depth may be deep from the point of view of short waves, but "shallow" according to very long waves. For example, the continental shelf is deep for the common gravity waves, but shallow for tides or tsunamis.

In shallow water, there is interaction between the water and the bottom. The bottom slows down the wave, and the wave moves around the bottom sediments. The bottom constrains the water's vertical motion much more than the horizontal motion. Therefore, the water's orbital motion is elliptical rather than circular, and these ellipses become smaller and increasingly flattened near the ocean bottom.

REFLECTION

When a wave runs into a wall, the water next to the wall has a special constraint. It cannot engage in the normal circular motion. Although the water can move sideways along the wall, it obviously cannot move in and out through the wall. This constraint requires that the incoming waves be reflected. Clockwise circular motion must be canceled by counterclockwise circular motion, which means that waves must be going in both directions. This is done automatically, because the wall automatically pushes back on the wave as hard as the wave pushes on the wall (Newton's second law of motion), so the wall reflects whatever comes in. As illustrated in Figure 9.14, the resulting motion occurs along the wall but not through it. If the wall does not extend all the way to the surface, then only a portion of the wave energy will be reflected, and the rest will continue on.

As you can imagine, there is a real interest in trying to minimize the wave energy reflected from harbor breakwaters. This is generally accomplished by making the breakwaters porous. Much of the incoming wave energy is absorbed, rather than reflected, as the water shoots into the structure through the holes and caves, and then trickles back down to join the ocean again.

WAVE DAMPING

Careful studies of waves indicate that free ocean waves can travel huge distances with little or no mea-

[3]You may wish to think about why this doesn't lead to waves going outward in *both* directions from the initial wave crest. Indeed it would, if the wave crest were initially standing still, but it wouldn't if the original wave crest were moving. The full answer has to take into account not only the interference of wavelets emanating from the wave's present position, but also those from all its previous positions. A mathematical treatment would show that the superposition of all rings of wavelets emanating from all the points along a wave crest results in constructive interference only in the forward direction.

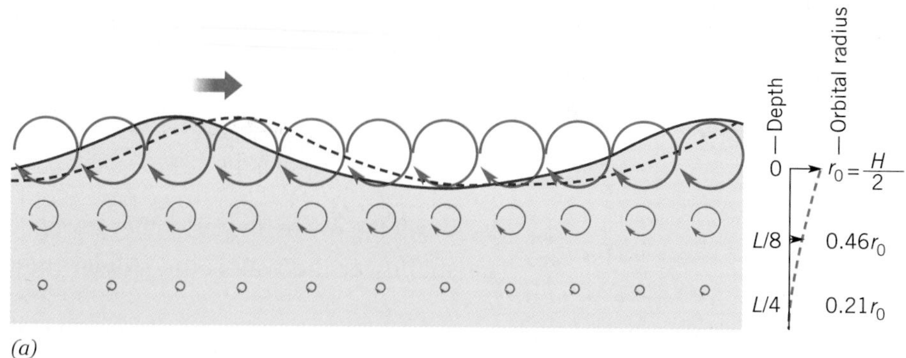

Figure 9.13
(a) Illustration of the orbital motion of a few representative water molecules as a deep water wave passes by. Dashed line indicates the position of the wave after each molecule has progressed through one-eighth of its orbit. The sizes of the orbits decrease with depth. At a depth equal to one-half the wavelength, the water motion is negligible. (b) Sketch of the horizontal and vertical spacings of a few representative water molecules as waves pass by. Seaweeds would have the orientations of the vertical dashed lines. The horizontal lines indicate the paths that would be followed by horizontally moving fish or submarines.

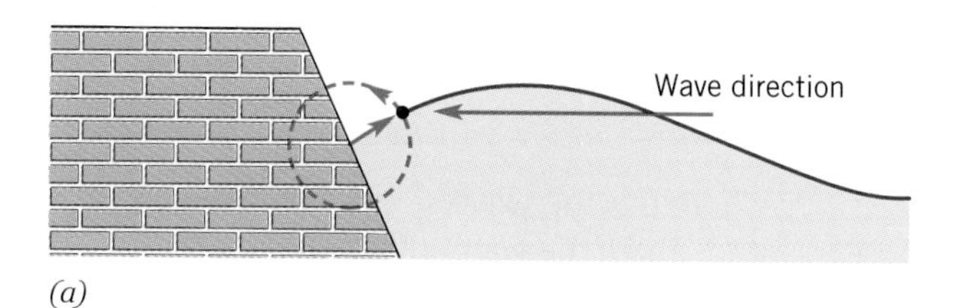

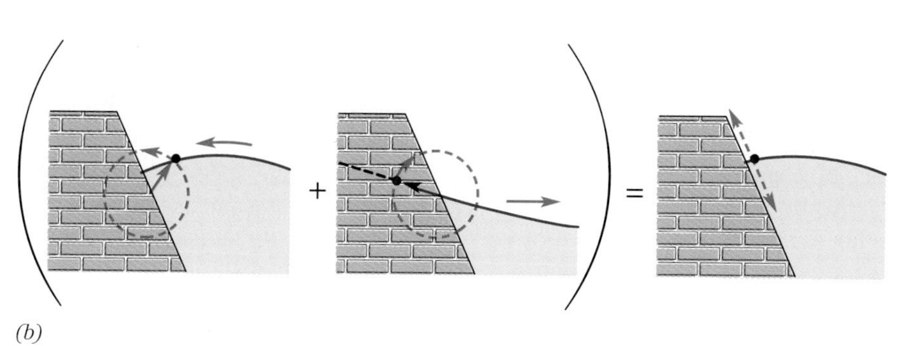

Figure 9.14
(a) When a wave runs into a wall, the normal orbital motion would require those water molecules at the edge to penetrate the wall half of the time and vacate the wall the other half. (b) However, the wall reflects the wave, and so the superposition of the orbital motions of the water molecules for incident and reflected waves yields linear displacement along the wall.

surable loss of energy. This is about as close to frictionless motion as we can get on Earth, and it is a very efficient means for transferring energy over large distances. Once generated, most ocean waves continue across the ocean and dump their energy on some distant coast. Waves may lose energy, however, if the water contains floating objects.

Waves may be reduced by anything that either removes energy, or interferes with the circular orbital motion of the water. Basically, anything floating in the water will work. Examples include seaweeds, debris, schools of fish, and ice. You may have noticed that it is much easier to carry cold drinks without spilling if they have ice chips floating in them. The floating ice helps calm the waves and prevent the liquid from spilling out.

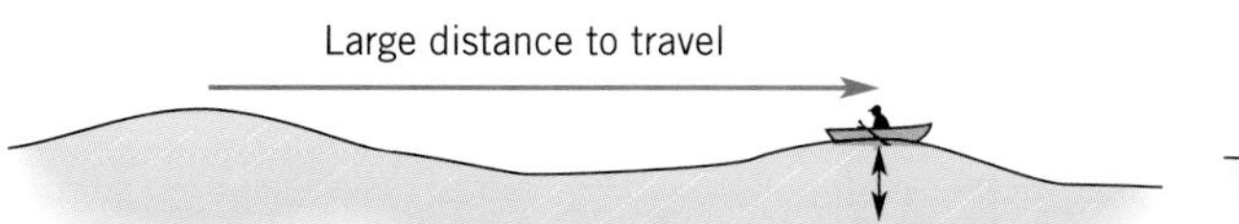

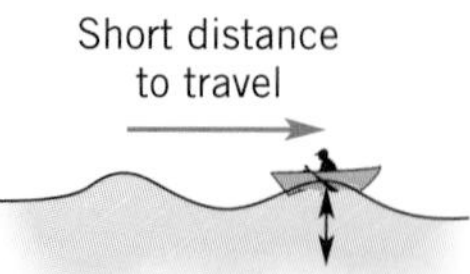

Figure 9.15

Longer waves travel faster than shorter waves because longer waves have farther to go to keep pace with the up and down undulations of the water's surface.

> It is much easier to carry cold drinks without spilling if they have ice chips floating in them. The floating ice helps calm the waves and prevent the liquid from spilling out.

Due to the inertia of things floating in the water, their up and down motion tends to lag behind that of the water. We say that they oscillate somewhat **out of phase** with the water. In addition, the rigidity of floating objects also tends to interfere with the fluid orbital motion of the water. For example, the water on one end of the object may be going up and forward, while that on the other end may be going down and back. A rigid object cannot do both at once. Consequently, both the inertia and rigidity of the floating objects tend to impede the oscillatory water motion and remove energy from the wave.

Floating objects are most effective in calming the waves that are the size of the objects or smaller. Small objects have little effect on large waves. Schools of small fish or small floating sticks may dampen waves smaller than a few centimeters long. The damping of the little ripples by floating debris creates smooth shiny areas on the water, called **slicks**. Pods of whales or a field of large icebergs may be able to calm waves that are much larger. Notice how smooth the water surface is in all the photos in this book that show fields of floating ice (Figures 4.6, 7.11c, and 8.22).

9.2 WAVE SPEED

WAVELENGTH AND WATER DEPTH

The speed with which free gravity waves travel across the ocean can be calculated using known mathematical relationships between forces and motion. Wave speed depends on two factors: the length of the wave and depth of the water.

1. The longer the waves, the faster they travel.
2. The shallower the water, the slower they travel.

Although the general solution is fairly complicated, it becomes simple in the two extremes of either very deep or very shallow water. For these two extremes, we have the following:

1. If the water is deeper than half the wavelength, then the bottom has essentially no effect. The wave speed depends only on the wavelength and is given by the formula

   $$v = 1.25\sqrt{L}$$

   where v is the wave speed in meters per second, and L the wavelength in meters.[4]
2. If the water is shallower than one-twentieth the wavelength, all waves travel at nearly the same speed. The shallower the water, the slower they go. The speed for these shallow water waves is given by

   $$v = 3.13\sqrt{d}$$

   where v is the wave speed in meters per second, and d the water depth in meters.

These two extremes are the basis of our definition of *deep water* and *shallow water* waves. The water is "deep" if the bottom has no effect on the wave motion, and it is "shallow" if the bottom completely controls the wave motion. Although the derivation of the complete formula for wave speed is beyond the scope of this text, the basic ideas are as follows.

Longer waves travel faster, basically because longer waves have a greater distance to travel between successive up and down undulations of the water (Figure 9.15). (Actually, the water surface undulates more slowly with longer waves, but the difference is not enough to make up for the difference in wavelength.) Long waves also "reach" deeper. That is, water motion extends deeper beneath a long wave than beneath a short one. Consequently, as waves come into shore, the faster longer waves run into the bottom first and are slowed down more. When the

[4] In some writings, the wave speed is given the symbol c for "celerity."

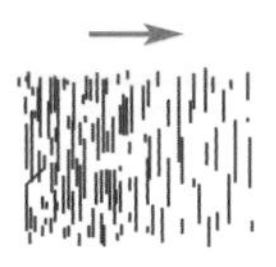

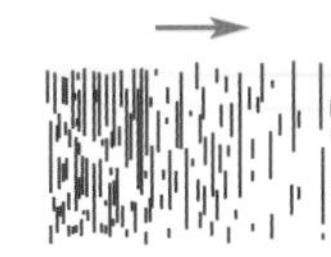

Figure 9.16
Illustration of a wave group crossing an ocean on 4 successive days, viewed from above. As the group progresses, the long wavelength components (being faster) take the lead and the slower, shorter components fall behind.

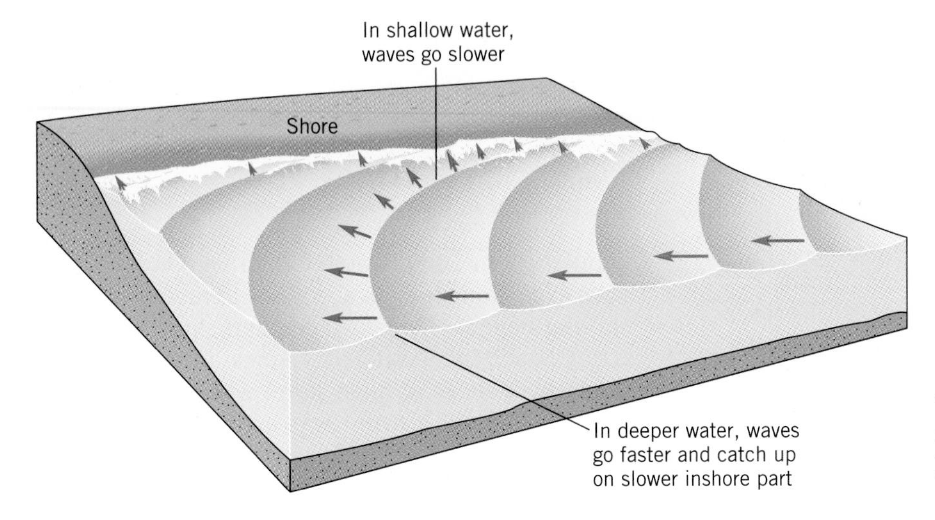

Figure 9.17
Refraction of waves coming in toward shore.

water becomes sufficiently shallow, the long fast waves have been slowed down so much that they are going the same speed as the slower shorter waves.

DISPERSION

The fact that longer waves travel faster has an interesting effect on the pattern of waves generated by storms at sea. A storm generates a rough and rugged ocean surface, which is a superposition of many different wave groups of various amplitudes and wavelengths, going different directions. As all these waves travel away from the storm, the longer waves travel faster and take the lead, while the slower shorter ones fall behind (Figure 9.16). Far from the storm center, all the waves in the region will be going in the same direction, and the waves will be sorted according to wavelength, with longer faster waves farther out in front. This gradual spreading and sorting of waves according to wavelength is called **dispersion**.

By the time the large storm waves have traveled a thousand kilometers or more from the storm center, they have become very well sorted, with all the waves in a region having long parallel crests and uniform wavelengths. We refer to these well-sorted uniform waves as **swell**. When the swell reaches our coasts, we can tell how far the storm center was out at sea by how far the long waves have gotten out ahead of the shorter ones.

You can observe dispersion by throwing some stones into a quiet pond. As the various waves travel outward from the original splash, the longer waves take the lead, forming into nice uniform rings of well-sorted waves as they go. The smaller ripples travel much more slowly and tend to remain back near where the stones struck the water. The original splash disappears, and the various waves sort themselves out as they travel. Storms at sea do the same, except that both the original "splash" and the resulting waves tend to be larger than those in a pond.

> You can observe dispersion by throwing some stones into a quiet pond. As the waves travel outward from the original splash, the longer waves take the lead, forming into nice uniform rings of well-sorted waves.

REFRACTION

The slowing of waves in shallow water causes them to change their direction of travel. The portion of each wave that reaches shallow water first slows down, and the portion still in deeper water catches up. Consequently, waves change their headings, turning toward shallower water where they go more slowly (Figure 9.17). This bending of waves toward

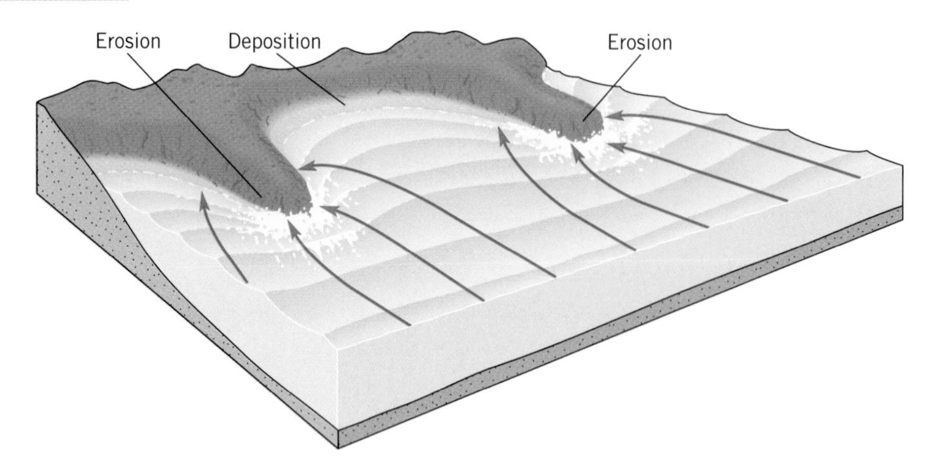

Figure 9.18

Incoming waves reach shallow water off the points first. Therefore, they bend toward the points as they come in. Solid lines indicate the direction of wave propagation (everywhere perpendicular to the crests) and divide the incoming waves into equal energy increments. It is seen that the energy of the incoming waves is concentrated on the headlands and the coves are relatively quiescent.

regions where they go more slowly is called *refraction*. Due to refraction, all waves seem to come nearly straight in toward shore, even though they seemed chaotic and were traveling in various directions when out at sea.

This behavior is not peculiar to ocean waves. All types of waves, and indeed all things in motion, always tend to turn toward where they go more slowly. While walking, if your right foot stops while the left foot keeps going, you turn toward the right. While driving, if the wheels on the right side slow down while those on the left side keep going, the car turns toward the right. Waves of visible light bend while traveling through lenses and prisms for the same reason.

On jagged coastlines, incoming waves encounter shallow water off the points (called **headlands**) first, so the waves bend toward these regions. Consequently, the wave energy is concentrated on the points, and the coves are quite calm by comparison. We say, "The points draw the waves." (See Figure 9.18.)

Some large waves, such as the tides and tsunamis, reach all the way to the deep ocean floor. Storm waves do not normally reach as deep, but some can reach the deepest parts of the continental shelf. Bumps and dips in the ocean floor can cause these large waves to change their course as they cross the oceans and shelves, directing the waves toward some coasts and away from others.

An interesting example of refraction occurred in April 1930 in Long Beach, California. On a day that was exceptionally calm according to records at neighboring beaches, breakwaters, and ships at sea, the Long Beach breakwater was struck and heavily damaged by a series of huge storm waves. This mysterious and peculiar event was eventually explained when a hump was found on the continental shelf south of the breakwater. Long period swell, coming from a storm in the Southern Hemisphere, was refracted while passing over this hump. Much as a lens focuses light, the incoming storm waves were focused on the tip of the breakwater. It received all the wave energy, and the neighboring shoreline received little or none.

WAVE GROUPS

Waves usually travel in groups. Familiar examples of small wave groups include the waves made by a moving boat (the boat's "wake") and the circles of waves made by throwing a stone into a quiet pond. Larger wave groups are generated by storms at sea. As a wave group travels across otherwise quiet water, the wave at the front of the group must set the newly encountered water into motion. This takes energy from the lead wave, and so it dies. On the other hand, stopping the circular orbital motion at the rear of the passing wave group yields extra energy, which appears as a new wave forming at the rear (Figure 9.19a).

Consequently, if you make a splash in a quiet pond and carefully watch the lead wave in the group, you will find that it quickly disappears. What was once the second wave in the group becomes the first. Soon, it too will disappear, and the once third wave will then take the lead. Similarly, if you watch the last wave in the group, you find that a new one soon appears behind it. As time goes on, more and more new waves appear at the rear. The overall picture is that as the group travels, individual waves spring up

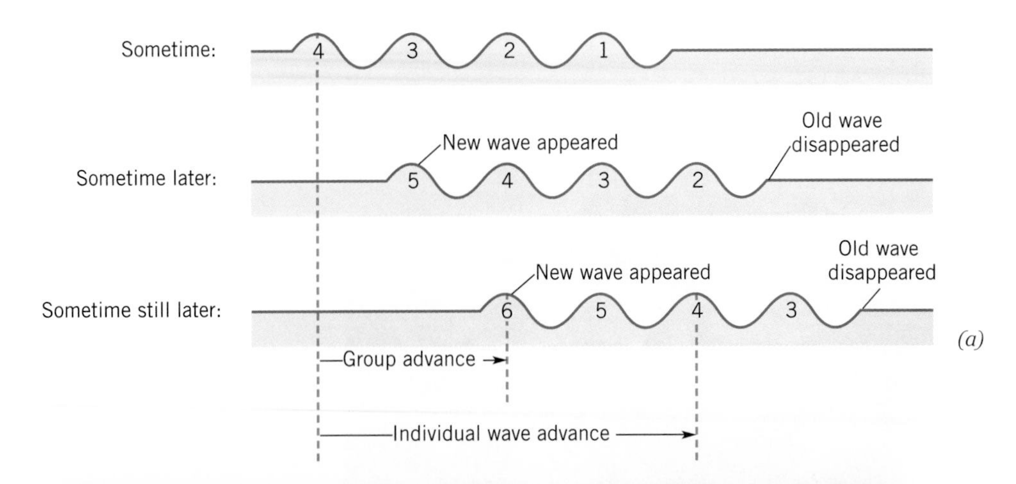

Figure 9.19

(a) Illustration of the motion of a wave group. The lead waves keep dying out, exhausting their energy by setting the newly encountered water into motion. New waves form at the rear, where extra energy is released as the water motion stops. In deep water, the group travels at half the speed of the individual waves. (b) The wake of a boat provides a way to study the behavior of waves in a group. Evolution in time can be determined in this still photo by noting that points in the wake farther from the boat have been underway longer. Following any one wave in the above group, you can see that it is born (point closest to the boat) at the rear of the wave group and dies (point farthest from the boat) at the front of the group.

at the rear, gradually work their way to the front, and then die.

This effect can be seen in still photos of ship wakes, such as Figure 9.19b. Let your eyes follow the wake that trails back behind the boat in this photo. Farther away from the boat are waves that have been under way for longer times. Consequently, going back away from the boat along any wave, you trace its development in time. You see that individual waves are formed at the back of the group (the inside of the wake) and die out at the front of the group (the outside of the wake).

For free deep water waves, the individual waves travel exactly twice as fast as the group, springing up at the rear of the group, traveling up to the front, and then dying.[5] We have to watch out for this factor when we try to predict when the waves from a distant storm might arrive at our coast. They take twice as long as we would predict using the wave speed and distance to the storm center.

[5]As the water shoals, however, the relationship between the speed of the individual waves and that of the group becomes much more complicated.

Figure 9.20

As waves come in towards shore, trailing waves catch up on leading waves, because the trailing waves are in deeper water and traveling faster. Wavelengths decrease and wave heights increase. Taller waves mean larger orbitals and faster orbital speeds for the water. Orbital speeds increase and wave speeds decrease. Eventually the orbital speed of the water is greater than that of the wave itself. The crest gets out ahead of its support, and the wave breaks.

9.3 BREAKING WAVES

WHY WAVES BREAK

As a series of waves approaches shore, the lead waves are in the shallowest water and going slowest. The rear waves gain on them, and so the distance between waves decreases. The water and energy of each wave become concentrated in a narrower zone, and so the wave grows taller. The water's orbital velocity increases, because taller waves require larger orbitals, which means the water has to go farther each time around. You notice this increased orbital velocity when standing in the water, where you are alternately pulled very strongly out to sea by the water in the trough and then pushed very strongly toward shore by the water in the crest.

At the same time that the orbital speed is increasing, the wave speed is decreasing, due the the shallower water. Eventually, the orbital speed is larger than the wave speed. The crest gets out ahead of the wave, and the wave breaks (Figure 9.20). A related factor that adds to a wave's instability is that as waves head toward shore, water beneath a crest is deeper than that beneath the trough in front of it. So the crest travels faster and tends to catch up on the trough. As a general rule, waves break when they reach a water depth, measured to the still water level, that is about 1.3 times the wave height. Consequently, by standing on shore and estimating the height of the waves where they are breaking, you can determine the water depth out there without getting wet.

Frequently, waves break more than once before reaching shore. Large heavy waves tend to drag sand from the beach and deposit it in **longshore bars**. The incoming waves encounter shallow water first over these longshore bars and break over the bars rather than on the beach. In this way, longshore bars protect the beaches from taking the full force of the waves during periods of heavy seas. You can detect the position of longshore bars by looking for offshore regions where the waves are breaking prematurely.

Water cannot support waves with steepness (H/L) greater than 1/7. Waves may break before their steepness reaches this value if subjected to other constraints, such as shoaling waters or the forceful impact of strong winds. But even in deep water, waves will break if their steepness grows too large. This can happen, for example, in a region where a storm is causing the waves to grow larger, or by the

Figure 9.21
Ships must head into breaking waves to take as much of the force as possible on the bow.

chance superposition of two waves in such a way that the resulting wave height is too steep.

WAVES OF TRANSLATION

Once a wave has broken, the water formerly in the crest spills down the front face of the wave and continues on across the water surface. These broken waves are called **waves of translation**, because of the sideways or "translational" motion of the water in them.

Waves of translation can be quite destructive. Ships and offshore structures can handle the cyclical orbital motion of water beneath unbroken waves, but when waves of translation slam into them, the moving mass of water can be overpowering. Ships head into large breaking waves in order to take as much of the force as possible on the bow (Figure 9.21). Catching large waves of translation broadside could result in severe damage or loss of the ship.

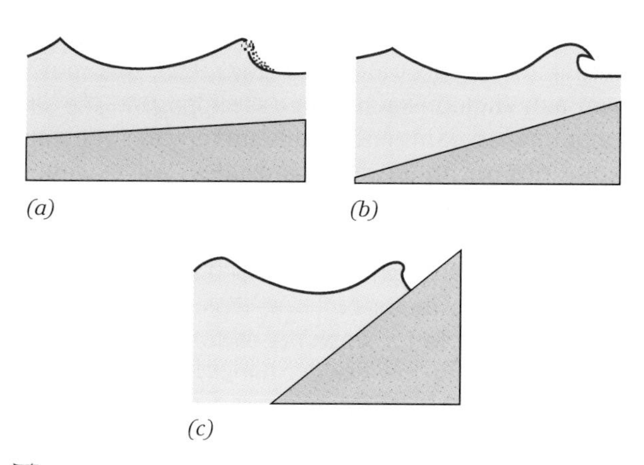

Figure 9.22
Illustration of the three different types of breakers. (a) Spilling breakers are generally found where there is a gently sloping bottom. (b) Plunging breakers are found where the bottom slope is moderate. (c) Surging breakers are found where the bottom slope is so steep that the wave doesn't break until it is right at the shoreline.

TYPES OF BREAKERS

Breaking and broken waves along the coast are colloquially referred to as **surf**, and the coastal region containing the breaking and broken waves is called the **surf zone**. In the surf zone, breaking waves are named according to the form they appear to take, or equivalently, the abruptness with which they break.

Where there is only very gradual slope to the bottom, incoming waves grow gradually toward breaking proportions. The wave breaks slowly and over a long distance, with the crest spilling down the face of the wave. These are called **spilling breakers** (Figure 9.22a). Where there is greater slope to the bottom, the front base of the wave may slow down so quickly that the crest gets out ahead of it. The crest curls over the front of the wave and plunges down toward its base, rather than spilling down its face. Such breakers are called **plunging breakers** (Figure 9.22b). Finally, the slope of the bottom may be so abrupt that a wave does not build up and break until it reaches the beach. That is, the broken wave falls directly onto the beach, rather than onto the water in front of the wave. These breakers that suddenly build up and break right at the shore are referred to as **surging breakers** (Figure 9.22c).

9.4 WIND-DRIVEN WAVES

Most common gravity-driven surface waves are generated by the wind. As the wind blows, tiny capillary waves spring up first, making the sea surface rougher

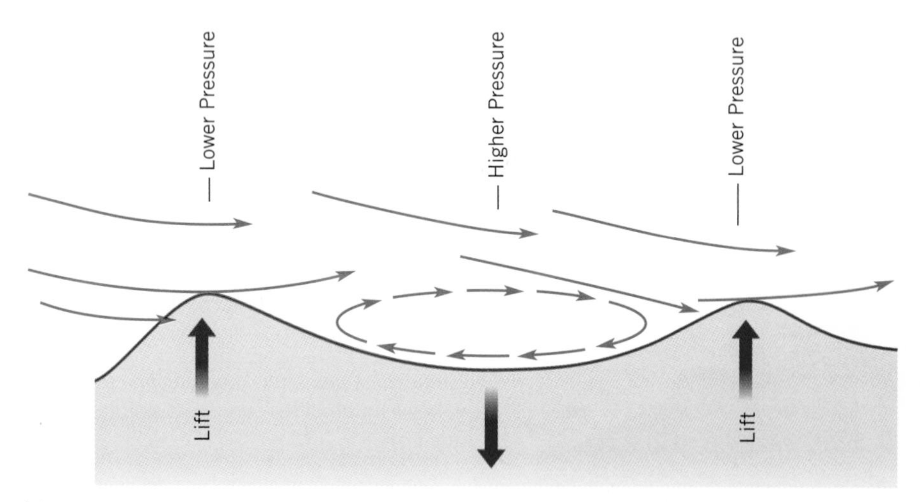

Figure 9.23

Fluids accelerate from higher pressures toward lower pressures. Therefore, their internal pressure is lowest where they are moving fastest (Bernoulli's principle). That is what provides lift to airplane wings. It also provides lift to wave crests, because the air speed over the crests is greater than that in the protected troughs. In addition, the wind transfers energy to the waves by pushing against the back of the crests, and eddies in the troughs may help reinforce the backward orbital motion there.

and increasing the friction with the wind. Small waves develop quickly, and larger ones more slowly. The energy transfer between wind and waves involves many mechanisms, including that which gives lift to airplane wings as the air flows over them, and that which causes flags to wave in the breeze (Figure 9.23). In addition, the wind pushes against the backs of wave crests, and eddies (i.e., swirls) in the troughs reinforce the down and back orbital motion of the water there.

The most rapid and violent build-up of seas involves both strong winds and breaking waves. When the waves break, they generate more waves, some of which are of longer wavelength. Thus, breakers provide another mechanism, in addition to the direct force of the wind, that helps the build-up of longer waves in storm centers. The unsorted, choppy mixture of waves in a region where the waves are building up is sometimes colloquially referred to as **sea**.

If the wind continues, the waves grow larger until the average wavelength is sufficient that the waves are moving as fast as the wind (Figure 9.24). When this stage is reached, the waves grow no more, and the state of the sea is referred to as **fully developed**. Of course, other things may happen to end the development of the waves before becoming fully developed. In particular, the wind may stop, or the waves may leave the windy region. Consequently, the average length of waves generated by winds depends on not only the wind speed, but also the wind's duration and the distance over which it blows, or the **fetch**. In ship logs, it is standard to describe the state of the sea surface according to a numerical scale, referred to as the **Beaufort scale,** on which the numbers range from 0 (no wind, complete calm) to 12 and above (hurricane conditions).

9.5 TSUNAMIS

In addition to the wind, waves may also be created by movements of the solid Earth. Just as the movement of your body generates waves in your bathtub, movement of the ocean bottom generates waves in the ocean. As illustrated in Figure 9.25, this movement may include such things as slippage of the sea floor along an earthquake fault, underwater volcanic explosions, or underwater landslides. Waves generated in this manner are called *tsunamis*, or *seismic sea waves*. Tsunamis usually have long wavelengths and small wave heights. Waves 200 kilometers long and less than a meter high are typical. It would be nearly impossible for ships at sea to detect the passage of such long, low waves beneath them.

The ocean is everywhere very shallow compared with the length of most tsunami waves, and so they are quite strongly affected by the bottom. Due to refraction, they are bent toward some regions and away from others by rises and dips in the ocean floor.

As is true for all waves, when a tsunami comes into shallow coastal waters, its wavelength decreases and its height increases. When it reaches shore, the

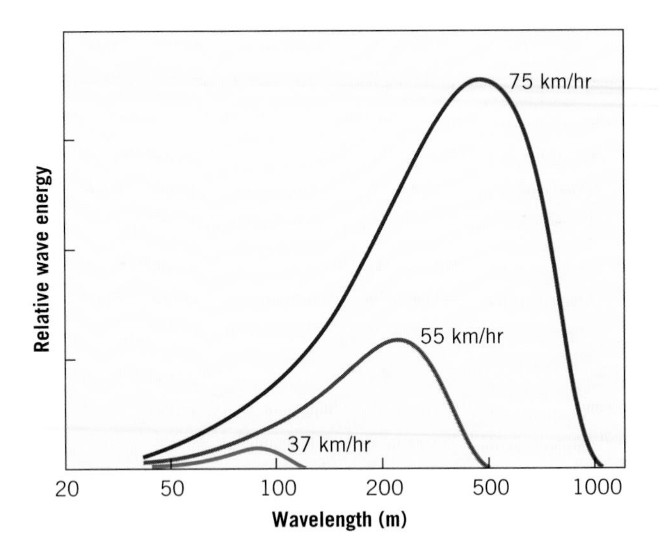

Figure 9.24

Plot of wave energy vs. wavelength for fully developed seas in winds of 75, 55, and 37 kilometers per hour, respectively. Stronger winds generate waves that are both longer and more energetic, on average.

wave height may have attained several meters. This is still very small compared with its wavelength, which would have been shortened to several tens of kilometers by then. If we were on the coast as a typical tsunami came in, we would not observe the tsunami as a huge overhanging breaking wave, crashing down on unsuspecting coastal villages, as some sensationalistic books and movies would have us believe. Rather, we would see a gradual rise and fall of sea level, with a period of typically 10 to 20 minutes.

> Just as the movement of your body generates waves in your bathtub, movement of the ocean bottom generates waves in the ocean.

Nonetheless, tsunamis can be quite dangerous. A few coastal areas are exceptionally susceptible to tsunami damage because of ocean bottom contours that tend to focus the energy of the incoming tsunamis. Particularly vulnerable are coastal regions that are low, flat, and normally do not experience large fluctuations in sea level. In these areas, entire towns and villages may be built near sea level. As a tsunami comes in, the rising water may flood these areas. The strong currents of the flooding waters and the ever-present wind-driven waves on the surface may smash into buildings that were not designed to withstand such battering. One of the most spectacular tsunamis of all time occurred with the eruption of Krakatoa in Indonesia in 1883. Some of the resulting waves reached heights of 40 meters in the shallow waters of neighboring islands, and the death toll in the low-lying coastal areas of these islands reached 36,000.

For those coastal areas that might be vulnerable to damage, tsunami prediction is an important but difficult endeavor. The detection of underwater seismic activity somewhere in the world does not mean that a tsunami has been generated. Even if one is under way, the waves are so long and low that they often cannot be detected until they reach shore somewhere. Refraction by sea floor features changes their directions. Even if we did know where they were headed, tsunamis travel too fast to warn and evacuate imperiled coastal populations. Wave speeds for tsunamis across the deep ocean are typically 750 kilometers per hour.

Many coastal populations are accustomed to large tidal ranges and do not build structures near the shoreline. These regions are usually safe from tsunami damage, unless there are exceptional circumstances, such as the tsunami occurring during a very high tide, or at the same time that a storm surge has raised sea level on the coast.

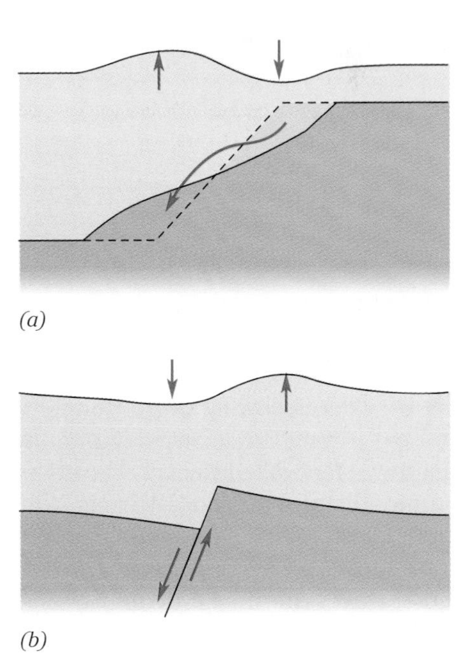

Figure 9.25

Two possible causes for tsunamis are underwater landslides (a) and slippage along a fault (b).

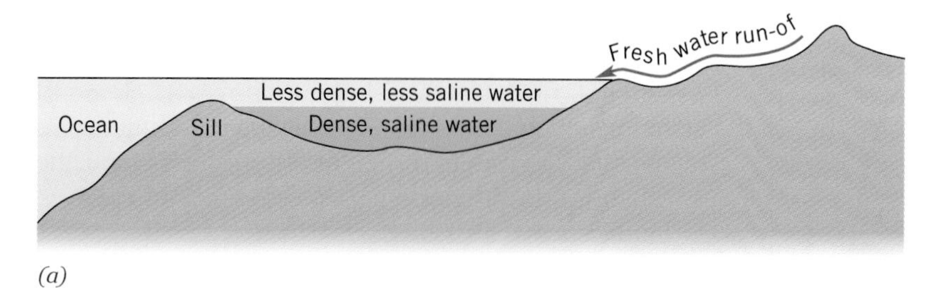

(a)

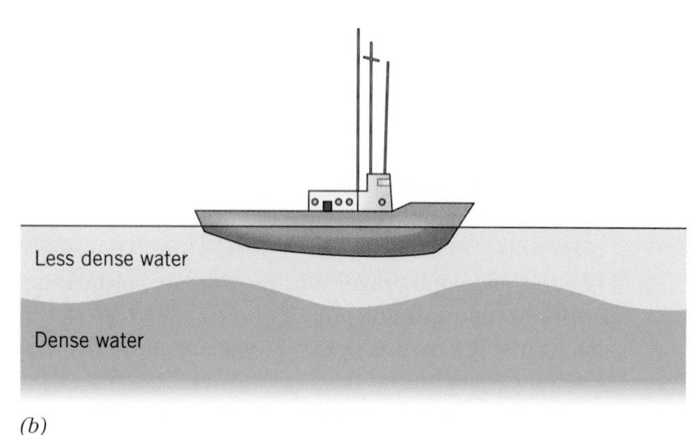

(b)

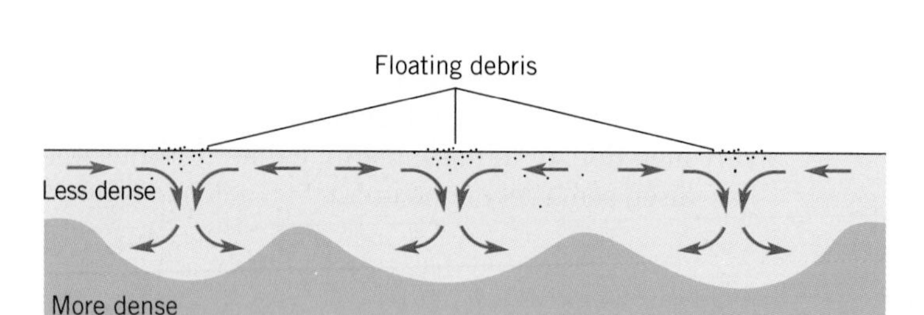

(c)

Figure 9.26

(a) Internal waves are common where light water flows out over denser water. This is especially common near the mouths of rivers and streams, and in fjords. Internal waves might be generated by the flow of the light surface layer or perhaps the motion of a slow-moving boat. (b) The boat handles sluggishly due to its invisible subsurface wake. (c) Floating surface debris collects over the troughs in the internal waves, as the surface water must flow toward the troughs to fill them in. This debris dampens the wind ripples, making visible slicks.

9.6 INTERNAL WAVES

In the ocean, subsurface *internal waves* form when one layer of water is flowing over or under another (Figure 9.26). Along the coast they often form in fjords (narrow steep-walled glacially carved inlets) or near the mouths of rivers and streams, where the runoff from rainstorms or snowmelt flows out across the ocean's surface.

Internal waves are driven by gravity and travel very slowly. You may have noticed that you fall much more quickly through air than water. You are much denser than air, so gravity has a correspondingly larger effect on you there. Gravity pulls denser things downward, and the greater the difference in density, the more quickly the denser things fall. For this reason, the speed of waves along the interface between two fluids depends on their relative densities. Surface waves travel fast, because water is much denser than air. By contrast, internal waves travel very slowly, because the densities of two water masses differ only very slightly.

The presence of internal waves can sometimes be detected by long parallel slicks on the ocean surface. As is illustrated in Figure 9.26c, the surface water collects over the troughs of the internal waves and is thinned over the crests. This motion of the surface water away from the crests and toward the troughs causes floating surface debris to collect over the troughs. This debris dampens the small surface ripples, making the surface smooth and glassy.

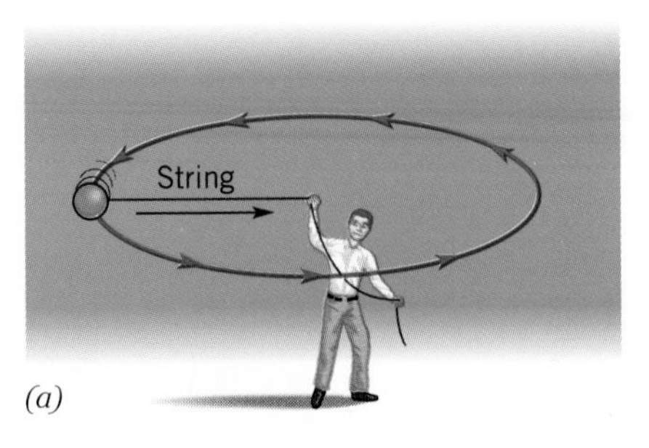

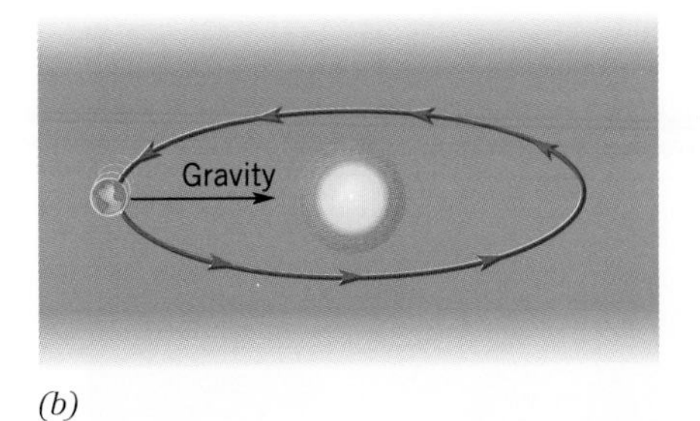

(a) (b)

Figure 9.27

(a) When you twirl a ball on the end of a string, the string pulls the ball toward your hand. If it were not for this, the ball would fly away. (b) As a planet orbits the Sun, gravity provides the inward force that keeps the planet in its orbit. Gravity pulls on the orbiting planets, just like the string pulls on the orbiting ball. Gravity also holds together the Earth and Moon as they orbit each other, just as if they were tied together by a string. (Actually, more like a rubber band.)

Table 9.1 Average strengths of the Sun, Moon, and Earth's own gravitational forces on the Earth's surface. Also listed are the differences in the gravitational forces from one side of the Earth to the other. (These differences are what produces the tides.) Forces are given in millionths of a Newton force per kilogram of mass.

Object	Gravity (μN/kg)	Near and Far Side Difference (μN/kg)
Earth	9,800,000	0
Sun	5,930	0.51
Moon	33	1.10

9.7 TIDES

The largest of the waves in our oceans are the tides, which are caused by the gravitational forces of the Moon and the Sun. One such wave stretches half-way around the Earth. Unlike the other types of waves we have studied, these waves are as regular and predictable as the sunrise tomorrow.

THE ASTRONOMICAL CAUSE

As we learned in Chapter 2, gravitational forces depend on mass and distance, increasing with mass and decreasing with distance. That is, the more massive the object, the greater its gravitational force. The greater the distance to the object, the smaller its gravitational force. For points on the Earth's surface, the Earth's own gravitational force is by far the most influential, and Earth has pulled itself into a spherical shape.

A few other objects also exert gravitational forces on the Earth, although these forces are very, very small compared to that of the Earth itself (Table 9.1). Of these others, the Sun and Moon are the most influential: the Sun because it is massive, and the Moon because it is near. Their gravitational force here is only a tiny fraction of the Earth's own gravity, so the distortion of the Earth's surface due to these objects is small.

When you twirl a ball on a string, the string pulls the ball toward your hand and keeps the ball in its circular orbit (Figure 9.27). In a similar fashion, the gravitational force between celestial objects is what keeps them in their orbits. It is what keeps the Moon in orbit around the Earth and the Earth in orbit around the Sun. This same gravitational force is also responsible for the tides, with the important aspect being that the strength of gravity between two objects decreases with increasing distance.

When you twirl a ball on a string, the string pulls the ball toward your hand and keeps the ball in its circular orbit. In a similar fashion, the gravitational force between celestial objects is what keeps them in their orbits.

We consider first the gravitational force of the Moon on the Earth.[6] Because the force of gravity between two objects decreases with increasing separation between them, the side of the Earth nearest the Moon gets pulled harder than the side away from the Moon. This results in some stretching of the Earth

[6]Of course, the Earth pulls on the Moon just as hard as the Moon pulls on the Earth. But because we wish to understand tides on the Earth, we study what the Moon does to the Earth, not what the Earth does to the Moon.

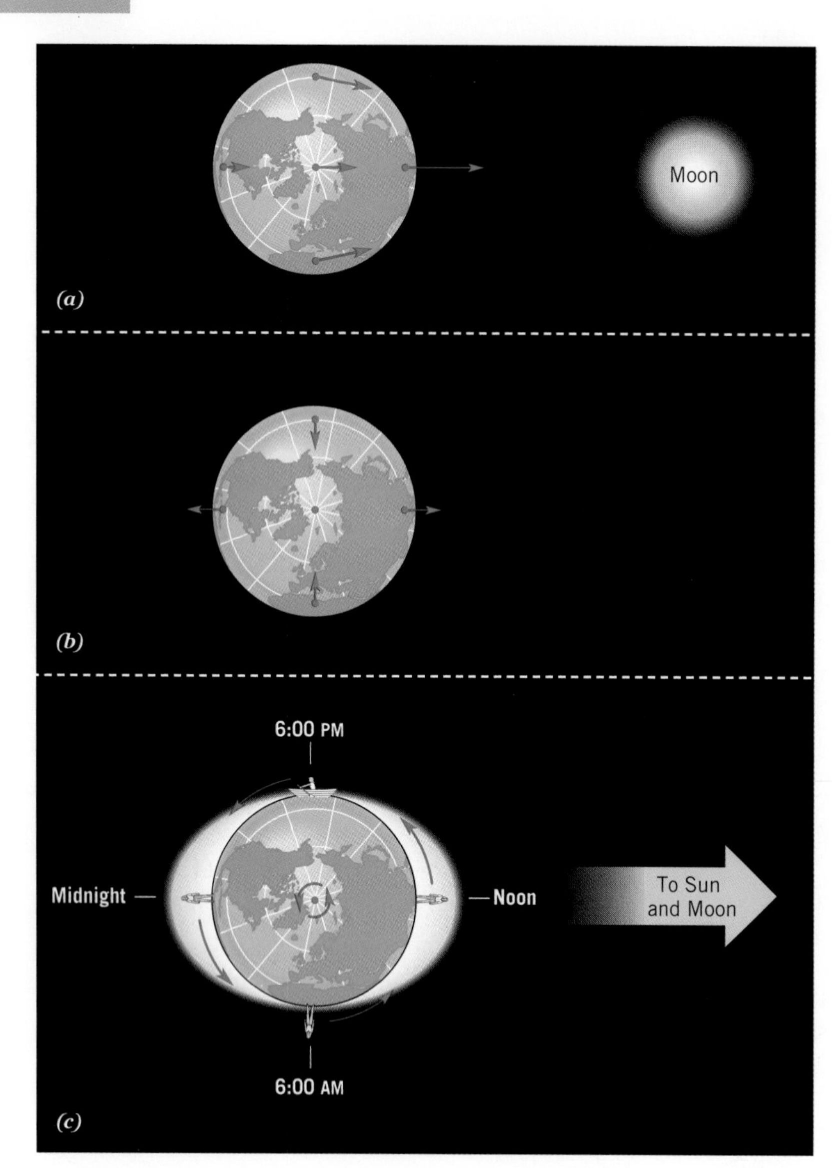

Figure 9.28

(a) Acceleration of various parts of the Earth toward the Moon (not to scale). Regions closer to the Moon are pulled harder. (b) Acceleration of various parts of Earth relative to the Earth's center. (c) View of the Earth from above the North Pole. The tidal bulges are aligned with the Sun and Moon. As the Earth spins, we are carried through these bulges, thus experiencing two high tides and two low tides per day. In the case illustrated here, we would experience the high tides at noon and midnight, and the low tides at 6:00 PM and 6:00 AM, as indicated.

(Figure 9.28). This *stretching force* is nearly 10 million times smaller than the Earth's own gravitational force, as can be seen in Table 9.1. Consequently, these tidal distortions of the Earth's shape are extremely small compared to the size of the Earth and greatly exaggerated in the drawings in this section.

The change in the strength of the Moon's gravitational pull from one side of the Earth to the other results in a **tidal bulge** on *both* sides of the Earth. Compared to points in the Earth's middle, points on the side nearest the Moon get pulled harder, and therefore stretched toward the Moon. Points on the far side get pulled more weakly and therefore get left behind by comparison. The two bulges are equal in size.

By analogy, consider a drag race among a dragster, car, and bicycle. They all start together and accelerate in the same direction. But from the car's point of view, the dragster shoots out in front and the bicycle falls behind. The amount by which they spread out depends not on how fast they accelerate, but rather on the *differences* in their acceleration. (If they all accelerated at the same rate, no matter how fast that is, they would remain together.) Similarly, the side of the Earth nearest the Moon gets pulled harder than the side away from the Moon. Therefore, relative to the rest of the Earth, the near side shoots out ahead, and the far side gets left behind. There is a tidal bulge on *both* sides.

Of course, there are tidal bulges in the solid Earth as well as the oceans. If the Earth were completely fluid and perfectly smooth, the bulges due to the Moon's gravity would have a height of about 2.8 meters. However, the oceans cover only parts of the Earth and are shallow, and the Earth beneath them is not very fluid. Typical heights of tidal bulges are only

Earth, as seen from Sun

Earth, as seen from Moon

Figure 9.29

View of Earth as seen from the Sun and Moon. From the Sun's perspective, the Earth is just a distant speck The Moon is 400 times closer, however, so from the Moon's perspective, the Earth appears much (i.e., 400 times) wider, taller, and thicker. You can see that the *difference* in the distances to the Earth's near and far sides is much greater from the Moon's perspective than the Sun's. That is why the tides due to the Moon are larger than those due to the Sun, even though the Sun's gravity is much stronger.

about 26 centimeters in the solid Earth and 70 centimeters in the oceans. There are similar tidal bulges in the atmosphere as well.

The Earth's daily rotation carries us through both tidal bulges each day, as illustrated in Figure 9.28c. From our point of view, we say that the tide "comes in" and "goes out" twice a day. From a broader perspective, however, it doesn't come in at all. Rather, it just sits there, and we are carried into it by the Earth's rotation. Actually, because the Moon is orbiting the Earth, it moves enough each day that it takes little more than one complete spin (24 hours) for us to catch up to it. Our tidal patterns have a period of 24 hours and 50 minutes, on average.

The Sun also creates tides on Earth. Although the Sun's gravity is much stronger than the Moon's, the *difference* in the Sun's gravity from one side of the Earth to the other is less than the difference in the Moon's gravity. The Sun is 400 times farther from us than the Moon. From the Sun's position, we are but a distant speck, both sides of which are nearly the same distance away (Figure 9.29). From the Moon's much closer perspective, however, we are huge, and our near side is considerably closer than the far side. Since it is the *difference* in gravity from one side to the other that causes the tides, the tides due to the Moon are actually larger than those due to the Sun. Tidal bulges caused by the Sun average only about 46% as high as those of the Moon.

MONTHLY VARIATIONS

The Moon orbits the Earth in $29\frac{1}{2}$ days (which is the original basis for the "moonth" on early Roman calendars). As the relative positions of the Sun and Moon in our sky change during the month, so do the relative orientations of their respective tidal bulges. As is seen in Figure 9.30, when the Moon is near its new or full phases, the Moon's tidal bulges line up with the Sun's, and the resulting tides are extra-large. However, when the Moon is near either the first or third quarter, the two sets of tidal bulges are at right angles to each other, and they tend to offset each other.

The extra-large tides associated with the new and full Moon are called **spring tides**. The extra-small tides that we experience when the Moon is near the first or third quarter are called **neap tides**. As the Moon orbits the Earth, periods of spring tides and neap tides alternate at about 1-week intervals, because it takes the Moon about a week to pass between quarter phases.

SEASONAL VARIATIONS

In addition to these variations in tidal patterns during the month, there are also predictable variations during the year. The cause of our seasons is that the Earth's spin axis is tilted by $23\frac{1}{2}°$ relative to the Sun. In December, for example, the North Pole is tilted away from the Sun, and the Northern Hemisphere receives relatively little sunshine. Six months later, however, Earth has traveled half-way around the Sun, the North Pole is tilted toward the Sun, and the Northern Hemis-

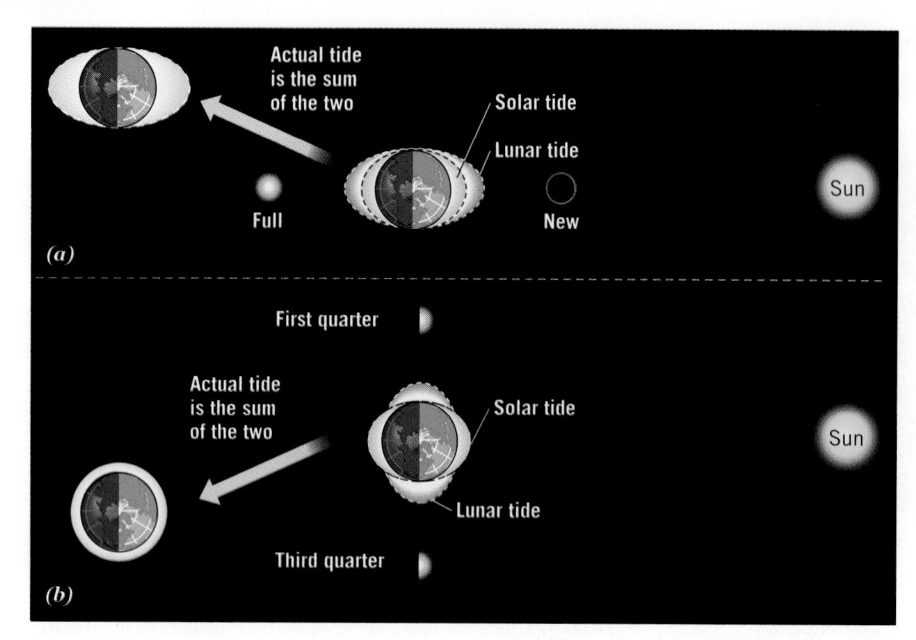

Figure 9.30
(a) When the Moon is new or full, the solar and lunar tides are in the same directions, giving the extra large spring tides. (b) When the moon is at first or third quarter, the solar and lunar tides are 90° to each other, tending to offset each other, and giving the smaller neap tides. Due to superposition, the actual tides we see are the sum of these two components. (For example, in the second case, we will see the water surface averaged out. The two components roughly cancel each other out: one's high on the other's low, and vice versa.)

phere receives relatively large amounts of sunlight. This is illustrated in Figure 9.31a.

To understand seasonal variations in the tides, consider the case of spring tides (new or full Moon) in the northern winter. As you can see in Figure 9.31b, the tidal bulges line up with the Sun and Moon, not the Earth's equator. Consequently, the bulge tends to be below the equator on one side of the Earth and above the equator on the other side. As the spinning Earth carries us through these two bulges each day, the two tides will not be equal. The nighttime tide will be larger in the Northern Hemisphere, and the daytime tide larger in the Southern Hemisphere. In summer, we will experience the opposite pattern, as illustrated in Figure 9.31c. (Notice that person A in this figure misses one of the tides each day, so this person experiences only one high tide per day rather than two.)

Recall from our earlier discussion of waves that any complicated signal can be written as a superposition of sine wave components. We can express the tidal pattern described above as the sum of **diurnal** (daily) and **semidiurnal** (twice-a-day) components. When there is only one high tide per day, the diurnal component dominates. When there are two nearly equal tides per day, the semidiurnal component dominates. Most places experience a combination of the two: There are two tides per day, but one is larger than the other. As we have seen, the diurnal component has a period of 24 hours and 50 minutes, and therefore, the semidiurnal period is 12 hours and 25 minutes.

In sum, the tilt of the spin axis means that we will be carried through different parts of the tidal bulges on opposite sides of the Earth. We may be carried under the peak of the bulge on one side, and under the tail of the bulge on the other. We may even miss one of the bulges altogether. This will depend on our latitude, the season of the year, and the phase of the Moon.

VARIATIONS OF THE TIDAL FORCES

Neither the Earth's orbit around the Sun nor the Moon's orbit around the Earth is a pure circle. Rather, both are ellipses. Consequently, the distances to the Sun and Moon vary, and therefore, their tidal influences vary. You may have noticed that sometimes the Moon seems larger than normal. This is not just your imagination. Sometimes, the Moon is indeed closer to us and therefore appears larger in the sky.

The distance between the Sun and Earth varies by about 3.5% during the year, resulting in an 11% variation in the height of the solar tides. There is even greater variation in the Earth–Moon distance, which changes by about 11% during the month and makes a 37% variation in the heights of the lunar tides here. Not only are the Moon's tides larger than those due to the Sun, but there is greater variation in them as well.

The Moon's orbit is tilted by about 5° relative to the plane of the Sun and the rest of the planets. Like a spinning top, it slowly wobbles with a period of 18.6 years. Although the Earth's spin axis is tilted $23\frac{1}{2}°$ relative to the Sun, its tilt relative to the Moon varies either way from this figure by about 5° (see Figure 9.32), causing corresponding changes in the orientations of the tidal bulges.

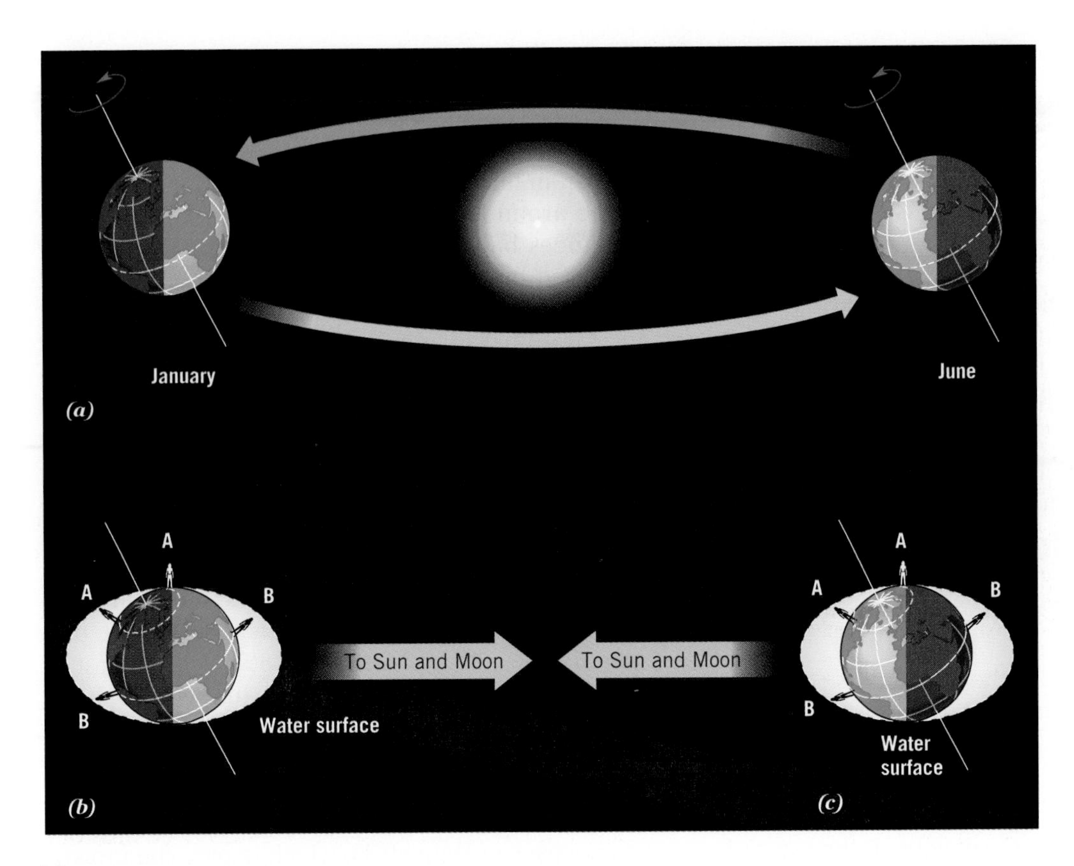

Figure 9.31

(a) The Earth spins about an axis that is tipped relative to the plane of its annual orbit around the Sun. In winter, the Northern Hemisphere is tipped away from the Sun and gets relatively little sunlight. By summertime, it is tipped toward the Sun, accounting for the warm weather this time of year. (b) Tidal patterns depend on not only the phase of the Moon, but also season and latitude. In the Northern winter, observer A gets only one high tide per day (at midnight) and observer B has two high tides per day (one at noon, and the other at midnight.) (c) In the Northern summer, observer A gets his high tide at noon instead of midnight.

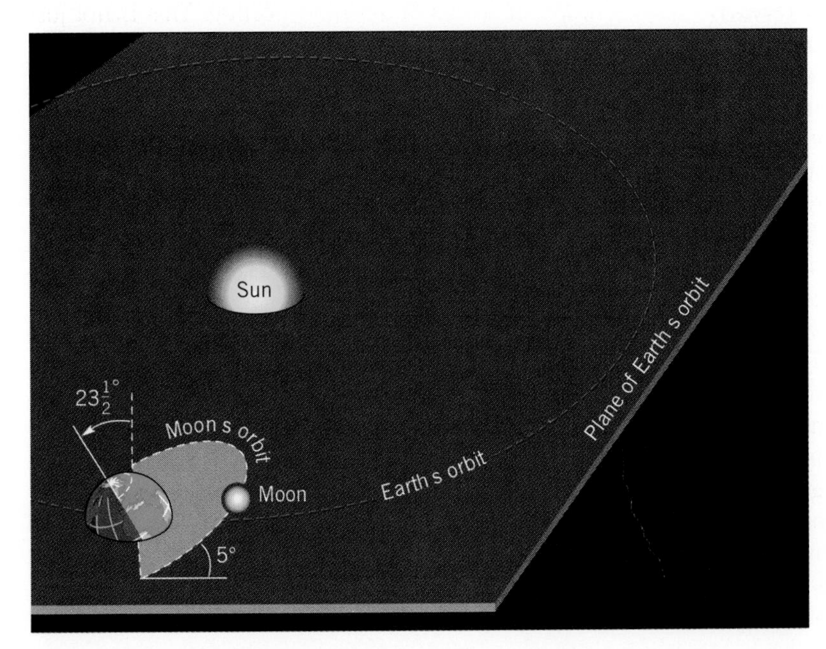

Figure 9.32

Sketch of the imaginary plane of the Earth's orbit. Relative to this plane, the Earth's equator is tilted by $23\frac{1}{2}°$, and the Moon's orbit by about 5°. Because of these tilts, the tidal bulges due to the Sun and Moon are not always centered over the equator. Sometimes they are above it, and sometimes below.

REVIEW

At this point, you have been introduced to enough complexity in the tide producing forces that a brief review is in order. With regard to the tidal influences of the Sun and Moon, the following should be highlighted:

1. The observed tide is a combination of that due to the Sun and that due to the Moon. Each produces tidal bulges on *both* sides of the Earth, and the Earth spins daily beneath these bulges.
2. The Moon's tidal bulges are a little over twice as large as the Sun's, on average.
3. The relative orientations of these two sets of tidal bulges vary during the month, giving alternately spring and neap tides at roughly 1-week intervals.
4. The orientations of the tidal bulges vary with the seasons, because the Earth's spin axis is tipped.
5. The distances to the Moon and Sun vary during the month and year, respectively, causing corresponding variations in the sizes of the tidal bulges.
6. The orientation of the Moon's orbit varies over an 18.6-year period as it "wobbles."

Up to this point, we have been studying what is called the **equilibrium theory** of the tides. It is essential that you understand the equilibrium theory. It provides not only the correct basic understanding of the fundamental cause of the tides, but also the underlying reasons for the daily, monthly, and seasonal variations in the tidal patterns that are observed. However, the theory is incomplete. It treats the Earth as if it were covered with uniform deep oceans and no continents. The real Earth does not quite fit this model, and the details of the real Earth provide additional complexities in the tidal patterns. Among the most important of these details are the presence of continents and the relative shallowness of the oceans.

CONTINENTS AND THE CORIOLIS EFFECT

The real tidal bulges cannot just sit still while the Earth spins beneath them, for a variety of reasons—the most obvious of which is that the continents get in the way. Not only does the spinning Earth carry people, but it carries the continents as well. When one of these continents runs into a tidal bulge, the tide rebounds back across the ocean, like a baseball off a bat. From our point of view, the tide follows the Sun and Moon westward across an ocean and runs into the continent on the western side. It rebounds from the continent, going back in the opposite direction. Usually, it arrives back on the eastern side of the ocean in time to catch the Sun and Moon their next time around. So from our point of view, the tide bounces back and forth across an ocean, getting dragged by the Sun and Moon going one way and then recoiling off a continent for the return trip.

Even more complexity is added by the Coriolis effect, which we studied in Chapter 2. Because the Earth is spinning, winds and waters traveling across the Earth's surface appear to curve to the right in the Northern Hemisphere and to the left in the Southern Hemisphere. So from our point of view here on the spinning Earth, the tides don't go straight back and forth across the oceans. Rather, they slosh around in circles. (See the discussion of tidal currents in Chapter 8.) If you compare the oceans to water in a pan, the tides would be more like water sloshing around the perimeter of the pan, rather than waves bouncing back and forth across it. This sloshing of water around a basin of any size is referred to as **amphidromic motion**.

You can see this circular sloshing of the tides in the oceans in Figure 9.33, which depicts the progression of the high tides in the various oceans. The spiderlike lines are called **cotidal lines**, and they indicate the locations of the high water levels at any given time. The numbers on these lines give the time in hours, so they show the time progression of the high tides. You can see that the tides slosh around central points where there is no motion, called **nodal points** (or sometimes *null points*, or *amphidromic points*). Just as with water sloshing in a pan, the tide's amplitude tends to get higher toward the outside of the basins and smaller toward the nodal points.

Not only does the spinning Earth carry people, but it carries the continents as well. When one of these continents runs into a tidal bulge, the tide rebounds back across the ocean, like a baseball off a bat.

You might notice that the water sloshes counterclockwise around basins in the Northern Hemisphere. This may seem backwards, considering that the water is circling to the right. Nonetheless, it is correct. Just as in the case of ordinary waves, the motion of a tidal wave is quite different from that of the water. The water goes in small circular orbits, while the wave crosses the entire ocean. Water curving to the

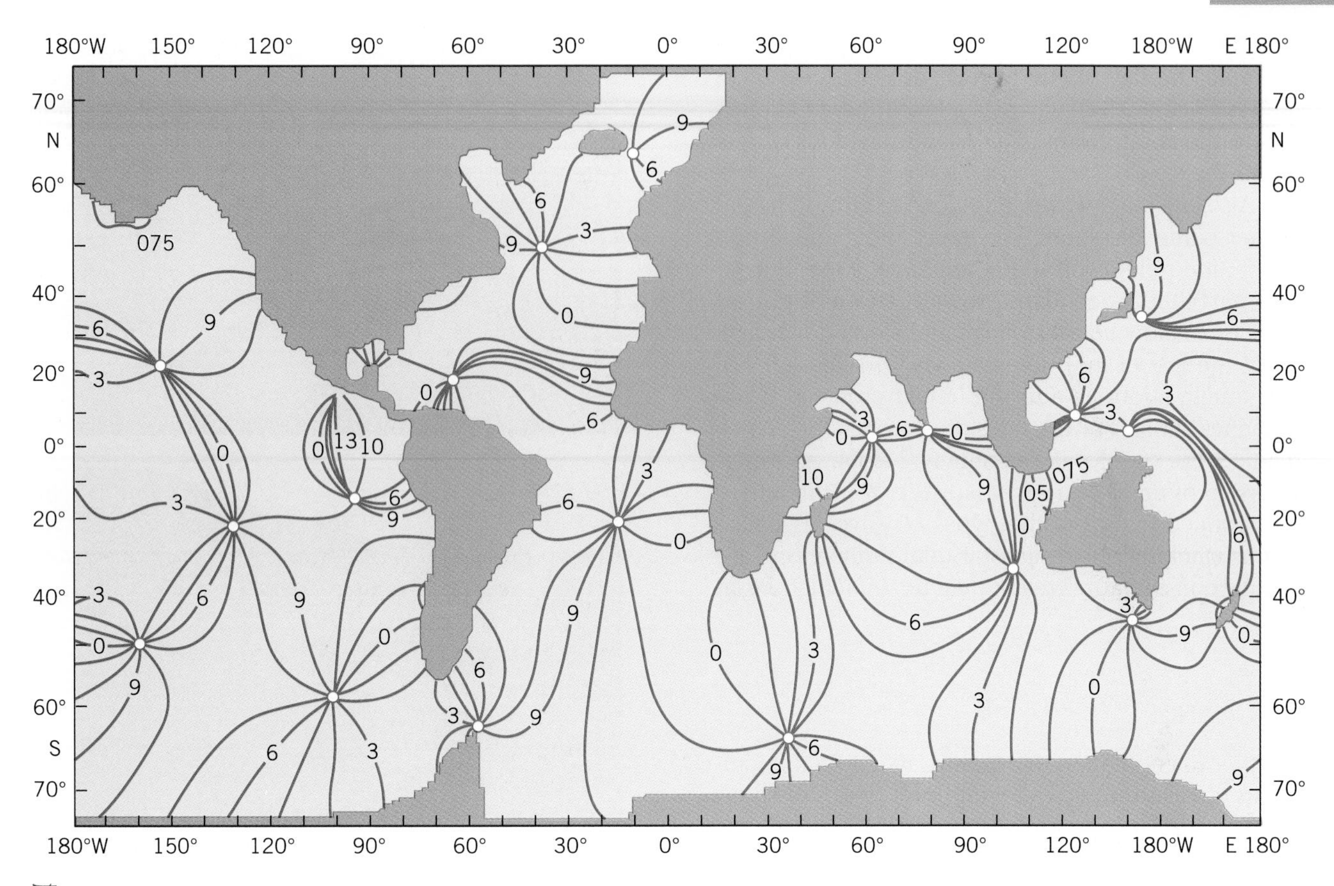

Figure 9.33

Cotidal lines for spring tides. These show the positions of local high tides at various times. The numbers on the lines indicate hours, with 0 corresponding to the time that the Moon passes over Greenwich, England. The amplitudes are not indicated, but they get larger with increasing distance from the nodal points. You can see that the tides tend to slosh in circles around the basins, going clockwise in the Southern Hemisphere and counterclockwise in the North. (From Dietrich et al., 1980, *General Oceanography*, 2nd ed., John Wiley & Sons, New York. After Pekeris and Accad, 1969, *Philosophical Transactions of Royal Society*, 265, 413–436.)

right tends to stack up on the right side of the basin as the tide progresses. Stacking up on the right side means that it will slosh counterclockwise around the basin. (Consider a bug crawling around a pan. If the bug keeps crawling up against the side of the pan on its right, it will end up circling left, going counterclockwise around the perimeter of the pan.)

SHALLOW WATER

Another important deviation in tidal patterns from the ideal case of the equilibrium theory is caused by ocean bottoms. It is easy to see that from a tide's perspective, the water is very shallow. One tide extends half-way around the Earth, or about 20,000 kilometers, and the oceans are only about 4 or 5 kilometers deep. So the speed with which these waves travel is completely determined by the depth of the oceans, according to the formula for shallow water waves. This speed is typically 750 kilometers per hour over the deep basins and 100 kilometers per hour over the continental shelves. This is not fast enough to keep up with the Sun and Moon in their daily trek across the sky.

One result of the tides' inability to keep up with the Sun and Moon is that the times of local high tides are not the same as the times when the Sun and Moon are directly overhead. On some coasts the tides may be hours ahead of the Sun and Moon, and on other coasts they may be hours behind.

Because the tides are in *shallow water* everywhere, the directions in which they travel are strongly affected by the contours of the ocean bottom, bending toward regions where they go more slowly. As tides cross the oceans, they can be focused by rises in the ocean bottom or the contours of the outer continental shelf, such that their energy is concentrated in some regions and directed away from others. For example, the Bay of Fundy gets tides in excess of 13 meters (Figure 9.34), whereas a few hundred kilometers to the south, Nantucket Island has tides of less than 1 meter.

Also like other waves, tides tend to grow in height as they come into shore. The lower and main portion of the tide reflects off the continental slope. But for that part that proceeds on toward shore, the front reaches shallow coastal waters and slows down first, while the rest of the wave catches up (Figure 9.35). The length shortens and the height grows as it approaches shore. For the tides, this process begins already at the base of the continental slope, so it progresses over a longer distance and is more pronounced than for other types of waves. Because of this stacking effect, coastal tidal fluctuations are usually considerably greater than tidal fluctuations at sea.

Figure 9.34
Tidal extremes in the Bay of Fundy, Nova Scotia.

RESONANCE

Another phenomenon that greatly affects tidal fluctuations is **resonance**. The width and depth of some parts of the oceans are such that the natural frequency with which the water would slosh back and forth on its own nearly matches the diurnal period of the Sun and Moon that drive the tides. When this happens, the amplitudes of the tides build up, and we have a very large seiche.

Just north of Antarctica, all oceans are connected at their southern ends. There are no continents to obstruct the tides as they travel westward around the globe down there. Furthermore, the Earth is not so far around at this latitude as it is at the equator, and the tides can keep up with the Sun and Moon. As a result, the tidal wave is quite pronounced. As it passes the southern end of the major oceans, it sends a tidal surge up each one. There is no corresponding phenomenon in the Northern Hemisphere because of the continents.

TIDES IN SEAS

Until they reached the British Isles, the Romans were not aware of tides and their relationship to the Moon. The tides in the Mediterranean Sea are so small that they almost go unnoticed. The Mediterranean is only one of a large number of similar enclosed or semienclosed seas that empty into the major oceans. In most of these, the tides are small, because the natural resonant sloshing periods are much shorter than the diurnal period of the Sun and Moon that drive them.

The floors of most sea basins are of oceanic crust, and so the depths of most seas are similar to that of the major oceans. Therefore, the speed with which tidal waves cross them is similar to the speeds with which they cross an ocean—typically 750 kilometers per hour. At this speed, a basin would have to be roughly 8000 to 9000 kilometers across for a wave's round trip to match the diurnal period of the Sun and Moon. Such large widths are typical for the major oceans, but are far larger than those for any sea.

Therefore, although the tides in some parts of the ocean can reach a resonance with the Sun and Moon, such resonance does not happen for most seas. It is like pushing a child on a swing at the wrong times. Not much happens. The water in seas sometimes sloshes due to tsunamis or wind-induced movement, but tidal forces usually don't do much.

Many seas communicate directly with the ocean over wide regions. Examples include the South China Sea, Gulf of Mexico, and Caribbean Sea. The tides experienced in these areas are primarily diminished remnants of the tides rolling in from the neighboring ocean—diminished because these seas are separated from the adjacent ocean by shallow sills (i.e., places

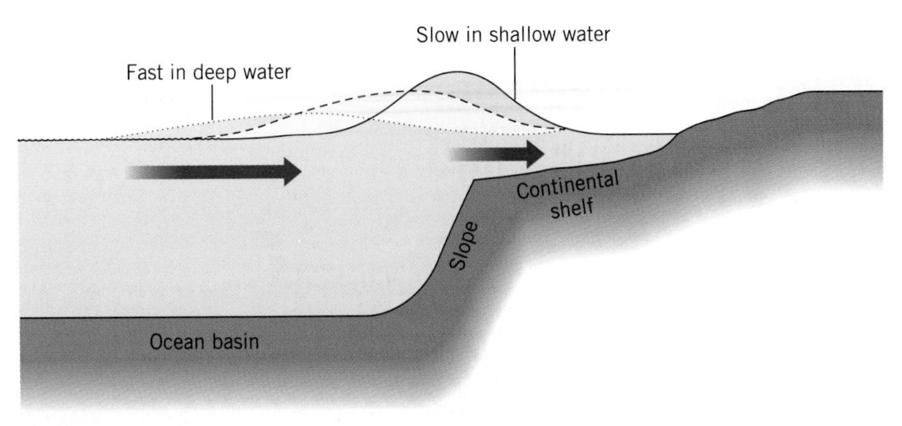

Figure 9.35

Illustration of the tide growing shorter in length and taller in height as it approaches the coast. The front reaches shallow water first and slows down, while the rear catches up. Although this particular illustration is for tides, the same happens to all waves as the water gets shallower. (Of course, the effect cannot begin until the water is shallower than half the wavelength, because the wave starts slowing down until it feels the bottom.)

where the ocean floor rises nearly to the surface) that tend to deflect incoming tidal waves.

Further reduction in the heights of the tides in these seas is accomplished by the depth contours of the ocean floor. We have seen in Section 9.2 that waves bend toward shallow water. Wave energy is concentrated on the headlands. On the large scale of a tide's perspective, shallow water is often encountered off the continents first. Therefore, tides are often directed toward these continental *headlands* and away from the protected and recessed seas.

Waves go more slowly in shallower water. A few seas are so shallow, and the tides travel so slowly, that the resonant sloshing frequency does correspond to the diurnal or semidiurnal period of the Sun and Moon. For example, if a sea were 200 kilometers long and only 50 meters deep, then the speed of a tide in this sea would be such that it could travel up and down its length in about 12 hours. The impulse provided by the rise and fall of the tides in the adjoining ocean would set up a resonance condition in this sea, and you would expect exceptionally large tidal fluctuations there. The Bay of Fundy has approximately the dimensions of this hypothetical sea, so resonance is one of the primary reasons for the large tidal range in this bay (Figure 9.35). The North Sea is another example of a shallow sea where resonance occurs. The speed of the tidal wave along its length is slow enough that it takes about 24 hours for one complete cycle along its length, which also matches the impetus of the tides in the adjoining ocean.

TIDE TABLES

Because of the complexities described above, the theoretical prediction of coastal tides is extremely difficult. As a result, tide tables are produced by extrapolation from past data instead of theory (Figure 9.36).

Earlier in this chapter, we learned that any complicated signal can be made from the superposition of sine wave components. For tides, the sine wave components are sometimes referred to as **partial tides**. Seven of these are usually dominant, and the superposition of these seven partial tides can correctly reproduce the variation of sea level due to the following:[7]

1. The solar diurnal variation, with a period of 24 hours.
2. The solar semidiurnal variation, with a period of 12 hours.
3. The lunar diurnal variation, with a period of 24 hours, 50 minutes.
4. The lunar semidiurnal variation, with a period of 12 hours, 25 minutes.
5. The solar annual variation with a period of 1 year. (This reflects the yearly variation in the distance to the Sun and the tilt of the Earth's axis relative to the Sun.)
6. The lunar monthly variation, with a period of $29\frac{1}{2}$ days. (This reflects the monthly variation in the distance to the Moon and the tilt of the Earth's spin axis relative to the Moon.)
7. A lunar precession variation, with a period of 18.6 years. (This reflects the wobble in the plane of the Moon's orbit.)

[7]There are many different but equivalent ways of doing this. One common method uses interference between waves of slightly different frequencies to reproduce the long-term variations. In this approach, the periods for some of the seven components differ from those listed here.

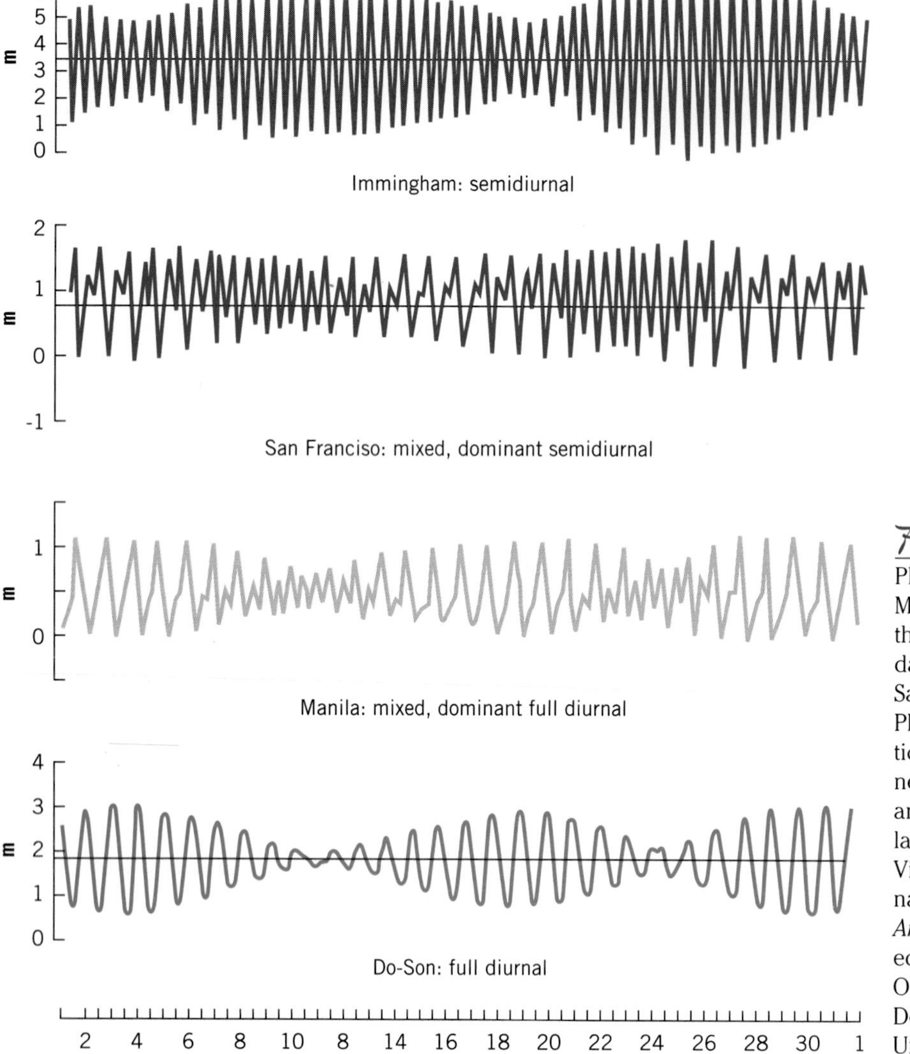

Figure 9.36

Plots of tides in various ports for March 1936. In Immingham, England, the semidiurnal component (twice daily) component dominated. Both San Francisco California and Manila, Philippine Islands showed mixed tides, with the semidiurnal component being larger for San Francisco and the diurnal (daily) component larger for Manila. At Do-Son, Vietnam, the tide was entirely diurnal. (From H. O. Bowditch, 1958, *American Practical Navigator*, rev. ed., U.S. Naval Oceanographic Office, Washington, D.C., or A. Defant, 1958, *Ebb and Flow*, University of Michigan Press, Ann Arbor.)

By using past tidal data, the amplitude and timing of each of these seven components can be adjusted appropriately for any particular coastal location. With these main sine wave components known, they can be added together to give the net tidal displacement for any future time. This is how most tide tables are made.

Summary

Wave Properties

Waves are generated by a variety of forces, the most common of which is the wind. All complex wave patterns result from the superposition of simple wave groups, moving independently of each other. The resulting displacement of the sea surface at any point is simply the sum of the displacements due to the individual waves. In some regions waves may interfere constructively and in other regions destructively.

Most common waves are driven by gravity, but for very small waves, surface tension provides the main restoring force. These capillary waves are important in the transfer of energy between wind and waves.

Every point on a wave crest acts as a source for generating wavelets going outward in all directions. This property allows waves to bend in behind obstacles. Water undergoes circular orbital motion as waves pass by. The size of these orbits decreases with depth, until at a depth of half the wavelength, the water motion is negligible. In shallow water, the orbits are elliptical, becoming more and more flattened near the ocean floor. When striking a flat surface, the wave must be reflected, because only the superposition of both a clockwise and counterclockwise orbital motion can yield linear motion for the water along the surface. Waves are calmed by anything floating in the water.

Wave Speed

Long waves travel faster than short ones. Long waves also reach deeper and run into bottom first. Consequently, long waves are slowed down most as they approach shore. In very shallow water, all waves travel at the same speed. The shallower the water, the more slowly they go. Waves bend toward where they go slowly. Among other things, this *refraction* explains why all waves in shallow water head nearly straight toward shore, regardless of what directions they were traveling at sea. It also means that bumps and valleys in the ocean floor can change the headings of large waves as they cross the ocean.

As waves travel from a disturbance, they become sorted, with the longer waves taking the lead. Waves travel in groups. The individual waves travel faster than the group, springing up at the rear and dying out at the front.

Breaking Waves

Waves break when they become too high to be adequately supported. As waves come into shallow water, their lengths shorten, heights increase, wave speed decreases, and orbital speed increases until the crest gets out ahead of the rest of the wave. Breaking waves can be placed into three categories according to how rapidly they build up and break. After a wave has broken, a wave of translation is formed. For these, the water is actually moving along with the wave, rather than undulating in cyclical orbits.

Wind-Driven Waves

Most common waves are driven by the wind, growing until the waves are sufficiently large on average that they are going roughly at the same speed as the wind. This fully developed state cannot be reached if the wind stops blowing or the waves leave the windy region first.

Tsunamis and Internal Waves

Tsunamis result from underwater land movements. They are typically very long, low waves and move across ocean basins at speeds of about 750 kilometers per hour. It is nearly impossible to tell if a tsunami is under way or where it is headed until it has reached a coast. Tsunamis are particularly hazardous to coastal regions unaccustomed to large variations in sea level.

Waves can propagate along the boundary between any two fluids. When between layers of water of different densities, they are called *internal waves*. Internal waves travel very slowly and can sometimes be detected by the long parallel surface slicks.

Tides

The largest waves on the ocean are the tides. They are caused by the gravitational force of the Moon and, to a lesser extent, the Sun. The force of gravity falls off with distance, which results in a stretching of the Earth. The side nearer the Moon gets pulled slightly harder than the middle, and the far side gets pulled slightly less. This results in tidal bulges on both sides of the Earth. The Earth spins beneath these tidal bulges, carrying us through both each day.

When the bulges caused by the Moon and those caused by the Sun are lined up with each other, we have spring tides, and when they oppose each other, we have neap tides. Among the factors that influence tidal patterns most are the phase of the Moon, latitude of the observer, tilt of the Earth's spin axis, and season of the year.

The travels of the ocean's tidal bulges around the Earth are obstructed by the continents, slowed by the ocean bottom, and curved by the Coriolis effect and contours in the sea floor. Complications such as these make detailed theoretical calculations of tides nearly impossible. Tide tables are constructed from information in earlier tidal records.

Tides in seas are generally very small, because the time it takes for a large wave to slosh back and forth is short compared with the 24-hour period of the Sun and Moon. Seas are also often protected from the tides of the neighboring ocean by shallow sills and refraction of the tides toward the continents. However, in a few shallow seas, the resonant sloshing frequency for the sea matches that of the tides, and resonance can make tides in these seas quite large.

Key Terms

orbital
internal waves
tsunami
tidal wave, or tide
free waves
forced waves
crest
trough
wave height
significant wave height

wavelength
wave steepness
still water level
amplitude
period
wave speed
wave group
deep water wave
shallow water wave
superposition
destructive interferenc
constructive interference
surf beat
standing waves
antinode
node
edge waves
seiche
resonant frequency
sine waves
restoring force
gravity waves
surface tension
capillary waves
Huygens' principle
diffraction
out of phase
slick
dispersion
swell
headland
longshore bars
wave of translation
surf
surf zone
spilling breaker
plunging breaker
surging breaker
sea
fully developed sea
fetch
Beaufort scale
tidal bulge
spring tide
neap tide
diurnal
semidiurnal
equilibrium theory
amphidromic motion
cotidal lines
nodal points
resonance
partial tides

Study Questions

1. Explain how surf beat is created.
2. What causes surface tension? What are some of the things you can think of which demonstrate that water molecules are very sticky?
3. What minimum wavelength would storm waves have in order to move sediments on the continental shelf at a depth of 100 meters?
4. When a wave runs into the edge of a swimming pool, how can the water move straight up and down along the wall if the water's orbital motion is circular? Use a sketch to help explain.
5. A smooth glassy water surface may indicate the presence of a subsurface school of fish or bed of seaweeds. Why?
6. How does wave speed depend on wavelength? Why? How does it depend on water depth, and why?
7. As they approach shallow water, longer faster waves slow down more than shorter slower ones, until all are going at the same speed. Why do the longer ones slow down more?
8. How are the well-sorted long parallel waves called *swell* created?
9. At the beach, why do the waves come straight into shore, regardless of their heading when at sea? On jagged coastlines, why are the coves so calm?
10. Why do waves become taller as they come in toward shore?
11. Why do waves break in shallow water? In deep water?
12. For *fully developed seas*, why do stronger winds make longer waves?
13. What might be some of the reasons why some coastal villages may suffer severe damage from a tsunami, and others none at all?
14. Why do slicks form over the troughs of internal waves? Why does floating debris make the water surface smooth and glassy?
15. The planets are moving at very high speeds as they go around the Sun. Why don't they fly off into outer space; that is, what keeps them going around the Sun? If the Earth is constantly pulled toward the Sun, why don't we run into it?
16. Why is there a tidal bulge on the side of the Earth nearest the Moon? Why is there one on the far side?
17. Why do spring and neap tides alternate at weekly intervals?
18. From our point of view, the Sun and Moon go around the Earth once a day, and the tide "comes in" and "goes out" twice a day. What really happens?
19. Why does the daily tidal pattern at any one coastal city vary from week to week? From season to season?
20. Briefly explain how the following three factors each cause the real tides to differ from the predictions of the equilibrium theory: (a) The continents, (b) the Coriolis effect, and (c) the shallowness of the oceans.
21. Why are tides small in most seas?

Critical Thinking Questions

1. If gravity were stronger, would waves travel faster or slower? Why?

2. Make a sketch of the breakwater and wave motion of Figure 9.12e. Explain why the waves behind the breakwater are going in a different direction than they are at sea. Do you suppose that the curved beach behind the breakwater is a result of the wave motion, or is the wave motion a result of the curved beach? Explain.
3. Calculate the wave speed for a tide or tsunami in a typical ocean basin with a depth of 4.5 kilometers. What would be its speed over the continental shelf where the depth is 100 meters? 1 meter? Answer these same three questions for a gravity wave with a 40-meter wavelength.
4. Make a sketch of a ship's wake as seen from above. Explain why the individual waves in the wake don't start at the ship and go outward forever.
5. Why are waves of translation more damaging to ships than large unbroken waves? Some naval engineers believe that the waves of the Great Lakes are more damaging to ships than those of the ocean. Can you think of a reason why this might be true? (*Hint*: Is there some reason why a larger fraction of the waves on the Great Lakes might be breaking waves?)
6. Suppose it is summer, the Moon is full or new, and you are in the Northern Hemisphere. From the equilibrium theory of the tides, what two times of day would you expect to experience high tides? Which of these two times would the high tide be highest? Why?

Suggestions for Further Reading

BROWN, JOSEPH. 1989. Rogue Waves. *Discover* 10:4, 46.

GARRETT, CHRIS. 1993. Tides and Their Effects. *Oceanus* 36:1, 27.

GIESE, GRAHAM S. 1993. Coastal Seiches. *Oceanus* 36:1, 38.

HAYES, BRIAN. 1993. Trails in the Trackless Sea. *American Scientist* 81:1, 19.

HOLLAND, WILLIAM R. 1987. Computer Modeling in Physical Oceanography from the Global Circulation to Turbulence. *Physics Today* 40:10, 51.

JELLEY, J. V. 1989. Sea Waves: Their Nature, Behavior, and Practical Importance. *Endeavor* 13:4, 148.

KOMEN, G. 1987. Interactions of Wind and Waves. *Nature* 328:6130, 480.

LOCKRIDGE, P. A. 1988. Tales of Shores, Ships, and Tsunamis. *Sea Frontiers* 34:5, 262.

LOCKRIDGE, P. A. 1989. Tsunami: Trouble for Mariners. *Sea Technology* 30:4, 53.

OPEN UNIVERSITY. 1989. *Waves, Tides, and Shallow Water Processes*. Pergamon Press.

Science. 1989. The Big Picture of the Pacific's Undulations. 243:4892, 739.

WALLNER, E. 1988. Ocean Tides. *Sky and Telescope* 75:2, 195.

WOLKOMIR, RICHARD. 1988. The Mechanics of Waves—and the Art of Surfing. *Oceans* 21:3, 36.

Although stormy seas are more energetic, gentle seas are more common. Consequently, storm erosion is occasional, but severe, with gradual recovery in between.

The coast feels the interplay between two different worlds.

Ten

THE COASTAL OCEAN

The coastal margins of continents are the most dynamic of all Earth's regions. The complex variety of coastal processes is of particular importance to us humans, because we so heavily populate coastal regions and rely on activities involving the coastal waters. We begin this chapter with a study of beaches, those very narrow and constantly changing zones of interaction between land and ocean. We look at natural processes that work to make and modify beaches, and then at some of the consequences of human interference with these processes. Then we look at processes that work on a much larger scale and over much longer time periods to shape the overall features of the continental margins. Finally, we study coastal waters, including those of the ocean adjacent to the continents and those found in more protected coastal embayments.

10.1 BEACHES

For most of us, our first encounter with the ocean is the **beach**, a place where waves crash, children frolic, and all are impressed by the ocean's power and beauty. The "beach" is the region of wave-worked sediments; it includes the entire active area where we might expect to find changes from month to month. Beaches exposed directly to heavy ocean waves might have a width of a few hundred meters, extending from several meters above the high tide level to several meters below low tide level. The much broader region, over which the proximity of the ocean has a noticeable effect on the climate and foliage, is called the **coast**. The coast typically extends several tens of kilometers inland.

We describe a beach in a variety of ways. First, we might describe its general topography or shape. For example, is it wide, narrow, steep, flat, jagged, or straight? Second, we might describe its location: Is it exposed to direct wave onslaught, or is it somehow protected? Is it on the main coast, in an estuary, or on a barrier island? Third, we describe the beach by the size and nature of its sediments. For example, are they fine muds, coarse gravels, cobbles, or sands? Are they made of quartz, coral, granite, basalt, or volcanic debris?

PARTS OF A BEACH

The beach is divided into three regions called the **backshore**, **foreshore**, and **offshore** regions, according to whether they are above the high tide level, between high and low tide levels, or out beyond low tide level, respectively (Figure 10.1). If the damp lower foreshore region is also relatively flat, it is referred to as the **low tide terrace**. The **beach face** is the exposed portion of the foreshore. The turbulent noisy region of foam and spray where waves are breaking and washing in toward shore is called the *surf zone*.

Some beaches are backed up by a sea cliff, some by dunes, and some have neither (Figure 10.2). Most have a **berm**, which is the dry sandy region where you sunbathe or have a picnic. The berm may have one or more **berm crest**, which is a gentle ridge or drop-off edge in the sandy berm. It indicates the highest point of normal wave activity. You can camp behind it with reasonable assurance that you won't wake up soaked in the morning. From the berm crest, the beach face slopes down to the **shoreline**, which is the actual line of contact between water and land. At the outer edge of the surf zone may be longshore bars, which are offshore reservoirs of beach sand (Figure 10.1). Sometimes these bars are so shallow that they are partially exposed at low tide, and sometimes they are at depths of 10 meters or more.

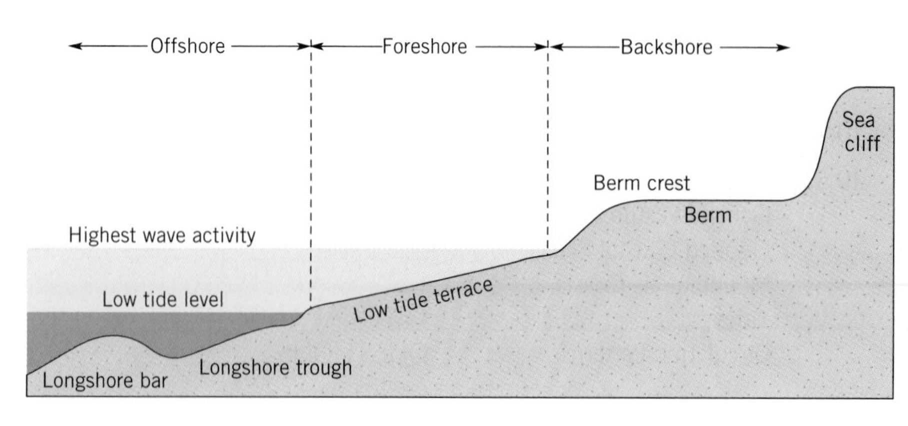

Figure 10.1

Cross section of a beach showing various features. Details vary from one beach to another.

(a)

(b)

Figure 10.2
Some beaches are backed up by sea cliffs *(a)*, some by dune fields *(b)*, and some by neither.

Figure 10.3
Beach sediments may include many materials. This beach is made of large rocks and seaweeds.

BEACH MATERIALS

The beaches popular with most bathers are sandy, but sand is only one of a wide variety of materials from which beaches can be made. Rocks, shale, shingles, cobbles, mud, and gravel are frequent beach materials. Rocks and other coarse sediments are not transported easily and therefore usually originate close to the beach where they are found. However, the finer sediments, such as sands and clays, might have come from far-away places, having been carried long distances by rivers, streams, and longshore currents.

The white sand on some Florida beaches is largely due to ground-up skeletons of coral and other marine organisms. Much of the yellow California sand comes from weathered granitic rocks (light-colored continental igneous rocks). The gray-green sands of the Pacific Northwest are derived from basalts (dark-colored oceanic igneous rocks), and the ground-up rocks of some volcanic islands yield a black beach sand. Organic materials such as dead seaweed or driftwood might contribute to the sediments in some places (Figure 10.3). Take a magnifying glass to the beach with you, look closely at some grains of sand, and guess where they came from.

In Chapter 5, we introduced both the standard for classifying sediments according to grain size (Table 5.2) and the method of describing the mineral content of those sediments (Section 5.1). Figure 5.10 illustrates how beach and shelf materials tend to depend on latitude, with finer sediments and coral at lower latitudes and coarser sediments at high latitudes where waves and weather are more severe.

WHY SAND STAYS ON THE BEACH

If you think about it, you may wonder why there is any sand on the beach at all. If waves can destroy sturdy rock outcroppings and sea cliffs, shouldn't they easily carry away the soft beach materials? Shouldn't the sand quickly work its way downhill off the beach and out onto the continental shelf?

Sand remains on the beach, because it is carried uphill toward shore by the unbroken incoming waves. As is illustrated in Figure 10.4a, the orbital motion of the water beneath unbroken incoming waves tends to carry the sand grains up and forward as they pass, but the "down and back" motions are obstructed by collisions with the bottom. In this way, sand grains leapfrog up the beach beneath incoming waves. Any beach sand that might stray out to sea too far would be carried back in by this mechanism. Of course, this inward sand transport can only happen out beyond the surf zone. Once the waves have broken, the water no longer undergoes orbital motion beneath the waves, so this "leapfrogging" process no longer works. The unbroken waves bring sand in as far as the surf zone, but once it reaches this point, different processes take over (Figure 10.4b).

Sand grains leapfrog up the beach beneath incoming waves. Any beach sand that might stray out to sea too far would be carried back in by this mechanism.

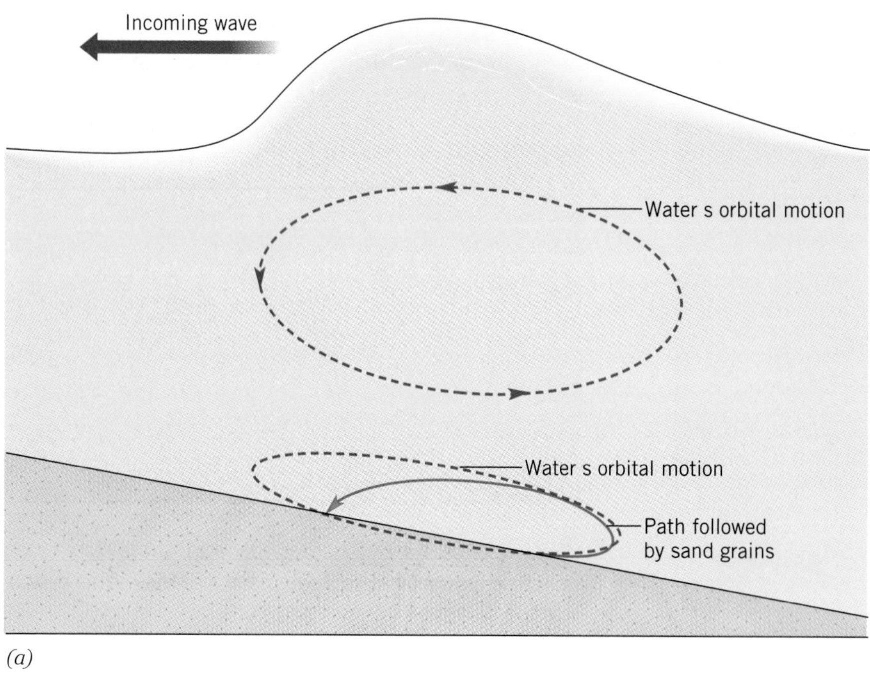

(a)

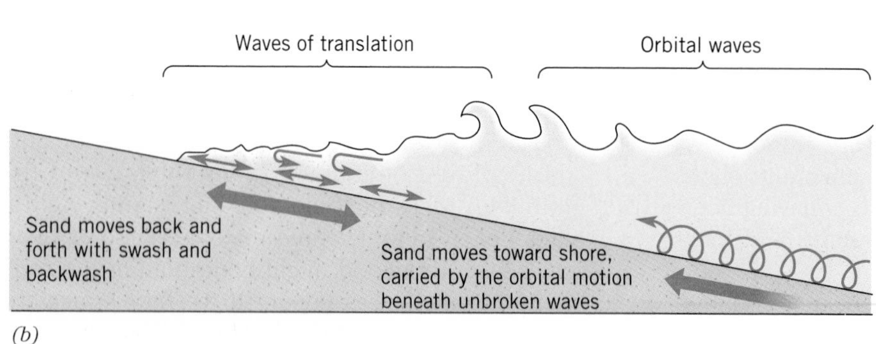

(b)

Figure 10.4

(a) Near the bottom, the water's orbits are flattened ellipses, even more flattened than shown in this drawing. When a wave passes overhead, the water drags some sand grains with it, going up and forward as the wave crest passes. The sand settles slowly through the water and strikes bottom ahead of its previous position. From there, it will be lifted again as the next wave passes, and the cycle repeats. In this way, the sand grains leapfrog toward shore. (b) Before the waves break, the sand leapfrogs toward shore with the passing of each wave, being carried up and forward by the water as each crest passes. This process ceases once the waves break, however. Beyond that point, the swash carries the sand up the beach, and the backwash carries it back down.

CHANGES IN THE BERM AND BAR

Within the surf zone, waves of translation are coming into shore and wash up the beach face in what is called the **swash**. Some of the swash returns back down the beach face in what is called the **backwash**, and some soaks in and percolates back down through the sand. Sand is carried up the beach with the swash and back down with the backwash. The net transport of sand within the surf zone depends on which of these two processes dominates.

In times of light wave activity, most of the swash soaks into the beach as it washes up. Consequently, when waves are small, there tends to be little backwash. Sand is carried up the beach by the swash and deposited when the swash soaks in (Figure 10.5a). The berm grows larger. On the other hand, when waves are heavy, the incoming swash finds the beach saturated with water from the previous large waves. Unable to soak in, some of the water from the swash must return seaward over the beach as a backwash. As it flows back down, the backwash gains in speed and turbidity. The sand it carries makes it denser, so it flows underneath the incoming waves, lifting them off the beach (Figure 10.5b). This prevents these incoming waves from dragging sand up the beach, while increasing the backwash's sediment transport down the beach. Therefore, in periods of heavy wave activity, the backwash dominates the sand transport. The berm shrinks as sand is taken away from it.

The sand carried by the backwash cannot go beyond the surf zone, because farther out are unbroken waves that carry the sand back in. Consequently, sand eroded from the berm during periods of heavy waves is deposited in longshore bars near the outer margin of the surf zone, but not beyond it. The longshore bars help protect the beach from further erosion, because they make the incoming waves break prematurely as they cross over the bars. You can determine the locations of longshore bars by noting where the waves are breaking.

Although stormy seas are more energetic, gentle seas are more common. Consequently, storm erosion tends to be occasional, but severe, with relatively long periods of gradual recovery in between.

Putting all these processes together, we see that the sand cannot normally go out beyond the surf zone, because the water motion beneath the incoming unbroken waves would carry it back in (Figure 10.4). Within the surf zone, sand is carried up the beach by the swash and down the beach by the backwash. In periods of small waves, the swash dom-

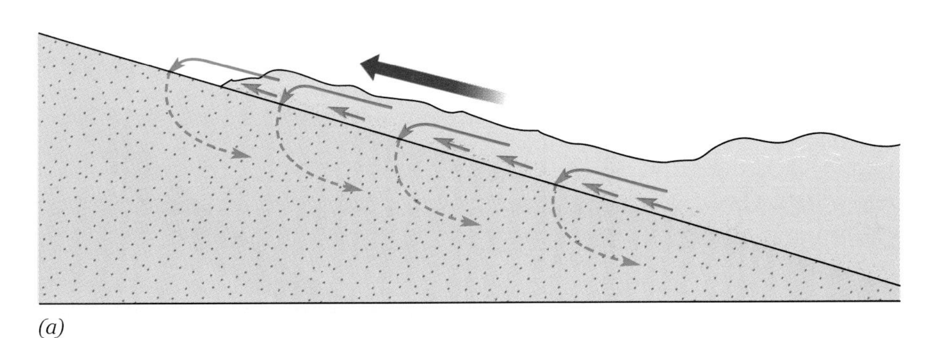

(a)

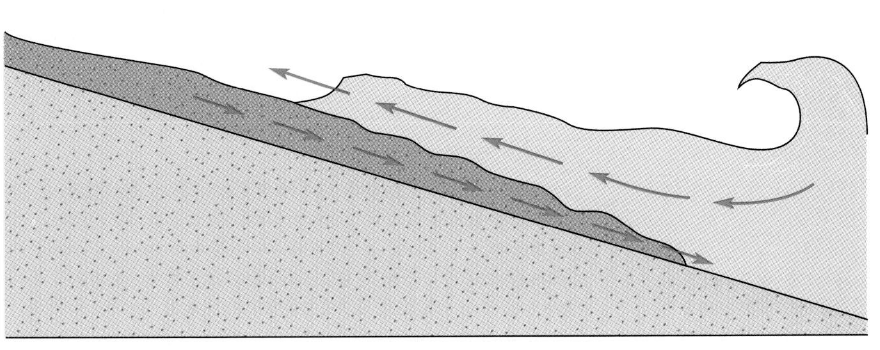

(b)

Figure 10.5

(a) In periods of light waves, the swash soaks in and percolates back through the beach, depositing its sediment load on the surface. The berm grows as the swash transports sand up the beach. (b) During heavy waves, the beach is saturated with water and little of the swash can soak in. The sand-laden backwash travels back down the beach face and under the successive incoming waves, carrying sand from the berm and depositing it in the longshore bar.

inates and the berm grows. In periods of heavy waves, the backwash dominates and the berm shrinks.

The two opposing processes aren't quite symmetric in that large storm waves can erode material from the beach much more quickly than the gentle waves of calm seas can replace it. Although stormy seas are more energetic, gentle seas are more common. Consequently, storm erosion tends to be occasional, but severe, with relatively long periods of gradual recovery in between.

During periods of heavy erosion, the removal of sand from the beach face often leaves a sudden drop-off or **scarp** on the beach near the high tide line. The erosion also tends to preferentially remove the finer sediments, leaving the coarser, rockier materials behind. We say that the remaining rockier beach is **armored**, and this rocky armor does indeed help protect the beach from further erosion.

Normally, wave activity is light during summer and heavy during the more stormy winter. Consequently, there tends to be a seasonal oscillation in the direction of sand transport, going up the beach in summer and down the beach in winter (Figure 10.6). That is, berms grow at the expense of longshore bars during the summer, and longshore bars grow at the expense of berms during the winter. If you have ever visited your favorite beach during the winter and found it rocky and void of the familiar large sandy berm, you needn't worry. The sand hasn't gone far. It is just out beneath the surf in longshore bars.

WAVES, GRAIN SIZE, AND BEACH STEEPNESS

There is a correlation between wave activity, beach slope, and sediment grain size on beaches. More exposed beaches tend to have steeper slopes and coarser sediments (Figure 10.7a). To understand the reason for this, consider what happens when wave activity is heavy. The turbulent water removes the finer sediments and carries them away until they can find a quieter environment to settle in. Consequently, the remaining beach sediments are coarse.

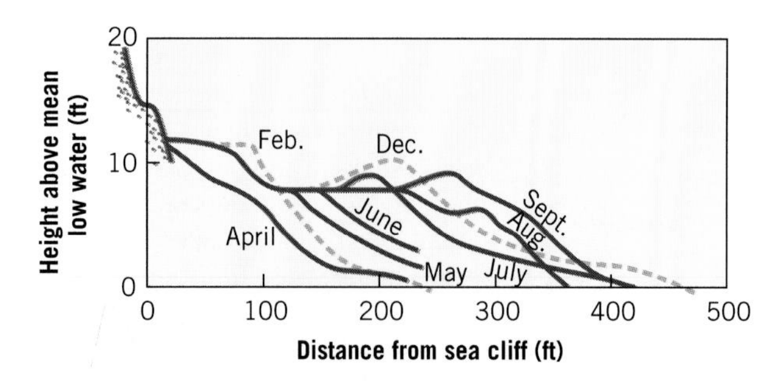

Figure 10.6

Cross sections of a beach at Carmel, California. Seasonal changes in the berm are indicated. (From Willard Bascom, *Waves and Beaches*, copyright © 1964 by Educational Services, Inc. Used by permission of Doubleday and Company, Inc.)

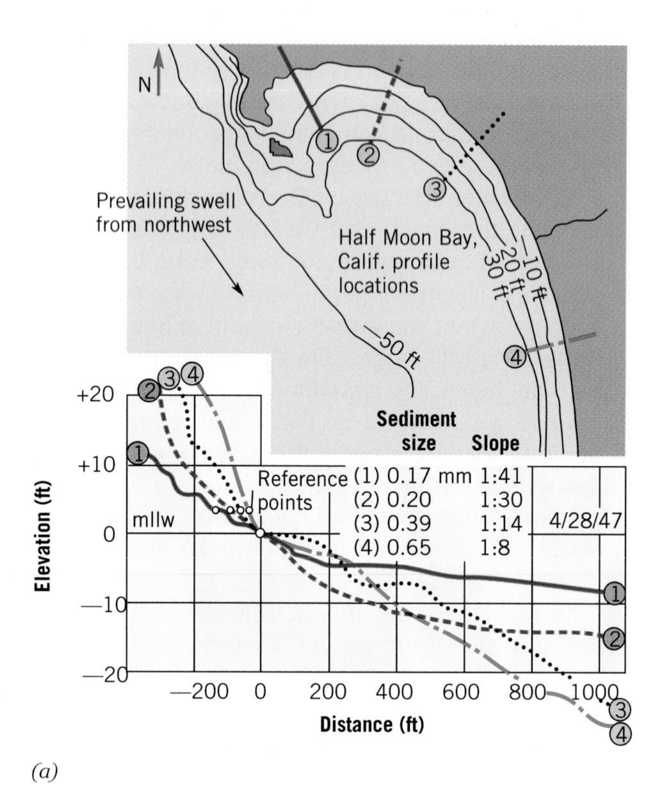

(a)

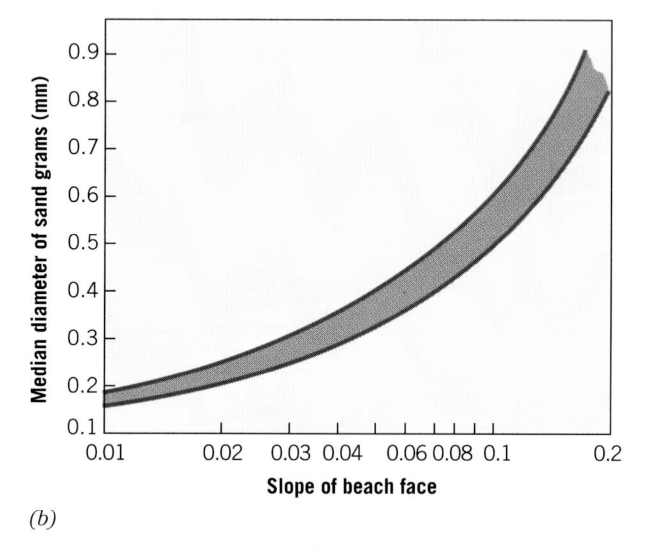

(b)

Figure 10.7

(a) Several beach profiles at Half Moon Bay, California. The more protected regions (numbered 1 and 2) have gentler slopes and finer sediments than the more exposed regions (3 and 4). (From Willard Bascom, *Waves and Beaches*, copyright © 1964 by Educational Services, Inc. Used by permission of Doubleday and Company, Inc.) (b) Plot of grain size vs. slope of beach face for mature beaches. Steeper beaches have coarser sediments.

Coarse sediments result in steeper beach faces because the swash readily soaks in and percolates down through the sediments, leaving a small or nonexistent backwash. The swash tends to push the sand (or gravel, etc.) up the beach face, making the berm larger and beach steeper. If there were a backwash, it would do just the opposite, dragging materials from higher up and depositing them lower down. Since the swash dominates on these coarser beaches, the beach face is steeper (Figure 10.7b).

MINOR BEACH FEATURES

In addition to the above overall beach processes and patterns, a close examination of beaches reveals a variety of smaller curious features, which change from beach to beach and time to time. Examples include diamond-shaped patterns and pinholes in the sand, tiny ridges of sand outlining the upper edge of previous swashes, ripples and tiny riverlets, gentle ridges and valleys along the shoreline, and a host of other possibilities. When you stroll along a beach, notice these small features and see if you can explain how they were formed.

10.2 LONGSHORE SAND TRANSPORT

In the previous sections, we studied how the broken waves carry sand up and down the beach. In addition, they carry sand along the beach, parallel to the shoreline, in what is called the **longshore transport**, or **littoral drift**.

LONGSHORE CURRENTS

Unless the crests of the incoming waves are exactly parallel to the shoreline when they break, the breaking waves roll in at a slight angle to the shoreline. This makes the swash wash up the beach at a slight angle, and so the water in the surf zone zigzags along the beach in the direction favored by the incoming waves (Figure 10.8). This component of the water motion parallel to the shoreline is called the **longshore current**. Because there is no net motion to the water beneath unbroken waves, the longshore current exists almost exclusively in the surf zone. Farther out may be other wind-driven surface currents, but these move comparatively slowly and have very little impact on the transport of sediments along the beach.

The water in the surf zone is quite turbid, and the longshore current carries these suspended sediments with it. The volume of sand transported in this manner can be quite impressive. For example, sand generally flows southward along both North American coasts at a rate of 500 to 2000 cubic meters of sand, or about 50 to 200 large fully-loaded dump trucks, passing any one point per day. This ongoing excavation is responsible for the formation of many of the large-scale coastal features and is a major problem for human activities that would impede this flow.

Sand generally flows southward along both North American coasts at a rate of 500 to 2000 cubic meters of sand, or about 50 to 200 large fully-loaded dump trucks, passing any one point per day.

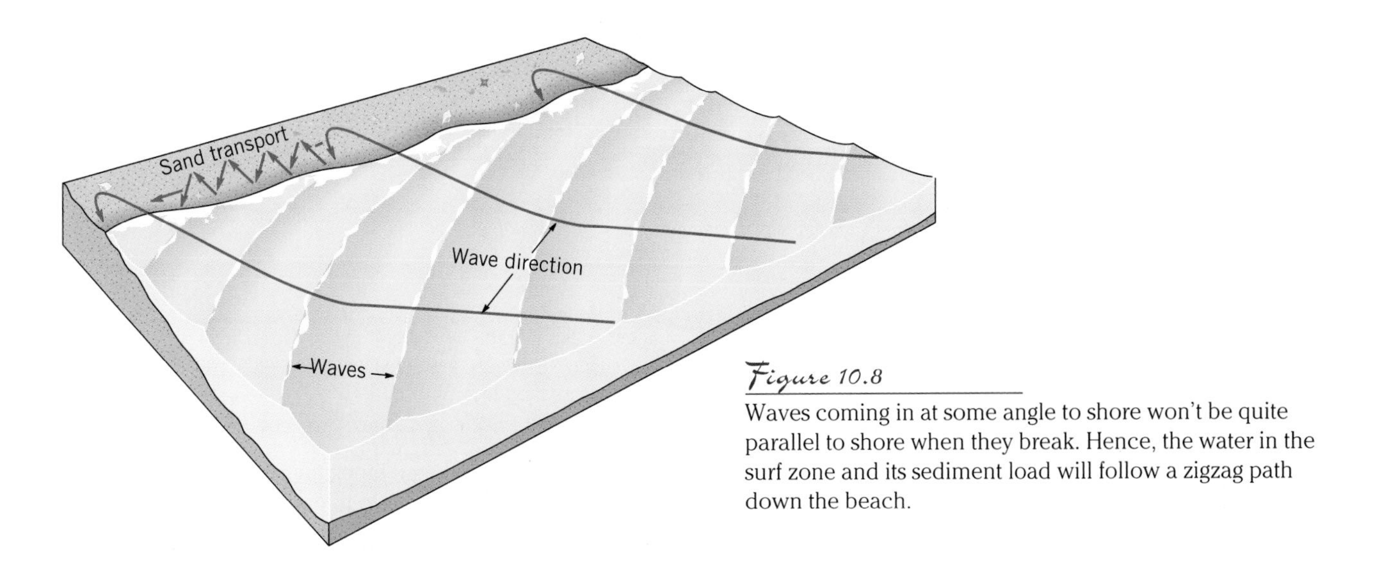

Figure 10.8
Waves coming in at some angle to shore won't be quite parallel to shore when they break. Hence, the water in the surf zone and its sediment load will follow a zigzag path down the beach.

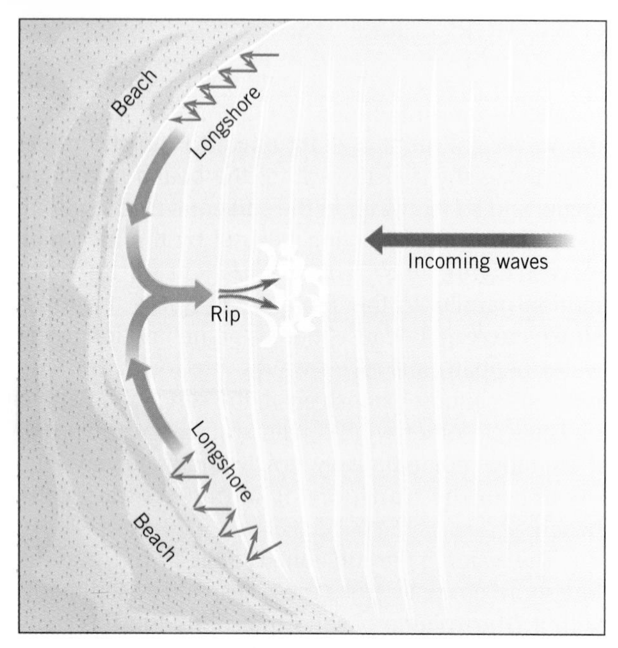

(a)

(b)

Figure 10.9

(a) Waves coming into a curved beach may generate longshore currents in the surf zone that converge at some point and return to sea in a rip current. (b) Rip currents may be spotted by the floating foam and debris as well as the suspended sediments that they carry back out to sea.

RIP CURRENTS

Where longshore currents converge, the accumulated water may flow back out to sea in very swift, narrow flows, called **rip currents**. Converging longshore currents are particularly likely on beaches where the shoreline has gentle curvature, as in a bay or cove (Figure 10.9), because the incoming waves drive the longshore currents in different directions on different parts of the beach. Rip currents are most likely to happen when wave activity is heavy and the backwash otherwise has difficulty getting out through the incoming waves. They can usually be spotted by the foam and debris that they carry away from shore with them.

Rip currents have claimed the lives of many swimmers who panicked when finding themselves being swept out to sea. Many of these lives could have been spared if they had just swum a few strokes sideways, parallel to the beach. Rip currents are always narrow and easily exited to the side.

BEACH COMPARTMENTS

The littoral flow of sand along some coasts is occasionally interrupted. In some places on the U.S. West Coast, sand flows into the upper ends of submarine canyons and disappears from the beaches entirely, eventually sliding down through the canyons and ending up in turbidite deposits on the deep ocean floor (Chapter 4). Headlands may also block the flow of sand, creating **dune fields** as the sand accumulates behind them (Figure 10.10). On the East Coast, the flow of sand frequently ends in the extension of barrier beaches, cusps, or in dune fields. It may flow

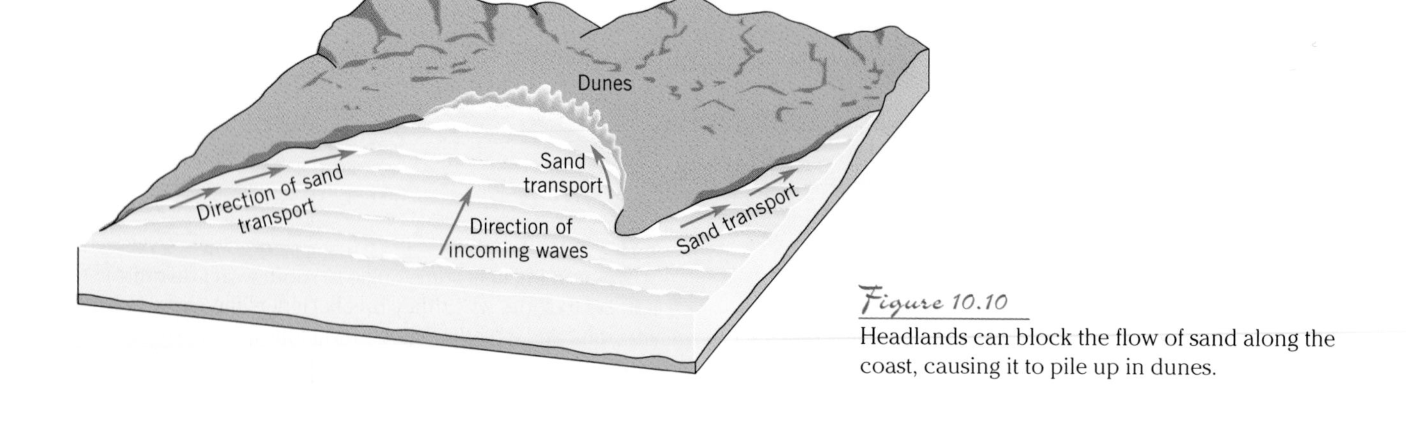

Figure 10.10

Headlands can block the flow of sand along the coast, causing it to pile up in dunes.

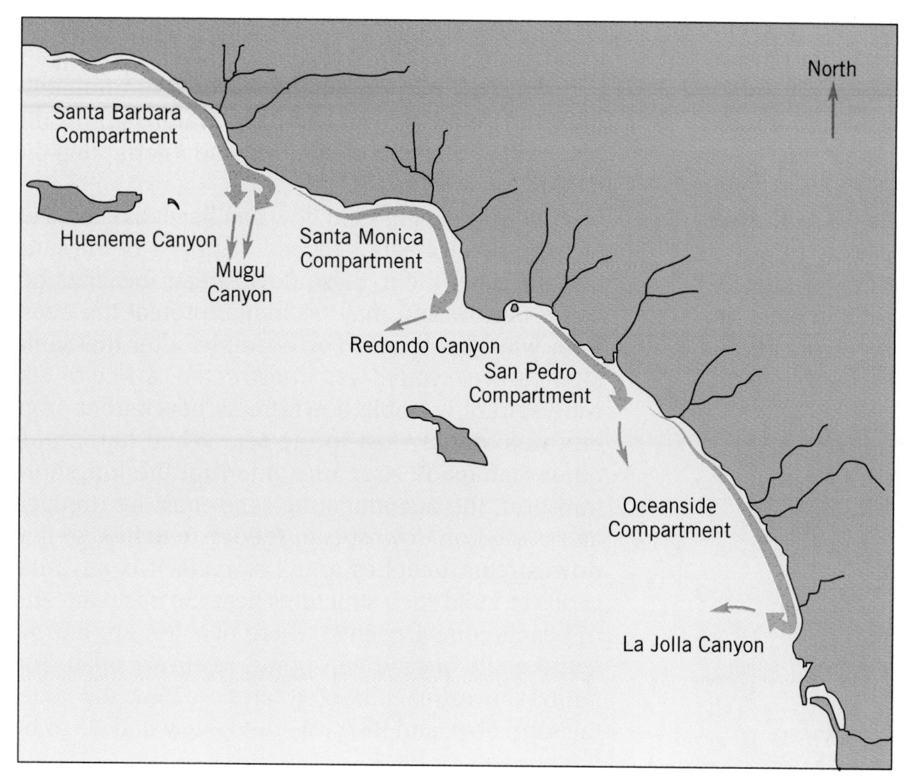

Figure 10.11

The southern California coast, extending from Santa Barbara to the Mexico border. In each beach compartment, the beach tends to be narrowest at the northern end. As the sand moves south along the shore, the beach becomes wider as more sand is added from sources such as rivers and the erosion of sea cliffs. At the southern end of each compartment is a submarine canyon into which the sand is dumped, disappearing from our beaches forever. Along other coasts, beach compartments may end in dune fields or the extension of barrier beaches.

past a channel or river mouth, where strong currents carry it out to deeper water.

These interruptions of the sand flow result in **beach compartments** (Figure 10.11). These are especially prominent on the rugged West Coast, where interruptions are many, and the sand is not so plentiful. Each beach compartment begins just "downstream" from an obstacle or interruption. At this point, the beach is narrow and rocky, because it is deprived of incoming sand from the upstream beaches. Further downstream along the beach, rivers and wave erosion add sediment to the beach, so it grows wider and wider until the next interruption blocks the sand flow and marks the downstream end of that compartment. Beyond that, the beach is again narrow and rocky, and a new beach compartment begins.

HUMAN INTERFERENCE

We sometimes try to protect certain coastal interests from waves or erosion using various kinds of rugged walls made of rocks, broken concrete slabs, or other heavy demolition debris (Figure 10.12). **Groins** extend straight out from shore and are intended to reduce erosion and encourage deposition by obstructing the flow of sediments along the beach. Onshore **seawalls** and offshore **breakwaters** may be built parallel to the beach to protect it from incoming waves and erosion. Sometimes, we reduce the need to dredge the mouths of harbors or rivers by building long **jetties** that block the flow of sand into these areas.

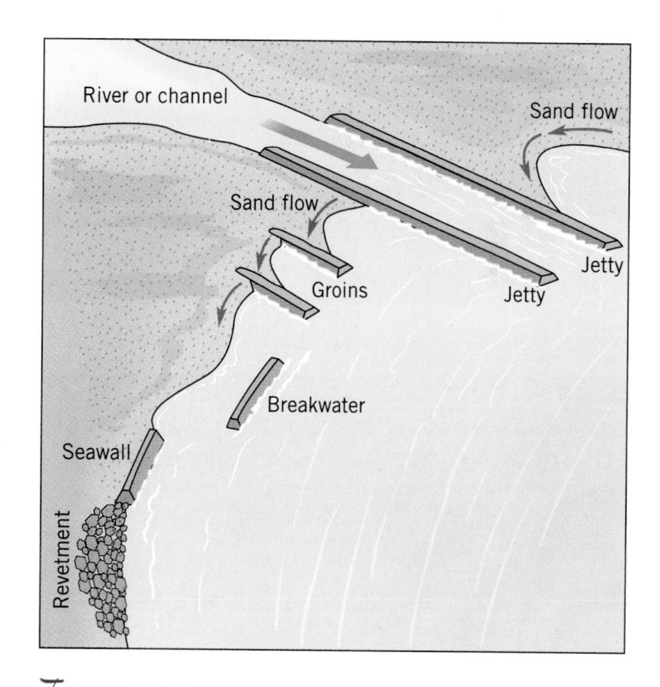

Figure 10.12

Some common types of barriers constructed to protect beaches.

Sometimes, we place **revetments** directly on the beach to protect it from waves and further erosion. This armor may take the form of solid retaining walls, or large chunks of rock or concrete placed directly on the beach. In some places, we combat beach erosion by adding sand to replace that which is being carried away. Such a program is called **beach nourishment**.

The blockage of sand flow by jetties is illustrated in Figure 10.13a, and the settling of sand in quiet protected waters behind a breakwater in Figure 10.13b. In Santa Monica, a breakwater was built offshore and parallel to the shore, with the hope that it might allow the sand to drift on through, while protecting anchored boats from incoming waves. Unfortunately, the sand settles out in the calm water behind the breakwater, filling in the harbor, and interrupting the littoral drift anyhow.

Interrupting the sand flow can have disastrous effects on downstream beaches. Deprived of the sand that once fed them, these downstream beaches become narrow and may no longer protect the coast from wave onslaught. For example, after the Santa Monica breakwater was constructed, a one-block-wide strip of valuable downstream beach-front property was quickly lost to the sea. When harbors or other man-made structures interrupt the longshore transport, the accumulating sand must be dredged and placed on downstream **feeder beaches**, so that downstream beaches aren't starved. It is advantageous to build such structures near the upstream end of beach compartments, where beaches are narrow and the rate of sand transport is relatively small. The farther downstream the structures are built, the larger the sand flow, and the more dredging will have to be performed.

Inland interference with rivers and streams may also deplete our beaches. When streams are dammed for water reservoirs or power plants, or when their channels are lined with concrete for flood control, the ocean beaches are deprived of the sediment that these streams once delivered. The result is narrower and less protective downstream beaches. Many beach-front homes have been damaged or destroyed, because man-made interference in feeder streams has depleted sand from protective beaches.

(a)

(b)

Figure 10.13

(a) Lake Worth Inlet, Florida. Sand flows from north to south (top to bottom) along the seaward side of the barrier island, but is stopped by the jetty. (b) Aerial photo of Santa Barbara Harbor. The sand flows from left to right in this photo. The accumulaton of sand behind the "upstream" left side of the breakwater has permitted the construction of buildings and a parking lot there. As can be seen in this photo, the sand now flows on around the breakwater and settles out in the mouth (right side) of the harbor, which is kept open by dredging.

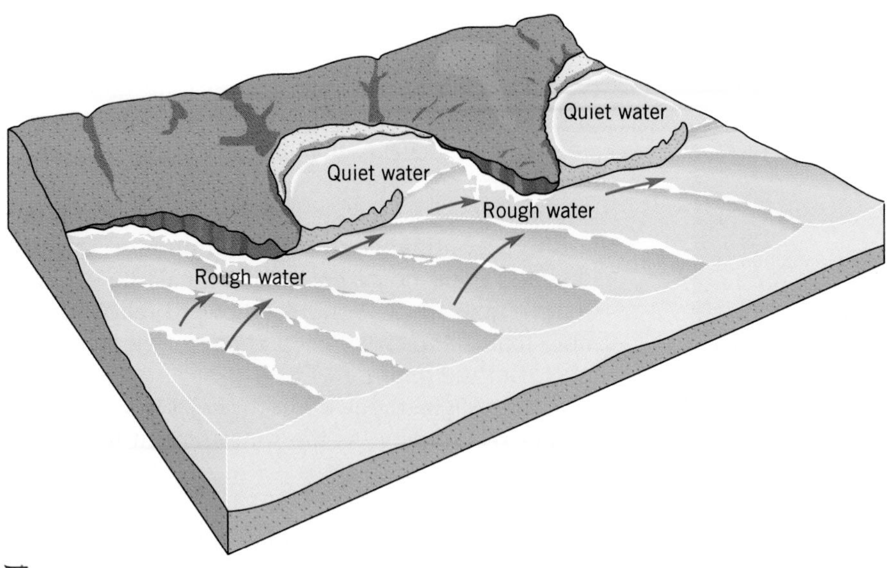

Figure 10.14

Where there are longshore currents, the sediment is kept in suspension in the rough water off the headlands, but is deposited in baymouth bars in front of the quiet coves.

> Many beach-front homes have been damaged or destroyed, because man-made interference in feeder streams has depleted sand from protective beaches.

STRAIGHTENING OF COASTLINES

The longshore sand transport is another process that contributes to the gradual straightening of coastlines. We have seen that incoming wave energy is concentrated on headlands, because that is where they encounter shallow water first. This causes erosion of the points and deposition in the neighboring quiet coves (Figure 9.18).

Additional straightening is due to the formation of **baymouth bars**. As sand travels down the coast, it remains in suspension as it passes the rough waters of the headlands, but then settles out in the quiet water of the neighboring bays (Figure 10.14). The baymouth bars formed in this manner extend from the headland in the direction of the longshore current.

Baymouth bars further protect the waters in the bays behind them. Sediments carried by the run-off from the nearby land settle out in these quiet waters, gradually filling them in. While still in transition, these regions become broad, shallow, muddy, marshy regions, which are rich in nutrients and have high biological productivities (Section 10.5).

10.3 PROCESSES AFFECTING OVERALL COASTAL FEATURES

The overall features of a coast depend on many long-term processes. These include such things as exposure to ocean waves and currents, the tidal range, the local sediment sources, and whether erosion or deposition dominates. Coastal features strongly reflect the local geography and coastal materials.

Coastal features also depend on whether sea level is rising or falling. The major changes in sea level relative to land involve two quite distinct types of processes that operate on quite different time scales. Tectonic processes cause changes over millions of years, and ice ages cause changes over tens of thousands of years.

COASTAL TECTONICS

Tectonic processes determine whether the coastal region is stretched, compressed, rising, or sinking. For example, for the past several million years, the U.S. West Coast has been rising, and so we can find the relics of former beaches and marine terraces high above the present beach (Figure 10.15). The uplifting has created coastal mountains, which sometimes carry a covering of loosely consolidated marine sediments. Many small streams feed into the ocean through the valleys, and in some higher latitudes, beautiful coastal valleys have been carved by glaciers.

By contrast, most of the U.S. East and Gulf Coasts have been slowly sinking over the past 150 to 180 million years, since the opening of the Atlantic Ocean. This sinking is partly due to the crustal materials becoming cooler and denser as they move away from the Mid-Atlantic Ridge, and partly due to the added

Figure 10.15

Relict beaches above the present beach indicate that this part of the coast has risen relative to sea level. Streams have carved gullies down through these terraces which mark the former beaches.

weight of sediments that rivers and streams have deposited on these margins. In some places, these sediments on the continental rise reach a thickness of 15 kilometers. On average, the rate of sedimentation has been nearly equal to the rate of sinking, so the two processes nearly offset each other.

In addition to the uplifting or sinking of continental margins, processes beneath the deep ocean floor can also cause large changes in sea level along the coast. For example, during periods of accelerated seafloor spreading, the oceanic ridge is broader and hotter, and therefore floating higher on the asthenosphere. This lifting of the central ocean floor creates a corresponding elevation of the ocean surface during these periods.

ICE AND ISOSTASY

Ice ages cause relatively short-term fluctuations in sea level that are superimposed on the long-term tectonic changes. During a typical ice age, the Earth's climate is cooled by about 5° C, and water removed from the ocean forms huge continental ice sheets. The last ice age peaked about 18,000 years ago, when the ice sheets extended south as far as the mid-North American Continent, and sea level was about 135 meters below its present level. This left most continental shelves exposed.

As the climate warmed, ice began melting back and the sea level rose. It rose quickly at first, but then nearly leveled off about 6000 years ago. During this past century, however, the rate of rise has accelerated and now appears to be rising at a rate of about 20 centimeters per century. This most recent rise is thought to come primarily from the warming and expansion of surface waters, rather than the further melting of ice.

While sea level is changing, the continental margins do not necessarily remain stationary. When the weight of the last ice sheet was removed from the northern continents, the land rebounded upward (isostatic adjustment). For the portion of the U.S. East Coast north of New York, this upward rebound was more than sufficient to keep ahead of the rising sea level. But the southern East Coast and Gulf Coast were not weighted down by ice sheets and therefore are not rebounding. As the water rises, it floods further inland in these regions.

If the climate continues to warm, the sea level will continue to rise for two reasons. One reason is the continued thermal expansion of water. For every 1°C increase in the average temperature of the oceans, the sea level rises about 2 meters. Continued warming would also cause the melting of more ice. If all the ice in the world were to melt (most of which is in Antarctica), the sea level would rise about 60 meters.

THE SEDIMENT SUPPLY

Among the long-term processes that determine the general features of a coast are erosion and deposition. Sediments are slowly depleted from beaches by several processes, most of which involve littoral transport. These include the flow of sediments out onto the shelf, down through submarine canyons, and onto the ocean floor, and the accumulation of sediments in local dune fields. Also included is the gradual grinding and flaking of sediments into finer particles, which may be more easily suspended and transported to deeper calmer waters or dissolved into the water.

To counteract these losses, there are several processes that contribute new sediments to beaches, among which is the transport of sediments to the coast by rivers and streams. Also, some sediments are created by direct wave erosion, as the incoming waves crash into solid outcroppings. Finally, some are swept landward from the continental shelf by the orbital motion of the larger incoming waves.

The tectonically active edges of continents, such as the West Coasts of the Americas, have narrow continental shelves and rugged sea floors. There are relatively small quantities of sediments, erosion is dominant, and the continental rise is essentially nonexistent. On the passive edges, by contrast, the shelves are wide, smooth, and well developed.

Sediments are thick and the continental rise is well developed. Deposition usually dominates on these coasts. In general, however, this deposition does not forestall the landward intrusion of the ocean as the sea level rises. Rather, as the sea level rises and the ocean moves landward, it simply sweeps these sediments with it, by processes described in Section 10.1.

BARRIER ISLAND FORMATION

At the peak of the last ice age, the continental shelves were exposed. Then, as the ice sheets melted back and the sea level rose, the waves swept shelf sediments upward with them as the shoreline rose to its present position. These relic sediments that were swept up from the continental shelf combine on our present coasts with sediments carried by rivers draining the inland watersheds and those currently being produced by wave erosion. These processes produced an extraordinarily large accumulation of coastal sediments on coasts with broad continental shelves, shallow coastal waters, and large drainage from the inland areas. The East Coast of the Unites States is one such area.

When these large accumulations of sand are exposed to the forces of the freshwater drainage pushing them outward from land, the waves pushing them inward from the ocean, and the longshore currents carrying them sideways along the coast, the result is often the formation of extremely long, thin exposed sandy barriers that parallel the coast, but that are separated from land by broad shallow bays. When these long, narrow exposed deposits are connected to the mainland, they are called **spits**, or *baymouth bars* if they front a bay. If they connect the mainland to an offshore rock or small island, they are called **tombolos**. If they are completely separated from the mainland, running roughly parallel to the coastline, they are called **barrier islands**, or *barrier beaches* (Figure 10.16a). Spits and barrier islands are prominent features along the U.S. East Coast.

Figure 10.16b displays the structure of a typical barrier island. The beach on the ocean side is an exposed sandy beach, molded by the processes described in Section 10.1. Behind the beach are dunes of loose sand that move gradually with the winds and rapidly with the waves and waters of strong coastal storms. The dunes offer protection to the barrier flats behind them, enabling some plants to take hold and further stabilize this region. Behind the barrier flat is a salt marsh where the barrier island meets the shallow embayment, called a **lagoon**, that separates it from the mainland.

Many barrier islands are slowly moving landward with the rising sea level, so they are overriding the

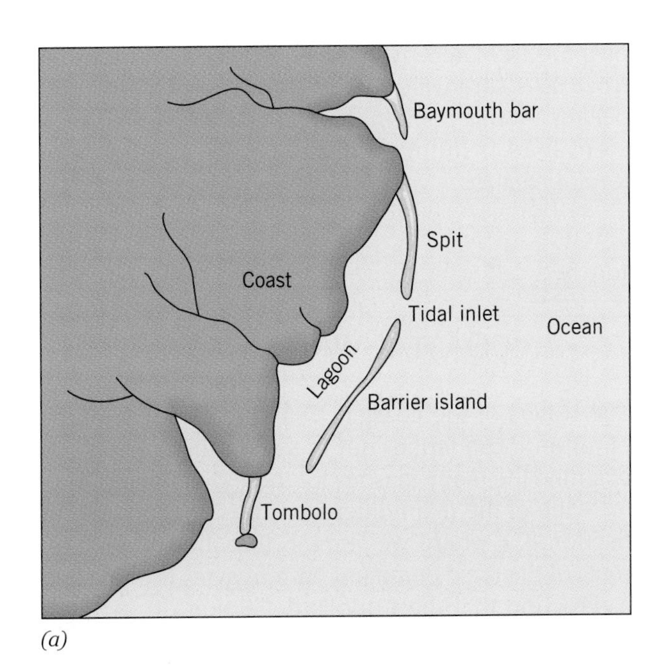

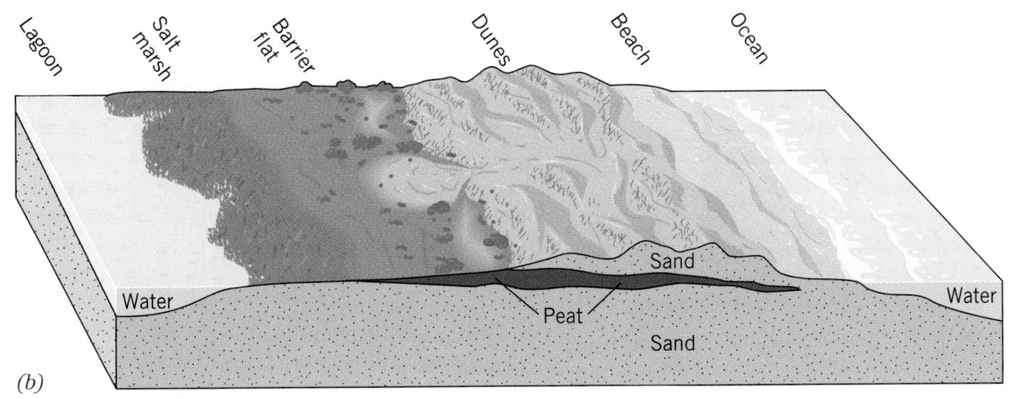

Figure 10.16
(a) Illustration of the names of various types of long, thin, exposed offshore deposits. (b) Cross section of a typical barrier island.

salt marshes on the landward edge as they move. Consequently, beneath many present barrier beaches are peat (i.e., partially decayed plant matter) deposits from the earlier marshes that have been overridden, and these deposits are exposed again on the seaward side as the barrier island moves on.

Barrier islands are interrupted by occasional channels, called **inlets**. Overwash from storm surges sometimes carves temporary inlets through the barrier islands, but these are usually short-lived, as the littoral drift fills them back in after the storm is gone. More lasting inlets are maintained by pressures from the landward side. In particular, as rivers and streams flow into the lagoons, the force of the accumulating waters produce and maintain inlets through which they drain out into the adjoining ocean. Some inlets are also maintained by the ebb and flow of the tides as the sea level rises and falls with the daily tidal cycle.

DELTA FORMATION

Many larger rivers carry sediments in such quantities that the longshore transport cannot keep up with the accumulating deposits. Consequently, where one of these rivers enters the ocean, we find a large low flat plain of sediment deposit, called a **delta** (Figure 10.17). The river flows across the delta through channels called *distributaries*. The delta deposits are mostly finer silts and clays. Where the deltas meet the ocean, however, the waves remove the finer sediments, leaving sands and other coarser sediments on the beaches.

Figure 10.17

NASA satellite photo of a portion of the Mississippi River delta.

When the river floods, the water rises in the distributaries and spreads out in a shallow layer over the adjacent *flood plains*. As the sediments settle out of these shallow flood waters, they create leveed banks along the distributaries and add fertile soil to the flood plains. Consequently, deltas provide fertile and well-irrigated agricultural land. Furthermore, the delta's channels and wetlands are rich in marine life, such as clams, crabs, shrimp, oysters, and fish. Although these features made them advantageous to early agricultural societies, the fact that they are low and vulnerable to coastal storms and flooding makes problems for some modern urban centers that are built on deltas.

Many large rivers do not enter the ocean directly and therefore do not form deltas. Important examples are the large rivers on the U.S. East Coast that flow into estuaries rather than directly into the oceans. At this point in the Earth's history, their sediment load is filling in these estuaries rather than forming deltas.

10.4 COASTAL WATERS

The waters of the coastal ocean are distinctly different from those farther out to sea. In this section, we examine the many reasons for these differences.

One of the main differences was encountered in Section 8.3, where we learned that the surface currents of the large dominant current gyres flow in such a direction that the Coriolis deflection drives these waters away from the neighboring continent (Figures 8.11 and 10.18). In some cases, the water that replaces them is upwelled from depth. This upwelling tends to dominate where continental shelves are narrow and the major surface currents flow slowly, such as along the West Coasts of the Americas. In other cases, the replacement waters arrive as countercurrents flowing down along the coast from higher latitudes. This tends to be the dominant replacement mechanism along the East Coasts where the oceanic surface currents are intense and the broad continental shelves prevent upwelling. Whether upwelled from depth or intruding down the coast from higher latitudes, the coastal replacement waters are quite different from the surface waters of the adjacent ocean. In general, they are colder, richer in nutrients, and support greater biological activity.

Other differences in the coastal waters are due to their shallowness and closeness to land. They are much more responsive to seasonal heating, cooling, and dilution by freshwater run-off from land. Regional variations are caused by plumes of low-salinity water from coastal rivers and streams.

As rivers and streams enter into the coastal ocean, they tend to flow along the coast in the direction favored by the Coriolis effect—to the right in the Northern Hemisphere and to the left in the Southern Hemisphere. In each case, this makes them tend to flow along the coast in the direction opposite to that of the main gyre that is farther out to sea.

Coastal waters also tend to be turbid, due to river discharge, waves, storms, and coastal currents. Finally, for reasons to be studied in Chapter 11, coastal waters are a great deal more active biologically compared to adjacent oceanic waters. The abundance of microscopic life, in particular, also adds to the turbidity and greenish color of these waters.

10.5 ESTUARIES

Fresh water from land enters the ocean through rivers, streams, and ground water, all of which flow through valleys. Where these valleys meet the oceans are embayments, which may be as narrow as the mouths of clean stream-cut channels, or as wide as

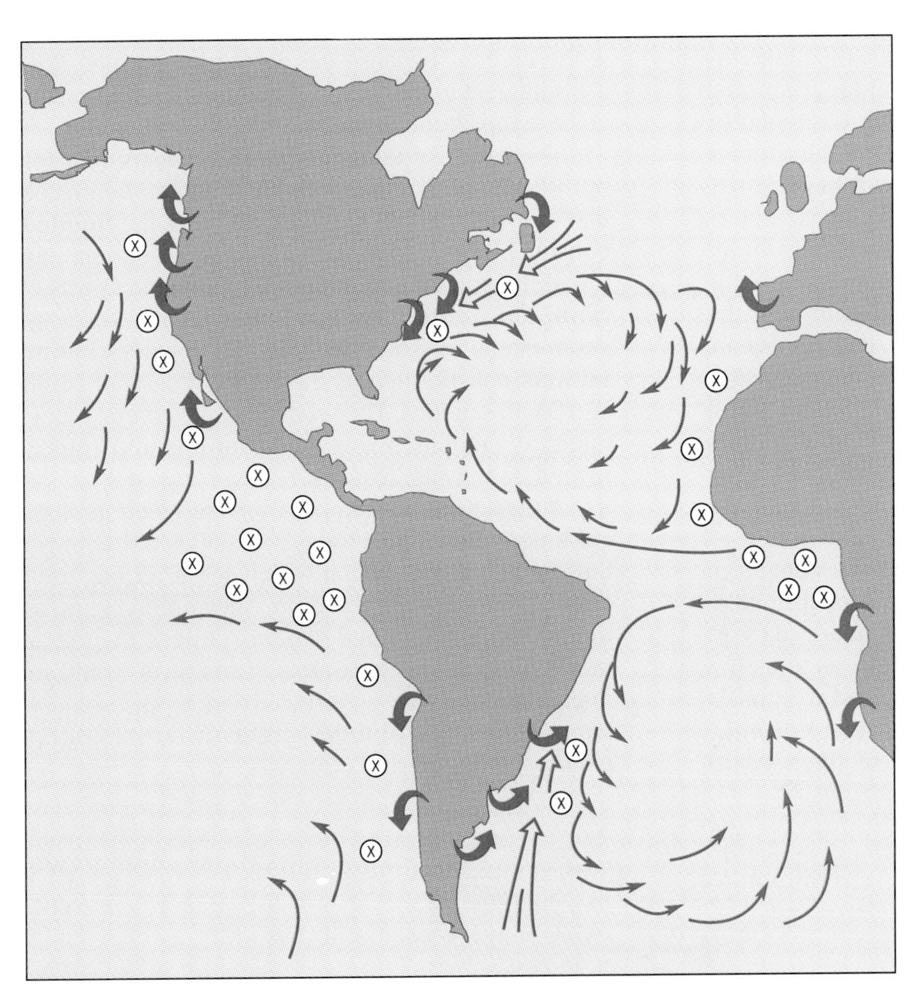

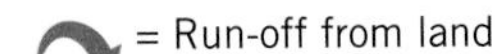

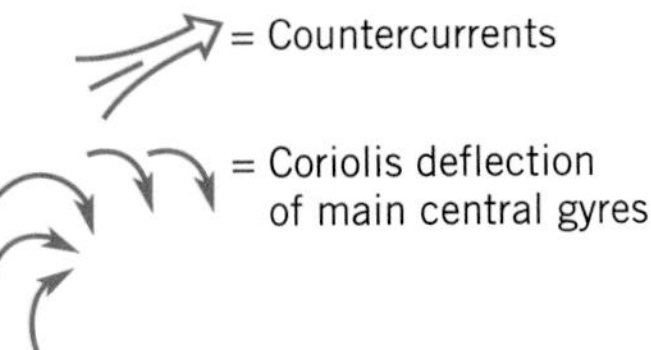

Figure 10.18

Coastal waters differ considerably from those farther out to sea. The Coriolis deflection of the main ocean surface currents draws these surface waters away from the continents, and they are replaced by coastal upwelling, coastal countercurrents, and river discharge.

the lagoons behind long barrier islands. These embayments, where fresh water from the land meets salt water from the ocean, are called **estuaries**.

GENERAL FEATURES

Because of their small size, shallow water, and closeness to land, all estuaries are heavily influenced by processes on the neighboring land, such as seasonal variation in temperature and run-off from the rains. For the same reasons, they are also heavily influenced by human activity, such as excavation, construction, and various forms of pollution.

In estuaries, fresh water from the land mixes with salt water from the oceans. The degree of mixing is heavily influenced by both the rate at which fresh water enters from rivers and streams and the volume of tidal flushing from the ocean. The rate of freshwater inflow is often expressed in terms of the **flushing time**, which is the amount of time it would take the inflowing fresh water to replace the entire freshwater content of the estuary. Tidal flushing is described by the **tidal volume** or *tidal prism* that is the average volume of water between the high and low tide surfaces (Figure 10.19). It corresponds to the volume of water entering and leaving the estuary during an average tidal cycle. The **head** of an estuary is where the fresh water enters from a river or stream, and the **mouth** is where the estuary empties into the ocean. An estuary may have many heads and many mouths. For example, many lagoons on the U.S. East Coast have several rivers and streams flowing into them from land and also several inlets through the barrier islands through which they communicate with the ocean.

Although estuaries contain both fresh water from land and salt water from the ocean, the relative influence of the two varies greatly from one estuary to another. Clean **stream-cut channels**, such as the mouths of the Mississippi, Congo, and Columbia Rivers, are dominated by forceful river currents. Oceanic influence is overshadowed by the seaward flow of fresh water. At the other extreme, some estuaries have only seasonal free-flowing streams; during parts of the year, fresh water enters through the slow flow of ground water only. In spite of this wide variation in estuary types, they all share two common features: They have some degree of isolation from the ocean and some dilution by fresh water from land.

The most commercially important and heavily populated estuaries are the mouths of major rivers. The forceful flow of the rivers keeps the main channels sufficiently deep for navigation by ocean-going ships, and the rivers themselves afford transportation to points farther inland. Tectonic and glacially formed embayments also are frequently sufficiently deep to offer convenient ports for ocean commerce. At the other extreme are the mouths of minor streams, from which the flow is not sufficient to prevent sediments from filling in the estuary. Such estuaries are too shallow to be of major commercial importance. Frequently, barrier beaches or baymouth bars form at the mouths, because the outward freshwater flow is insufficient to flush the sediments carried along the coast in the littoral drift.

The nature of estuarine systems reflects the geology of the coasts. On this basis, most can be placed into two broad categories. One includes the **flooded coastal plains**, broad flat land areas extending into broad flat continental shelves. The Atlantic Coasts of North and South America, Europe, and Africa are good examples. The other category includes **mountainous coasts**, such as the Pacific Coasts of North and South America. These have rugged coastal topograhies, continuing below the water line into equally rugged underwater features.

All estuaries are temporary features on geological time scales. All of today's estuaries are younger than 18,000 years, during which sea level has risen about 135 meters. Furthermore, they are filling in, as the inflowing rivers and streams deposit their sediments in these quieter waters. If sea level would remain constant, the lagoons, wetlands, and shallow backwaters of all of today's estuaries would disappear in a few thousand years.

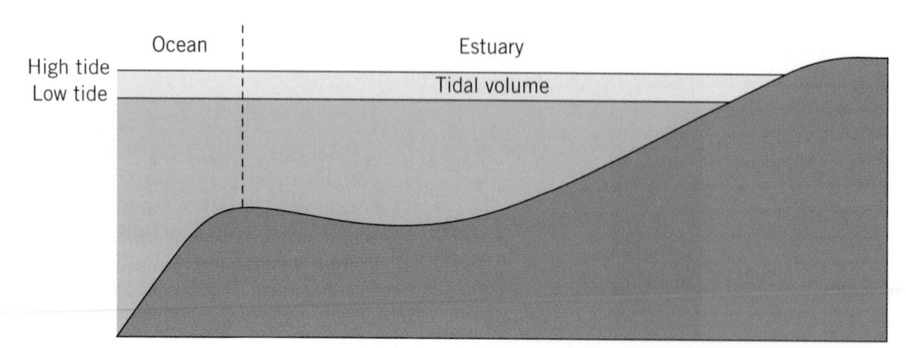

Figure 10.19
The tidal volume (or tidal prism) of an estuary is the average volume of the water contained between the high and low tide levels.

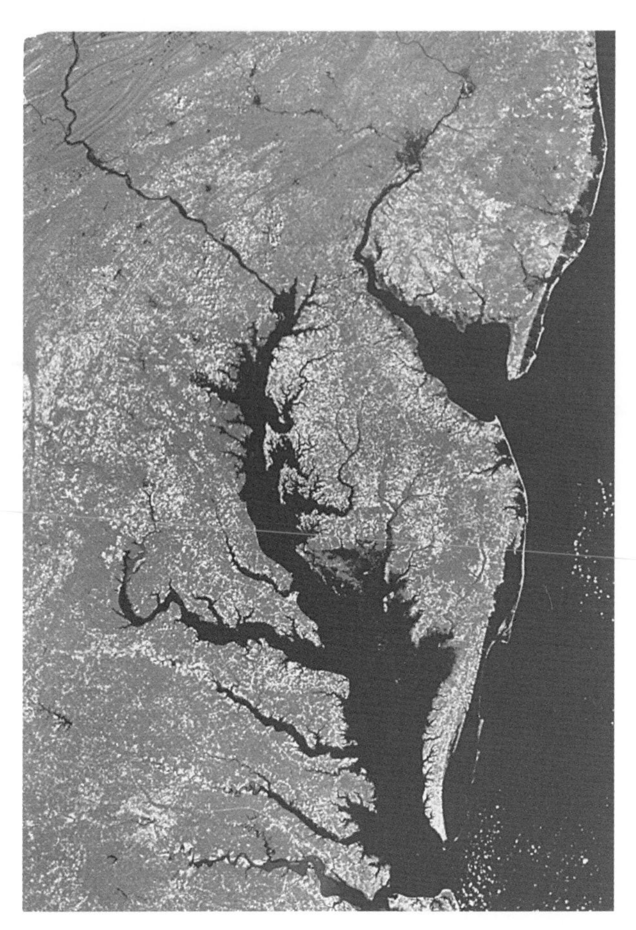

Figure 10.20

High altitude photo showing some drowned river valleys along the East Coast of the United States, including Delaware Bay (above) and several drowned river valleys in the Chesapeake Bay system. Also visible are the very thin barrier islands along the coast, with streams emptying into the lagoons behind them.

FLOODED COASTAL PLAINS

Much of the U.S. East Coast is a flooded coastal plain. During the last ice age, much of the wide eastern continental shelf was exposed as an extension of the continent. Rivers and their tributaries cut valleys across these plains and out to the edge of the shelf where they entered the ocean. Then as the ice melted, sea level rose and flooded over portions of these coastal plains. The rising waters extended inland further over the lower areas, creating the broad **drowned river valleys** that we find today (Figure 10.20).

On flooded coastal plains, deposition dominates and sand is plentiful. Furthermore, the water is shallow, and there are little or no underwater features that would interfere with the littoral transport of sand. These conditions nourish the growth of extensive

Figure 10.21

A cove between rocky headlands on the Hawaiian coast. Due to a relatively small supply of sediments, mountainous coasts have small and narrow beaches. Interesting and varied coastal rock structures result from differing exposure to waves, weather, ground water, and different materials in the original rock.

long, narrow offshore sand deposits that are excavated and maintained by the longshore currents. Many of these deposits are exposed above the water, such as spits and barrier islands, and are constantly being reshaped and repositioned by the waves and winds. Such changes can be gradual, taking place over years, but storms can cause major overnight modifications, including displacement or disappearance of the entire deposit.

The lagoons behind these spits and barrier islands are shallow, so they heat up quickly, cool down quickly, and their salinities vary with changes in the weather. They include extensive salt marshes that support vigorous biological activity, based on the high plant productivities in these nutrient-rich areas. It is common along flooded coastal plains for drowned river valleys to empty into lagoons rather than directly into the ocean. Barrier islands may protect the lagoons from oceanic influences so well that wind-induced currents and seiches may cause greater changes in sea level along lagoon shores than the tides.

MOUNTAINOUS COASTS

Mountainous coasts are rough, irregular (Figure 10.21), and have coastal waters that often get deep very rapidly. Sand is less plentiful than it is on the coasts of flooded coastal plains. Although spits and baymouth bars are fairly common (Figure 9.12e), there is not an extensive system of barrier islands as is common for flooded coastal plains. One reason for

Figure 10.22
The steep walls of this fjord are typical of glacially cut valleys.

Figure 10.23
Tidal bore moves slowly upstream along the Petitcodiac River, New Brunswick.

this is that the mountainous barricade blocks the flow of rivers to these coasts and thereby deprives the coast of a major source of sediments. Other impediments to the formation of extensive barrier beaches are the deeper water and obstructions to longshore transport, such as submarine canyons or gullies cutting into the shelf or rocky headlands jutting out into the oceans. Without the protection of barrier beaches, mountainous coasts are more exposed to direct attack by waves, so erosional features such as sea cliffs and rocky headlands are common.

The bays of mountainous coasts are relatively narrow, corresponding to the irregular topography. Because of their comparatively small size, these bays are generally more heavily influenced by the tides than the large lagoons of flooded coastal plains. Marshes are alternately inundated and exposed as the tide rises and falls.

The mountain-building processes sometimes cause crustal blocks to fall below sea level. Examples are San Francisco Bay (California) and the Strait of Juan de Fuca (Washington state). Such **tectonic-formed estuaries** are deeper than lagoons and can usually accommodate ocean-going ships without dredging. At higher latitudes, we find coastal **fjords**, where the slow but forceful flow of glaciers has carved impressive U-shaped valleys down through the mountains leading to the coast (Figure 10.22). Many of these glaciers have melted back since the last ice age, and the rising ocean has moved into these magnificent steep-walled estuaries.

TIDES

The protected waters of estuaries often communicate with the open ocean through relatively small openings. Large volumes of water must flow in and out through these channels during the rise and fall of tides, resulting in strong currents and interesting navigational hazards.

The period of time in the tidal cycle when water is flooding into the estuary is called **flood tide**, and the period of time when it is emptying is **ebb tide**. A *stand of the tide* or **slack water** refers to a period when water is neither entering nor leaving. Because of the time it takes for the estuary to fill and empty, the tidal cycle in an estuary usually lags behind that in the neighboring ocean.

Most estuaries are shallow, which means that the travel speed of an incoming tidal wave is relatively slow. Furthermore, the inward flow of the tide is opposed by the outward flow from the one or more streams that empty into the estuary. As a result, tidal waves frequently progress very slowly upstream along an estuary. Before one has run its course, the next one may be coming in. For example, tides travel so slowly up the Chesapeake Bay that there are often remnants of three different tidal waves at various points along its course simultaneously.

Sometimes, there is a stand-off. The incoming tide makes very little progress against the outflowing current, while the tide continues to rise. This can create a wall of water, called a **tidal bore**, that moves slowly upstream against the current (Figure 10.23). Among the most spectacular of these are those found on the Amazon River, where they can reach a height of 5 meters, and on the Tsientang River in China, where they may reach a height of 8 meters. Most tidal bores, however, are less than a meter in height. Bores can be hazardous, due to the sudden rise in water level as they pass, and they might capsize small boats that are not prepared for them. To most residents along the shores of the appropriate river mouths, however, they are little more than a small daily nuisance, and for some surfers, they present an opportunity for recreation.

CIRCULATION PATTERNS

As the human population increasingly crowds the oceans' shores, estuaries are becoming increasingly threatened. Because estuaries are small, shallow, sensitive, and delicately balanced, human activities are much more influential than they are on exposed coasts, where the powerful unrelenting ocean forces easily overpower temporary human intrusions. The study of estuarine circulation is necessary in order to minimize the impact of construction and improve our understanding of such issues as water quality, sediment transport, and biological activity.

We begin our study of circulation in estuaries with the **salt wedge** model. This model applies best to estuaries where the outward flow of fresh water is sufficiently strong to prevent ocean water from entering along the surface, and where the estuary is sufficiently deep that surface waves and turbulence have little effect on mixing the deeper waters. Estuaries that fit this model are called *salt wedge estuaries*. In these, the fresh water from the stream flows out along the surface, and a wedge of denser salt water flows in along the bottom (Figure 10.24a). Along the interface between the two waters, salt water from below mixes with fresh water above and is carried back out beneath the surface as brackish water (i.e., less salty than pure sea water).

Although there is always a tendency for the denser salt water to enter along the bottom and the lighter fresh water to flow out along the surface, the simple salt wedge picture is often considerably modified by a number of factors. Wind, waves, weather, and currents create turbulence and mixing. The relative influence of the flow of fresh water from the land vs. the flow of tidal currents from the ocean varies greatly from one estuary to another, as does the depth of influence of various surface processes.

The salt wedge model is most accurate in the deep stream-cut channels of large rivers, where tidal influences are overshadowed by the freshwater flow and surface effects are minimized by the great depth. On the other extreme, the salt wedge model is quite difficult to apply to broad, shallow lagoons, where salt water intrusion is obstructed by barrier islands

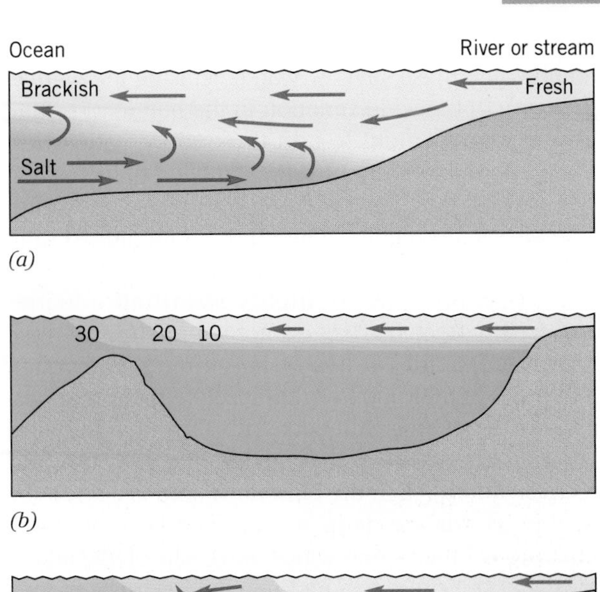

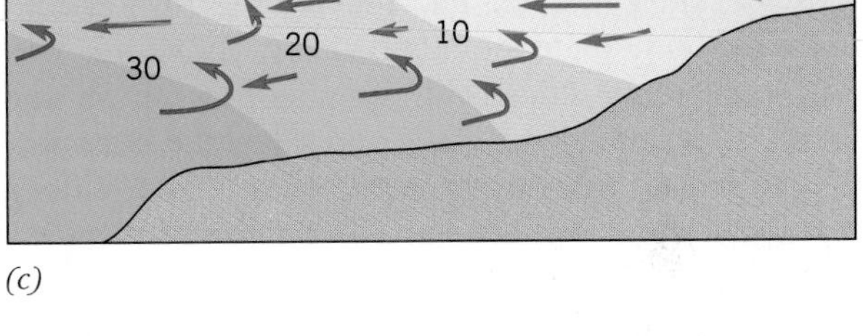

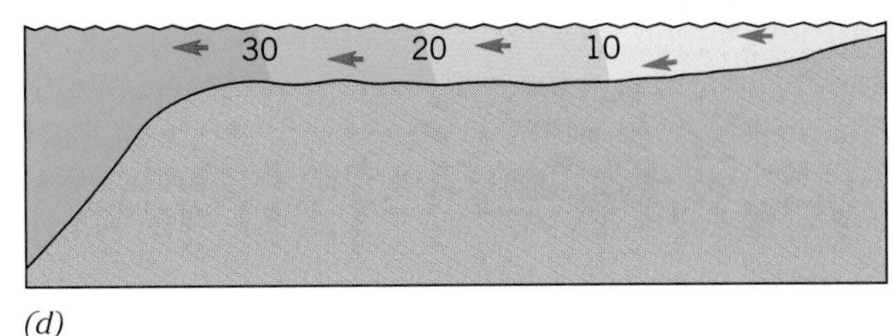

Figure 10.24
Cross sections of estuaries (with huge vertical exaggeration) showing typical salinity profiles. (a) Salt wedge estuary. The salt water intrudes in a wedge along the bottom, mixes with the fresh water along the interface, and exits as a brackish mixture. The freshwater flow is sufficiently strong to prevent salt water intrusion at the surface. (b) Highly stratified estuary. (c) Partially mixed estuary. (d) Well-mixed estuary. The contour lines marked 10, 20, and 30 indicate the salinity in parts per thousand.

and wind-induced waves and current patterns thoroughly mix these shallow waters, erasing the wedge. Because of the wide variation in the nature of estuaries, it is convenient to add three more categories, based on the amount of vertical mixing of the estuarine waters. Besides salt wedge estuaries, the other categories are highly stratified, partially mixed, and well-mixed estuaries (Figures 10.24b, c, and d).

The deep waters of **highly stratified** estuaries are well protected. They do not experience the mixing of strong tidal or freshwater currents. They often have sills at their mouths that insulate their deep waters from oceanic influence and flushing. They are also sufficiently deep that mixing due to winds and waves cannot reach the bottom. A fjord is a good example. Fjords are deep, surrounded by mountains that protect them from winds, have sills of glacial deposits at their mouths, and receive fresh water only from the immediate mountain valley. Therefore, their waters are calm and highly stratified, with fresh and brackish water at the surface and salt at the bottom.

Partially mixed estuaries are sufficiently deep that waves and winds cannot ordinarily mix the waters all the way to the bottom. The force and courses of the tidal and freshwater currents provide moderate but incomplete mixing. The salinity increases from top to bottom, but the stratification is not as great as that of fjords. As with the well-mixed estuaries, there will be changes in salinity both along and across across the estuary, due to such factors as the courses followed by the incoming and exiting tidal and freshwater currents. Also, the Coriolis deflection of the intruding salt and fresh waters causes lateral variations in salinity. In the Northern Hemisphere, both are deflected to the right as they enter, so that when facing outwards toward the ocean, there will be a tendency for the water on the right to be fresher, and that on the left saltier. Examples of partially mixed estuaries include Puget Sound and San Francisco Bay.

Well-mixed estuaries tend to be broad and shallow. Wind and waves can mix the waters vertically all the way down, so that there is little or no vertical stratification. That is, there are no layers of differing salinity. The salinity changes are lateral, rather than vertical, varying from place to place depending on the shape of the estuary and its flow patterns. Examples of well-mixed estuaries are the Chesapeake and Delaware Bays, and for that matter, all the lagoons on the U.S. East Coast.

Although most estuaries have some sort of bar at their mouths, some would not be an estuary except for the bar. That is, a barrier island may extend along a portion of a coast where there is no permanent stream entering the ocean, or any significant indentation in the coastline. Such **bar-built estuaries** display a wide variety of characteristics. In some, the rate of evaporation exceeds the freshwater input, so that the estuarine water is actually saltier and denser than sea water. It flows out of the estuary along the bottom and the sea water enters along the surface, which is just the opposite of the flow patterns in most estuaries. These are called "negative" or **inverse estuaries**. Laguna Madre on the Texas Gulf Coast is one such example.

WETLANDS

Many estuaries have extensive shallow areas that are covered by water at high tide, but exposed when the tide is out. These areas are called **wetlands**. They are composed of muddy flat **tidal flats** and a network of channels that carry water to and from the wetlands as the tides come and go.

Salt-tolerant grasses called *marsh grass* grow in the wetlands of tropical and temperate latitudes. These grass-covered **salt marshes** are typically 5 to 10 times more productive than agricultural crop lands, and the high productivities support active animal populations. Important wetland products include clams, scallops, oysters, and fish. The high productivities also result in a rapid accumulation of organic detritus in the sediments. Consequently, wetland sediments are organic muds, rich in partially decayed organic matter and in fine clays carried by the inflowing rivers and streams.

At latitudes lower than about 30°, mangrove trees and bushes sometimes replace the marsh grasses as the dominant wetland plants, and we call these wetlands **mangrove swamps**. The upper portions of the mangrove roots are exposed, making these trees look like they have legs. In fact, over periods of years, mangroves can slowly move by growing new roots on one side and discarding those on the other. Both marsh grass and mangrove roots tend to trap sediments and thereby give extra stability to tidal flats that would otherwise quickly erode with storms and tides.

Most pollutants enter the estuary with the fresh water from land and tend to settle out in the wetlands along with the fine sediments. Therefore, in addition to producing food, the wetlands also serve the greater estuarine area by removing pollutants and purifying the water.

THE BIOLOGICAL COMMUNITY

Most estuaries are exceptionally rich in biological activity and crucial to the lives of many organisms. Most nutrients enter at the head with the freshwater run-off from land, and the intruding salt water wedge tends to keep them there. In addition, most estuaries are sufficiently shallow that sunlight penetrates to the bottom. The combination of sunlight and nutrients promotes vigorous plant growth.

The plants, in turn, provide food and shelter for local animals. On land, grazing animals tend to have their offspring in the spring when grass is tender and feeding is easy. Similarly, many marine animals tend to have their offspring in estuaries, where feeding is easy. Estuaries provide breeding grounds, nurseries, and juvenile habitats for a large number of animal species, many of which spend their adult lives in the open ocean. In addition, the abundant shallow water vegetation provides protection for juveniles and structural support for tiny mollusks, crustaceans, and larvae of many species. Although estuaries otherwise have relatively little impact on the larger ocean, their biological impact is significant.

THE HUMAN INTEREST

Not only are estuaries particularly important to the lives of many marine organisms, but they also attract human populations as well. Conflicts of interest between the two groups arise, and they are becoming more critical as human populations continue to grow.

The majority of the world's metropolitan areas are centered on estuarine systems, for reasons that are easily understood. The protected waters provide ports for ocean-going vessels. Connections with rivers provide routes for further commercial connections with inland areas. This access to goods and commerce spawns industry and related services. In addition, the biological productivity of these estuaries supports healthy fishing industries. Thus, commerce, industry, and fishing together provide a strong economic base for growing human populations.

The conflicts of interest between human and marine populations tend to take several forms. Humans tend to crowd the water's edge, draining or filling tidal flats and marshes for human construction, or dredging them for commerce (Figure 10.25). In the United States, over 50% of the coastal wetlands have been lost since the Pilgrims arrived, and this loss continues to grow at an average rate of 1300 square kilometers per year. These wetland areas are the most biologically productive of all oceanic regions. Many are several times more productive than the finest agricultural crop lands and support much of the marine life in deeper parts of the estuary. Dredged channels change the current patterns and muddy the water to the point where many native organisms can no longer survive.

Figure 10.25

Tidal flats and coastal salt marshes, which are important components in the ecology of estuaries and the surrounding ocean, are being filled and built upon.

10.6 SEAS

We now turn our attention to seas—bodies of water larger than estuaries but smaller than oceans. Unlike estuaries, the water in seas is truly oceanic. It is salty and well mixed. Most of the major seas include basins that are underlain by oceanic crust, and their greatest basin depths are sometimes comparable to those of the oceans. However, on average they are shallower. Because seas are smaller than the oceans, the continental shelves of the adjacent land mass usually encompass a larger fraction of their total area. In fact, some, such as the North Sea, Baltic Sea, and Yellow Sea, have no deep water at all. Those seas that are completely surrounded by land are called *mediterranean*, and those that are on continental margins, being separated from the neighboring ocean by land and shallow sills, are called *marginal*. An example of a mediterranean sea is the European Mediterranean Sea, and an example of a marginal sea the Caribbean.

Roughly half the world's coasts border on seas, rather than directly on an open ocean. If you look at a map of the world, you will see that this is true for nearly all of Asia, most of Europe, and half of North America. Although seas encompass only about 10%

Table 10.1 Relative Areas and Average Depths of the Oceans and their Adjacent Seas. Total World Ocean Area is 362 Million Square Kilometers.

Ocean or Sea	Percent of World Ocean Area (%)	Mean Depth (km)
Pacific Ocean	45.92	4.19
Asiatic Mediterranean (between S.E. Asia and Australia)	2.51	1.25
Bering Sea	0.63	1.49
Sea of Okhotsk	0.38	0.97
Yellow and East China Seas	0.33	0.27
Sea of Japan	0.28	1.67
Gulf of California	0.04	0.72
Total Pacific	50.09	3.94
Atlantic Ocean	23.91	3.74
Arctic Ocean	2.62	1.33
American Mediterranean (Gulf of Mexico and Caribbean Sea)	1.20	2.16
Arctic Mediterranean (between N.W. Territories and Greenland)	0.77	0.39
Mediterranean	0.69	1.50
Black Sea	0.14	1.19
Baltic Sea	0.11	0.10
Total Atlantic	29.44	3.30
Indian Ocean	20.28	3.87
Red Sea	0.13	0.54
Persian Gulf	0.07	0.10
Total Indian	20.48	3.84
World totals	100.00	3.73

Source: Data from Menard and Smith, 1966, *Journal of Geophysical Research* 71, 4305.

of the total area of the oceans (Table 10.1), they are snuggled up against land masses, so they dominate the coasts of many continents.

SURFACE CURRENTS

In shallow seas that are fairly isolated from oceanic influences, overall surface current patterns reflect the prevailing winds. Compared to oceans, the currents in seas are more strongly constrained by adjacent coastlines and more affected by water run-off from adjacent land. For reasons we have already seen, tidal currents are generally not as influential in seas as they are in the open oceans, but because of resonance effects, the North Sea and Gulf of California are exceptions.

Some seas, such as the Gulf of Mexico, Caribbean Sea, and Asiatic Mediterranean (between Southeast Asia and Australia), communicate well enough with the open ocean that the dominant surface currents are driven by ocean processes. Others, like the Arctic Ocean and Black Sea, are so isolated that the ocean has no direct influence at all.

CIRCULATION IN DEEPER BASINS

Because of the variety of influences exerted by winds, neighboring oceans, and continents, each sea has its own peculiar set of surface water properties and motions. Superimposed on these particular local patterns, however, are some general patterns influencing the circulation in seas with deeper basins. These general patterns are of two types, depending on whether or not the fresh water removed through evaporation exceeds that received through precipitation.[1]

Where evaporation is greater than precipitation, the surface water becomes saltier and sinks (Figure 10.26a). The sinking surface water carries dissolved oxygen that supports a relatively large population of deep animals. New surface water flows into the sea to replace that which is sinking, and the sinking waters continually flush the bottom water back into the neighboring ocean. The average length of time the water stays in the sea, between entering at the sur-

[1]The term *precipitation* includes precipitation received both directly and as run-off from a nearby land mass.

face and being flushed back out from the bottom, is called the **residence time**. The Mediterranean is an example of this kind of sea (Figure 10.26b), and its water has a residence time of about 80 years. Heavy evaporation makes the eastern end about 15 centimeters lower than the level at the Strait of Gibraltar.

A different situation exists in seas where precipitation exceeds evaporation. The lighter, fresher water remains on the surface. The bottom water does not get flushed, and it tends to become stagnant and dead. Once animal life has used up the dissolved oxygen, it does not get replenished by more from the surface. Furthermore, beyond a depth of a hundred meters or so, there is not enough sunlight for plants. The deep waters, then, are relatively barren of life. The situation is depicted in Figure 10.26c. Examples of this type of sea include the Black Sea and Arctic Ocean.

SEAS OF THE ATLANTIC

We now briefly describe the major seas of the world (Figure 10.27 and Table 10.1), organized according to the ocean with which they communicate. Seas of the Atlantic Ocean include the Mediterranean Sea, Caribbean Sea, Gulf of Mexico, and Arctic Ocean.

The Mediterranean Sea has two basins, roughly 4 kilometers deep, separated by a sill that is about 0.4 kilometer deep at the Strait of Sicily. The basin on the east has several narrow ridges and deeps, and the basin on the west is fairly flat. This pattern indicates that the Mediterranean may be compressed in the east (Africa getting closer to Europe) and stretched in the west (Africa getting farther away). The very thick salt deposits underlying the Mediterranean indicate periods of heavy evaporation and isolation from the Atlantic. The deposits are much thicker than could be explained if the Strait of Gibraltar closed off and the entire Mediterranean evaporated dry. Therefore, the equivalent of many complete evaporations must have occurred.

The Caribbean and Gulf of Mexico are sometimes referred to as the *American Mediterranean*. The Caribbean is confined to the south and west by the coasts of South and Central America and the Yucatan Peninsula, to the north by the Greater Antilles (Cuba, Hispaniola, Puerto Rico, etc.), and to the east by the Lesser Antilles. It communicates with the Atlantic primarily along the northern boundary of the South American coast and with the Gulf of Mexico through the Yucatan Straits. The Caribbean floor is complex, being divided by several ridges into four basins of

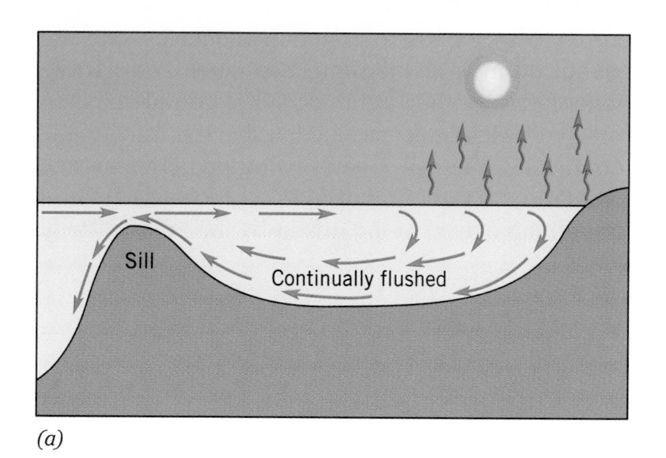

(a)

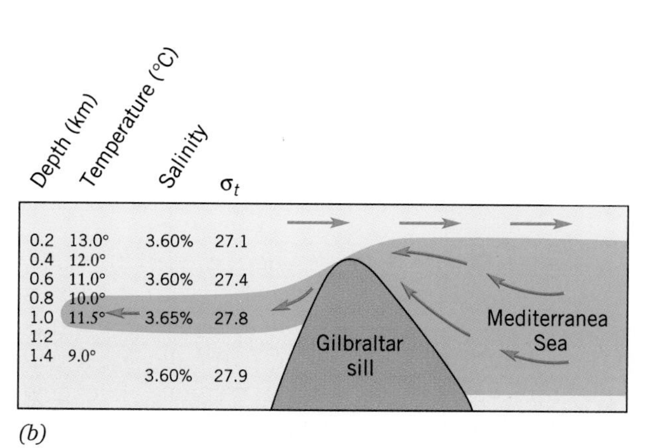

(b)

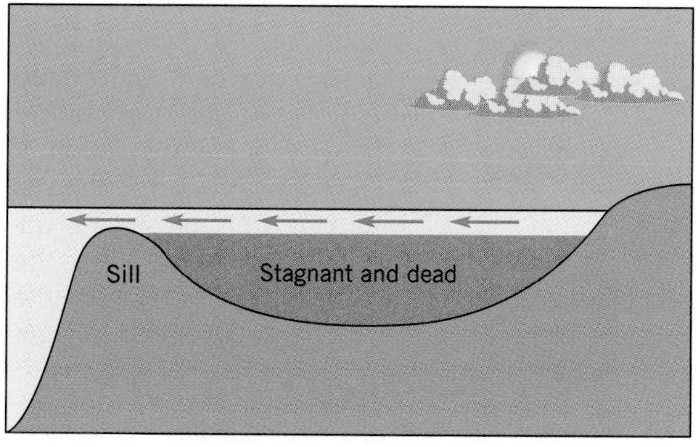

(c)

Figure 10.26

(a) In seas where there is large evaporation, surface water becomes saline and sinks, continually flushing out the deeper waters. (b) Spilling over the Gibraltar sill, Mediterranean water finds equilibrium about 1 to 1.5 kilometers down in the Atlantic. At this depth, it is both warmer and saltier than the surrounding waters of the Atlantic. (After P. H. Kuenen, 1963, *Realms of Water*, John Wiley & Sons, New York.) (c) Where precipitation is large, the surface water remains fresher and lighter than deeper water, and there is little vertical mixing. Once the oxygen is depleted from the deeper water, it is not replenished from the surface, and so it is lifeless.

about 4-kilometer depth each. This complexity probably reflects compression of the small plate that underlies this sea.

The Gulf of Mexico's floor is simple by comparison, having just one central basin that is about 3.5 kilometers deep. It communicates with the Caribbean through the Yucatan Straits and with the Atlantic through the Florida Straits. Both the Gulf of Mexico and Caribbean Sea are separated from the Atlantic Ocean by sills that are about 1 kilometer deep, which allow both surface currents and some deeper waters to enter and leave.

The dominant surface current in the Caribbean Sea and Gulf of Mexico is a continuation of the North Equatorial Current in the Atlantic, some of which enters the Caribbean along the northern edge of South America. This *Caribbean Current* flows on across the Caribbean and into the Gulf of Mexico between Cuba and the Yucatan Peninsula. This accumulation of surface water in the Gulf of Mexico creates a dome of water, whose central elevation is about 10 centimeters higher than that in the neighboring Atlantic. In addition to the inflowing Caribbean Current, the Gulf of Mexico also receives freshwater run-off from two-thirds of the continental Unites States.

The current from the Gulf of Mexico then flows "downhill" into the Atlantic, following a curvy path dictated by the coastline with the help of other influences such as the Coriolis effect and winds. It is rather diffuse in the western part of the Gulf, but intensifies as it approaches the straits of Florida. As it circulates clockwise around the Gulf of Mexico, it is called the *Loop Current*, but its name changes to the *Florida Current* as it flows around the Florida peninsula and back into the Atlantic. There it joins other northward-flowing waters to form the Gulf Stream.

The Arctic Ocean is the world's largest sea. It is bordered by extensive continental shelves, and the central basin is divided into two parts by a narrow ridge. Because the Arctic Ocean is completely surrounded by land and continental shelves (truly "mediterranean"), the deep water is trapped. The surface circulation occurs primarily in a clockwise fashion around the ocean, being driven by the polar easterly winds.

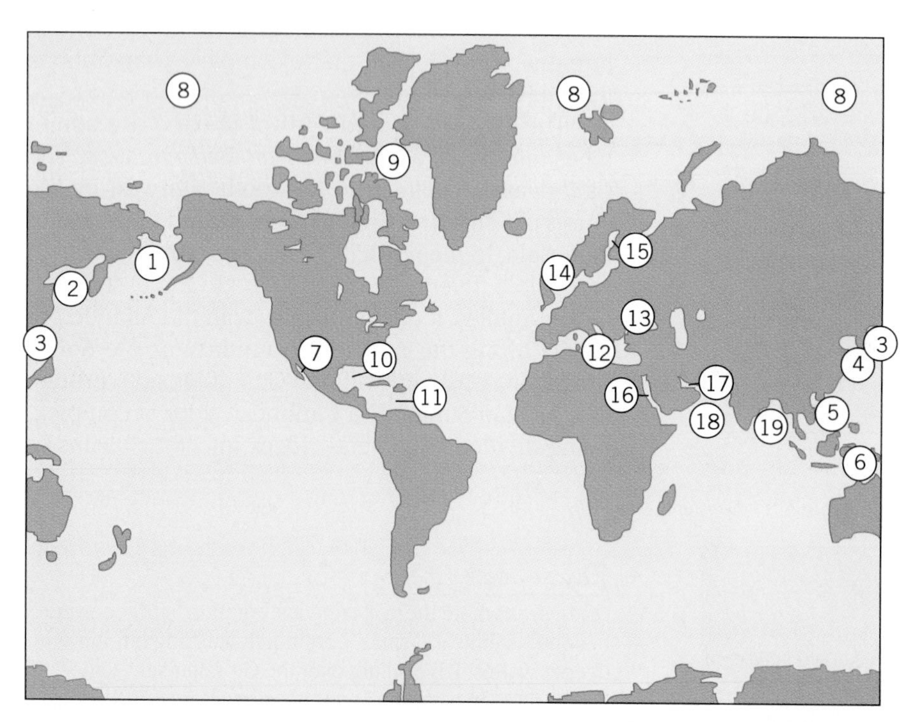

Pacific Ocean

1. Bering Sea
2. Sea of Okhotsk
3. Sea of Japan
4. Yellow Sea
5. South China Sea
6. Other Seas of the Asiatic Mediterranean
7. Gulf of California

Atlantic Ocean

8. Arctic Ocean
9. Arctic Mediterranean
10. Gulf of Mexico
11. Caribbean Sea
12. Mediterranean Sea
13. Black Sea
14. North Sea
15. Baltic Sea

Indian Ocean

16. Red Sea
17. Persian Gulf
18. Arabian Sea
19. Bay of Bengal

Figure 10.27

Map of the world indicating the locations of the major seas.

SEAS OF THE PACIFIC

Important seas of the Pacific Ocean include the Gulf of California, Bering Sea, and the string of seas that run down the East Asian Coast. These include the Sea of Okhotsk, Sea of Japan, Yellow Sea, and South China Sea.

The Gulf of California is long and narrow, reflecting the tectonic process that is tearing Baja California away from the mainland. It is only about 4-million-years-old and has several small shallow basins, reflecting its youth and tectonic activity. The Colorado River empties into the northern end, and the southern end is at the Tropic of Cancer. Circulation in this narrow sea is seasonal with the winds, and so are the upwelling and fishery patterns. The tides seem to resonate in this sea; the tidal range is about 1 meter at the mouth but up to 10 meters at the northern end.

The floor of the Bering Sea is dominated by the broad Alaskan continental shelf, but there is a small basin in the southwestern corner. The surface currents tend to flow counterclockwise, as is appropriate for the surface winds at these latitudes. The Polar easterlies drive the surface water northwest through the Bering Straits and into the Arctic Ocean, where they join the westward-flowing surface waters there. They are replaced by waters flowing into the Bering sea from the Pacific Ocean at the southwest corner. Most of the Bering Sea is covered by ice in the winter. In late spring, the ice melt creates a sharp halocline, and the bloom in microscopic organisms begins in the marginal zones between ice floes. As long as there is floating ice that dampens the waves, vertical mixing is reduced, and both the halocline and bloom remain near the surface.

The series of seas along the East Asian Coast lie between the mainland and offshore volcanic ridges. We have seen how the subduction of the Pacific Plate creates these volcanic features, which include Kamchatka, the Japanese Islands, and the Philippine Islands. The combination of the southward-flowing countercurrents, wind-driven surface currents, and an erratic coastline yield a fairly complex pattern of surface currents in these East Asian seas.

SEAS OF THE INDIAN OCEAN

The Red Sea is a new ocean opening up as Africa tears away from Asia. It is less than 20-million-years-old. Because its floor is a new hot, buoyant oceanic ridge, its maximum depth is only about 2.3 kilometers. There are some very hot salty brine pools on its floor, due to water exiting cracks in the hot ridge rocks. It is also in a very hot arid part of the Earth, so its surface waters tend to be warm and salty too. These are conditions favored by corals, so the shelves carry extensive coral reefs.

The Arabian Sea and Bay of Bengal are not as individually distinct as other seas, because they are not isolated from the main ocean by sills or narrow channels. The surface circulation in these seas is seasonal, being dominated by the seasonal Asian monsoon winds, but also influenced by the equatorial currents in the Indian Ocean. Surface currents are driven northeast by the summer monsoons and southwest by the winter monsoons, which cause a seasonal rise and fall in the coastal sea level of about 1 meter. The north end of both these seas receives considerable freshwater run-off from the continent. The Indus river flows into the Arabian Sea. The Ganges and Brahmaputra rivers flow into the Bay of Bengal, creating a huge delta and delivering a sediment load that dominates the floor of the entire sea.

Summary

The coast is a very dynamic place. It responds quickly to changes in wind, weather, waves, sea level, and other oceanic and continental processes.

Beaches

A beach may be made of a variety of materials. Even the composition of the same beach changes with position and the season. Sand stays on the beach, because the orbital motion of incoming unbroken waves carries the bottom sediments up and forward as each wave passes overhead. Within the surf zone, the swash carries sand up the beach, and the backwash carries it back down. In periods of light waves, the swash dominates and the berm grows. When wave activity is heavy, however, the backwash dominates, and sand is removed from the berm and deposited in longshore bars. Coarse beach materials and steep beaches are associated with heavy wave activity.

Longshore Sand Transport

Because incoming waves are often not yet quite parallel to the shoreline when they break, there is a zigzag motion of the water in the surf zone. This results in a net longshore current along the beach. Rip currents are found where longshore currents converge.

The longshore current carries sediments with it, and this results in a surprising amount of sand transport along our beaches. This littoral drift is compartmentalized. The beaches at the upstream end of any compartment are narrow, but they grow wider further downstream due to the addition of more sediments. A compartment ends with an obstruction, such as a headland or submarine canyon. The sand usually ends up in dune fields, extensions of barrier beaches, or by getting flushed out to deeper water or down some submarine canyon.

Human interference with the longshore flow of sand can have disastrous effects, including the rapid erosion of downstream beaches. Along jagged coastlines, the sand tends to stay in suspension past the points and deposit in baymouth bars in front of the bays.

Processes Affecting Overall Coastal Features and Coastal Waters

Along flooded coastal plains, abundant sediments and shallow waters support the formation of extensive barrier islands protecting shallow lagoons. Drowned river valleys are also common. Mountainous coasts change more abruptly, estuaries tend to be smaller, and sediments are not so abundant. Tectonic-formed estuaries and fjords are among the kinds of estuaries found on these coasts. Coastal waters are distinctly different from those farther out to sea for a variety of reasons.

Estuaries

The flood and ebb of tides in estuaries are generally delayed relative to those of the neighboring ocean. Strong tidal currents often flow through tidal inlets. Tides entering estuaries from the ocean are slowed by shallow water and outflowing currents. In estuaries, fresh water enters at the head and mixes with the salt water from the ocean. There is a wide variation in types of estuaries, varying from those that are deep and highly stratified, to those that are shallow and vertically well mixed. Estuaries are places where nutrients accumulate. This fosters flourishing plant life, which, in turn, supports vigorous animal populations, including juveniles of important marine species. Estuaries are also places where human populations concentrate.

Seas

Seas are very large bodies of water. Some are heavily influenced by neighboring oceans, and others are not at all. Surface currents are driven by a variety of forces. The general deep circulation pattern in seas depends on whether evaporation exceeds precipitation. If it does, then the surface water is continually sinking and flushing out the bottom water. If not, then the surface water remains on the surface, and the deep water stagnates. The Mediterranean, Caribbean, Gulf of Mexico, and Arctic Ocean are important seas of the Atlantic. The Gulf of California, Bering Sea, and string of seas along the East Asian Coast are important seas of the Pacific. The main seas of the Indian Ocean include the Red Sea, Arabian Sea, and Bay of Bengal.

Key Terms

beach
coast
backshore
foreshore
offshore
low tide terrace
beach face
berm
berm crest
shoreline
swash
backwash
scarp
armored beach
longshore transport
littoral drift
longshore current
rip current
dune fields
beach compartment
groin
seawall
breakwater
jetty
revetment
beach nourishment
feeder beaches
baymouth bar
spit
tombolo
barrier island
lagoon
inlet
delta
estuary
flushing time
tidal volume
head
mouth
stream-cut channel
flooded coastal plain
mountainous coast
drowned river valley
tectonic-formed estuary
fjord
flood tide
ebb tide
slack water
tidal bore
salt wedge
highly stratified
partially mixed
well-mixed
bar-built estuary
inverse estuary
wetlands
tidal flats
salt marshes
mangrove swamps
residence time

Study Questions

1. Why does sand stay on the beach? What seasonal changes would you expect on a sandy beach, and why?

2. Why do protected beaches tend to have finer sediments than unprotected beaches? Why do beaches with coarser sediments tend to have steeper slopes?
3. With the aid of a sketch, show why rip currents might be more common on a curved beach than a straight one.
4. How does the refraction of incoming waves help straighten a jagged coastline? How does the longshore sand transport also help?
5. Large changes in sea level are caused primarily by what two types of processes? What are the time scales associated with each?
6. What processes remove sand from our beaches? What processes add sand to the beaches?
7. Barrier islands are found on coasts with what characteristics? Where did the sand on the U.S. East Coast come from (several possible sources)?
8. Describe some of the origins of coastal waters, and explain why they are different from those surface waters farther out to sea.
9. What are some of the reasons why you would expect the temperature, salinity, and turbidity of estuarine waters to vary more than those of the open ocean?
10. What kinds of estuaries are common along flooded coastal plains? How are they formed?
11. How would you expect the beaches and estuaries along mountainous coasts to differ from those of flooded coastal plains?
12. With the help of a diagram, describe the salt wedge model of estuarine circulation.
13. Why are most estuaries exceptionally productive biologically?
14. What are some of the economic motives for humans to congregate around estuaries?
15. Explain how the relative rates of precipitation and evaporation determine the overall circulation patterns in a sea.
16. Describe the flow of surface currents through the Caribbean Sea and Gulf of Mexico.

Critical Thinking Questions

1. During a storm, a barge full of sand overturns near the shore but out beyond the breaking waves. Describe what happens to that sand, and where it ends up. A week later, the storm waves have passed, and gentle waves return. What happens to that sand then?
2. Other things being equal, would it be wisest to build a harbor at the upstream end, downstream end, or a central portion of a beach compartment? Why? If we need to dam up a stream for a water reservoir, we would be wisest to pick a stream flowing into which part of a beach compartment? Why?
3. Explain how a breakwater built offshore and parallel to shore can interrupt the longshore transport of sand.
4. What is an inverse estuary? Would you be more likely to find one where the climate is warm and dry, or where it is cold and wet? Why? Would an inverse estuary more likely be deep or shallow? Why?
5. Discuss some of the ways that human activities impact estuaries.

Suggestions for Further Reading

BIRD, E. C. F. 1985. *Coastline Changes: A Global Review*. John Wiley & Sons, New York.

BRINK, KENNETH H. 1992. Coastal Physical Oceanography. *Oceanus* 35:2, 86.

DOLAN, R. AND H. LINS. 1987. Beaches and Barrier Islands. *Scientific American* 257:1, 68.

HECHT, J. 1988. America in Peril from the Sea. *New Scientist* 118:1616, 54.

KETCHUM, B. H., ED. 1983. *Estuaries and Enclosed Seas*. Elsevier Scientific Publishing, New York.

LASCOMBE, HENRI. 1990. Water, Salt, Heat, and Wind in the Mediterranean Sea. *Oceanus* 33:1, 26.

LAWRENCE, S. 1987. How to Feed a Beach; Where the Sand Comes from and Where It's Going. *Oceans* 20:2, 42.

MORRISSEY, S. 1988. Estuaries: Concern over Troubled Waters. *Oceans* 21:3, 22.

PILKNEY, ORRIN H. 1990. Barrier Islands; Formed by Fury, They Roam and Fade. *Sea Frontiers* 36:6, 30.

REYNOLDS, E. V. 1988. Beachless Summer. *Discover* 10:1, 38.

WANLESS, H. R. 1989. The Inundation of Our Coastlines. *Sea Frontiers* 35:5, 264.

Daily production by microscopic marine plants is typically 40 to 50% of their total mass, and the average life expectancy of a marine plant is just a few days.

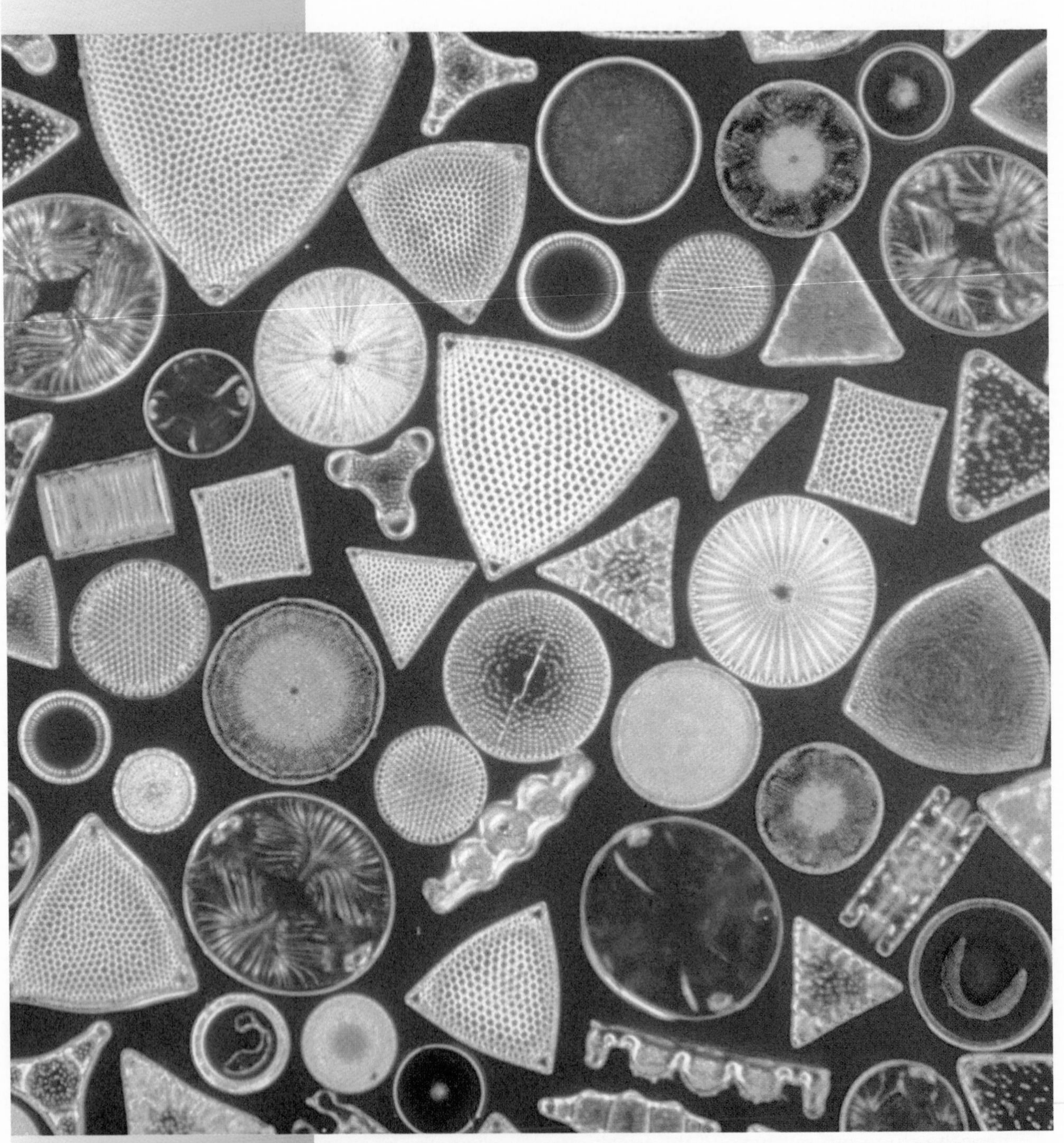

Magnified view of the external skeletons of various diatoms (microscopic plants).

LIFE AND THE PRODUCTION OF FOOD

The same surface water that gives the Earth its unique appearance and moderate climate also nurtures an elaborate and intricate web of life. Over the ages, the organisms and their interrelationships have become extremely complex. But the basic biological processes, and the role of water in these processes, have remained unchanged.

Virtually all life on Earth is supported by plants (Figure 11.1), which have the ability to store the energy of sunlight in the food they produce. Plants tend to produce food in excess of their own immediate needs, which has enabled them to survive hardships such as severe winters, prolonged darkness or drought, or excessive grazing by animals. This excess food also supports a variety of animals, such as ourselves. We animals derive the energy for our activities from the food that was originally produced by the plants. Animals that feed directly on plants are called **herbivores**, and animals that feed on these herbivores and other animals are **carnivores**. Each level through which the food is passed is called a **trophic level**, with the plants being the zeroth trophic level, the herbivores the first, and so on.[1]

Although plants use energy from the Sun, a very interesting but minor amount of food production in the ocean is carried out by bacteria that use certain forms of chemical energy. In order to include these organisms in our discussion of food production in the oceans, and avoid some controversy regarding whether certain kinds of microscopic food producers should really be called "plants," we often refer collectively to all organisms that produce food as the **primary producers** and the food (i.e., organic matter) they produce as **primary production**. The rate at which food is produced is referred to as **productivity**.

On land, most plants are large, with sizes measured in centimeters or meters. By contrast, the overwhelming mass of primary producers in the ocean are microscopic single-celled organisms (Figure 11.1b). The **cell** is the fundamental biological unit of all organisms, large or small. A cell is separated from its environment by a **cell membrane**, through which it both receives the materials it needs and discards all its waste products. Within each cell is encoded in chemical complexes all the information needed to guide the cell to carry out all its functions.

The sizes of these tiny single-celled marine organisms are measured in millionths of a meter. In the metric system, the prefix for a millionth is "micro" and has the symbol μ, so a millionth of a meter is a **micron** or **micrometer** and has the symbol μm. The size of a typical single-celled marine plant is in the range of 5 to 50 micrometers, which is roughly the width of a hair. We will learn the reason for these small sizes later in this chapter.

All life in the ocean is dependent on food production. When productivity is high, there is vigorous biological activity of many forms, and when productivity is low, there is little biological activity of any kind. Therefore, to understand the overall patterns of life in the ocean requires a basic understanding of food production by plants. For this reason, we begin our study of marine biology by looking at the very basic processes involved in food production and consumption. This will provide the foundation for a more detailed study of marine organisms in the following two chapters. In Chapter 12, we look at the different kinds of marine organisms and how they function, and in Chapter 13, we examine the various marine environments and the lifestyles of their inhabitants.

[1]For the numbering of trophic levels, we use the convention of J. H. Ryther, 1969, *Science* 166:3901, 72–76.)

11.1 ORGANIC SYNTHESIS

We begin our study of marine biology with very basic processes. In particular, we wish to know how food and other organic matter are produced, how energy is stored within them, and how this energy is released when needed by an organism at a later time.

PHOTOSYNTHESIS AND RESPIRATION

Through a process called **photosynthesis**, plants convert incoming solar energy into chemical energy that is stored in the organic matter that plants produce. This energy is released when the organic matter is burned through **respiration**, either by the plants themselves or other organisms at higher trophic levels to whom this organic matter is passed. In the oceans, very few rays of sunlight penetrate beyond the first hundred meters of water, so marine plants can survive in the surface waters only.

A few bacteria can manufacture food using energy from inorganic chemical reactions rather than sunlight. The production of food in this manner is called **chemosynthesis**. Although it is the mechanism for food production in the vent communities of the dark ocean floor, chemosynthesis accounts for only a very minor fraction of the total food produced in the oceans.

Scientists normally measure productivity in terms

(a)

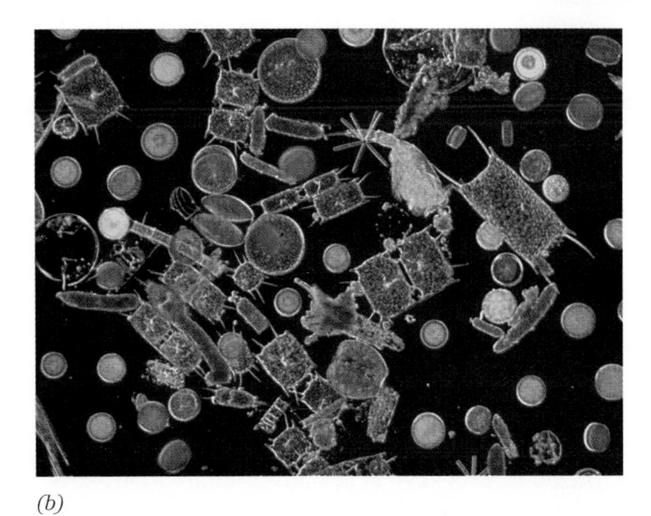

(b)

(c)

Figure 11.1

Plants produce the food that supports all life on Earth. Virtually all primary productivity on land is accomplished by large plants (a) such as these. This is in stark contrast to the ocean environment, where virtually all primary productivity is accomplished by microscopic single-celled plants (b) such as these. Seaweeds (c) such as these require shallow water for sunlight penetration and firm substrate for anchorage by their holdfasts. Such rocky coastal areas are only a tiny fraction of the total ocean area, so seaweeds make up only a tiny (although quite visible) fraction of the total marine plant population.

of grams of carbon that have been converted into organic matter per square meter of ocean area per year. Typical values lie in the range between about 50 and 300 gC/m^2/yr.[2] Interestingly, the productivities of land plants fall within nearly the same range, with deserts being near the lower end and irrigated agriculture near the top.

CARBOHYDRATES

The immediate products of photosynthesis are molecules made up of hydrated carbon atoms, called **carbohydrates** (to **hydrate** means to add water). Figure 11.2 illustrates what happens on a molecular level. The energy from sunlight is used to remove oxygen (O_2) from carbon dioxide (CO_2), leaving bare carbon (C) atoms. The carbon atom by itself is quite reactive, so in order to be safely stored for later use, it is "placated" by attaching water molecules to it. Symbolically, this process may be represented as

$$CO_2 + H_2O + [\text{energy}] \longrightarrow C(H_2O) + O_2$$

The end product of photosynthesis, then, is free oxygen (O_2) and carbohydrates. In the ocean, the released oxygen dissolves in the water. The carbohydrates produced by plants are slightly more complex than the single hydrated carbon atom, $C(H_2O)$, which is used here for illustrative purposes. For example, the "sucrose" sugar molecule is six hydrated carbon atoms ($C_6H_{12}O_6$), and the molecules of common table sugar are made of 12 carbon atoms hydrated with 11 water molecules ($C_{12}H_{22}O_{11}$).

Energy that is stored in hydrated carbon atoms can be released later by burning, or **oxidizing**, the carbon. This is what happens when gasoline burns in your car engine, for example (Figure 11.3). The slow controlled burning of carbon within the cells of organisms is called *metabolism* or *respiration*. It can be represented symbolically as

$$C + O_2 \longrightarrow CO_2 + [\text{energy}]$$

[2]The extremes range from less than 1 gC/m^2/yr to over 1250 gC/m^2/yr.

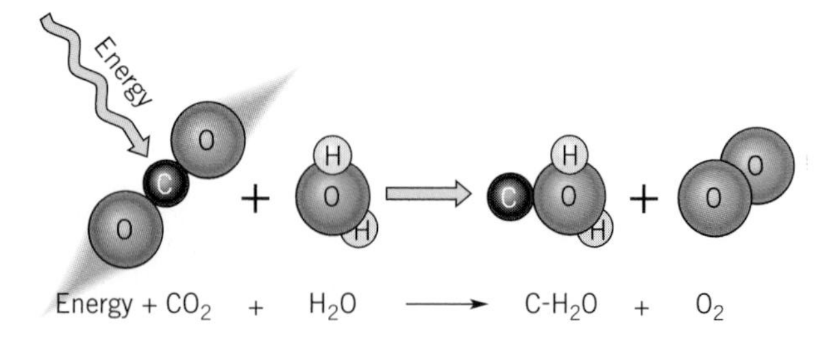

Figure 11.2

In photosynthesis, solar energy is used to remove oxygen (O_2) from carbon dioxide (CO_2). Then water (H_2O) is attached to the carbon atom (C) to form hydrated carbons, or *carbohydrates*, $C(H_2O)$. The solar energy is now stored and ready to be released when the carbohydrates are burned.

or, if we include the water attached to the carbon in the carbohydrate,

$$C(H_2O) + O_2 \longrightarrow CO_2 + (H_2O) + [\text{energy}].$$

Notice that this energy-release process is just the reverse of the energy-storage process of photosynthesis. You can feel the energy released when standing next to a roaring fire, or sitting next to a student who has run all the way to class, burning carbohydrates in doing so. It is the constant controlled burning of these carbohydrates that keeps your body temperature above that of your surroundings and provides the energy for all the things you do. All this energy came originally from the Sun and was stored for us in the food produced by plants.

Figure 11.3

The energy for these lights, these automobiles, and for the people who live here all comes from hydrocarbons that were originally produced by plants.

MORE COMPLEX MOLECULES

Although carbohydrates provide the basic energy source, more complicated molecules are also needed by organisms to help them carry out their various functions. For example, particular catalysts, called **enzymes**, are needed within the cells to help carry out particular chemical reactions, such as photosynthesis or respiration. Special materials may also be needed for structure, muscle contraction, circulation, nerve signals, food storage, reproduction, and so on.

Most of these more complicated organic molecules are constructed from large numbers of carbohydrates, with the addition of small amounts of other materials. Most fall into the general categories of **lipids** (e.g., waxes and fats) or **proteins**. Like carbohydrates, these materials can also be oxidized (i.e., burned) to release energy, although carbohydrates are usually burned first. Scientists normally describe the very complex protein structures in terms of *amino acids*, which are the basic building blocks from which they are constructed. A protein molecule may contain several thousand of these building blocks, each of which is much more complicated than a typical carbohydrate molecule.

The complexities of life on Earth are directly linked with the complexities of these organic molecules, which in turn are due to the versatility of the carbon atom. Each carbon atom can share four electrons with neighboring atoms. This property enables carbon atoms to form very large molecules, such as long chains of carbon atoms, or **carbon chains**, to which other things are attached. (See Figure 11.4.)

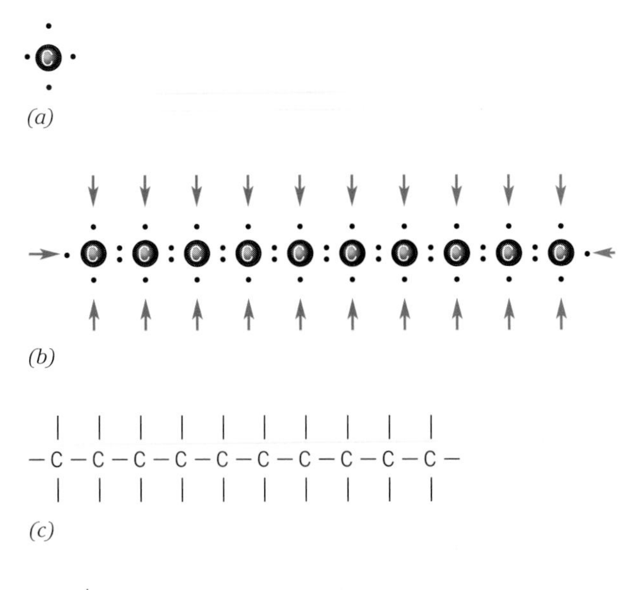

Figure 11.4

The chemical versatility of the carbon atom makes it an ideal basis for the large, complicated organic molecules of life. (a) The carbon atom has four electrons in its outer shell, which is four short of the eight and four more than the zero needed for the preferred ideal gas electronic configurations (Chapter 2). Consequently, it eagerly shares its four electrons with those of other atoms, making chemical bonds. (b) One thing the carbon atom can do is form long chains. Each carbon atom shares an electron with the carbon neighbor on each side, which leaves two electrons for other bonds (indicated by arrows). (c) In carbon chemistry, the sharing of electrons with a neighbor is indicated by a line. So this chain is the same as the one above. (d) Another common configuration of carbon atoms is the hexagonal ring.

OTHER INGREDIENTS

Unlike their land counterparts, marine plants are bathed in unlimited supplies of both water and dissolved carbon dioxide, which are the two main ingredients in the synthesis of organic matter. All the other ingredients together make up only a few percent of organic matter by weight (Table 11.1).

Of these remaining ingredients, nitrogen and phosphorus are usually in highest demand. Nitrogen is particularly important in the formation of amino acids and other complex proteins. Although the nitrogen molecule N_2 is the major constituent of our atmosphere and also is amply abundant in sea water, this form is not usable by most plants in photosynthesis. They require it in an oxidized form, normally as nitrate (NO_3^-), sometimes as nitrite (NO_2^-), or in some cases as ammonia (NH_3). By contrast, phosphorus is relatively scarce in the biosphere, but in the ocean it is normally encountered in the oxidized phosphate (PO_4^{3-}) form that plants need.

Together, nitrogen and phosphorus make up only a few percent of organic matter. Within typical plant material, carbon, nitrogen and phosphorus are found in the ratio

$$C:N:P = 100:17:3 \qquad \text{by weight,}$$

or

$$C:N:P = 100:15:1 \qquad \text{by number of atoms}$$

However, they are essential. Their scarcity in the marine environment severely restricts plant productivity. Consequently, the addition of small amounts of ni-

Table 11.1 Ingredients in the Production of Organic Matter, Listed According to Fraction by Weight of Typical Plant Materials

Material	% of Total Weight
Major Ingredients	
Oxygen	62–66
carbon	18–21
hydrogen	8–10
Minor Ingredients	
1. Nitrogen, phosphorus	1–5
2. Calcium, potassium, sulfur, sodium, chlorine, magnesium	0.05–1
3. Boron, manganese, zinc, silicon, cobalt, iodine, fluorine, iron, copper, etc.	Less than 0.05

trate or phosphate may stimulate a large amount of biological activity. For this reason, they are referred to as *nutrients*.

Many other materials are needed in even smaller quantities. These include some common and easily accessible salt components, such as potassium (K), calcium (Ca), sodium (Na), sulfur (S), and chlorine (Cl), whose abundance in sea water is more than sufficient to meet plants' needs. But there are some trace materials in sea water whose abundance is sometimes not sufficient to meet the particular needs of certain plants. When this happens, the lack of the particular trace material can restrict the plant's productivity just as much as lack of nutrient nitrogen or phosphorus. This particular vital and restrictive material is called a **micronutrient** for that particular plant. For example, the microscopic plants called *diatoms* require silicate (SiO_4^{4-}) for the production of their tiny silica (SiO_2) external skeletons, or *exoskeletons* (chapter opening photo and Figure 11.1b). Silicate is not very abundant among the materials dissolved in the ocean, and the lack of it can be as limiting to diatom productivity as nitrate or phosphate. Consequently, silicate is a micronutrient for diatom productivity.

THE EQUATION OF LIFE

The **equation of life** summarizes the above description of the production and use of food on Earth:

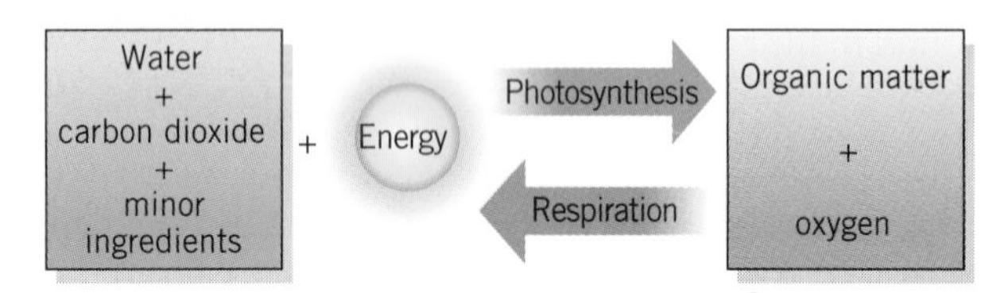

In photosynthesis, the energy used is sunlight. When plants and animals respire, the energy released is used in their own biological processes, such as locomotion, tissue production or repair, digestion, body heat, or reproduction.

11.2 NUTRIENTS

The equation of life represents a cycle for the materials involved—mostly carbon dioxide and water, but with small amounts of nutrients and other materials. These are incorporated into organic matter during photosynthesis and then are returned to the environment during respiration, whereupon the cycle can be repeated.

NUTRIENT SCARCITY

In surface waters, where light is available for photosynthesis, it is the scarcity of nutrients that usually limits productivity. Once consumed, nutrients are unavailable for use by another plant until the original organic matter decomposes. Thus, the productivity of an area depends on the nutrients getting returned to the environment and recycled as quickly as possible.

Surviving organisms have become proficient at collecting and hoarding needed scarce materials.Just as we humans must store the fall harvest to survive the winter, plants tend to hoard scarce materials in order to survive periods when these materials are unavailable. Unfortunately, this crucial ability to hoard also increases the scarcity. The more that is hoarded by some, the less that is available for others. As we learned in Chapter 6, the ability to hoard also frequently extends to trace materials that are not needed and may even be harmful to the organisms. Even materials that are needed in small quantities may be toxic in large quantities. In polluted coastal waters, relatively small concentrations of these harmful materials may cause widespread devastation in biological communities (Figure 6.10).

Just as we humans must store the fall harvest to survive the winter, plants tend to hoard scarce materials in order to survive periods when these materials are unavailable.

RELEASE FROM ORGANIC DETRITUS

The scarcity of nutrients is made even worse by the fact that acquired nutrients are only very reluctantly returned to the environment. When in need of energy, organisms tend to burn the carbohydrates first and the lipids next. Only when in dire need of energy and their carbohydrate and fat stores have been largely exhausted, do they begin oxidizing protein, in which the bulk of the nitrogen and phosphorus is stored.

Some nutrients become available through the excretions of organisms, especially when feeding is heavy and digestion incomplete. About half the nitrogen excreted by animals is in the form of ammonia, which tends to dissolve directly from the excretions and get back into the food cycle rather quickly. However, many nutrients must await death and decomposition of the organism, or decomposition of the excretions, before they are converted into a form usable by most plants.

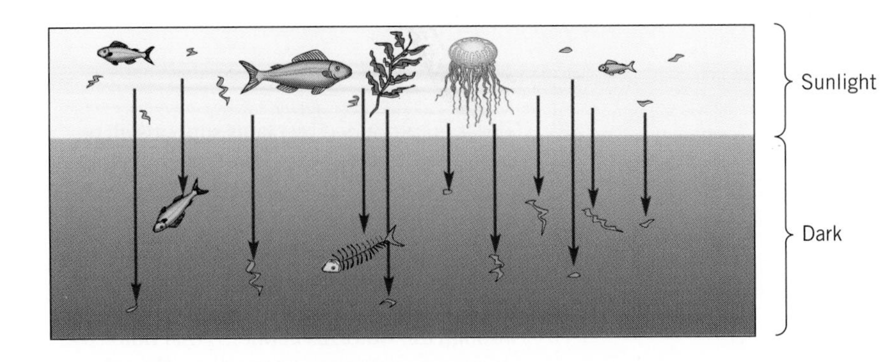

Figure 11.5

The remains of surface organisms (mostly microscopic) sink into deeper waters before decomposing. In this manner, nutrients are removed from the sunlit surface waters and released in the dark deeper waters. But because sunlight is needed for photosynthesis, these nutrients cannot be used again for producing organic matter until some process returns them to the surface.

The release of nitrate is particularly slow, requiring a three-step process involving different bacteria at each step. Much of the dead and decaying organic matter, called **detritus**, has sunk below the productive sunlit surface waters before the nitrate is released (Figure 11.5). It is of no use in these dark deeper waters, because sunlight is needed for photosynthesis. It may be many, many years before it can return to the surface, where plants can make use of it again.

Phosphate, on the other hand, requires only a one-step bacterial process to be released from organic detritus. Consequently, phosphate is released relatively quickly, often before the detritus has sunk beneath the sunlit surface waters, so it can be recycled quickly to the next generation of plants. Furthermore, there are other sources of phosphate in the oceans besides the decay of organic matter. For these reasons, it is usually nitrate, rather than phosphate, that most severely restricts primary productivity in the ocean.

In summary, of all the materials needed in photosynthesis, nutrients are in the smallest supply relative to their demand. All life in the oceans is dependent on food produced by plants. The ability of plants to produce this food is entirely dependent on getting the nutrients out of the detritus and back into the surface waters where they are again available for photosynthesis. Therefore, in dealing with the productivity of an area, we are particularly concerned with the **nutrient cycle**, especially the nitrate cycle.

11.3 MEASURING PRODUCTIVITY

We now combine this background on the basic chemistry of food production with our knowledge of ocean circulation that we learned in previous chapters. Together, they should help us understand the overall patterns of primary production in the oceans, as determined from a variety of measurements. We first study these measurement techniques and then examine the production patterns that they reveal.

LIFE SAMPLES

Since plants support the entire biological community, we can recognize the productive regions of the oceans by the abundance of life of all forms (Figure 11.6). For example, the distribution of fishermen or pelicans would be some reflection of primary productivity, because they depend on fish, which in turn depend on the production of food by plants. Likewise, the distribution of worms or bottom scavengers would reflect the productivity of the surface waters above. Higher productivity means more organic detritus falling to the bottom, which supports more animal life on the ocean floor.

For accurate measurements of primary productivity, however, scientists need something better than a simple measure of total biological activity (e.g., total mass of live organisms per cubic meter of water). Most animals engage in a variety of behaviors that affect their distribution. Porpoises, salmon, and sea gulls, for example, don't spend all their time feeding. They have other activities that remove them from the good food areas. Consequently, when samples of marine life are taken, they are usually done with other objectives in mind, such as determining the extent of a species, its relative importance in the local biological community, its migratory habits, and so on.

Even sampling the plants themselves is not particularly accurate, because plants in some waters may be very productive, whereas those in other areas may be inactive or even dormant. Nonetheless, there is a strong tendency for the abundance of plant life to correlate with productivity. Consequently, one common way to estimate the productivity of a region is to count its plants, nearly all of which are microscopic and single-celled. These plant counts are usually recorded in terms of number of plant cells per liter of sea water. Of course, we can't really count the number of microscopic plants in a liter, because the numbers are too large and the plants are too small.

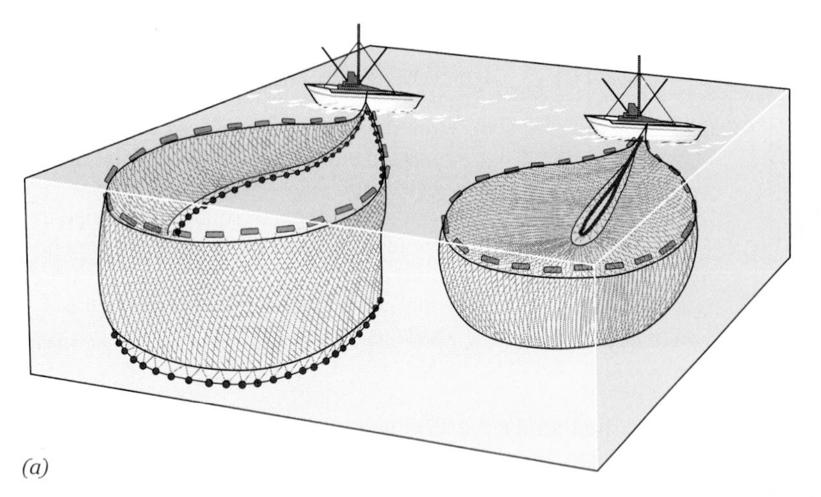
(a)

Figure 11.6

The food produced by plants supports all organisms. Therefore, the productivity of plants in the surface waters is reflected in the abundance of organisms of all types, including fish, seagulls, and fishermen as in photo (b) and including microscopic organisms that are sampled by dragging very fine "plankton nets" through the water, as in photo (c). Photo (b) shows a commercial purse seiner. The seine is laid around a school of fish. A "purse string" in through the bottom is drawn to prevent escape, and then the net is hauled in (a).

(b)

(c)

Rather, we count the number in a single drop of water on a microscope slide and multiply this by the number of drops in a liter. Alternatively, we can force a liter of sea water through a fine filter and divide the mass of the plant matter collected by the average mass of a single microscopic plant. Sometimes, the abundance of plant life can be inferred from the concentrations of certain chemicals associated with the plants, such as these described in the next section.

SATELLITE MEASUREMENTS

A high-tech method for estimating plant productivities uses Earth-orbiting satellites to determine the abundance of chlorophyll, which is the material that gives photosynthetic plants their characteristic green color. When observing an area of the ocean surface, the satellite measures the intensities of certain colors that are enhanced by chlorophyll. From the amount of enhancement of these particular colors, we can determine the amount of chlorophyll present.

This method has the same disadvantage as the method of counting microscopic plants, because although it gives an indication of the amount of plant material present, it does not directly indicate how active these plants are in producing food. Furthermore, only the plants in the top 2 meters of the ocean can be observed from these satellites. On the other hand, this method has the very important advantage that measurements can be made over the entire ocean surface and several times a day. Therefore, it is excellent for measuring global patterns and changes in those patterns.

Figure 11.7 gives satellite data on the abundance of chlorophyll. These results correspond well with more direct measurements of productivity, both on land and in the ocean.

DISSOLVED OXYGEN

A more direct method of determining the productivity of microorganisms in an area is to monitor the dissolved oxygen in a water sample in a closed clear bottle that is lowered to the desired depth. Oxygen is released by photosynthesis and removed by respiration. We then place an identical water sample in a second bottle, the walls of which are painted black

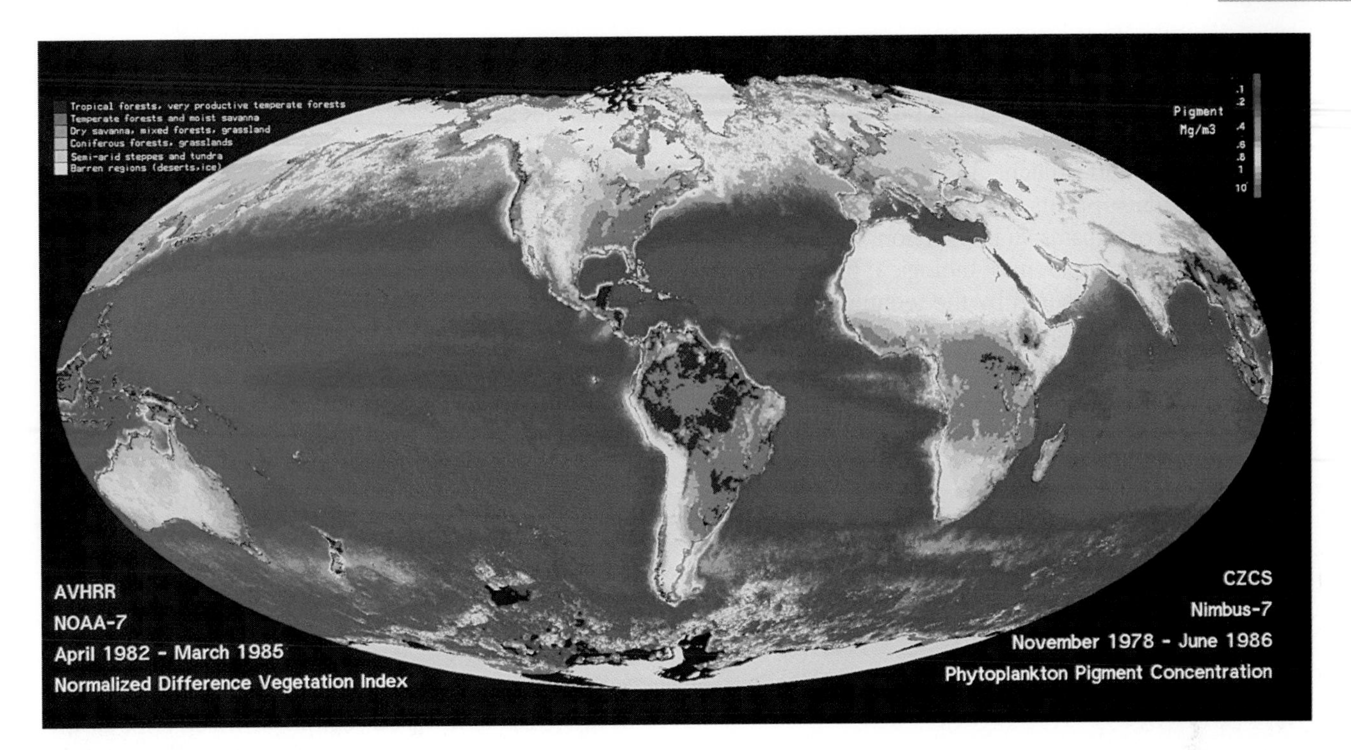

Figure 11.7

Plant productivities, as inferred from satellite data on chlorophyll concentrations. On land, the darkest greens indicate the highest productivity. In the ocean, reds (followed by yellow) indicate the highest productivity, and violet the least. On land, high productivities are correlated with rainfall and sunshine. In the oceans, they are associated with continental shelves and upwelling, where nutrients get recycled quickly.

so that no sunlight can get inside (Figure 11.8a). In total darkness, no photosynthesis at all can occur, so the rate of depletion of dissolved oxygen in this second bottle is due to respiration alone. By comparing the net change in oxygen in the clear bottle to that lost to respiration in the dark bottle, we can determine how much photosynthesis had occurred.

For example, suppose we measure a net gain of 5-milliliters of oxygen in the clear bottle and a depletion of 3-milliliters in the black bottle. That tells us that the 5 milliliter gain was *in spite of* a 3-milliliter loss to respiration. Photosynthesis actually produced 8-milliliters of oxygen, 3-milliliters of which was consumed by respiration, leaving us with the measured net gain of 5 milliliters. By analogy, if the number of students in a room increases by five, while three students leave, then we know that eight have entered. This **light–dark bottle** system was used to obtain the data displayed in Figure 11.8b, which shows the productivity of diatoms in surface waters for different times of day and different seasons.

CARBON-14

Another method for measuring the productivity of a sample involves using a radioactive isotope of carbon, C^{14} (carbon-14). A measured amount of carbon dioxide, made ("labeled") with C^{14}, is dissolved in the water. After a certain amount of time, some of this carbon will have been assimilated into new organic tissue. The rest is then driven off (e.g., by heating). From the measured radioactivity remaining in the sample, we know how much of the radioactive carbon has been incorporated into the plant tissues, and therefore, we know the rate at which the production of food has occurred.

11.4 GENERAL PATTERNS IN PRODUCTIVITY

Photosynthesis requires both sunlight and nutrients, but these two resources tend to be found in different places. In the surface waters where sunlight is plentiful, nutrients are usually scarce. And in deeper waters where nutrients accumulate, sunlight is lacking. Consequently, most of the ocean's waters are rather barren and lifeless—comparable to deserts on land.

In some surface regions, however, nutrients are plentiful. These tend to be either regions where the water is shallow (e.g., continental shelves) and the nutrients cannot sink far, or regions where deep water rises to the surface, bringing their nutrients with them. In either case, plants thrive on the combination of nutrients and sunlight, and the food produced by these plants supports a vigorous community of many kinds of organisms. In this section, we try to understand the overall pattern of biological productivity in the oceans as measured by the techniques of the previous section. To do this, we must understand where we would find large concentrations of nutrients in the surface waters. We now turn our attention to this question.

UPWELLING

As we have seen, upwelling of nutrient-rich water occurs frequently along continental margins, where winds drive surface currents out to sea and deeper waters surface to replace them. It also occurs in some places at sea where surface currents diverge, again meaning that deeper waters are rising to replace those departing at the surface. (See Figure 11.9.)

The overall large-scale patterns of upwelling and downwelling in the oceans are understood rather simply in terms of the major current gyres and their Coriolis deflection. As we learned in Chapter 8, the central portions of the main current gyres in all oceans are regions of converging surface waters and gentle downwelling. Conversely, the outer perimeters of the gyres tend to be regions of divergence at the surface and upwelling from beneath (Figures 8.11 and 10.18). These regions of upwelling along the gyres' outer perimeters include the **equatorial divergence** along the equator and **coastal upwelling** along the continental margins (Figure 11.10a). At high southern latitudes, the prevailing westerlies drive the Antarctic Circumpolar Current, or West Wind Drift, completely around the globe. The northward Coriolis deflection of this current creates the

P—R R

(a)

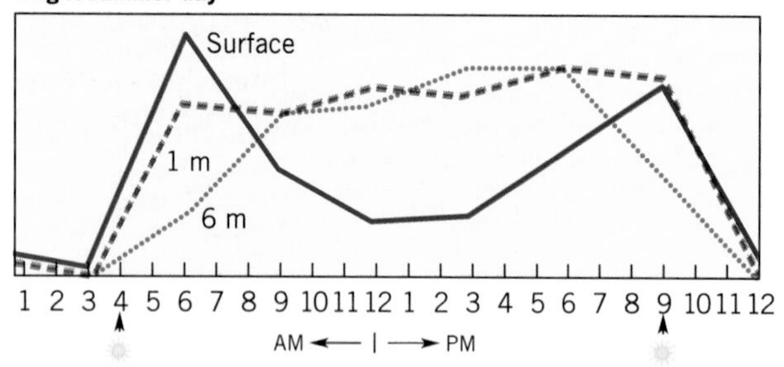

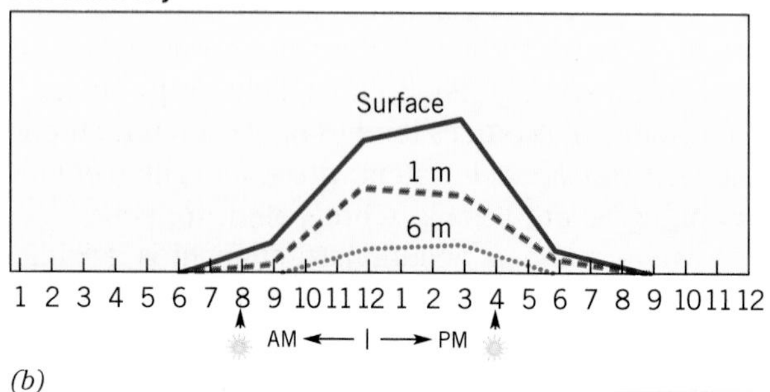

(b)

Figure 11.8

(a) Two bottles are needed when measuring oxygen production to determine the rate of photosynthesis. In the dark bottle, only respiration is occurring, so the measured oxygen depletion in this bottle yields the rate of respiration *R*. In the clear bottle, both respiration and photosynthesis are occurring, so the measured increase in oxygen concentration represents the difference between the oxygen produced in photosynthesis and that consumed in respiration (*P–R*). Since the two bottles give us *R* and *P-R*, respectively, we can add the two together to get *P*, the rate of photosynthesis. (b) Productivity of diatoms at about 60°N latitude, as measured by the two-bottle method. The times of sunrise and sunset are indicated. Measurements were made at the surface and at depths of 1 and 6 meters below the surface, as indicated. How would you explain the midday drop in surface productivity during a bright summer day? (From A. P. Orr and S. M. Marshall, 1969, *The Fertile Sea*, Fishing News Ltd.)

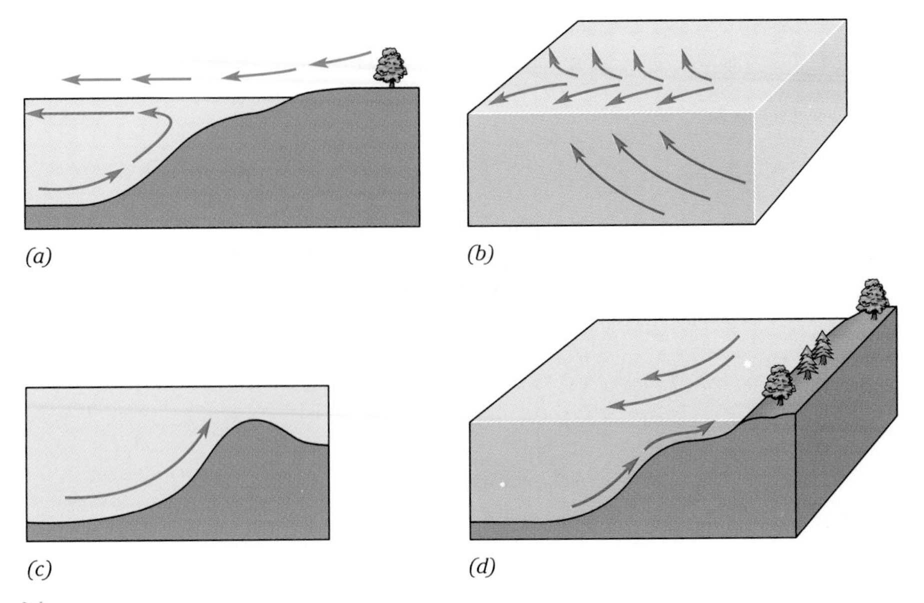

Figure 11.9

Illustration of some upwelling mechanisms. (a) In the tropics, trade winds drive the surface waters out to sea away from the western continental margins. The departing surface waters are replaced by deeper waters. (b) Where surface waters diverge, such as at the equatorial or Antarctic divergences, deeper waters must come up as replacement. (c) Reefs, islands, and rises in the ocean bottom may force deep currents upward. (d) The Coriolis effect may divert currents out to sea, requiring replacement by deeper waters.

Antarctic Divergence, with upwelling everywhere along its southern boundary.

Upwelling is especially pronounced along the eastern margins of oceans in tropical regions. Trade winds blow the surface water from east to west across the tropical ocean, driving the surface waters away from the western continental boundaries. To replace them, deeper waters rise from below. This happens in the Pacific off Peru, and in the Atlantic off equatorial Africa, making these areas particularly productive and good fishing grounds (Figure 11.10b).

In addition to these large-scale patterns of upwelling and downwelling in the oceans, there are also some more local effects (Figure 11.9). In some regions, deep ocean currents are forced upward as they encounter rises in the ocean floor, reefs, or a portion of a continental slope. Land masses and continental boundaries frequently guide or deflect major ocean currents in such a way as to force convergence in some places and divergence in others. In some areas, surface currents are driven by seasonal winds, like the monsoons, so that regions of upwelling and downwelling change with the seasons.

TURBULENCE

In shallow coastal waters, turbulence tends to bring sinking nutrients back to the surface. It also makes coastal waters more turbid with suspended sediments and other debris, thereby reducing the penetration of sunlight. Normally, the gain in productivity due to the increase in nutrients more than offsets the loss due to diminished sunlight, so turbidity is usually associated with a net increase in productivity.

The most common turbulent mechanism involves waves, which mix the upper 100 meters or so. This is not enough to retrieve materials from the deep ocean depths, but it is enough to keep the waters of the continental shelves well mixed. This accounts for the high biological productivity of the shelves, particularly in tropical and temperate latitudes where sunshine is plentiful (Figure 11.11).

Surface currents also produce turbulence. Most extend only to depths of a kilometer or so, which is not deep enough to retrieve nutrients from near the deep ocean floor. However, the Antarctic Circumpolar Current does reach all the way down to the ocean floor in many places and therefore continually brings a supply of nutrients back to the surface. These waters, along with the neighboring waters of the Antarctic Divergence, are therefore particularly productive biologically. They support a huge crop of tiny shrimplike krill, which are a favorite food for much of the world's whale population.

Our interest in upwelling and turbulence is that these mechanisms bring nutrients to the sunlit surface waters and therefore support increased biological productivity (Figures 11.10a and 11.10b). The

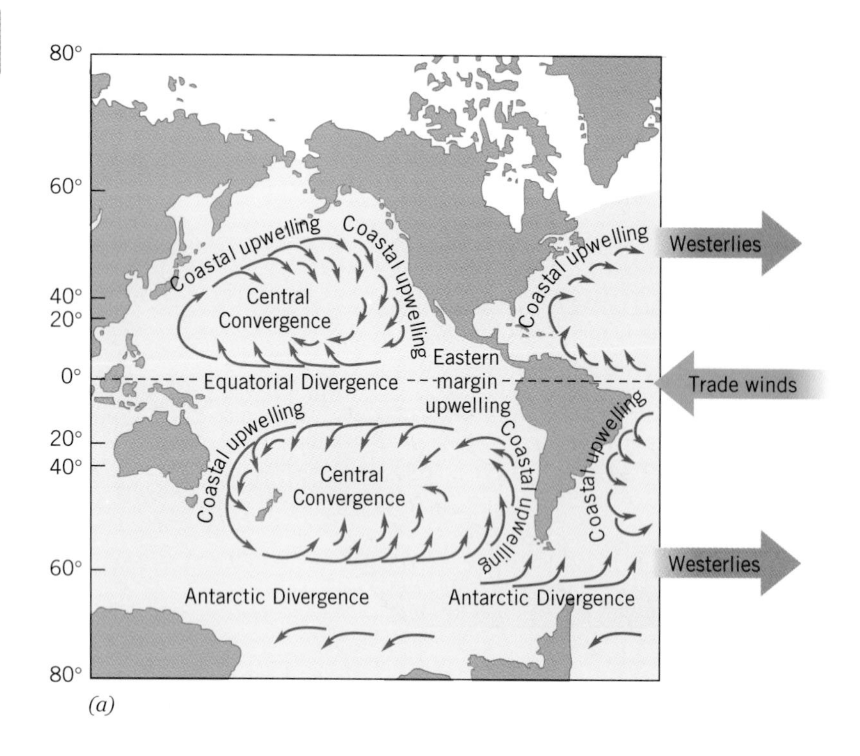

(a)

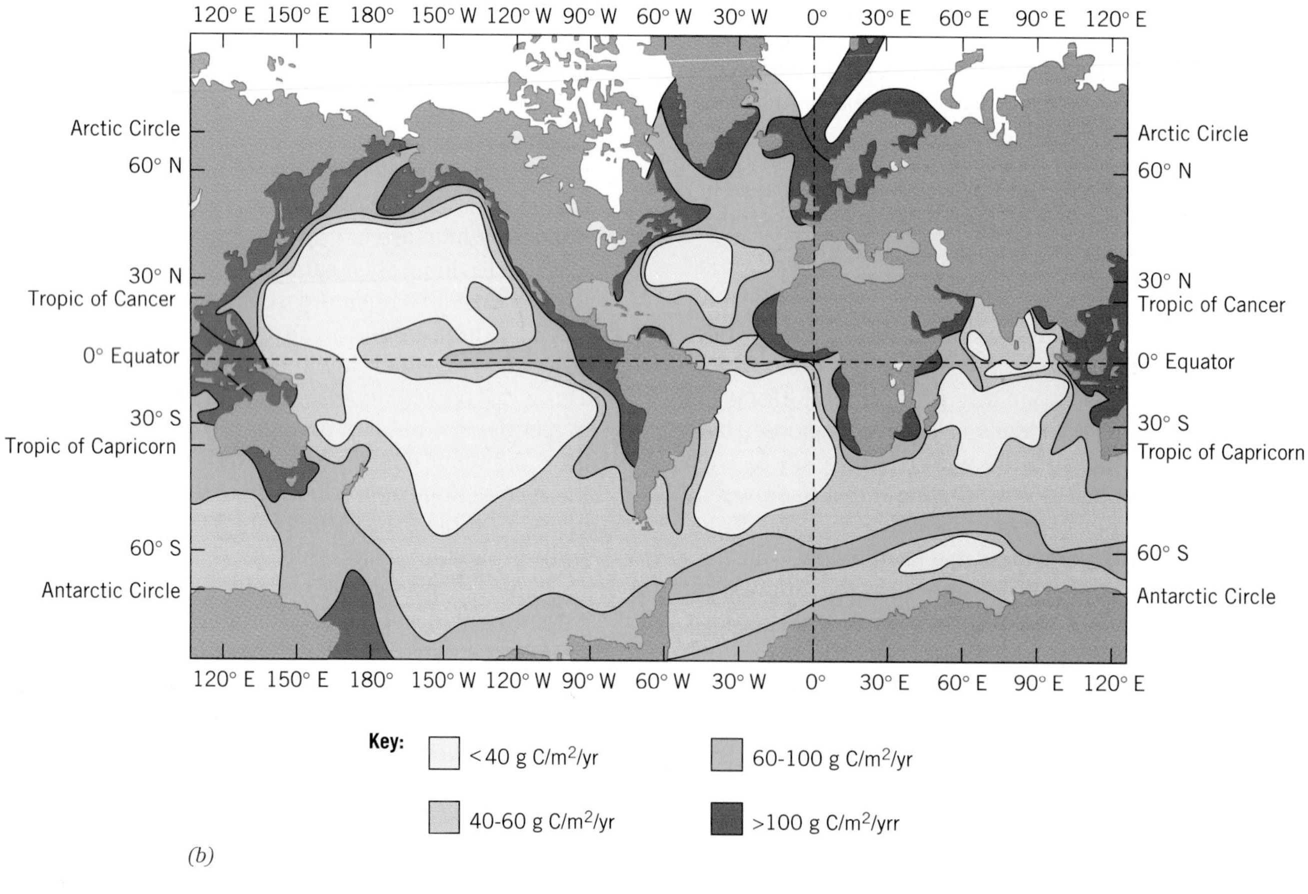

(b)

Figure 11.10

(a) The Coriolis deflection of the surface currents of the main gyres forces the surface water away from the perimeters, which causes upwelling along the coasts, along the equator, and at high southern latitudes. The trade winds also cause extensive upwelling in the eastern tropical oceans, as they blow surface waters out to sea. (b) Measured plant productivities, with darker shadings indicating higher productivities. The continental shelves and upwelling regions clearly have the highest plant productivities, due to the availability of nutrients. (After A. Couper, ed., 1983, *The Times Atlas of the Oceans*, Van Nostrand Reinhold, New York.)

Figure 11.11

Due to the availability of nutrients and sunlight, coastal waters such as these support a healthy biological community.

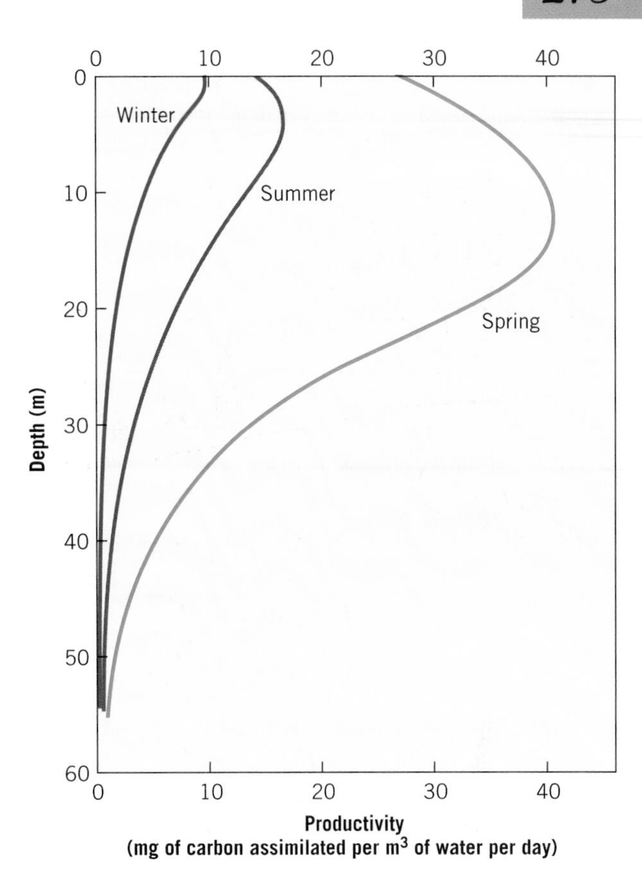

Figure 11.12

Typical variations in primary productivity with depth in the coastal waters of the Northeast Pacific Ocean. (Based on data from G. C. Anderson,1964, *Limnology and Oceanography*. **9**; 294.)

patterns of plant productivities are reflected in the abundances of organisms at all higher trophic levels, including the harvests of our commercial fisheries. We now summarize these patterns.

As we might expect, the very highest levels of plant productivity are found in the eastern tropical oceans, due to strong wind-induced upwelling and abundance of sunshine. We find slightly reduced, but still quite high plant productivities in regions of equatorial upwelling, coastal upwelling, and the Antarctic Divergence. In each of these areas, either the strength of the upwelling, or that of the sunshine, is somewhat reduced from that in the very highly productive eastern tropical oceans. Waters over the continental shelves also tend to be relatively rich in nutrients and therefore biologically productive. By contrast, the large central regions of the oceans tend to be relatively unproductive, due to convergence of surface waters, gentle downwelling, and a resulting depletion of nutrients.

PLANT EFFICIENCIES

Although scarcity of sunlight can limit plant productivity, especially in polar regions where nutrients are plentiful, it is not true that more sunlight necessarily means more food production. Perhaps it goes without saying that on the average, the organisms that get along best are those that do best under average conditions. For example, those plants that need very bright sunlight will find themselves in trouble on cloudy days, or when they are not right at the water's surface. Consequently, the majority of productive plants in the ocean have the highest efficiency under average conditions, rather than under the extremes of very bright or very dark. When sunlight and nutrients are abundant, maximum productivity does not occur right at the surface, but rather many meters below (Figure 11.12).

TYPICAL PRODUCTIVITIES

The productivity of plants in the surface waters restricts the amount of animal life that can exist at all depths. Where productivity is greater, correspondingly more respiration can be supported.

The average productivity for the entire ocean is about 50 grams of carbon being assimilated into organic matter per square meter of ocean area per year. Typical values range from five or six times this amount in upwelling areas, to somewhat less than this amount in the open ocean tropical waters, where nutrients are scarce. Under the Arctic ice cap, productivity is only about 1 gram of carbon assimilation per square meter of ocean per year, due to scarcity of sunlight there.

Table 11.2 Average Plant Productivities and Total Annual Plant Production in Upwelling, Coastal, and Nonupwelling Open Ocean Areas

Area	Productivity ($gC/m^2/yr$)	Percent of ocean area	Percent of total ocean production
Upwelling	300	0.1	0.5
Coastal	100	9.9	18.0
Open ocean	50	90.0	81.5

Table 11.3 Comparison of Productivities in Land and Ocean Environment

	Average Productivity ($g\ C/m^2/yr$)	Fraction of Earth's Surface	Total Production (tons of carbon assimilated/yr)
Land	160	28%	25 billion
Ocean	50	72%	20 billion

A study by J. H. Ryther[3] looked at average plant productivities in upwelling, coastal, and nonupwelling open ocean areas. The results are summarized in Table 11.2, where you can see that the ratios of plant productivities in these areas are

Upwelling:coastal:open ocean = 6:2:1

Because the open oceans encompass 90% of the total ocean area, the total amount of organic matter produced in this expansive region is largest, even though the productivity per unit area is the least.

COMPARISON TO LAND

The average productivity of the *open ocean* area is comparable to that of deserts on land, whereas the productivities of the coastal and upwelling regions are comparable to those of continental forests and agricultural crop lands. Although deserts make up only a small fraction of land areas, the corresponding ocean areas constitute 90% of the total. The overwhelming bulk of the ocean, then, is a biological desert.

For this reason, the average productivity per square meter on land is more than three times that in the oceans. In fact, even though the land has only 28% of the Earth's area, the total food production on land is slightly greater than that of the oceans (Table 11.3). The total amount of carbon assimilated into organic matter per year by plants is about 25 billion tons on land and 20 billion tons in the oceans.

At first thought, it should be surprising to us that the average productivity of the marine plants is so low. After all, conditions on the continents are much harsher than those in the ocean. Most land plants must develop extensive root systems in order to obtain water and extensive leaf systems to get carbon dioxide and oxygen. Ocean plants, in contrast, are continually bathed in these materials and do not need these complicated special systems. They are microscopic and very efficient. Why, then, is the productivity of the ocean so low?

The primary reason for the ocean's low productivity is its depth. Nutrients released by decaying detritus have difficulty returning to the surface waters where they may again be used by plants. On the continents, by contrast, detritus remains on or near the surface where the nutrients get recycled relatively quickly. Clearly, if we hope to develop aquaculture in response to the world food problem, we are going to have to find ways to either keep the nutrients in the sunlit surface waters, or recycle them more quickly (Chapter 14).

11.5 GEOGRAPHICAL AND SEASONAL VARIATIONS

In the preceding section, we learned that the surface waters of coastal and upwelling regions tend to be rich in nutrients, whereas the extensive central regions of the oceans are nutrient-deficient. This pattern is reflected in the pattern of plant productivities. But sunshine is the other important ingredient in photosynthesis, and it causes important seasonal and ge-

[3] J. H. Ryther, 1969, *Science* **166:**3901, 72–76.

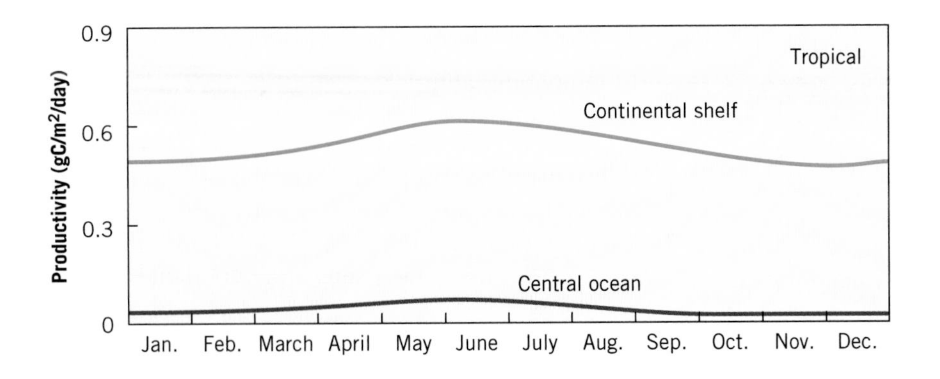

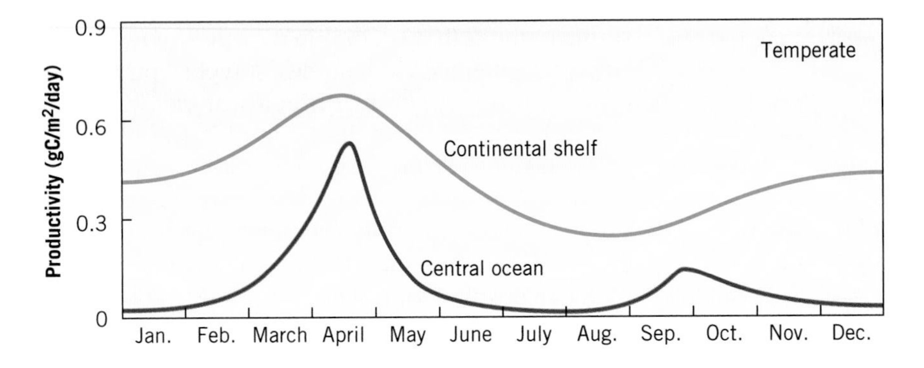

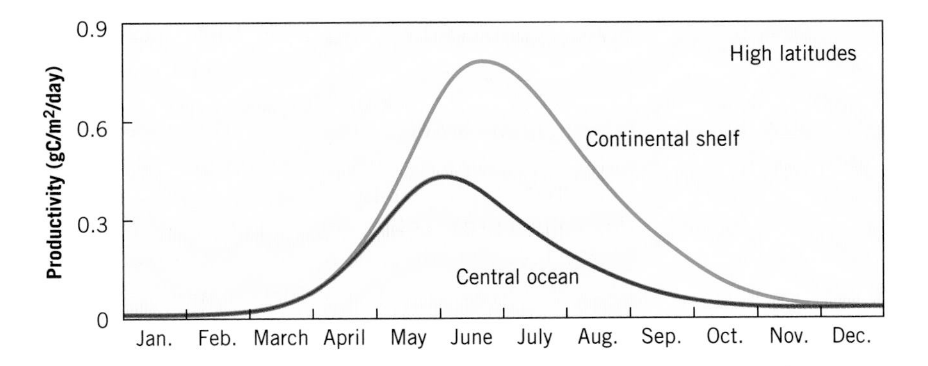

Figure 11.13

Typical seasonal variations in productivities for various ocean regions in the Northern Hemisphere. Availability of sunlight is more restrictive in high latitudes, so seasonal variations are more pronounced. Availability of nutrients is more restrictive in central ocean surface waters, especially where they are warm and do not mix with the cooler deeper nutrient- rich waters. (Although these patterns are typical, there are large variations due to factors we have already studied.)

ographic variations in the pattern of productivities. This is illustrated in Figure 11.13 and described in the paragraphs that follow. Throughout this section, a recurring theme is that warm surface water does not mix with cold deep water below. As organic detritus sinks into colder deeper water, nutrients are lost to the surface and do not get recycled. The thermocline that separates warm surface waters from cool deeper waters in temperate and tropical latitudes also prevents the recycling of nutrients.

TROPICS

In the tropics, sunshine is plentiful. Nonetheless, scarcity of nutrients keeps plant productivities quite low throughout most of the tropical oceans. Solar heating makes the surface water considerably warmer and less dense than the water below. The thermocline is permanent and prevents the recovery of nutrients from the deeper waters. Therefore, throughout most of the tropical ocean, surface waters and waters derived from them (e.g., the Gulf Stream) tend to be a beautiful deep blue. This identifies them as being biological deserts. Biologically active waters would be filled with tiny microorganisms and organic debris, which would give them a turbid green or yellow-green appearance.

However, there are a few regions of the tropical oceans where nutrients are plentiful. The combination of nutrients and sunlight make these regions very productive. As we have seen (Figure 11.10b), these areas include the eastern margins of tropical oceans, where trade winds blow the surface waters away from land, resulting in the upwelling of nutrient-rich deep waters from below. Upwelling also occurs along the Equatorial Divergence that continues on across the ocean. There are also local areas of upwelling, where reefs or rises force deeper currents to the surface. Finally, in the tropics as at all latitudes, the shallow waters over the continental shelves tend to be rich in nutrients, because decaying detritus cannot sink so deep that the released nutrients cannot be retrieved by waves and other turbulent mechanisms. Consequently, in the tropics we find very high productivities in the following limited areas: eastern ocean margins, equatorial upwelling, reefs and rises, and continental shelves. But away from these limited regions, the tropical ocean is relatively barren.

POLAR REGIONS

In polar regions, it is usually scarcity of sunlight, rather than scarcity of nutrients, that restricts biological productivity. The Sun is low in the sky at best, spending most of the time either just above the horizon or just below it. Furthermore, extensive ice cover often screens out much of the already meager sunlight. With the exception of some limited regions of convergence and downwelling, nutrients are plentiful throughout most polar surface waters due to seasonal mixing (Chapter 7, Figure 7.3). Each winter the frigid surface water sinks, forcing upward nutrient-rich water from below.

Because nutrients are plentiful, the availability of sunshine is crucial. Productivity is high during the months of summer sunshine, and it is very low during the darker winter months. This supports an animal population that feeds vigorously during the summer months and then goes nearly dormant for the rest of the year. Some migratory species, such as whales and salmon, include high latitudes on their summer feeding agenda, but go elsewhere for the rest of the year.

TEMPERATE LATITUDES

In temperate latitudes, there is gentle convergence and downwelling in the large central portions of all oceans and coastal upwelling along the margins. Consequently, plant productivity throughout the central regions of the oceans is relatively low, but relatively high along the coasts. Superimposed on this overall pattern are interesting and important seasonal variations.

We have seen that primary productivity is usually restricted by scarcity of nutrients in tropical waters and reduced sunlight in polar waters. It should be no surprise, then, that a combination of the two is involved at intermediate latitudes. A typical pattern for primary productivity in surface waters over deep ocean basins in these temperate latitudes is illustrated in Figure 11.14 and goes as follows.

In winter, nutrients are more plentiful, because surface waters are cool and either sink or mix with the nutrient-rich deep waters below. However, productivity is low due to the reduced sunlight. As the days begin to lengthen, the increased sunlight causes an early spring explosion of plant life, called a **bloom**, that reaches a peak and starts to decline in the late spring and early summer. The bloom begins at the surface and then proceeds to greater depths as the nutrients at the very surface are depleted.

There are several causes for the late spring decline. One is that the nutrient supply cannot support unlimited growth. Another is the proliferation of the tiny animal **grazers** that graze on the myriads of tiny plants. Probably the most important, however, is the heating of the surface waters by increasing daily sunshine. This creates a thermocline that prevents the return of nutrients from deep waters. Hence, as the surface nutrients are consumed, they do not get replaced, and so productivity declines.

In late fall, the daylight hours shorten and the surface water cools off, allowing it to mix with deep

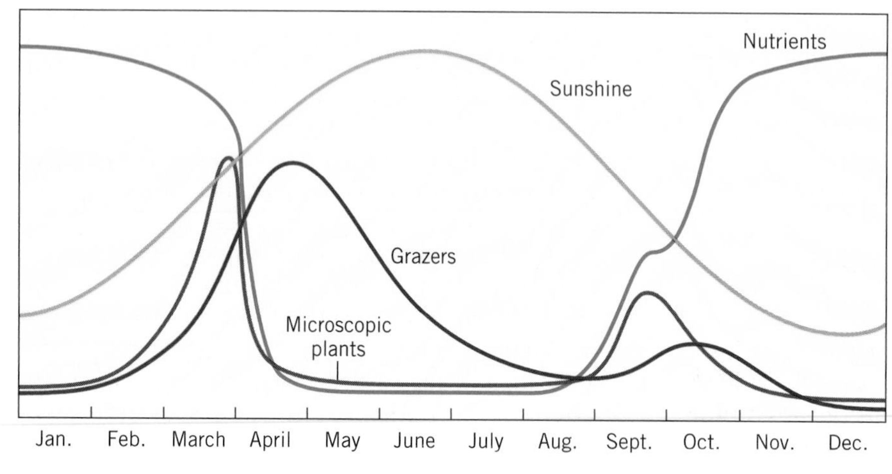

Figure 11.14
Typical seasonal variations in the abundance of sunlight, nutrients, microscopic plants, and grazers in surface waters at temperate latitudes.

waters. The nutrients return to the surface, causing a fall bloom in plant life. But this bloom is short-lived, because the days are getting short and sunlight is insufficient to sustain vigorous plant life. The winter lull in productivity returns and the cycle repeats itself.

Over the continental shelves, scarcity of nutrients is not such a problem. They cannot escape to deep water, because the shelves are shallow. Sunlight can penetrate nearly to the bottom in most regions, and turbulence due to waves and currents keeps the waters well mixed. There is still a shortage of nutrients in the summer, simply because the spring bloom consumes much of them, and the smaller summer waves do not create as much turbulence. But the development of a summer thermocline is not a problem like it is over deep ocean basins, so the seasonal fluctuations in productivity are not as pronounced near the coast (Figure 11.13).

VARIATION WITH DEPTH

In winter, productivity tends to be highest very near the surface, because the sunlight is meager. That is also where the spring bloom begins, but it then moves to deeper water as sunlight becomes more plentiful and surface nutrients are consumed. Figure 11.12 shows seasonal variations in productivity with depth for temperate coastal waters.

The depth beyond which there is not enough sunlight for the plants to sustain themselves is called the **compensation point**. Beneath this depth, plants cannot photosynthesize enough to meet their own respirational needs. Consequently, they cannot really be considered "producers" in this deeper environment. The depth of the compensation point depends on many things, of course. It depends on the plant species, as some are more efficient in their use of meager sunlight than others. Also, it depends on the availability of nutrients. But it is most sensitive to the intensity of sunlight and therefore related to the season of the year and the turbidity of the water. As a rule of thumb, the compensation point is found at a depth where the light intensity is about 1% of that at the surface, which is typically about 100 meters in clear ocean water and 20 meters in turbid coastal waters.

11.6 THE FOOD WEB

In the preceding sections, we studied the production of food. We now continue our study of basic biological processes in the ocean by looking at what happens to the food after it has been produced.

THE FOOD SUPPLY

There are many very interesting and intricate interdependencies among marine organisms, involving such things as providing support or protection, supplying vitamins, antibiotics, structural materials, and the removal of parasites or wastes. But the most important of these interdependencies involves the supply of food.

Less than 2% of the ocean is sufficiently shallow and has sufficiently firm substrate to accommodate large attached plants. Consequently, seaweeds such as those we find attached to rocky shores (Figure 11.1c) or washed up on our beaches do not represent a very large component of the plant population. Most of the ocean's primary productivity is carried out by microscopic single-celled plants.

Most of the organic matter produced by these plants is stored and subsequently oxidized by the plants themselves. But some is passed on to higher trophic levels. In general, the individuals at each trophic level tend to be larger and more complex than those at lower trophic levels on which they feed. But this isn't always true. For instance, simple microscopic parasites may feed on large fish or mammals. Often, the trophic level is a matter of maturity rather than breed. The larvae or juveniles of larger carnivores may be microscopic and feed directly on the plants.

A general rule of thumb is that each trophic level uses up about 90%[4] of the food energy it consumes, on average, and stores only about 10% for passage up to higher trophic levels (Figure 11.15). That is, most of the food materials consumed by any organism stop there and do not get passed on further. They are used for such things as the function of its muscles or organs, or the production of nonnutritional materials, such as chitin, bone, scales, and so on. But some goes into growth or reproduction in such a way that there is more mass available to the predators of the next trophic level.

After materials have been metabolized, the organism returns the waste products back into the marine environment, where they can be reused by plants to start the cycle all over again. Some of these, such as carbon dioxide, are ready for reuse immediately, but others, such as nutrients, tend to remain bound up in chemical complexes, which must await the work of bacterial decomposers before being released in a usable form (Figure 11.15).

[4]This 90% figure is an average. Various measurements have given ranges from about 77% to as high as 97%, depending on such factors as the particular species, age, and feeding environment.

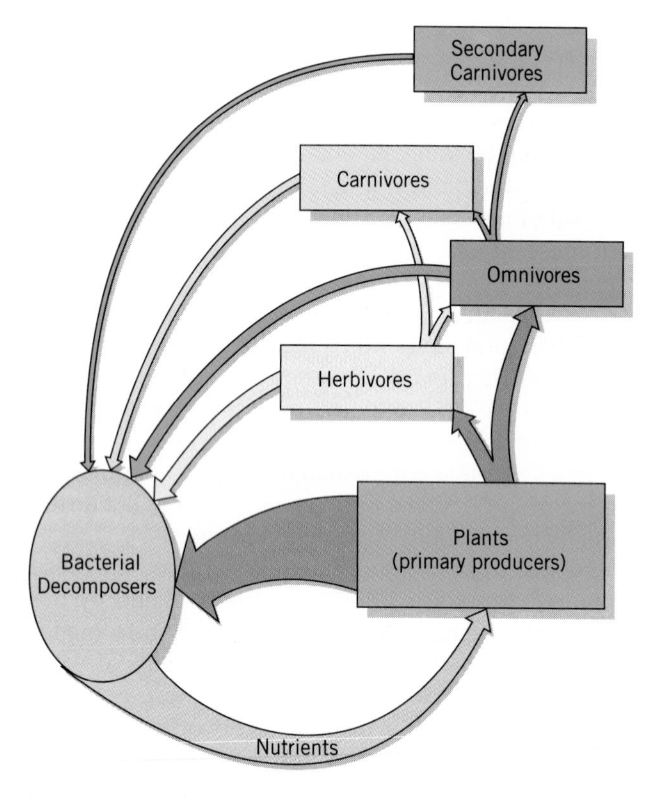

Figure 11.15

The food produced by the plants supports all organisms. The organisms at each trophic level use most of the food they consume for their own activities, but pass a small fraction onto the next trophic level. After the energy is used, bacteria decompose the dead and discarded materials, and the nutrients are returned to the plants to begin the cycle again.

ORGANIC DETRITUS

Because the primary producers support the entire marine biological community, you might think that the bulk of the organic materials in the ocean is contained in microscopic plant life. This is only partly true. These plants do contain most of the *living* organic matter, but the overwhelming bulk of the organic matter in the oceans is no longer alive.

Since the total mass of plants stays constant over the years, for every new plant produced, on average one dies. Daily production by microscopic marine plants is typically 40 to 50% of their total mass and the average life expectancy of a marine plant is just a few days. Production and consumption go on at alarming rates in the oceans. Imagine what it would be like if land plants doubled their mass every few days and were consumed at comparable rates by land animals!

> Daily production by microscopic marine plants is typically 40 to 50% of their total mass and the average life expectancy of a marine plant is just a few days.

Vigorous microscopic biological activity quickly converts organic materials produced by plants into tiny fragments of organic materials of other forms. On average throughout the oceans, about 99.8% of all organic matter exists in the form of dissolved or dead particulate matter (Table 11.4). These include protein, lipid, and carbohydrate molecules, tiny fragments of plants and animals, wastes, dead or dormant microscopic organisms, and other organic detritus. Only about 0.2% of all organic matter is actually in living organisms, most of which are microscopic plants (Figure 11.16). Consequently, it is much more common for **food chains** (i.e., the sequence of trophic levels) to begin with bacteria and microscopic animals feeding off dead organic detritus, than feeding directly off living plants.

In any case, rapid production and consumption in surface waters result in large quantities of organic detritus at various states of decomposition and at all depths. Of course, it is most heavily concentrated at the surface, but consumption by microscopic animals causes an interesting dip in its concentration in the 50- to 300-meter depth range (Figure 11.17).

Table 11.4 Estimates of Amounts of Organic Matter of Various Forms, at all Depths, Averaged Throughout the Oceans (Abundances in Grams of Carbon per Square Meter of Ocean Surface.)

Type of organic matter	Abundance ($g\ C/m^2$)	% of total
Dissolved organic molecules	2000	98.3
Particulate	30	1.5
Microscopic plants (phytoplankton)	4	0.2
Microscopic animals (zooplankton)	1	0.05

Source: After data by Ryther and Menzel, 1965, *Deep-Sea Research* 12, 200; and Michael J.Kennish, ed., 1989, *Practical Handbook of Marine Science*, CRC Press, Boca Raton, FL 287.

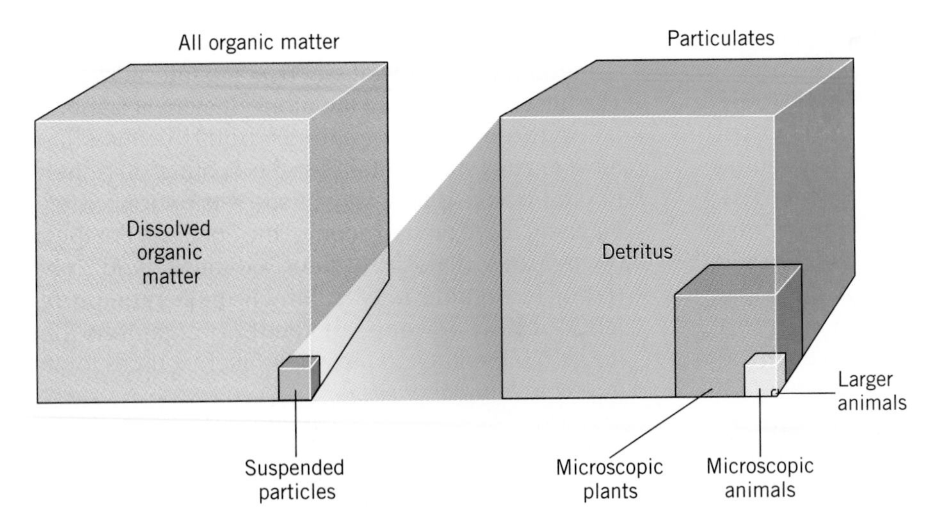

Figure 11.16

Schematic illustration of the relative amounts of organic matter in various forms. Dissolved organic matter is 98.3% of the total, and particulate detritus 1.5%. The live microscopic plants make up only 0.2%, and the live microscopic animals only 0.05% of the total. The larger swimming animals are only a trace.

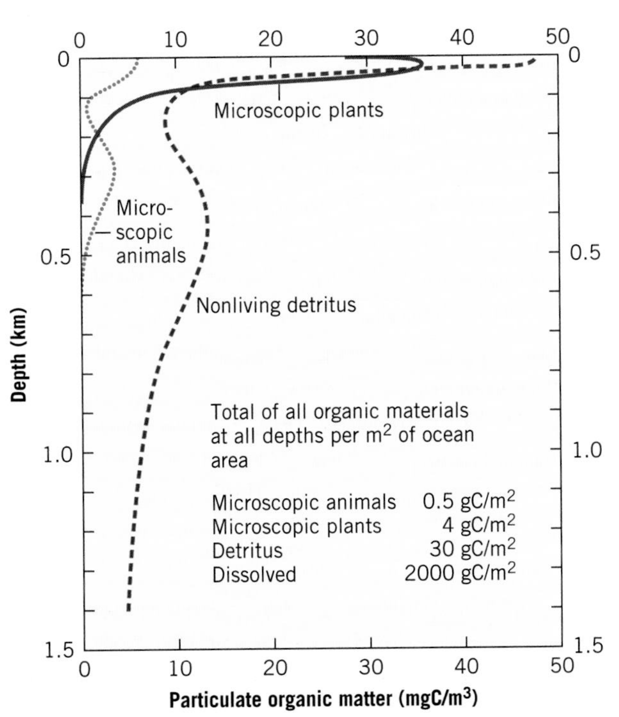

Figure 11.17

Typical distribution of microscopic particulate organic matter in the oceans, measured in grams of carbon per cubic meter of water. (The total dry weight of the organic matter is roughly five times this amount, and the actual live or *wet* weight even more.)

The highest concentrations occur in the top 100 meters, where most of the food is produced. Live microscopic plants are almost entirely confined to these waters. Living microscopic animals are found mostly near their food source in these surface waters, but are distributed over greater depths due to vertical migrations.

Gradual sinking and mixing cause the nonliving particulate organic matter to be distributed to much greater depths. Concentrations of dissolved organic matter are far too large to be shown on the scale of this graph.

Estimates of the total amount of organic matter at all depths in each of these categories are also given for comparative purposes.

FOOD CONCENTRATION

The animals that feed on tiny plant life and suspended particles of organic matter are mostly microscopic **multicellular** (i.e., many-celled) organisms (Figure 11.18). They have special adaptations for filtering these tiny food particles out of the water. They serve the larger community by collecting the food contained in these dispersed particles for easier passage up the food web. For example, we could not hope to gain our sustenance by filtering these fine dispersed particles from the water. Nor could a pelican, nor a tuna fish. We could, however, feed on anchovies, which feed on tiny animals, which feed on these tiny food particles. One ecological function of

Figure 11.18

Tiny microscopic animals, such as these copepods and worm and barnacle larvae, feed on the microscopic plants and other organic particles.

each trophic level is the collection and concentration of food for passage onto higher levels.

The abundance of dissolved and suspended organic materials in the ocean's surface waters varies greatly from one region to the next, but concentrations of a few parts per million by weight are typical. They tend to be **surface-active**, which means that they tend to stick to any solid surfaces found in the marine environment. You have probably noticed that rocks, sticks, pilings, boats, and such soon acquire a slimy coating after being put in the water. In time, the slimy coating attracts bacteria and algae, followed by barnacles, mussels, and a host of other organisms intent on making the food-rich surface their new home (Figure 11.19).

Rocks, sticks, pilings, boats, and such soon acquire a slimy coating after being put in the water. In time, the slimy coating attracts bacteria and algae, followed by barnacles, mussels, and a host of other organisms intent on making the food-rich surface their new home.

In addition to its accumulation on solid surfaces, some organic materials accumulate in "slicks" on the water surface, accompanied by driftwood and other floating debris. Like tiny microscopic animals, these floating materials also concentrate the surface-active dissolved and suspended food particles for passage up the food web.

Figure 11.19

Suspended organic matter tends to be surface active, forming a slimy coating on rocks, pilings and other solid surfaces, and serving as the basis for the development of a larger community, such as this one photographed during a very low tide.

Suspended sediments also serve as nuclei for the accumulation of surface-active organic materials. The finer sediments are the more efficient at accumulating these organic coatings, both because they have greater surface area for the coatings to adhere to, and because they stay in suspension longer, giving them more time to acquire the coatings. For these two reasons, deposits of finer sediments tend to be richer in organic matter, which makes them preferred regions for **deposit feeders** (organisms that ingest sediments) and also more likely sources for future petroleum deposits.

OXYGEN DEPLETION

The abundance of oxygen in the ocean is as important to the activities of respiring organisms as the abundance of food. These organisms need oxygen to release the energy contained in the food they produce or consume. Most organic matter is consumed and oxidized within 500 meters of the ocean surface, before settling deeper. Photosynthesis provides more than enough oxygen to support respiration in the upper 100 meters or so, but immediately beneath that, photosynthesis cannot occur, so it cannot replace the oxygen that is consumed through respiration. This is the reason for the *oxygen minimum layer* that we encountered in Chapter 6 (Figure 6.9).

The depletion of oxygen in the oxygen minimum layer is heightened by the tendency for microscopic grazers to spend most of their time in this region. Most rise to feed at night and then return to the dark waters below during the day. Staying out of the sunlight helps protect them from predators. But this behavior also tends to remove nutrients from the surface waters, and their respiration further reduces the oxygen concentrations in the oxygen minimum layer. At still greater depths, animal activity diminishes, and the waters generally contain sufficient oxygen to support the relatively small amount of respiration in these regions.

Much of the remaining organic matter that escapes consumption in the upper 500 meters eventually reaches the ocean floor, where it collects and supports animal and bacterial activity there. As a consequence of this consumption of organic matter in the sediments, the supply of food usually diminishes with depth rather quickly on a scale of centimeters. Percolation, diffusion, and the activities of burrowing organisms tend to keep the upper few centimeters of sediments bathed in water containing dissolved oxygen. Below these surface layers, however, there is an increasing tendency for the water to become trapped within the sediments without replacement. The decay and digestion of organic material deplete the supply of both food and dissolved oxygen.

BACTERIA

Of all the microscopic single-celled organisms in the ocean, bacteria are among the simplest and smallest. Typical sizes are under 5 micrometers, whereas the sizes of the single-celled marine plants lie typically in the range of 5 to 50 micrometers. Although bacteria are vigorously active, their total biomass is small relative to that of other organisms. Not only are bacteria tiny, but they also belong to the category of the simplest and oldest forms of life on Earth (Chapter 12). They can, and have, survived for ages here without help from any of the more complicated organisms. The reverse is definitely not true, however. We more complicated organisms could not survive without the help of bacteria.

Bacterial decomposers are instrumental in the recycling of materials through the food web (Figure 11.15). They take the organic detritus and waste products from each organism and transform them back into forms needed by plants for photosynthesis. At each trophic level, the amount of material meeting this fate is considerably larger than the amount of material passed on to the next trophic level. The most easily decomposed materials are carbohydrates. The decomposition of lipids and proteins is slightly more complicated due to the release of nutrients and salts tied up in these compounds. Many structural materials, such as cellulose, chitin,[5] and bone, decay very slowly.

The bacterial decomposition process is similar to that labeled *respiration* in the equation of life of Section 11.1. But some bacteria carry the process further. Not only do they release carbon dioxide (CO_2) and water (H_2O) in the process of oxidizing the organic compounds, but they also oxidize the nitrogen and phosphorus bound up within these compounds to produce nitrate (NO_3^-) and phosphate (PO_3^{3-}), which are so desperately needed by plants.

Some bacteria can continue to decompose organic matter in anoxic (oxygen-depleted) environments, either by removing the oxygen they need from other compounds or using other oxidizing agents (i.e., chemical agents that accept electrons, like oxygen). These bacteria dominate only where there is no free oxygen dissolved in the water, and they are called **anaerobic bacteria**. One major source of oxygen for anaerobic bacteria is the sulfate ion (SO_4^{2-}), and bacteria that use this source are called *sulfate reducing* bacteria. When oxygens are removed from the sulfate, the remaining sulfur ion tends to combine with hydrogen ions in the water to produce hydrogen sulfide (H_2S) gas, which gives off a characteristic smell associated with bogs, stagnant water, and rotten eggs. Anoxic environments are found in stagnant deep water and within heavy organic deposits, where dissolved oxygen is consumed and not replaced.

For water trapped within the bottom sediments, decay and respiration tend to deplete the supply of dissolved oxygen. Anaerobic bacteria may continue the decomposition, however, as long as other oxidizing agents and organic matter are both available. In a recent study of drilling samples, bacterial activity was found at a depth of several hundred meters into the sediments. When the abundance of organic materials in sediments is particularly large, bacteria may deplete the oxidizing agents before the organic matter is completely decayed. These remaining organic materials then still retain some stored energy. Continued burial and the passage of millions of years may turn these materials into petroleum deposits that will help fuel the energy needs of some society in the distant future (Chapter 15).

Near the hydrothermal vents of the dark deep ocean floor, some species of bacteria can use chemical energy to produce food. This chemosynthesis by bacteria is similar to photosynthesis by plants, but with unoxidized chemicals replacing the sun as the energy source. These chemicals enter the water as it percolates though cracks in the hot rocks below the sea floor, producing relatively large amounts of hydrogen sulfide, various unoxidized metals, and various metal compounds that are oxygen-deficient. All these materials can be oxidized (i.e., burned) by the bacteria to provide energy. The bacteria store some of this energy in organic materials that they produce. These bacteria, then, are the primary producers in vent communities (Figure 1.20b).

11.7 THE CELL

In this chapter, we have been studying the very basic processes involved in the production and consumption of food. We now turn our attention to the question of *how* the organisms accomplish these things. How does a plant know that it is supposed to manufacture food? Once it has the raw materials, how does it know what to do with them? Why can't a

[5]Pronounced "kite-in", this is a tough material that forms the exoskeleton of insects, crustaceans, and some other invertebrates. Cellulose is a fibrous structural material that provides stiffness, such as that of woody stems.

grain of sand or drop of water do the same? What tells each part of each organism what it is supposed to do, and how to do it? Scientists are finding answers to these questions in the microscopic makeup of living things.

As we learned at the beginning of this chapter, the basic unit of any organism, large or small, is the biological cell. Different cells do different things. Whether it is a tiny single-celled organism all by itself, or a small part of a larger organism, each cell has a set of duties it must perform if the organism is to survive. These duties are carried out through physical and chemical processes involving the molecules within the cell. Among these are the chemical interactions carried out during photosynthesis or respiration.

THE TRANSPORT OF MATERIALS

Out of the cell's immediate environment and through the cell membrane must come the materials needed by the cell to carry out its functions. If it is involved in photosynthesis, it must have water, carbon dioxide, and nutrients coming in. If it is involved in respiration, then it must be appropriately supplied with organic materials and oxygen. In addition to having appropriate raw materials coming into the cell, waste products must also be carried back out.

Because of its excellent ability to dissolve materials, water usually serves as the medium that transports these raw materials and waste products. The fluid within a cell is water-based, and most cells require a water-based external environment as well. The reason your body fluids are made of water, rather than air or some other fluid, is due to its excellent solvent properties.

The transport of materials in and out of the cell is accomplished primarily by diffusion (Section 6.8), which is the tendency for molecules to go from regions of higher concentration toward regions of lower concentration. By analogy, if you have lady bugs in one room and house flies in another and you open the door between the two rooms, you will soon have both lady bugs and house flies in both rooms. The random motion of the insects tends to take them from the room of higher concentration toward the room of lower concentration, until their concentrations are evened out. Similarly, as raw materials are used up within a cell and their concentration diminishes, new raw materials diffuse through the cell membrane, restocking the depleted supply inside. Likewise, as waste products build up within the cell, they become more concentrated than in the outside medium, so they diffuse out.

> By analogy, if you have lady bugs in one room and house flies in another and you open the door between them, you will soon have lady bugs and house flies in both rooms. The random motion of the insects tends to take them from the room of higher concentration toward the room of lower concentration, until they are evened out.

In many organisms, cell membranes facilitate the passage of certain materials and allow some materials to pass in one direction more easily than the other. Although diffusion is the basic transporting mechanism, these discriminating cell membranes can cause the concentrations of some materials to be considerably different within the cell than outside.

Many large multicellular organisms isolate their internal body fluids from the external environment.

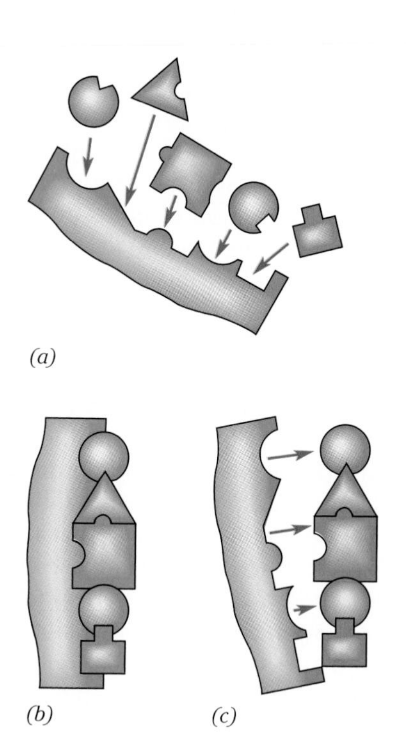

Figure 11.20

Schematic illustration of how an enzyme or RNA molecule can serve as the scaffolding on which materials are assimilated. (a) Random thermal motions eventually bring each of the needed raw materials to the appropriate position on the scaffolding. (b) These materials then interact with each other to form larger molecules. (c) The newly formed material then peel off the scaffolding, leaving it free to start the process again.

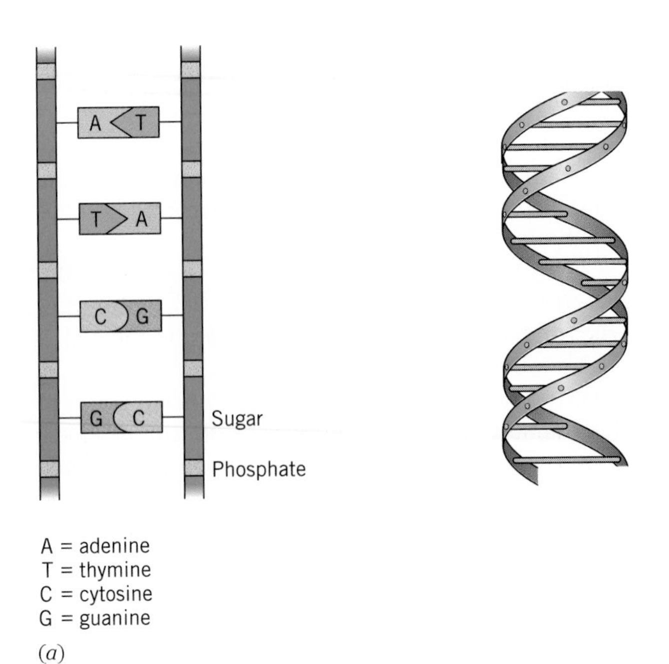

(*a*)

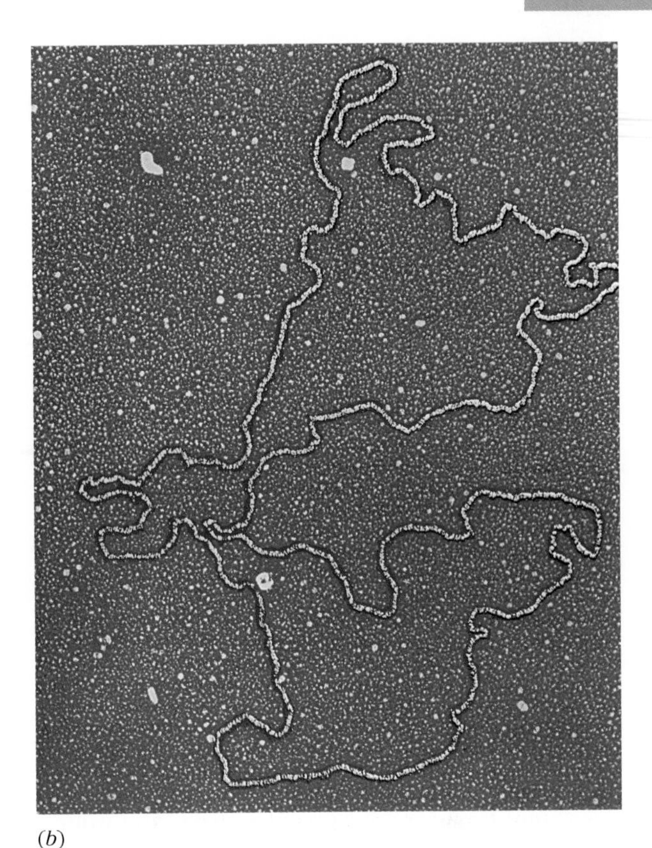

(*b*)

Figure 11.21
(a) DNA is like a microscopic ladder that has been twisted into the form of a double helix. The edge of each element of the ladder is a sugar, with phosphate connecting adjacent sugars. The coded information that regulates the cell is carried in the "rungs" of the ladder. There are four different kinds of rungs. One of each is illustrated in the drawing. The instructions for the cell are carried in the ordering of these rungs. (b) Electron microscope photo of DNA.

Communication between the two is allowed only through certain special multicellular membranes, such as lungs, gills, and guts. These membranes regulate the composition of internal body fluids so as to make the best possible environment for the function of the cells. For example, your body has lungs and a gut, which isolate your body fluids from the air you breathe and the food you eat. These membranes closely regulate the materials entering and leaving your body fluids, so that the cells within your body are continually bathed in a fluid that provides the optimum environment for them.

HOW SIMPLE CELLS DO COMPLICATED THINGS

Once it has the raw materials, how does a cell know what to do with them? For example, how does it know how to generate food stores, energy, add to its own growth, or reproduce itself? Scientists are still far from a complete understanding of how cells carry out all of their special functions, but the following gives a brief, simplified description of a few techniques that the cells employ.

Within a cell are enzymes that facilitate certain chemical reactions. Each enzyme has a chemical structure that makes it attractive to certain molecules. Its own thermal motion, as well as those of the other molecules in the cell fluid, causes it to undergo countless collisions each second as it moves around within the cell. Occasionally, it encounters the kind of molecule that is attracted to that enzyme, and so it sticks. This molecule's chemical nature may be changed through interaction with the enzyme directly, or it may wait until the enzyme has encountered more molecules of particular kinds, with different molecules sticking in different places on the enzyme (Figure 11.20). These molecules may then interact with each other, often with the help of some interactions with the enzyme itself. This new substance is then released, leaving the enzyme to start the process all over again. The molecules produced in these enzyme-induced reactions may be used by the cell directly, or may be used to form larger, more complicated molecules.

For complex functions, such as the manufacture of complicated proteins, there are large **RNA** molecules within the cell that serve as the scaffoldings on which the necessary raw materials are brought together. The process is similar to that carried out by many enzymes, but on a much larger scale. Various materials are attracted to various portions of a RNA molecule, so the ordering of the components along the RNA molecule determines the order of the ingredients in the new molecule being produced (Figure 11.20). Some of these ingredients are produced by enzymes, some are collected and transported by enzymes, and some just run into the RNA all by themselves. When the ingredients are finally assimilated, they interact with each other, often aided by certain enzymes present, and then peel off the parent RNA molecule.

You can see that for a cell to function properly, it not only needs the correct raw materials, but it also must produce the appropriate enzymes and RNA molecules. The information needed to do this is contained in long, thin protein molecules called **DNA** (Figure 11.21). Except for bacteria and their close relatives, the DNA for all other organisms is contained within a separate region of each cell called the **cell nucleus** (Figure 11.22). By the ordering of the various chemical groups along its length, a DNA molecule can produce the needed enzymes, RNA molecules, and it can even reproduce itself, all through processes similar to those described for enzymes and RNA molecules above. That is, DNA serves as the scaffolding on which many of these things are assimilated. Consequently, through the ordering of the chemical groups along DNA molecules, they contain all the information necessary to run the cell activities.

Any mutations of the DNA will cause corresponding changes in the cell's functions. Of course, most of these changes will be harmful, reducing the organism's ability to cope and survive. Occasionally, some mutations will be helpful, and the beneficiary organisms will have an advantage over others and perhaps eventually displace them.

To summarize, the behavior of a cell is influenced by its outer environment, which determines what materials are available to diffuse into the cell. It is also influenced by the cell membrane, which may have the ability to selectively concentrate some of these materials within the cell. Furthermore, there are enzymes and RNA molecules within the cell that facilitate the chemical reactions needed for the cell to perform its functions. The information necessary to produce these enzymes and RNA molecules is carried in DNA molecules, which are also able to reproduce themselves.

CELL EFFICIENCY

A cell is in contact with its environment only along its outer surface. Consequently, both the rate at which raw materials can get into a cell and the rate at which the waste products get out depend on the surface area of the cell. The larger the area, the more exchange can take place. However, the amount of raw materials *required* by a cell depends on its volume; bigger cells need more materials. The most efficient cell would require little, but get it very easily. In

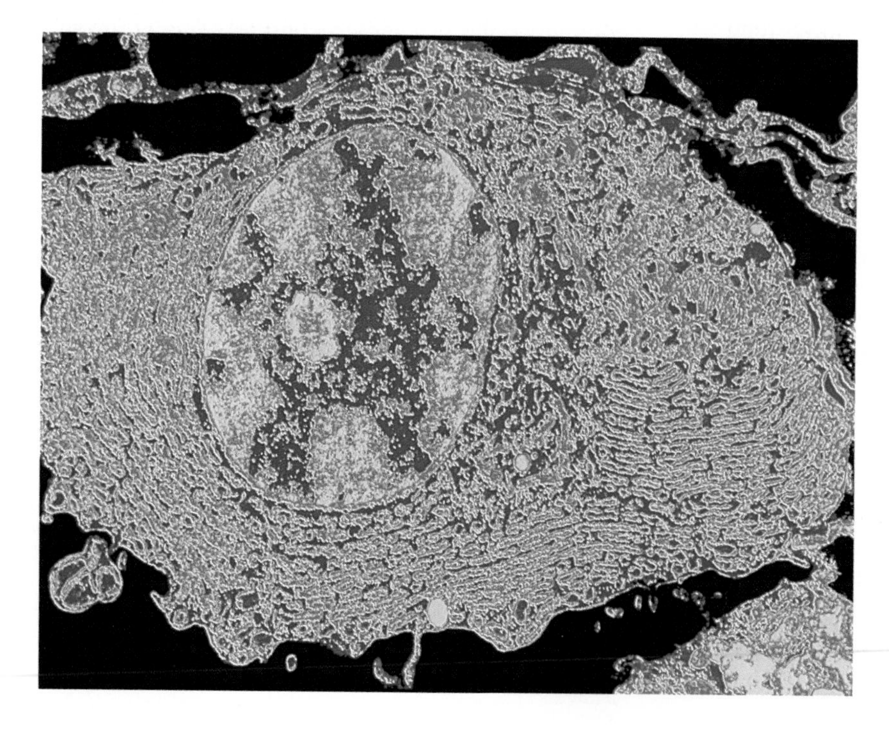

Figure 11.22

Transmission electron micrograph of a slice through a human cell. The large central oval-shaped region is the cell nucleus where the DNA is located. (Magnification about 2000 times.)

other words, the most efficient cell would have a small volume and large surface area.

Similar reasoning applies to the removal of waste products. The smaller the volume, the fewer waste products would be produced inside. The larger the surface area, the more quickly and easily these would diffuse out. So in getting rid of wastes, small volumes and large surface areas would also be best. (See Table 11.5.) If we are interested in **cell efficiency**, we must ask what characteristics would make a cell's surface large in comparison to its volume.

To answer this question, notice that whenever you break something, you expose new freshly made surfaces without changing the total volume. Continued breaking produces smaller pieces and more exposed surfaces. Smaller pieces mean a larger surface in comparison to the volume. From this analogy, we can see that the most efficient cell would be one that is as small as possible, yet still holds all the materials needed to perform its functions.

> Whenever you break something, you expose new freshly made surfaces without changing the total volume. Continued breaking produces smaller pieces and more exposed surfaces.

There is also a second important reason for the efficiency of a cell to depend on small size. The **diffusive transport** (i.e., transport by diffusion) of materials in water proceeds very slowly over large distances. For example, the characteristic time for the diffusion of a dissolved salt in water is about 1 second for a distance of 0.1 millimeters (the size of a large cell), 1 day for a distance of 3 centimeters, and 1 year for a distance of only 60 centimeters (Figure 6.13). Clearly, if the cell relies on diffusion for the transport of materials within it, then it must be small.

MICROSCOPIC MARINE LIFE

The advantages of efficiency in a nutrient-poor environment is the reason why the bulk of marine plant life is microscopic. Likewise, the overwhelming bulk of animal life in the oceans is also microscopic. The scarcity of food puts a premium on efficiency for the animals, as well. Contrast this picture to life on land. As you look around you, it is clear that by far the majority of life on land is large. Although there is ample microscopic life around, its total mass is very small in comparison to that of the large organisms. Just the opposite is true in the oceans.

Table 11.5 Characteristics of an Efficient Cell

An Efficient Cell Would	Implication
Require very little incoming nourishment	Small volume
Produce very little outgoing wastes	Small volume
Get its raw materials easily	Large surface area
Get rid of its wastes easily	Large surface area

Summary

Organic Synthesis and Nutrients

Plants have the ability to convert solar energy into stored chemical energy, which can later be released through the oxidation (burning) of the organic materials they produce. First produced are carbohydrates. These can be converted into more complex molecules through subsequent chemical reactions, which require small amounts of other materials. Although most of these other materials are sufficiently abundant in the ocean, nutrients such as nitrogen and phosphorus are not. Nutrient scarcity is aggravated by hoarding and the tendency for nutrients to be carried below the photic surface waters. The productivity of a region depends on the nutrients being released from the organic detritus and recycled.

Measuring Productivity

The productivity of a region is reflected in the abundances of organisms of all kinds there. It can be determined more accurately by counting the microscopic plants, measuring dissolved oxygen concentrations, measuring the rate of assimilation of carbon-14, or from satellite measurements of chlorophyll.

General Patterns in Productivity

Nutrients that sink into deeper water may return to the surface in areas of upwelling, or where turbulence extends sufficiently deep to retrieve them, such as over the continental shelves. Areas of upwelling include the eastern parts of tropical oceans, where trade winds blow surface waters out to sea, the perimeters of the large current gyres, caused by the Coriolis deflection toward the centers of these gyres, and reefs and rises that force deep currents upward. But throughout most oceanic regions, surface waters are lacking in nutrients. The total amount of organic matter produced annually by marine plants is less than that by land plants.

Geographical and Seasonal Variations

In most of the tropical ocean, primary productivity is restricted by the scarcity of nutrients. However, in shallow water and upwelling areas, the abundance of nutrients fosters vigorous plant activity. In polar waters, reduced sunshine is the constraining factor. In temperate latitudes, there are seasonal variations: Diminished sunshine constrains productivity in the winter, and scarcity of nutrients constrains productivity in the summer. Plant efficiencies are highest in moderate sunlight, and maximum productivity usually occurs several meters below the surface.

The Food Web

The entire biological community depends on the food produced by plants. At each trophic level, about 90% of food energy is metabolized, and only 10% is stored for passage up to higher trophic levels. Most of the organic material in the oceans is nonliving. Suspended organic detritus tends to stick to solid surfaces, including the surfaces of suspended sediments.

Tiny grazers serve the greater biological community by collecting and concentrating the tiny particles of organic matter, both living and nonliving, for passage up to higher trophic levels. Metabolism of this organic matter produces an oxygen minimum layer beneath the photic surface waters, where the lack of sunlight prevents plants from replenishing the oxygen lost to respiration. Nutrients are released from waste products by bacterial decomposers.

The Cell

The fundamental unit of all living organisms, large or small, is the biological cell. Whatever the cell's particular functions, raw materials must come to the cell through the cell membrane, and waste products must be carried away. The transporting medium is water, and diffusion is the process primarily responsible for the transport. Discriminating cell membranes and regulation of body fluids by special tissues, such as gills, lungs, or guts, may help improve the cell's living environment. Enzymes facilitate certain chemical reactions within the cell, and RNA molecules serve as scaffoldings on which complex molecules are assembled. The cell's activities are ultimately controlled by DNA molecules in the cell nucleus. Small size is advantageous in improving the ability to get needed raw materials, eliminate wastes, and transport materials around within the cell. The importance of efficiency is the reason why the bulk of all marine life is microscopic.

Key Terms

herbivore
carnivore
trophic level
primary producers
primary production
productivity
cell
cell membrane
micron
micrometer
μm
photosynthesis
respiration
chemosynthesis
carbohydrate
hydrate
oxidize
enzymes
lipids
proteins
carbon chain
micronutrients
equation of life
detritus
nutrient cycle
light–dark bottle system
Equatorial Divergence
coastal upwelling
Antarctic Divergence
bloom
grazers
compensation point
food chain
multicellular
surface-active
deposit feeder
anaerobic bacteria
RNA
DNA
cell nucleus
cell efficiency
diffusive transport

Study Questions

1. Where does the energy for life come from?
2. In the production of carbohydrates by marine plants, what are the needed raw materials? What is the waste product released? What are some of the additional materials that may be needed in the conversion of carbohydrates into more complex molecules?
3. Discuss some of the factors that contribute to the nutrient scarcity in surface waters.
4. How do upwelling and turbulence help increase the productivity of surface waters? Give some examples of places where each might be effective.
5. Discuss ways that the rate of primary productivity of a region might be determined. Why might the distribution of pelicans or fishermen be an indication of productivities? Why is the dark bottle needed when monitoring dissolved oxygen concentrations?
6. Why are most tropical waters unproductive biologically? Is this also true of shallow tropical waters, such as reefs or continental shelves? Why, or why not?

7. Polar waters are usually extremely productive, but just for a few months each year. Which months do you think that would be, and why?
8. Explain the seasonal variation in productivity in temperate latitudes.
9. Does maximum productivity usually occur right at the surface, where sunlight is most abundant, or somewhat deeper? Why?
10. How does the total annual food production by marine plants compare with that on land? Why is the annual productivity in the ocean so low?
11. Discuss the passage of food energy through the food web. At each trophic level, how much is used by that organism and how much is passed on to higher levels, on the average? What happens to the waste products?
12. Of the organic materials in the ocean, how much is contained in living plants? How much in other living organisms? How much as dissolved or suspended particulate matter?
13. Discuss two ways that suspended food particles are collected and concentrated for passage on up the food chain. Why would you expect finer sediments to be richer in organic matter than coarser ones?
14. Explain diffusion. How is it related to the transport of materials to, from, and within a cell?
15. How do enzymes, RNA, and DNA help the cell carry out its functions?
16. What are the advantages of small size for a cell? Is most of the ocean's food production carried out by large plants or microscopic ones? Why?

Critical Thinking Questions

1. Estimate the thickness of a page in this book in microns. (*Hint*: Measure the thickness of 100 pages and divide by 100.) How does the thickness of a page compare to the size of a bacterium? Or a microscopic plant that is 50 microns across?
2. Write down the chemical equation for the burning of a molecule of a simple sugar, $C_6H_{12}O_6$, in oxygen, O_2.
3. If organic detritus floated, how would that affect primary productivity in the oceans?
4. Hawaii is located in the central Pacific Ocean where plant productivities are very low on average, yet the Hawaiian coastal waters are full of life. Why?
5. Suppose you had a large coastal pond of sea water in which you were going to raise anchovies for sale as food. A friend suggests that you stock your anchovy pond with sea bass that feed on the anchovies and harvest the larger bass instead. Which would produce more food altogether? By about how many times?

Suggestions for Further Reading

COHEN, J. E. AND C. M. NEWMAN. 1988. Dynamic Basis of Food Web Organization. *Ecology* 69:6, 1655.

DRING, M. J. 1982. *The Biology of Marine Plants.* Edward Arnold.

FALKOWSKI, P. 1988. Ocean Productivity from Space. *Nature* 335:6187, 205.

FRIEDEN E. 1972. Chemical Elements of Life. *Scientific American* 227:1, 52.

HAYWARD, THOMAS L. 1994. The Shallow Oxygen Maximum Layer and Primary Production. *Deep Sea Research* 41:3, 559.

JOYCE, G. F. 1989. RNA Evolution and the Origins of Life. *Nature* 338:6212, 217.

KARL, D. M. ET AL. 1988. Downward Flux of Particulate Organic Matter in the Ocean: A Particle Decomposition Paradox. *Nature* 332:6163, 438.

KRAUT, J. 1988. How Do Enzymes Work? *Science* 242:4878, 533.

RYTHER, J. H. 1969. Photosynthesis and Fish Production in the Sea. *Science* 166:3901, 72.

TAPPER, R. 1989. Changing Messages in the Genes. *New Scientist* 121:1657, 53.

YOUVAN, D. C. AND B. C. MAIRS. 1987. Molecular Mechanism of Photosynthesis. *Scientific American* 256:6, 42.

Many dinoflagellates are bioluminescent. They can make the water seem to glow at night when agitated, such as in a ship's wake or when you step on a wet beach.

Life on a coral reef.

Twelve

MARINE ORGANISMS

In the preceding chapter, we learned that nearly all marine life is ultimately dependent on the food produced by plants in the surface waters. Where sunlight and nutrients are plentiful, plants are productive. The food they produce supports large populations of organisms of many types. Where sunlight or nutrients are scarce, plant productivities are low, and life is sparse.

These general considerations explain the overall patterns for the distribution of life in the oceans, but they do not give detailed insights into the organisms themselves. In particular, we are interested in what the organisms look like, how they function, the environments in which each might be found, what regulates their populations, and the way organisms interact with each other and their environments to form communities. These topics are studied in this and the following chapters.

12.1 CLASSES OF MARINE ORGANISMS

It helps our study of the various inhabitants of the Earth if we can put them into categories based on broad general characteristics. Of course, there are many different possible classification schemes. We could distinguish among them on the basis of color, length, weight, habitat, and so on. But the idea forming the basis of the more common detailed classification schemes is that the development of the individual organism reflects the history and development of its ancestors. As a result, most organisms in a given category have some developmental similarities and probably have had similar ancestry.

All organisms are divided into a few broad categories, called **kingdoms**, according to broad general characteristics. The members of any one kingdom are then subdivided according to more specific characteristics into **phyla**.[1] The members of any one phylum are then further subdivided into *classes*, according to even more specific characteristics, and so on. The order of the various steps of classification, from the largest, broadest groups to the most specific categories, is as follows:

1. Kingdom
2. Phylum
3. Class
4. Order
5. Family
6. *Genus*
7. *species*

Scientists rely heavily on Latin for the various designations. The genus is a capitalized noun and the species is an adjective; both are usually italicized. Examples are given in Table 12.1.

Because of the complexity of life on Earth, no classification scheme is without ambiguity. It is relatively easy for us to divide the familiar large land organisms between just two kingdoms, plants and animals. But in the marine environment, most organisms are microscopic and the distinction is much more difficult. For this reason, most classification schemes now have two kingdoms to accommodate single-celled organisms, in addition to one kingdom for the multicellular plants, one for the multicellular animals, and one for the fungi (Table 12.2). We now go through these five kingdoms and describe some of the more prominent marine members of each, which are listed in Table 12.3.

12.2 SINGLE-CELLED ORGANISMS

GENERAL

Single-celled organisms are extremely prominent in the oceans. In fact, they compose the vast majority of the ocean's living biomass. Some are producers, some are consumers, and some are decomposers. All single-celled organisms are divided into two kingdoms according to whether or not they have a distinctive cell nucleus, which is separated from the rest of the cell by a nuclear membrane and contains DNA and some of the other cell regulators.

Those that do not have a cell nucleus are called **procaryotes** and belong to the kingdom **Monera**. These very primitive organisms were the earliest form of life on Earth, and for the greater part of our planet's history, they were probably the *only* form of life here. Scientists believe that the earliest procaryotes appeared about 3.8 billion years ago, well within the first billion years of Earth history. This kingdom has two major subdivisions, both of which include organisms that are still very important members of the Earth's biological community. These two divisions are the bacteria and photosynthetic cyanobacteria, also called *blue-green algae* (Figure 12.1).

[1] Some use the term *division* in place of "phylum" for this level in the classification of plants.

Table 12.1 Examples: Classification of Certain Organisms

Organism	Diatom	Giant Kelp	Killer Whale	Human
Kingdom	Protista	Plantae	Anamalia	Anamalia
Phylum[a]	Chrysopyhyta	Phaeophyta	Chordata	Chordata
Class	Bacillariophycene	Phaeophycae	Mammalia	Mammalia
Order	Centrales	Laminariales	Cetacea	Primate
Family	Chaetoceraceae	Lessoniaceae	Delphinidae	Hominidae
Genus	*Chaetoceros*	*Macrocystis*	*Orcinus*	*Homo*
Species	*decipiens*	*pyrifera*	*orca*	*sapien*

[a]Some would use *division* here for the plants.

Table 12.2 The Five Kingdoms

Kingdom	Characteristics
Monera	Single-celled, no cell nucleus
Protista	Single-celled, cell nucleus
Fungi	Fungi
Plantae	Many-celled plants
Animalia	Many-celled animals

Table 12.3 Classification for Some of the Principle Organisms in the Ocean

Kingdom	Phylum	Class	Examples
Monera	Schizophyta		Bacteria
	Cyanophyta		Blue-green algae
Protista	Chrysophyta		Diatoms, coccolithophores, silicoflagellates
	Pyrrophyta		Dinoflagellates, zooxanthellae
	Protozoa		Foraminifera, radiolaria
Fungi	Mycophyta		Fungi, lichens
Plantae	Chlorophyta		Green algae
	Phaeophyta		Brown algae
	Rhodophyta		Red algae
	Tracheophyta		Marsh grass, eel grass, mangroves
Animalia	Porifera		Sponges
	Cnidaria (or Coelenterata)	Hydrozoa	Portuguese Man of War
		Scyphozoa	Jellyfishes
		Anthozoa	Corals, anemones
	Bryazoa		Moss animals
	Ctenophora		Comb jellies
	Plathyhelminthes		Flatworms
	Nematoda		Round worms
	Annelida		Segmented worms (polychaetes)
	Chaetognatha		Arrow worms
	Branchiopoda		Lamp shells
	Mollusca	Polyplacophora	Chitons
		Gastropoda	Snails, limpets
		Bivalvia	Clams, mussels, scallops, oysters
		Cephalopoda	Octopuses, nautiluses, squids
	Arthropoda		
	Subphylum: Chelicerata		Horseshoe crabs
	Subphylum: Crustacea		Barnacles, copepods, krill, shrimp, crabs, lobsters, isopods, amphipods
	Echinodermata	Stelleroidia	Starfish, brittle stars
		Echinoidea	Sea urchins, sand dollars
		Holothuroidea	Sea cucumbers
		Crinoidea	Sea lilies
	Chordata		
	Subphylum: Urochordata		Tunicates, sea squirts
	Subphylum: Vertebrata		
		Agnatha	Jawless fish: lampreys, hagfishes
		Chondrichthyes	Cartilaginous fish: sharks, skates, rays
		Osteichthyes	Bony fishes
		Reptilia	Snakes, turtles, caymans
		Aves	Birds
		Mammalia	Whales, otters, seals, sea lions, sea cows

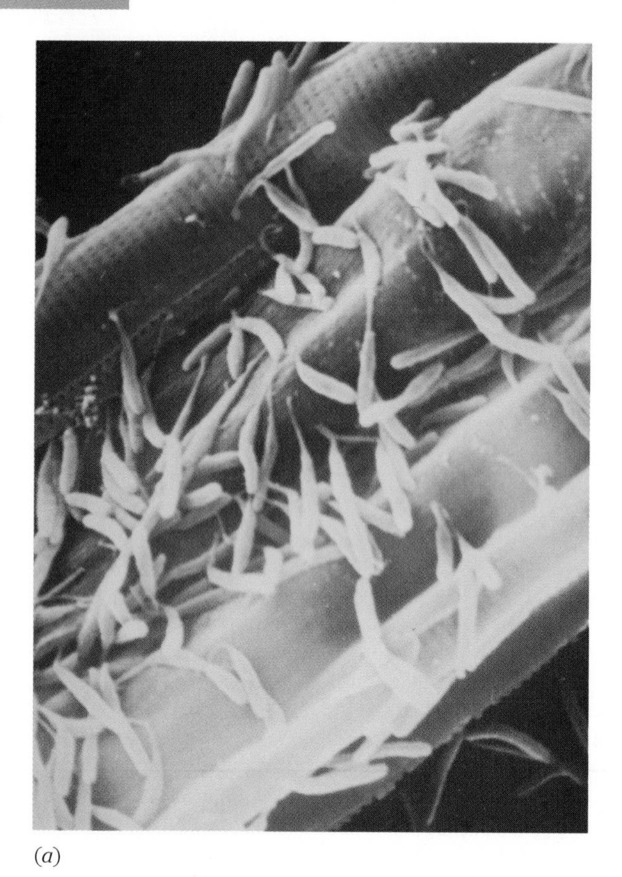

(a)

(b)

Figure 12.1

(a) Microphoto of tiny marine bacteria (each about 1 micrometer across) attached to the larger diatoms. (b) Microphoto of two species of blue-green algae (cyanobacteria) whose cells grow together to form filaments (magnified about 3000 times).

Organisms whose cells have distinctive cell nuclei are called **eucaryotes**. For these, the DNA and some other specialized regulatory materials are contained within the cell nuclei. The eucaryotes are more advanced and are relative newcomers on Earth. All multicellular organisms are eucaryotes. Single-celled eucaryotes make up the kingdom **Protista**. The most productive of the single-celled plants in the oceans belong to this kingdom, and the dominant members belong to four groups called coccolithophores, silicoflagellates, diatoms, and dinoflagellates. Single-celled animals, called protozoans, also belong to this kingdom.

Now let us look in more detail at the most important marine members of these two kingdoms of single-celled organisms. Their most important characteristics are summarized in Table 12.4.

BACTERIA

As we learned in Section 11.6, bacteria are extremely tiny one-celled organisms, typically less than 5 micrometers across. They are essential to life on Earth, because they decompose organic matter and release the nutrients for use by the next generation of plants. Marine bacteria are most densely populated in the bottom sediments and on surfaces, such as the surfaces of fish, boats, seaweeds, pilings, and suspended sediment grains. These are the places where surface-active organic materials collect, on which bacteria feed.

In spite of their important role as decomposers and recyclers, the total biomass of bacteria is small compared with the total mass of the microscopic single-celled plants. Consequently, bacteria are not a major food source in the ocean, in general, except for the communities that surround deep sea hydrothermal vents. Bacteria do supplement the diets of some larger single-celled organisms, such as some dinoflagellates and protozoans, some microscopic multicellular animals, and some deposit feeders, who ingest the bacteria along with the organic sediments.

Some bacteria and some cyanobacteria are **nitrogen-fixing**. That is, they produce ammonia and nitrate directly from nitrogen in the environment. This process introduces new nutrients into the life cycle to supplement those already there, such as those released as organic matter decomposes.

BLUE-GREEN ALGAE

The other phylum under the kingdom Monera are **blue-green algae**. These are sometimes called *cyanobacteria* because most have a cyan (blue-

Table 12.4 Characteristics of the Dominant Single-celled Marine Organisms

Type	Typical Size (μm)	Skeletal Material	Where Dominant
Bacteria	<5	None	Sediments and surfaces
Producers (plants)			Sunlit surface waters only
Blue-green algae	5	None	Nowhere, but tend to grow on surfaces
Coccolithophores	3–10	Calcium carbonate	Warm open ocean (tropical and subtropical)
Silicoflagellates	5–40	Silica	Cool open ocean (polar and subpolar)
Diatoms	20–80	Silica	Cool, nutrient-rich (upwelling, polar, and coastal)
Dinoflagellates	10–50	Cellulose or none	Warm quiet waters, wherever the others are scarce
Consumers (animals)			
Protozoans			
Radiolarians	50–500	Silica	Surface waters and sediments
Foraminifera	100–1000	Calcium carbonate	Surface waters and sediments

green) coloration due to the chlorophyll-a pigment in their photosynthetic parts. Some have other pigments as well, such as the species that gives the Red Sea its name and is also sometimes responsible for a reddish tinge in the waters of the Gulf of California. Some species of blue-green algae can use both nitrogen fixation to make their own nutrients and photosynthesis to manufacture their own food. These abilities give them very simple nutritional requirements and make them the most independent organisms on Earth. Consequently, they are found in a wide variety of environments, including some that seem extremely inhospitable.

Blue-green algae tend to be a little larger than bacteria, although still tiny compared to cells of eucaryotes. Dimensions of around 3 to 5 micrometers are typical. Like bacteria, they also tend to be found on surfaces, although free-floating forms are also common. Some grow side by side and form filaments. The most familiar blue-green algae are those that form the slimy scum and films on rocks and other surfaces in stagnant water and those that fog up the glass of an aquarium. You can easily grow some by putting dirt in a jar of water and letting it sit in the sun.

In keeping with their tendency to grow on surfaces, some species form sticky coatings on certain shallow regions of the ocean floor. Sediments then stick to this coating, forming a layered mat with algae covered by fine sediments. The algae then grow up through the sediment layer, forming another sticky algal layer, to which still another layer of fine sediments stick. The structure keeps building upward, with alternate layers of algae and sediments. Over the years, the covered algal layers decompose, leaving the peculiar layered rock formation, called a **stromatolite**. The discovery and dating of ancient stromatolites gives evidence that these very simple plants were present very early in the Earth's history.

The most familiar blue-green algae are those that form the slimy scum and films on rocks and other surfaces in stagnant water and those that fog up the glass of an aquarium.

COCCOLITHOPHORES AND SILICOFLAGELLATES

In spite of their simplicity, early development, and early dominance on Earth, blue-green algae are no longer the most important primary producers in the oceans. Now, the bulk of the ocean's primary production is carried out by coccolithophores, silicoflagellates, diatoms, and dinoflagellates. These all are single-celled marine plants that have cell nuclei. Therefore, they are eucaryotes and belong to the kingdom Protista.

The distribution of these microscopic single-celled plants in the oceans is variable, depending on light intensity, nutrients, intensity of grazing, currents, and many other factors. As we have seen, there is a deficiency of nutrients through most of the ocean's surface waters, and so small size and efficiency are particularly important to most marine plants. Therefore, the majority of all food production in the ocean is accomplished by tiny single-celled plants that are

(a)

(b)

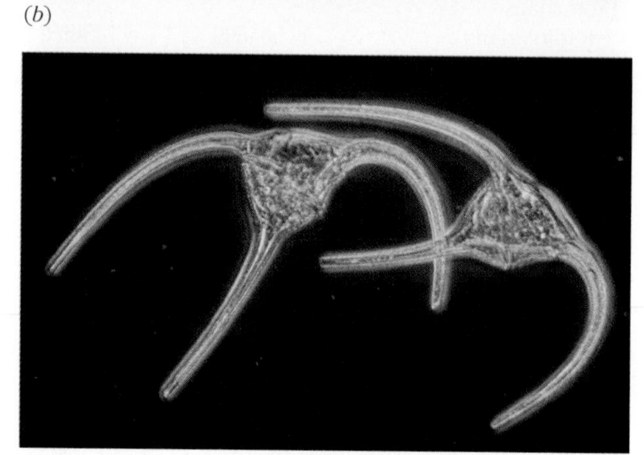
(c)

Figure 12.2
(a) Parts of coccolithophore skeletons, magnified about 14,000 times. The individual skeletal plates are called coccoliths, and their discovery in bottom sediments was our first clue that these very important and extremely tiny producers existed. (b) Several species of diatoms. (c) Dinoflagellate algae.

smaller than 10 micrometers. This small size has enabled some of them to avoid detection until relatively recently.

Of these tiny plants, **coccolithophores** (Figure 12.2a) have exoskeletons made of tiny **calcite** (calcium carbonate) plates and are found mostly in warmer tropical waters. Various species of tiny plants with exoskeletons of silica (SiO_2) are together called **silicoflagellates** and found mostly in cooler waters at higher latitudes. Calcium carbonate dissolves in colder waters, which is probably the reason that the silicoflagellates take over there. Both coccolithophores and silicoflagellates have tiny hairlike whips called **flagella** that give them some minor mobility. Their small size tends to make these tiny plants efficient producers and slow sinkers. Therefore, in nutrient-deficient or nonturbulent waters, such as those of the expansive central temperate and tropical oceanic regions, they have a definite advantage. Most of the food production in these waters is done primarily by the extremely tiny coccolithophores.

DIATOMS

The larger **diatoms** (Figure 12.2b) measure typically 20 to 80 micrometers across and generally have a golden brown color. Consequently, they are sometimes called *golden algae*. They thrive in waters of the continental shelves, regions of upwelling, and at higher latitudes where there is better mixing and larger nutrient concentrations.

Throughout most of the ocean, the diatom is handicapped by its larger size, both because of decreased efficiency and faster sinking. Efficiency is not so important, however, in the nutrient-rich coastal and upwelled waters. Furthermore, danger of extinction through sinking is diminished by the shallowness, turbulence, and mixing of coastal waters, and water motion in upwelling waters. The supply of sunshine and nutrients enables those that remain near the surface to reproduce rapidly and replace those lost to depth.

Because they dominate these nutrient-rich waters, diatoms are associated with the ocean regions of highest productivity per square meter. However, coastal and upwelling areas are only a small fraction

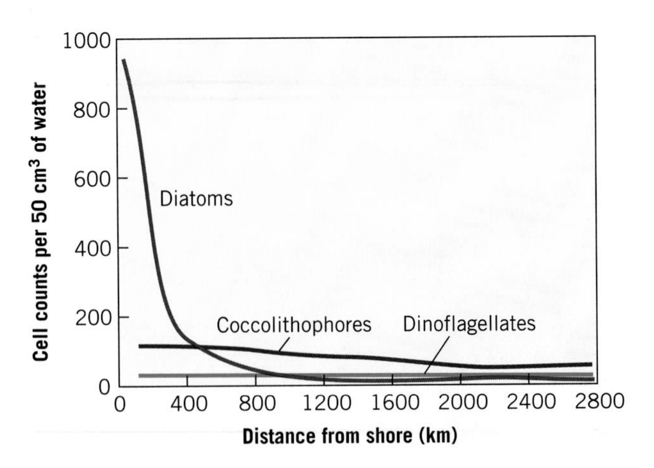

Figure 12.3

Distribution of diatoms, coccolithophores, and dinoflagellates with distance from land, going northeast toward the central North Atlantic from the Venezuelan coast. Diatoms often dominate in nutrient-rich coastal waters, but the smaller coccolithophores and dinoflagellates do comparatively better in the nutrient-deficient oceanic waters farther out. (Based on data from E. M. Hulburt, 1962, *Limnology and Oceanography* 7, 307–315.)

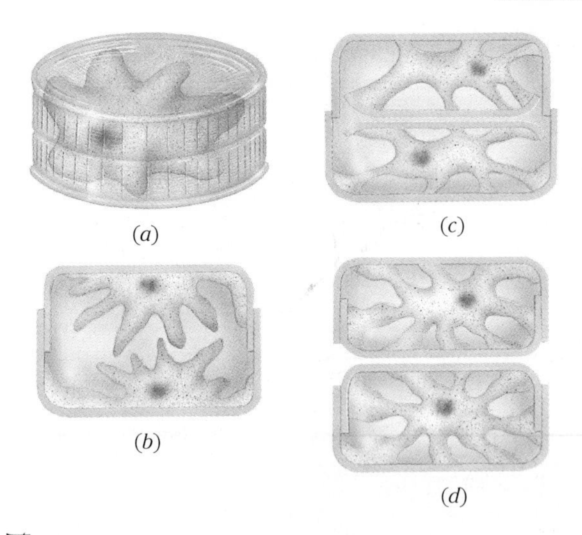

Figure 12.4

Illustration of successive stages of cell division for a diatom.

of the ocean, so the total amount of food produced by diatoms is less than that of the coccolithophores. The preference of diatoms for nutrient-rich coastal waters is illustrated in Figure 12.3, which shows a typical variation of microscopic plant populations with distance from shore.

Diatoms display a wide variety of geometrical shapes, such as tiny hockey pucks or pincushions, that are sometimes stuck together in short chains.

Each diatom has a perforated silica exoskeleton, called a *frustule*, that allows it access to nutrients and other needed materials in the environment without exposing itself to small predators. Also, the rough spiky shell increases friction with the water. This is important to the denser-than-water diatoms, which continually face the problem of sinking too rapidly out of the upper sunlit waters. They also produce oils to help neutralize their buoyancy, partially compensating for the dense protective frustules. Diatoms display a wide variety of geometrical shapes, such as tiny hockey pucks or pincushions, that are sometimes stuck together in short chains. These shapes tend to be either centric (i.e., cylinder or wheel shaped) or pinnate (i.e., cigar of football shape) (Figure 12.2b). Centric diatoms tend to live in the water, whereas pinnate diatoms are found mostly on the shallow ocean bottom or attached to surfaces.

Diatoms have the ability to produce thick-walled **resting spores**, which enables them to wait out extended periods in an inhospitable environment. When conditions again permit, the resting spore will develop into a diatom, and normal activity is resumed. This is clearly an advantage in polar latitudes, where organisms must wait out the long winter of darkness.

Under favorable conditions, diatoms can reproduce as rapidly as twice per day, quickly producing a diatom *bloom*. They reproduce by cell division. As is illustrated in Figure 12.4, their frustules come in two parts, one fitting over the other like the cover of a shoe box. When the cell divides, each new cell takes half of the former frustule and must manufacture the other half itself. The new half is always the inner half, so there is a tendency over the generations for the frustules to become smaller and smaller. Eventually, this housing is too small, and the diatom discards it entirely, produces a new more expansive frustule, and the process begins again.

In some places, the tiny intricate and porous exoskeletons of diatoms sink to the bottom and form deposits called **diatomaceous earth**, which has many commercial uses. These include fine filters, deodorizing and decoloring agents, cleaners, polishers, and paint removers. When absorbed into diatomaceous earth, the very touchy explosive, nitroglycerine, becomes the "safe" explosive dynamite. This discovery was the basis of the Alfred Nobel fortune, from which the annual Nobel Prizes are paid.

Figure 12.5
Lichen covered boulders. Lichens involve symbiotic relationships between a fungus that provides structure and moisture, and a microscopic algae that provides food.

DINOFLAGELLATES

Dinoflagellate algae (Figure 12.2c) generally have two pronounced flagella that run through grooves in the cell wall. Many have exoskeletons made largely of cellulose. Thus, they are more mobile and more buoyant than diatoms, and they therefore have an advantage in calm or downwelling water. Because dinoflagellates don't have a mineral skeleton, they don't contribute to the sediments. Some can photosynthesize and others cannot. Some have the ability to absorb nourishment from water when removed from sunlit regions. Many dinoflagellates are bioluminescent. They can make the water seem to glow at night when agitated, such as in a ship's wake, in the surf, or when you step on a wet beach. Some develop important symbiotic relationships with other organisms, such as the zooxanthellae that live within the tissue of corals.

Many dinoflagellates are bioluminescent. They can make the water seem to glow at night when agitated, such as in a ship's wake, in the surf, or when you step on a wet beach.

There is a tendency for dinoflagellates to be most heavily concentrated very near the surface, whereas other productive microscopic plants are most highly concentrated somewhat below the surface. The reason for this probably is related to the extra mobility of dinoflagellates, which allows them to stay closer to the bright sunlight. Dinoflagellates are versatile and found in a large variety of ocean waters, but tend to be more numerous where it is warmer.

Dinoflagellates may temporarily become the dominant producers when there is stagnant water or gentle downwelling, which is so detrimental to the other less mobile algae. During dinoflagellate blooms, some dinoflagellates give the water a reddish color, and some are poisonous to humans. The feared *paralytic shellfish poisoning* is associated with these **red tides** in some coastal areas.[2] Shellfish feeding on these algae are not affected, because they have nonacid digestion. But the human stomach contains acids. When we eat shellfish that have fed

[2]Reddish water can also be caused by a species of blue-green algae, which is not poisonous.

on these dinoflagellates, our stomach acids react with some of the dinoflagellate materials to produce toxins that attack our nervous systems. Depending on the amount consumed, the effects can be symptoms similar to flu or drunkenness, or in more severe cases can cause paralysis, starting with the lips, followed by the stomach, total paralysis, and then possibly death.

PROTOZOANS

Protozoans are single-celled animals that can be found either on the ocean floor, or floating in the water. They are most common in warm surface waters and environments containing decomposing organic matter. Some graze on diatoms and other phytoplankton, and some rely heavily on the direct absorption of organic molecules for their nutrition. The survival of most species is quite sensitive to environmental conditions, which enables us to learn about environmental conditions long ago by studying their skeletal remains in ancient sediments.

Many of the more plentiful kinds of protozoans secrete exoskeletons for protection, but can extend their protoplasm through perforations in these skeletons in order to capture food. Important examples include foraminifers, which have calcareous skeletons, and radiolarians, which have intricate silica skeletons (Figure 5.12). Calcium carbonate dissolves better in cooler deeper waters. Therefore, among the sea floor sediments, the calcareous foraminifer skeletons are more common in waters shallower than about 4 kilometers, and the siliceous radiolarian skeletons are more common in the deeper waters.

12.3 FUNGI

Fungi are decomposers. In the oceans, bacteria handle the decomposition chores very efficiently. Although microscopic fungi are found in a wide variety of marine environments, they constitute only a very tiny fraction of the ocean's living biomass. In general, they are less prominent in the ocean than on land. The most familiar marine representatives are lichens in the intertidal zone (Figure 12.5), in which the fungus has a symbiotic relationship with a type of blue-green algae. The fungus helps retain water for the algae when exposed and the algae help provide nourishment for the fungus.

12.4 MULTICELLULAR PLANTS

Because of their mild and moderated environment, marine plants tend to be much simpler than those on land. No mosses or ferns are found in the ocean. Flowering and seed-producing plants, so common on land, are found only occasionally in shallow coastal waters. Examples are marsh grasses and mangrove trees. The overwhelming majority of marine plants are much simpler than these and are referred to as **algae**.

As we have already seen, the most productive of these algae are the microscopic, free-floating, single-celled plants studied in the previous section. Because they are microscopic, we don't notice them in spite of their large numbers and importance. But there are some larger, multicellular algae as well. As you might guess, these larger algae are not as efficient, and they therefore are found only in nutrient-rich waters. In fact, they are almost wholly confined to the regions of the ocean that are very near shore. These anchored coastal *seaweeds* are more familiar to us, both because of their closeness to land and their size (Figure 11.1c).

SEAWEEDS

Seaweeds are algae that cling to rocks and other solid surfaces with **holdfasts**. Some may have central trunks, or **stipes**, extending upward, from which branch leaflike **fronds**. Seaweeds do not have vascular systems that transport water and nutrients as do rooted land plants. The holdfasts serve only to anchor the plants and are not roots. Coastal areas having stable or rocky bottoms make up less than 2% of the total ocean area, so even if they were comparably efficient, the anchored algae would contribute considerably less to the total ocean productivity than planktonic algae. Many also offer support, protection, and sometimes food for various smaller marine creatures. Because of their size and proximity to land, they can be harvested. Some are used in food, food additives, emulsifiers, and various drugs.

Green algae (phylum Chlorophyta) grow in leafy, matted, or filamentous forms. They are found attached to pipes, pilings, or other objects that have been sitting in shallow water (Figure 12.6a). The leafy forms are called *sea lettuce* because of their appearance. Unlike lettuce, however, most green algae have absolutely no stiffness or crispness. They rely entirely on the buoyancy of the water for their support, and they collapse completely if removed from water. Some warm water species of green algae se-

(*a*)

(*b*)

(*c*)

Figure 12.6
Examples of: (a) green algae (codium), (b) brown algae (knotted rockweed), and (c) red algae (Irish moss on a mussel shell).

crete calcium carbonate, which provides some rigidity and contributes to the building of coral reefs.

Brown algae (phylum Phaeophyta) are purely marine and the most advanced of the various forms of algae (Figure 12.6b). Familiar examples are kelp and sargassum, which we frequently find washed up on our beaches. Many brown algae have gas-filled bulbs that hold their leafy fronds near the surface, where the sunlight is stronger. They tend to be large, with some species growing to greater than 20 meters in length.

Sargassum along coasts in the Caribbean Sea is sometimes torn from its footings and carried by ocean currents into the central convergence of the North Atlantic Ocean. Huge collections of this seaweed there give this region the name *Sargasso Sea*. The floating sargassum weed can grow and reproduce while floating, and provides a home to many small organisms in this central ocean region.

Red algae (phylum Rhodophyta) are mostly marine and smaller than the dominant brown algae species. Red algae are the most colorful, varied, and widespread of all seaweeds (Figure 12.6c). Some grow in branching filaments and secrete calcium carbonate, which makes them look and feel like colorful branching corals.

Both red and brown algae tend to grow in deeper water than green algae, which should make sense to you if you note that the color of an object is the color that it reflects. Green algae reflect the middle part of the solar spectrum and therefore are not as efficient at absorbing sunlight as brown and red algae. This is particularly crucial in the deeper coastal waters, where sunlight is particularly meager and only the blue-green part of the spectrum reaches all the way to the bottom.

VASCULAR PLANTS

As we have seen, primitive plants were growing in the oceans billions of years ago, fairly close to the Earth's beginnings. However, plants did not begin to colonize land until about 450 million (0.45 billion) years ago, which is understandable in terms of the relatively harsh conditions encountered on land. Because of the need to cope with these conditions, land plants are relatively large and complicated. The dominant plants are vascular, flowering, and seed-

Figure 12.7
Marsh grass is a vascular plant, and is shown here at low tide.

producing plants (Phylum Tracheophyta). The vascular system is needed to transport water and nutrients between the roots and leaves.

A few of these more advanced plants colonize the ocean's margins—particularly the shallow estuarine waters. But no species of vascular plant is found in purely oceanic regions, where they cannot compete with the simpler and more efficient marine plants. Marine plants, after all, can concentrate on production and needn't divert significant resources to support and the transportation of materials as land plants must. The coastal marine vascular plants include marsh grass (Figure 12.7), which is frequently found in shallow brackish estuaries at latitudes below about 60°, and mangroves (Figure 5.11), which are prominent in shallow brackish waters at latitudes below about 30°. Marsh grass and mangroves owe their survival to their tolerance of salt and ability to aerate their roots via special tubes.[3]

12.5 MULTICELLULAR ANIMALS

Unlike plants that need sunlight, animals of one kind or another can inhabit all regions of the ocean; they are limited only by the supply of food and oxygen. Of course, the particular species of animal varies from one region to another. There is sufficient dissolved oxygen for animal life throughout most ocean regions, but food is most concentrated near the surface, where it is produced in photosynthesis. Consequently, most marine animals live within a few hundred meters of the surface. Some detritus accumulates on the ocean bottom, and so the bottom supports a collection of animals as well. Throughout most of the ocean, these two regions are separated by 3 or 4 kilometers of relatively barren waters, but over the continental shelf, the water is shallow and these two regions merge (Figure 11.11).

Animals were in the oceans long before life of any kind appeared on land. Consequently, compared to land animals, marine animals have had a longer evolution and display a wider variety of forms. All animal phyla are represented in the oceans, but not all are represented on land. Let us now consider some of the animals that are more prominent in the oceans, organized according to phyla. We begin with the phylum containing the simplest animals and proceed in order of increasing complexity.

SPONGES

Sponges (phylum Porifera) are porous filter feeders. They are multicellular animals that have few natural enemies, and they can be found attached to the sea floor at all depths (Figure 12.8). Sponges have very little organization among the cells, and their structural support is furnished by small spicules distrib-

[3]All plants need oxygen when they respire, and land plants get their oxygen from the air, rather than water.

Figure 12.8

Yellow sponges on a reef. Sponges filter their food particles from the water that passes through them.

(*a*)

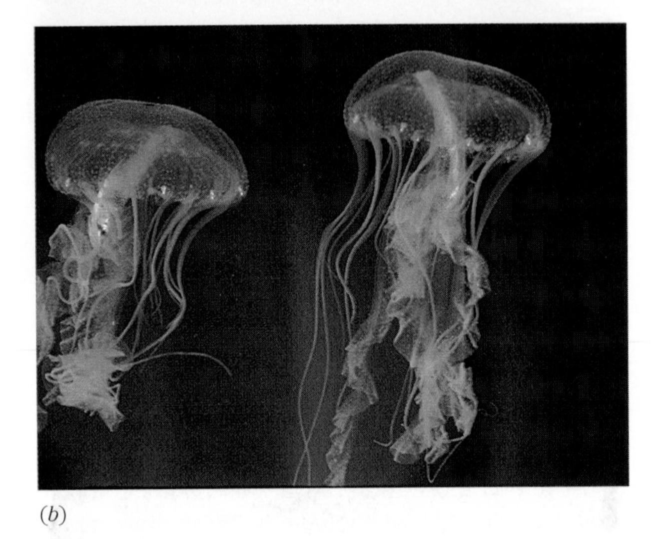

(*b*)

Figure 12.9

Coelenterates: (a) An anemone, and (b) two jellyfish, each consuming a small fish that it has captured.

uted fairly randomly throughout their volume. Sponges feed by filtering microorganisms and organic detritus from water passing through their porous bodies. This explains why they are full of holes. Water must pass through them in order for the food particles to be filtered out. Furthermore, only those cells near the surface of a sponge have any chance of access to such a food particle. Deep interior cells would get no food and die of starvation, thereby leaving another hole.

COELENTERATES

Coelenterates[4] (phylum Coelenterata, or often Cnidaria) are very simple animals, consisting essentially of an open gut that is surrounded by tentacles (Figures 12.9a and 12.9b). All coelenterates feed on meat, and all have stinging cells in their tentacles. The sting paralyzes small prey, and the tentacles help bring the prey into the gut. There is not as much specialization among the cells as there is in higher animals, and so each cell must serve a variety of functions. For example, there are no pure muscle cells. Therefore any motion, such as that of a tentacle, is accomplished by cells that must serve other functions as well.

There are a variety of members in this phylum. Some are fixed to the bottom and others are free floaters. This phylum also displays considerable **polymorphism**, or the ability of a single species to appear in a variety of forms. For example, successive generations of a species may alternate between attached and free-floating forms. Examples of coelenterates attached to the bottom are corals and **anemones** (Figure 12.9a), which form the largest class within this phylum. Anemones' open guts are surrounded by tentacles and face upward. They resemble flowers, 1 to 10 centimeters in diameter, and are popular and colorful inhabitants of tide pools.

[4]Pronounced "sill-en'-ter-ate."

Children enjoy agitating the center of the anemones with their fingers and feeling the tentacles close in. This is how anemones capture small prey, but as yet there has not been an anemone victorious over a finger. Food particles falling on individual tentacles are individually brought into the gut, although the tentacles can operate in unison in attempts to capture somewhat larger prey.

A coral is also like a tiny anemone—a miniature open gut surrounded by delicate tentacles facing outward to capture tiny food particles that are drifting by. But many of the very important corals have other interesting adaptations. Many excrete calcium carbonate and grow in close colonies, thus forming the hard coral skeletal structure found on coral reefs (Figure 4.22f). These structures contain tiny holes into which the corals can withdraw for protection. They often remain withdrawn during the daytime when predators could otherwise see and feed on them, but extend their tiny tentacles out through the holes to feed at night. Many reef-building corals also live in close symbiotic relationship with a dinoflagellate algae, zooxanthellae. The coral provides the support to keep the algae in the sunlight, while the algae provides nourishment to the coral and helps in the production of calcium carbonate.

Figure 12.10

Two ctenophores. Ctenophores are sometimes called "comb jellies" or "sea walnuts."

Another class of coelenterates includes jellyfish (Figure 12.9b). In contrast to anemones, they float through the water and their tentacles hang downward. Also in contrast to anemones, the tentacles of some species of jellyfish have registered minor victories over humans, sometimes inflicting pain to unsuspecting swimmers. Many jellyfish can swim slowly by rhythmic pumping of their gut. Some sink slowly through the water, capturing prey as they fall, and then "pump" back to the surface to start again. Another class includes the *Portuguese man of war* and *by the wind sailor* (velella). These resemble jellyfish, but the various parts of an individual are fairly independent, making the organism more like a close colony of individual animals with a special division of labor among them. Bryozoans (moss animals) appear similar to small anemones or corals, but bryozoans have a distinction that places them in a different phylum altogether. Although all coelenterates release undigested materials simply by opening up their tentacles, bryozoans have a separate anus, which allows these materials to pass out the side.

CTENOPHORES

Ctenophores[5] (phylum Ctenophora) look like small, new improved versions of jellyfish (Figure 12.10). A typical ctenophore is the size and shape of a walnut, but smooth, transparent, and has eight rows of tiny hairs along its side that beat in unison for minor propulsion. Some have no tentacles. Of those that do have tentacles, some have no stinging cells. Because they are small and transparent, ctenophores are not as familiar to the average person as the more visible coelenterates. But ctenophores have an important impact on the marine biological community due to their heavy predation on microscopic animal life. All ctenophores are bioluminescent, and so these small, symmetrical, graceful, transparent creatures are quite visible and strikingly beautiful in dark waters. They are sometimes given the common names *comb jellies*, because of their eight combs of beating hairs, or *sea walnuts*, because of their size and shape.

WORMS

There are many different phyla of worms. Compared to coelenterates and ctenophores, worms have more advanced and closed digestive tracts, and tend to be

[5]Pronounced "ten'-a-fores."

(a)

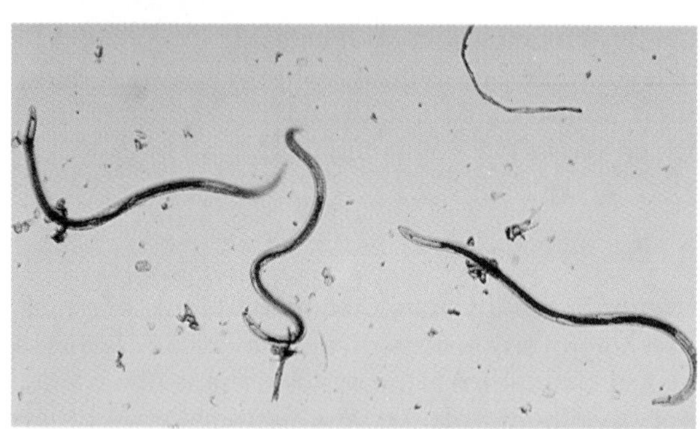

(b)

(c)

(d)

(e)

Figure 12.11

(a) A flatworm (phylum Plathyhelminthes) (b) A round worm (phylum Nematoda) (here magnified about 40 times). (c) A segmented worm (phylum Annelida). (d) Featherduster tube worm (phylum Annelida). (e) An arrow worm (phylum Chaetognatha).

smaller, thinner, and more mobile. Whereas coelenterates and ctenophores are purely aquatic and primarily marine, most worm phyla are found in a wide variety of environments. We very briefly describe a few kinds of worms found in the oceans.

Marine **flatworms** (phylum Plathyhelminthes) (Figure 12.11a) are flat, typically a centimeter or so in length, and have the simplest digestive tract of all worms. It has a simple gut with a single opening that must serve as both a mouth and an anus. The digestive tract serves more to distribute the food material to the cells, than to digest it. Marine species are mostly deposit feeders, inhabiting the rocks or burrowing in the mud, but some are parasitic on larger organisms. Flatworms move with the help of tiny hairs covering their bodies.

Round worms, or **nematodes** (phylum Nematoda or Aschelminthes), are microscopic worms, whose digestive tract runs the length of the worm and has an opening at both ends (Figure 12.11b). Nematodes have no hairs for propulsion like flatworms and rely on almost random wiggling for propulsion. Although unfamiliar to us because of their microscopic size, these primitive animals are surprisingly abundant in all life environments on Earth. In the oceans, they are particularly abundant in the bottom sediments, where they feed on microscopic detritus.

Earthworms and freshwater leaches are familiar **segmented worms** (phylum Annelida), sometimes called *annelids* (Figure 12.11c). Most marine species live on or in the ocean bottom, although a few species have developed special appendages to facilitate swimming. Some filter-feeding annelids build tubes attached to the bottom that they live in, spreading flowerlike tentacles from the top, with which they capture small food particles (Figure 12.11d). Some pump water through their burrows, so they can filter food from the water without being exposed to predators at the surface. Some are deposit feeders that ingest sediments rich in organic matter. A familiar deposit feeder on land is the earthworm.

Arrow worms (phylum Chaetognatha) live in the water, unlike the flatworms, nematodes, and segmented worms that live primarily on the sea floor. The adults are 1 to 4 centimeters long, transparent, have fins for locomotion, and are fast swimmers. They are sometimes called *bristle-jawed worms*, because they have movable bristles around the mouth, which they use to seize tiny prey and other food particles (Figure 12.11e). Adult arrow worms are carnivores, because their diet includes microscopic animals. This sets them apart from most other worms, which feed on microscopic plants and detritus.

MOLLUSKS

Mollusks (phylum Mollusca) all have soft bodies, a soft "foot" used for movement, and many secrete calcareous protective shells. There are a variety of mollusks, separated into several classes. We mention the most important marine classes here. A few are illustrated in Figure 12.12.

Chitons[6] are small, flat mollusks that have eight calcareous plates protecting their back. They live in shallow water or intertidal regions and feed on algae that they scrape from the rocks as they slowly creep along.

Gastropods are limpets,[7] snails, nudibranchs, or slugs that also move about on their "foot." In a few species, the foot has evolved into a means of propulsion through the water. Pteropods are tiny free-floating relatives of snails, which feed on floating organic matter, including microscopic animals. But most gastropods are confined to the ocean floor, consuming whatever appropriate organic material may be found there. Those of you familiar with aquariums may know that when gastropods are put inside, they creep over the glass, eating the algae and keeping the aquarium glass clean. Some gastropods are carnivorous. Some snails can bore through the shells of other mollusks to feed on the soft meat inside. **Nudibranchs** are very colorful carnivorous slugs that feed on such things as sponges, coelenterates, or even hapless worms.

Those of you familiar with aquariums may know that when gastropods are put inside, they creep over the glass, eating the algae and keeping the aquarium glass clean.

Bivalves, such as mussels, clams, scallops, and oysters, have hinged shells. The adults of this class live on or in the ocean bottom, or attached to solid surfaces. Like gastropods, they can use their foot for propulsion along the bottom. They feed by filtering bits of organic matter from water siphoned through tubes. Animals that filter food particles from the water are called **filter feeders**.

In another class of mollusks, **cephalopods**, the "foot" has developed into a ring of tentacles with suction cups that surround the mouth. Each cephalopod

[6]Pronounced like "kite" with an "n" added at the end.

[7]Limpets are sometimes called *Chinese hat snails*, because their shell is the shape of a low, flattened cone, resembling a tiny Chinese straw hat.

(a)

(b)

(c)

(d)

Figure 12.12

Mollusks: (a) An octopus. (b) Two nudibranchs. (c) A colony of mussels on a rock. (d) A chiton.

has eyes, a chewing beak, and the ability to squirt a jet of water. This group includes the octopus, squid, and nautilus. The adult giant squid is the largest invertebrate (i.e., no segmented backbone) animal on Earth, reaching lengths greater than 15 meters.

ARTHROPODS

Arthropods (phylum Arthropoda) are the largest group of animals on Earth, both in terms of number of different species and numbers of individuals. Arthropods have segmented bodies, jointed legs, and external skeletons ("arthro" refers to joints, and "pod" to feet). In addition, they usually have two pairs of antennae and a pair of mandibles (i.e., jaws). The majority of all land animals are arthropods that belong to the class Insecta, and the majority of all marine animals are arthropods that belong to the class Crustacea. We call the organisms in these classes *insects* and *crustaceans*, respectively. Over 75% of all animal species belong to this phylum, and the fraction of individuals is even greater. In some sense, arthropods rule the Earth, both land and sea.

Crustaceans are extremely varied, and so here we just describe some of the more prominent groups of marine crustaceans. The most numerous are the many various species of **copepods** (Figure 11.18), which resemble tiny shrimp that are typically 0.2 to 2 millimeters long. They make up the bulk of the animal mass in the oceans. They feed primarily on microscopic plants and particulate organic matter. They are the most important single link between the microscopic plants and larger animals in the ocean food chain. Copepods mature rapidly, and at higher latitudes they go dormant and sink into deep water to wait out the dark winter months. Another subclass of crustaceans includes the *euphasids* or **krill**, which are similar to copepods, but somewhat larger (Figure 12.13). As adults, krill range in length from 1 to 5 centimeters and are particularly abundant in cold waters of very high latitudes, where they are a favorite food of some whales.

Isopods are common small crustaceans that are flattened from top to bottom (like the tiny armadillo-like sow bugs you sometimes find under rocks) and

(a)

(b)

(c)

Figure 12.13

Antropods. (a) Krill are larger cousins to the copepods, typically 1 to 5 cm in length, and are abundant in productive sub polar waters. (b) Barnacles, such as this gooseneck barnacle, are similar to copepods in their larval stage. But then they attach themselves to a solid surface and produce a hard protective home inside which they stay, capturing their food particles as they drift past the small opening. (c) Crabs are decapods.

amphipods are flattened from side to side. Various species of both are found in a wide variety of marine environments, but many live on the ocean floor. You might see examples of each in shallow coastal waters when you are at the beach. A familiar amphipod is the *sand hopper*, which appears to beach goers like a low-flying insect that feeds on seaweed washed up on the beach.

Barnacles (Figure 12.13b) belong to another subclass and seem to bear absolutely no resemblance to the other crustaceans, all of whom seem obviously related to shrimp or crabs of various sizes. The link is that barnacles have a larval stage, and these tiny larvae do look like the other crustaceans. They are also mobile, in contrast to the adult barnacle, which accounts for the reason why barnacles can appear on such unlikely surfaces as boat hulls, whales, crabs, or sharks. The adult barnacle secretes a shell around itself after attaching itself to a surface. Its feathery legs reach out through a hole in this shell and create water motion to help it capture tiny particles of food that might float by. It is said that a barnacle lays on its back and kicks food particles into its mouth.

Lobsters, crabs, shrimp, and prawns (Figure 12.13c) belong to the order of crustaceans called **decapods**, whose names refer to their 10 legs. Crabs have the ability to regenerate outer parts of limbs that they have lost in combat.

ECHINODERMS

Echinoderms[8] (phylum Echinodermata) are spiny-skinned radially symmetric bottom dwellers with an internal skeleton. All members of this phylum are exclusively marine, and most have five-sided symmetry. One class includes starfish (see Figure 12.12c.), one class sea urchins and sand dollars, and one class sea

[8]Pronounced “ee-kine’-o-derms” and means “spiny-skinned.”

Figure 12.14
Echinoderms, including a sea urchin (spiney), a red starfish, and a dark red sea cucumber in the foreground.

cucumbers (Figure 12.14). To see the five-sided symmetry of a sea cucumber, you must look at the skeletal framework along its length, rather than from top down.

Most echinoderms are filter feeders, but many starfish are predators, feeding on such prey as oysters and clams (Figure 12.12c). Suction cups on their arms enable them to grasp and pull open the bivalves. Then they extend their own stomachs inside the shell to digest the soft tissue there. One species of starfish, called the *crown of thorns*, feeds on coral and has caused considerable destruction on some reefs in the Pacific.

Some echinoderms have peculiar ways of dealing with hardship or injury. When attacked, the sea cucumber can discharge some of its entrails, leaving a meal that, it is hoped, satisfies the predator. Later, it can regenerate replacement parts. The starfish can regenerate missing parts as well. At one time, oyster and clam fishermen would cut up the starfish they captured with the hope of killing this predator of the clam and oyster beds. Unfortunately, this did not accomplish what was intended.

CHORDATES

Chordates[9] (phylum Chordata) are animals that have gill slits and a cartilaginous skeletal rod at some stage in their development. Most marine chordates are large fast swimmers. However, one important primitive marine chordate, called a **tunicate** (Figure 12.15), has very limited mobility. Tunicates are transparent sacklike animals that resemble slim ctenophores. They can grow up to 8 centimeters in length and have an internal, thin, nonsegmented skeletal rod. There are openings in both ends of the sack, so that food particles pass in one direction through the primitive digestive tract. Tunicates attached to the sea floor are called *sea squirts*, and those that live in the water are called *salps*. Salps move with weak jet propulsion, forcing water out of one end of the sack. Some species form colonies in chains, or tube shapes.

The most common and well-known members of the phylum Chordata belong to the subphylum Vertebrata. A **vertebrate** is characterized by having an internal skeleton with a spinal column of vertebrae, a brain, red blood, and two pairs of appendages. People, of course, belong to this subphylum. (Did you know that you had gills once during your development?)

The most primitive vertebrates belong to a class of jawless fish that includes the sea lamprey (Figure 12.16), which has no scales or jaws and lives by attaching itself to larger creatures and sucking its nutrition from the body of the host. Hagfish are similar, but feed on animal carcasses.

Cartilaginous Fish **Cartilaginous fish** belong to another class (Figure 12.17). They have vertebrae and skeletons of cartilage, sharp pointed scales that do not overlap and give the skin a sand-paper feeling, mouths on the underside of the head, and no gill covers. Familiar members of this class include sharks, skates, and rays.

Cartilaginous fish have no air bladders, so they all sink when not swimming. Rays and skates are flat, spend most of their time on the bottom, and hide from predators by covering themselves with sedi-

Figure 12.15
A colony of tunicates, which are small primitive chordates, each having a cartilaginous skeletal rod.

[9]Pronounced "core'-dates."

(a)

Figure 12.16
(a) Photo of the suction mouth of a sea lamprey.
(b) Sketch of a sea lamprey (top) and hagfish (bottom). Actual lengths are about 60 centimeters.

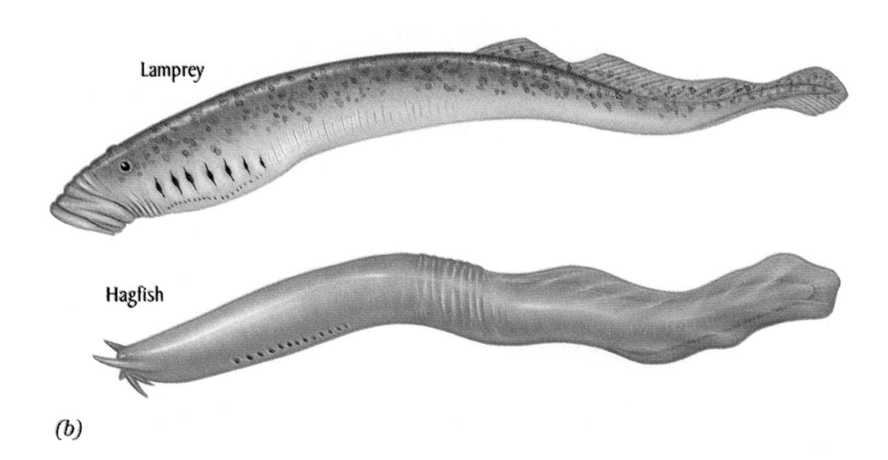

(b)

ments. Consequently, they are hard to see, except for their beady eyes and occasional movement of their gills. Most are harmless, but some skates and rays can give small electrical shocks to fend off would-be predators. One species of ray, a *sting ray*, defends itself with a barbed spike beneath its tail.

Sharks stay off the bottom by swimming continuously. This also prevents suffocation, because unlike bony fish, most sharks cannot pump water through their gills and therefore rely on their motion through the water for this purpose. Their tail fin is crescent-shaped, with an elongated upper portion. This shape of the tail fin, along with stiff pectoral fins and their body contours, help provide lift and prevent sinking as they swim.

Sharks first appeared in the oceans about 450 million years ago, so they have survived longer than most animals presently on Earth. Obviously, they must be well adapted to cope with life. Sharks have extremely good senses of smell and hearing, being able to smell traces of blood in the water at large distances and hear a fish that is in trouble and thrashing in the water. The only natural predator for a shark is a larger shark. Sharks do attack humans, but relatively rarely. Only about 100 shark attacks are recorded in the world per year, and few of these are fatal. The largest sharks are the whale shark and basking shark, both of which feed on copepods and other tiny animals. The whale shark can be up to 18 meters long and can weigh as much as two African elephants.

Although most marine organisms have enormous spawns (i.e., eggs or offspring produced in a batch), sharks, skates, and rays tend to produce only a few fertilized eggs, which are sufficiently well protected to ensure relatively high survival rates.[10] Unlike the eggs of bony fish, shark eggs are fertilized within the

[10]Marine mammals also give birth to relatively few, well-protected offspring, as do land mammals.

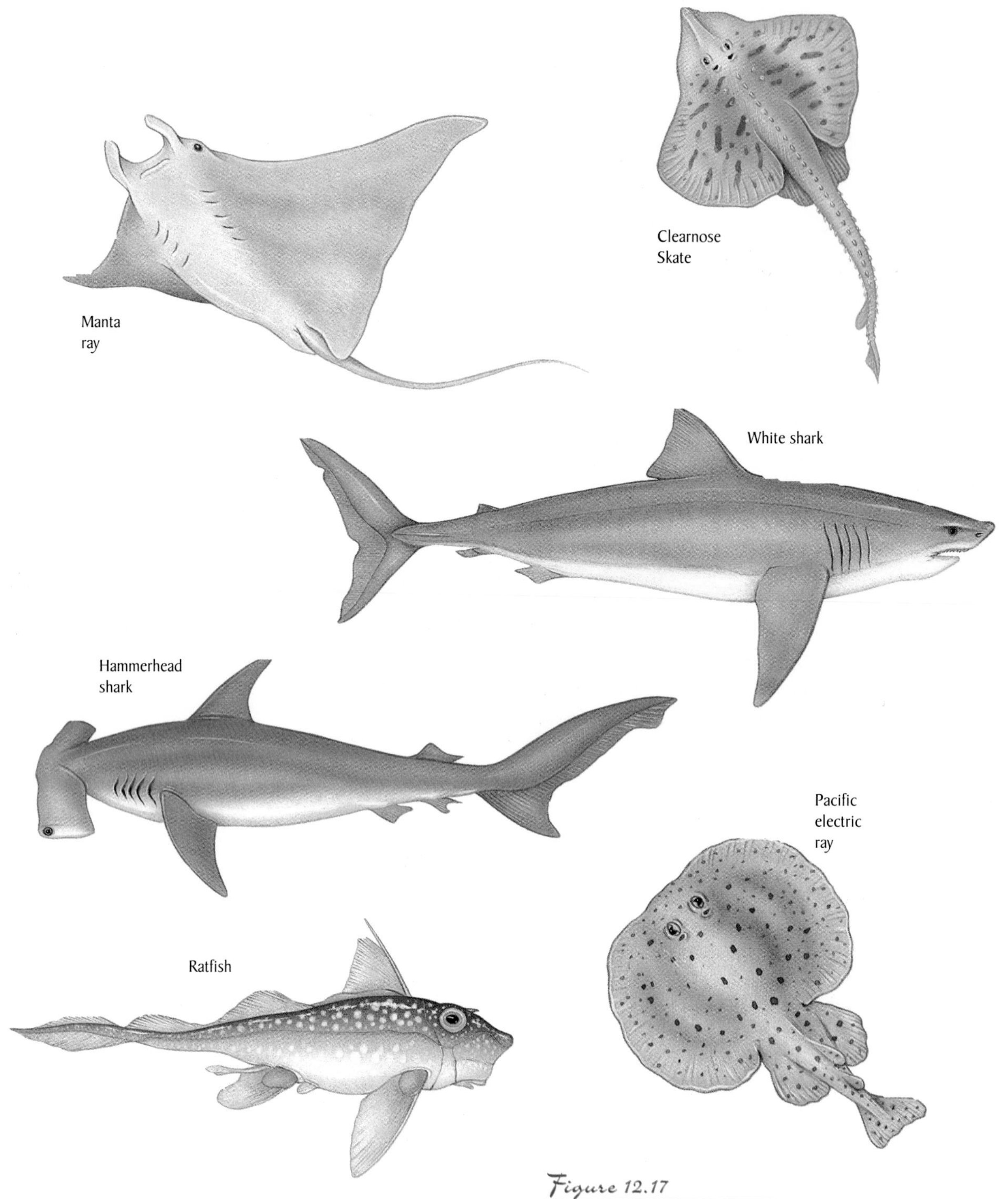

Figure 12.17

Cartilaginous fish. All sink slowly when not swimming. Note how the pectoral (side) fins have developed on the skates and rays, which spend most of their time on the bottom.

mother, and most sharks give birth to live pups. Some release their eggs in protective leathery sacks. They don't take care of their young, however, and might even eat them.

Bony Fish **Bony fish** differ from cartilaginous fish in that they have bony vertebrae and skeletons, overlapping scales, gill covers, and a mouth at the very front of the body. Some feed on seaweeds, some on tiny free-floating organisms, some on other fish, and some on bottom dwellers. Some have gas-filled air bladders to help neutralize their buoyancy.

(a)

(b)

(c)

Figure 12.18

The shape of the caudal (tail) fin is made for speed on cruising fish like the tuna or jack(a) and for acceleration on fish that lunge for their prey, such the grouper (b). The shape of the pectoral (side) fins also vary depending on the need. Many reef fish (c) use the pectoral fins for pinpoint navigation, fluttering them so fast they are a blur. Some have other uses, such as the flying fish that use them as sailing wings.

Some migrate large distances for feeding and spawning. Typical spawns involve many thousands or even millions of eggs, of which only two per lifetime must survive to replace the adults, on average. Clearly, the survival rate for youths is small. Youths need different foods than the adults, and some adults must migrate to ensure that the spawn will hatch in appropriate feeding grounds. Salmon and Atlantic eels migrate between fresh and salt waters. The salmon spawn in fresh water and mature in salt, and the Atlantic eels do the opposite.

Fish have a variety of shapes, depending on their needs. Some, such as tuna and mackerel, spend most of their time cruising at fairly high speeds. Albacore have been clocked at 75 kilometers per hour. These cruising fish cover much distance per day and thereby encounter lots of prey, although their success rate in capturing the prey they encounter is typically only around 10 to 15%. These cruisers have tall, thin crescent-shaped caudal (i.e., tail) fins that are good for speed, but not maneuverability or acceleration (Figure 12.18). Others, like bass or barracuda, have long and thin bodies, and have shorter and fatter caudal fins that provide good acceleration. These do not cover as much territory per day as the cruisers, but because of their powerful acceleration, they have a much higher success rate at capturing prey that they target, typically being around 70 to 80%. Compared to others, cruising fish tend to have higher metabolism and slightly elevated body temperatures. Their muscles have a higher oxygen content and a reddish color, compared to the white meat of fish that rely on acceleration or mobility for catching their sustenance.

Some fish, such as those living on reefs, rely on mobility instead of speed or acceleration for food gathering and protection. Many of these have tall and laterally compressed bodies, like a pancake on edge, and rely on active pectoral fins (below and behind the mouth) for pinpoint navigation (Figure 12.18c). Pectoral fins have become winglike in flying fish, thin and peglike for some who stand on the bottom, and

Figure 12.19
Marine reptiles include sea turtles and sea snakes.

long, broad, and wavy for skates and rays (Figure 12.17).

Among the common commercial fish species, the small fish that are netted in huge schools include herring, anchovy, menhaden, and sardines, all of which feed directly on tiny free-floating organisms (Figure 11.6a). The larger commercial fish that are often landed individually include mackerel, tuna, and swordfish (Figure 12.18a). Commercial species that lay flat on the ocean floor include flounder, halibut, and sole, and common commercial rockfish include cod, perch, and snapper.

Reptiles and Birds Reptiles are another class of vertebrates whose members are particularly varied. All reptiles breathe air, have scaly skin, are cold-blooded, and most lay eggs that have been fertilized within the female, although some lizards and snakes give birth to live young. Eggs are laid on land. In contrast to fish eggs, reptilian eggs have leathery protective shells with amniotic fluid inside, which helps protect the embryos from the harsh land environment. Most reptiles feed on whatever they can catch.

A few lizards, such as the marine iguana and Komodo dragon, live in certain coastal waters, as do relatives of alligators and crocodiles. Two additional classes of marine reptiles are turtles and snakes (Figure 12.19). Adult sea turtles weigh up to a ton and live more than 100 years. There are about 50 species of sea snakes, many of which are extremely poisonous. But fortunately for us, they are not aggressive and they have small mouths.

Birds evolved from reptiles during the age of the dinosaurs. The evidence of their reptilian ancestry is seen in the scaly skin on their legs, but the body scales have evolved into feathers. Like reptiles, birds breathe air and lay eggs with amniotic fluid and protective shells. But unlike their reptilian ancestors, they are warm-blooded. Many birds, such as the pelican, cormorant, seagull, and penguin, are dependent on the ocean for their food. Some can spend their entire lives at sea, needing land only for nesting.

12.6 MAMMALS

GENERAL

Mammals are another important class of chordates, and like fish, reptiles, and birds, they also belong to the subphylum of vertebrates. Mammals are warm-blooded, breathe air, have hair, give birth to live

(a)

(b)

(c)

Figure 12.20

Marine mammals. The sea otter (a) and polar bear (b) belong to the order Carnivora, and the dolphin (c) is a whale, belonging to the order Cetacea.

young,[11] and have mammary glands to nourish the young. Like birds, mammals tend to stay near or above the surface in shallow coastal waters, making them quite visible to us. Nonetheless, because their total biomass is very small compared to other marine organisms, and because they generally feed on fish and shellfish at the highest trophic levels, they play a relatively minor role in the overall marine ecosystem. They do cause some problems for certain commercial fisheries, however, because their feeding is in coastal waters and comes from the species that humans also like to harvest.

Of all the creatures of the oceans, mammals seem to have captured the most human interest. Perhaps it is because they are the largest and most intelligent of the marine animals, or perhaps humans feel some kindred spirit, having evolved from the same branch of the evolutionary tree. Because of this interest, we include here an entire section on marine mammals.

Life began on Earth about 3.8 billion years ago, but didn't start moving onto land until less than 500 million (0.5 billion) years ago. Although some land mammals were around as early as 200 million years

[11]With the exception of freshwater monotremes, such as the platypus.

ago, we find no evidence of *marine* mammals of any kind earlier than 50 million years ago. We conclude that all present ocean mammals evolved from land animals. That is, marine mammals demonstrate a *return* of some species to the oceans.

Marine mammals belong to the orders **sirenia**, **carnivora**, and **cetacea**. The sirenians are possibly the source of the mermaid myths. They include the dugongs and manatees, which are large slow-moving docile vegetarians and highly endangered inhabitants of warm and brackish coastal waters. They are commonly called *sea cows*. Cetaceans are whales. Carnivorans include a large variety of familiar land and sea animals. But the important marine carnivorans are the sea otter, polar bear, and an important suborder called the **pinniped**. Pinnipeds include seals, sea lions, walruses, and related animals. Pinnipeds spend nearly all their lives in the ocean, but come ashore sometimes for sun, rest, or breeding. Cetaceans and sirenians, however, never come ashore.

CARNIVORANS

The sea otter and polar bear are quite similar to terrestrial mammals, such as weasels and bears (Figure 12.20). They have four limbs for walking, like land mammals, but otters have webbed hind feet that help

(*a*)

Figure 12.21

Some pinnipeds. Sea lions (a) and walrusses (b) can fold their hind flippers beneath them and use then to help with travel and stability on land, whereas seals (c) cannot.

(*b*)

(*c*)

them swim. Unlike other marine mammals, the sea otter and polar bear rely primarily on their heavy fur to insulate them from the cold ocean waters, rather than a heavy layer of blubber beneath the skin.

The word "pinniped" means "feather-footed," and the reason for this name is evident to anyone who has seen a seal, sea lion, walrus, or their relatives (Figure 12.21). All four limbs of a pinniped have the same bone structure as the limbs of land mammals, but there are flippers instead of paws at the ends. That is, the "toes" on all four legs have evolved into long webbed flippers. This development facilitates their swimming, but makes it rather awkward or difficult for them to maneuver on land when they come out on coastal rocks or ice to sun themselves, rest, or reproduce. True seals have no external ears and cannot use their flippers to much advantage on land. Rather, they must slither. Sea lions, in contrast, can use their flippers to help them move on land, which is usually a lunging motion, with their front flippers helping to hold them upright. These are the "seals" of circus acts.

WHALES

Even though the total mass of all whales in the oceans is minuscule compared to the total mass of microscopic organisms, individual whales are much more conspicuous due to their large size (Figure 12.22). Whales are the most highly modified of the mammals that have returned to the sea. Whereas sea otters, polar bears, and pinnipeds still bear close resemblance to land carnivores, such as weasels, bears, and dogs, whales bear relatively little outward resemblance to their closest land counterparts.

Because they live in water and swim, whales might seem like large fish. But they are no more closely related to fish than you are. They differ from fish in the same fundamental ways that you and other mammals do:

1. They are very intelligent.
2. They breathe air, having lungs instead of gills.
3. They have hair—not nearly as much hair as do other mammals, even less than humans, and mostly on their heads. But it is hair, and not scales.
4. They give birth to live young.
5. They nurse their young, providing milk from mammary glands.
6. They have a relatively long life expectancy—20 to 40 years being typical.
7. They are warm-blooded.

The whale skeleton is similar to that of other mammals, and they are most closely related to hoofed mammals, like cattle and deer. The hind legs are missing, however. The two tiny detached hip bones are useless inheritances from generations long past. The whale's front legs have been molded into flippers, and the joints are inflexible except at the shoulder. Whales have streamlined bodies, but many of the larger species are rather slow swimmers. Unlike fish, their tail fins (called **flukes**) are horizontal, and they propel themselves by moving this fin up and down, rather than back and forth. The fluke is at the end of the tail, some two dozen vertebrae beyond where the pelvis once was. So the whale's fluke has an entirely different origin than the back flippers of the pinnipeds.

Because they live in water and swim, whales might seem like large fish. But they are no more closely related to fish than you are.

Whales breathe through blowholes on the tops of their heads. They seem to have no sense of smell and little or no taste. Their eyesight is generally good, although useless in dark deep waters, in turbid waters, or at night. However, their senses of touch and hearing are excellent. Whales are social creatures, usually traveling in groups, or **pods**. They communicate with a wide variety of sounds and songs and appear to keep in touch over large distances, sometimes exceeding 50 kilometers. For navigation they use **echolocation**. They make clicking sounds and listen for the echoes from the sea floor, sea surface, obstacles, other members of their pod, or fish or schools of tiny prey. The whale's clicks seem to originate in the forehead area, where there is a complicated system of air passages. Some of these clicks are audible to humans, but many are ultrasonic.

All whales are of two basic kinds, generally labeled as **toothed whales** and **baleen whales** (Figure 12.22). In general, toothed whales are smaller and faster. The largest is the sperm whale, which can grow up to 18 meters long. Other well-known toothed whales include beaked whales (up to 12 m long), belugas (5 m), narwhals (5 m), porpoises (2 m), dolphins (9 m), and killer whales (10 m).

Baleen whales include most of the larger whales that are heavily hunted, such as the right (up to 18 m long), gray (15 m), humpback (17 m), sei (20 m), fin (25 m), and blue whales (35 m). Baleen whales have no teeth. Instead, there are hundreds of long thin platelets, called *baleen*, hanging from the upper jaw.

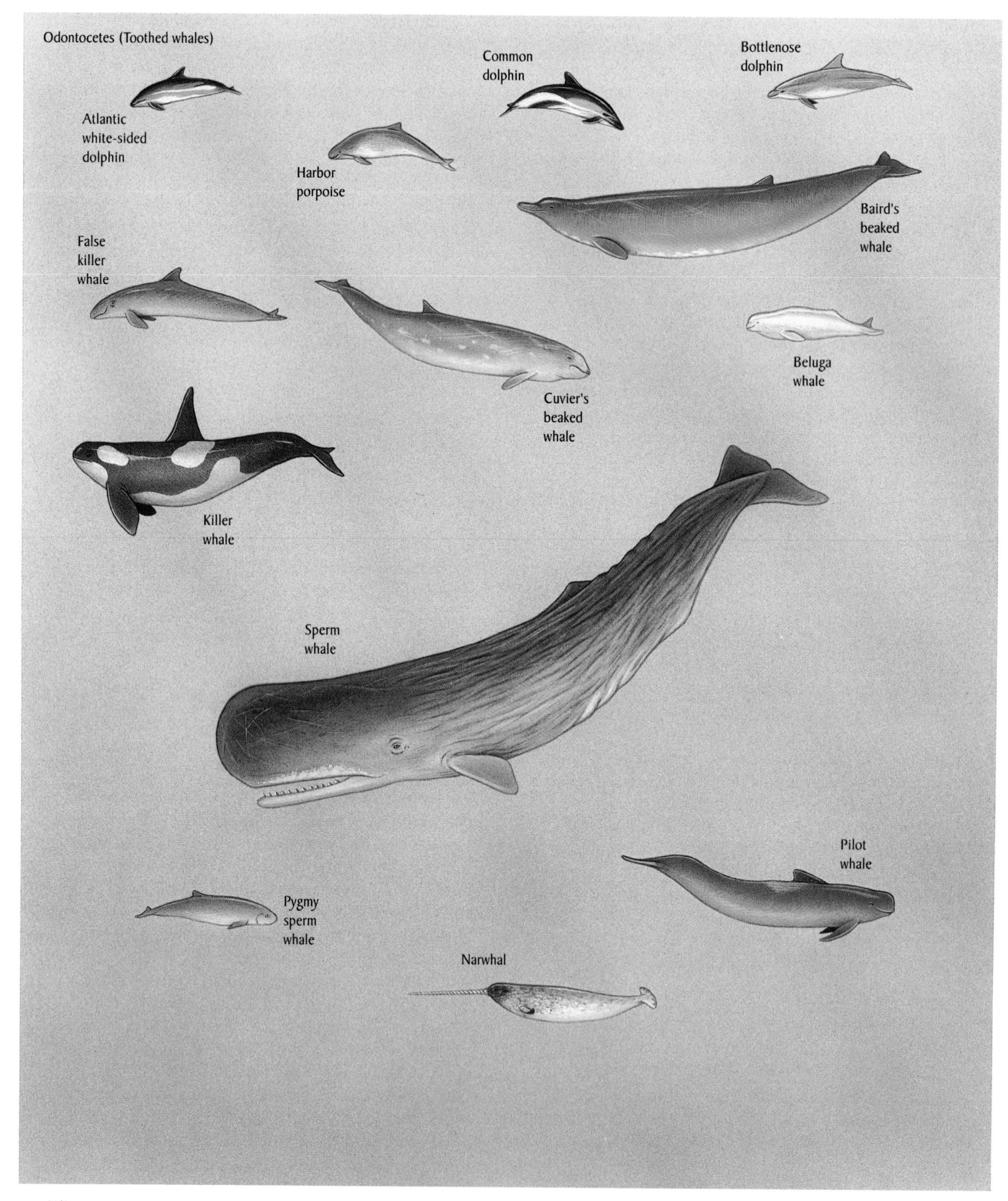

Figure 12.22

Some of the major kinds of whales.

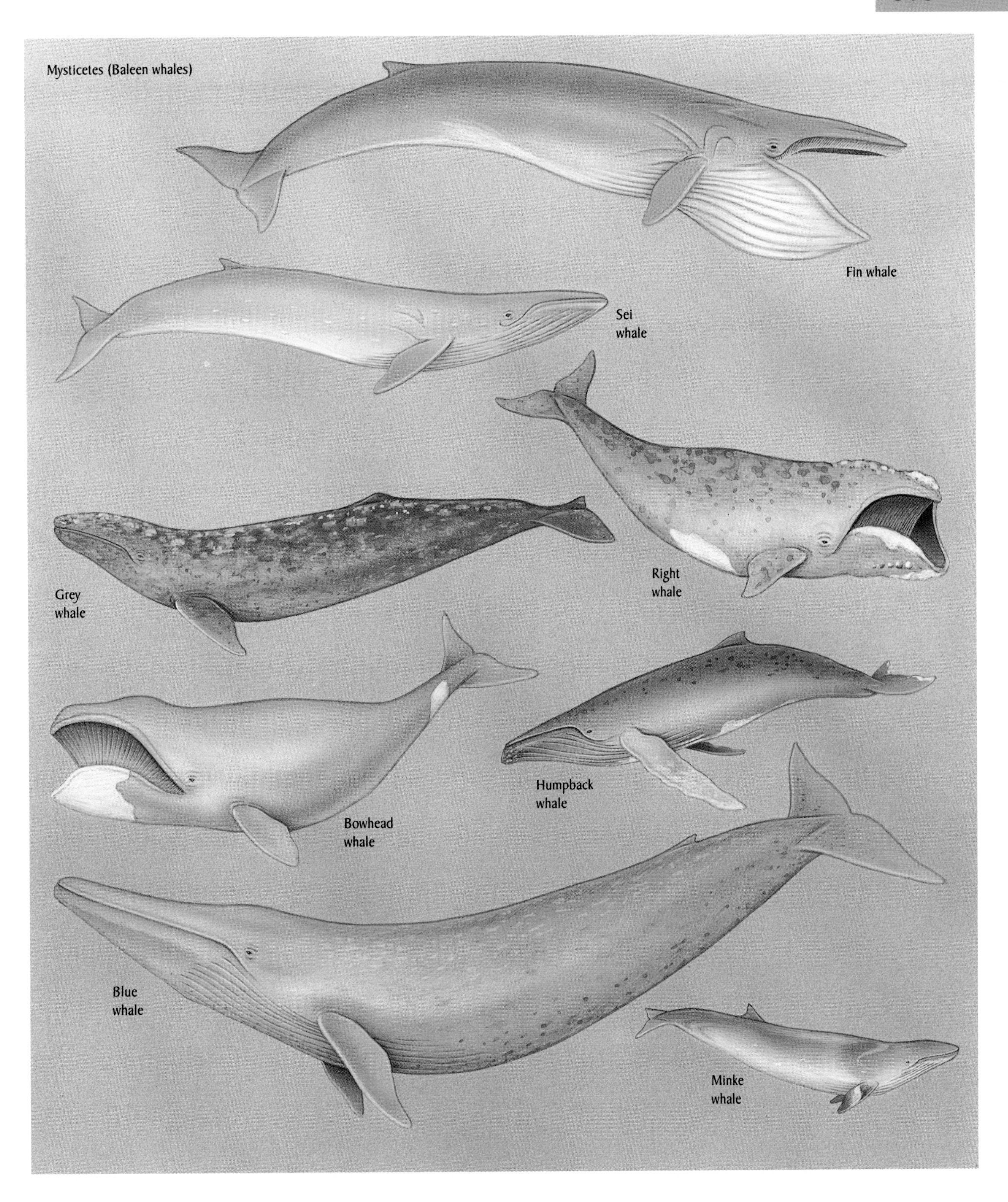
Mysticetes (Baleen whales)
Fin whale
Sei whale
Right whale
Grey whale
Humpback whale
Bowhead whale
Blue whale
Minke whale

Table 12.5 International Whaling Commission Estimates of the Populations of Various Commercial Whale Species

Species	Original Population	Population in 1980	Percent of Original
Sperm	2,400,000	2,000,000	83
Finback	543,000	123,000	23
Minke	295,000	280,000	95
Sei	254,000	51,000	20
Blue	226,000	13,000	6
Humpback	146,000	4,400	3
Right	120,000	3,100	3
Bryde's	92,000	92,000	100
Gray	20,000	11,000	55
Bowhead	20,000	2,000	10

This flexible, fingernail-like material was once called *whalebone* and was used in umbrellas, corsets, and fishing poles. Whales use the baleen as a sieve to filter their tiny food from the water. When they locate a school of krill or other tiny animals, they gulp in huge amounts of water and then force it out through the baleen that acts as a sieve to filter out their food. The largest mammal on Earth is the adult blue whale, which grows up to 35 meters in length and 120 tons in weight. This is much larger than the largest dinosaurs. Such large animals could not live on land, because bony skeletons could not support the great weight.

Whales have been hunted for centuries. The Norwegians began whaling in the late 800s and the Basques in the 900s. These were coastal operations. The *great age of whaling* developed in the 1700s and 1800s, when large numbers of European countries engaged in extended whaling expeditions. The harpoon gun was introduced in 1868, and shortly thereafter came exploding harpoons. The combination of this technology and increased whaling intensity severely reduced the populations of many major species by the early 1900s.

In addition to the meat, whale oil was extracted by heating the blubber. The oil was sold for use in lamps and for cooking. Today, petroleum products, natural gas, and electricity have largely replaced this use, but still the growing human population and improved technological ability to harvest whales have led to increasing harvests throughout most of the nineteenth and twentieth centuries. Whale harvests reached their peak in 1966, when 66,000 animals were harvested. The overkilling in these years created a sharp decline in subsequent harvests, as many species were decimated (Table 12.5). Several have not yet rebounded. Although no species has yet gone extinct, the gray whale is now gone from the North Atlantic.

Summary

Organisms are classified according to common characteristics that imply possible common ancestry.

Single-Celled Organisms

Single celled organisms, which are so important in the marine environment, are divided between two kingdoms based on cell structure. Marine plants tend to be very simple in comparison to their land counterparts.

Bacteria and blue-green algae are the earliest and most primitive forms of life on Earth, having no nuclear membrane to isolate the DNA and other cell regulators within the cell. The very tiny coccolithophores and silicoflagellates are the most efficient and prolific producers in the oceans. In nutrient-rich waters, where efficiency is not at such a premium, diatoms predominate. Dinoflagellates have an advantage in very calm waters or where there is gentle downwelling.

Fungi and Multicellular Plants

Fungi are less prominent in the marine environment than on land. Large multicellular plants are found nearly exclusively in shallow coastal waters, where sunlight penetrates to the bottom. These include green, brown, and red algae. Some advanced vascular plants, such as various marsh grasses and mangrove trees, grow in shallow protected coastal swamps and marshes.

Multicellular Animals

There are a wide variety of animals in the ocean. In order of increasing complexity, the more prominent phyla are sponges, coelenterates, ctenophores, various worms, mollusks, arthropods, echinoderms, and chordates. Most large fast swimmers in the oceans are vertebrates, which include various primitive members, cartilaginous and bony fish, reptiles, and mammals. Most marine animals reside within a few hundred meters of the ocean's surface, because that is where the food is produced. Some food escapes consumption at the surface and settles to the ocean floor, supporting a community of organisms there.

Sponges are loose colonies of individual cells that make porous structures. Food particles are filtered out as water passes through the pores. Jellyfish, anemones, and corals are all coelenterates, a phylum of animals that have a ring of tentacles surrounding an open gut. Ctenophores are like small, transparent, bioluminescent improved versions of jellyfish. There are several phyla of worms represented in the oceans, most of which live on the sea floor. Flatworms have a primitive digestive system and move by tiny hairs. Nematodes are tiny roundworms that live mostly in the bottom sediments. Most segmented worms are either deposit feeders or filter feeders. Arrow worms are tiny swimmers and capture tiny food particles that are floating in the water.

Snails, slugs, nudibranchs, clams, and mussels are all examples of mollusks, a phylum of animals that have a soft foot used for locomotion and sometimes carry one or two calcareous shells. The octopus, squid, and nautilus are mollusks with eyes, a chewing beak, and tentacles with suction cups.

Arthropods are the most numerous animals on Earth, both in terms of the number of species and the number of individuals. They include insects on land and crustaceans in the ocean. They all have jointed legs, segmented bodies, antennae, and an external skeleton made of a tough material. Important subgroups of crustaceans include the various kinds of copepods, euphasids, isopods, amphipods, barnacles, and decapods. Sand dollars, sea urchins, sea cucumbers, and starfish are all echinoderms, a phylum of animals with tough skins and radial symmetry.

Chordates are the most advanced animals. They have an internal skeletal rod and gills at some stage in their development. The most prominent subphylum are vertebrates, animals with a segmented spinal column. The most primitive vertebrates are hagfish and sea lampries. Sharks, rays, and skates have cartilaginous skeletons and must either swim or sink. Bony fish differ from cartilaginous fish in that they have bony skeletons, overlapping scales, gill covers, and a mouth at the very front of their body. The fin and body shapes vary greatly depending on whether the fish requires speed, acceleration, maneuverability, and so on. Some reptiles and birds are also represented in the oceans. These vertebrates breathe air and nest on land, which requires their eggs to have hard shells and amniotic fluid inside.

Mammals

All marine mammals evolved from land mammals that returned to the sea. The mammals have hair, breathe air, are warm-blooded, and give birth to live young that they nurse. Three orders of mammals are represented in the oceans: sirenians, carnivorans, and cetaceans. Carnivorans are sea otters, polar bears, and pinnipeds ("feather-footed" carnivores, such as seals, sea lions, walruses, etc.).

Cetaceans are whales and purely marine. They are structurally similar to other mammals. The front limbs have developed into flippers, and all joints have become rigid except for the shoulder. The hind limbs have disappeared, and the tail has become very muscular with horizontal flukes at the end. All whales can be divided into two groups. Toothed whales are generally smaller, faster, and are active predators. Baleen whales filter their planktonic nourishment from the water.

Key Terms

kingdom
phylum
procaryote
Monera
eucaryotes
Protista
nitrogen-fixing
blue-green algae
stromatolites
coccolithophores
calcite
silicoflagellates
flagella
diatoms
resting spores
diatomaceous Earth
dinoflagellates
red tides
protozoans
algae
holdfasts
stipes
fronds
green algae
brown algae
red algae
sponges
coelenterates
polymorphism
anemones

ctenophores
flatworms
round worms
(nematodes)
segmented worms
arrow worms
mollusks
chitons
gastropods
nudibranchs
bivalves
filter feeders
cephalopods
arthropods
crustaceans
copepods
krill
isopods
amphipods
barnacles
decapods
echinoderms
chordates
tunicates
vertebrates
cartilaginous fish
bony fish
mammals
sirenia
carnivora
cetaceans
pinnipeds
flukes
pod
echolocation
toothed whales
baleen whales

Study Questions

1. What is the difference between procaryotes and eucaryotes? Which are you?
2. What characterizes members of the kingdom Monera? What two major subdivisions are there within this kingdom?
3. Why do succeeding generations of diatoms get smaller and smaller? How is this problem corrected?
4. What is paralytic shellfish poisoning? Why does it affect humans and not shellfish?
5. Why do organisms with calcareous skeletons tend to be found more in tropical waters than polar waters?
6. How do sponges get their food? Why are they full of holes?
7. What do corals, jellyfish, and sea anemones have in common?
8. What are two different feeding methods displayed by marine segmented worms? How do arrow worms feed?
9. What are the main characteristics shared by all arthropods? What are the two largest classes of arthropods?
10. Describe mollusks, arthropods, and echinoderms. Give some examples of each. Which are related to insects? How can you tell?
11. Which animal phylum do you belong to? What must you have had at some stage in your development? When?
12. How do sharks differ from bony fish?
13. Give some examples of reptiles and mammals. What can you think of that might distinguish between these two?
14. What are some of the differences between bony fish that are good cruisers and those that are good accelerators?
15. Why do we think that marine mammals came from land?
16. What makes pinnipeds different from other carnivorans, such as weasels and dogs?
17. What structural differences make the tail fluke on a whale different from the rear flippers on pinnipeds?

Critical Thinking Questions

1. Why do you suppose that during periods of red tides, the breaking waves tend to glow, just like the wet sand when you step on it?
2. What are some differences between diatoms, coccolithophores, and dinoflagellates? In what kinds of oceanic regions do coccolithophores and dinoflagellates have an advantage over diatoms? Why?
3. See if you can think of two different ways that various mollusks might get their food. Do the same for arthropods and echinoderms.
4. What characteristics guarantee that you belong to the same subphylum as a shark? What subphylum is this?
5. When scuba divers go deep beneath the surface, the increased pressure forces air from their lungs into their blood. If they then rise to the surface too fast, the reduced pressure causes the air to bubble back out of the blood, producing a very painful and sometimes fatal condition called the bends. Whales have lungs and go for some very deep and prolonged dives (much deeper and longer than scuba divers), from which they can return to the surface very quickly. What do you suppose whales do to prevent the bends?

Suggestions for Further Reading

CARON, DAVID A. 1992. An Introduction to Biological Oceanography. *Oceanus* 35:3, 10.

DARLING, J. D. 1988. Whales, an Era of Discovery. *National Geographic* 174:6, 872.

FRENCH, KIMBERLY. 1990. Dangerous Liaisons? *Omni* 12:9, 28.

GRASSLE, J. F. 1988. A Plethora of Unexpected Life. *Oceanus* 31:4, 41.

HAMMER, W. 1984. Krill, Untapped Bounty from the Sea. *National Geographic* 165:5, 627.

MORSE, A. N. C. 1991. How Do Planktonic Larvae Know Where to Settle? *American Scientist* 79:2, 154.

SANDERSON, S. AND R. WASSEISUG. 1990. Suspension Feeding Vertebrates. *Scientific American* 262:3, 96.

SHERMAN, K. AND A. F. RYAN. 1988. Antarctic Marine Living Resources. *Oceanus* 31:2, 59.

SUMICH, J. L. 1992. *An Introduction to the Biology of Marine Life*, 5th ed. Wm. C. Brown.

WHITEHEAD, H. 1985. Why Whales Leap. *Scientific American* 252:3, 84.

WIEBE, PETER H. 1994. Visualizing Life in the Ocean Interior. *Oceanus* 35:3, 100.

The emphasis on small size and efficiency in most oceanic regions leads to a long food chain.

Spotted eagle rays in the Caribbean Sea.

Thirteen

ENVIRONMENTS AND LIFESTYLES

e began our study of marine biology by examining the very fundamental life processes and the ingredients they require. We saw how variations in the supply of these ingredients are reflected in the overall patterns of life in the ocean. We next studied the many different kinds of marine organisms, both producers and consumers, and how each functions.

We now look at the various marine environments, how organisms cope with problems encountered, and which organisms are best suited for survival. We also study the various kinds of marine communities, including how the organisms earn their livelihood through interactions with their surroundings and each other.

13.1 THE MARINE ENVIRONMENT

It is no surprise that life first developed in the oceans and only recently moved onto the land. By comparison, the land environment is very harsh. Land creatures must deal with gravity, dehydration, large temperature variations, and many other hardships not normally encountered by life in the oceans. Life on land has had to develop specialized systems and appendages in order to cope with the environment. For example, the ability to grow thick winter coats of fur, strong legs, complicated root systems, or fruits are a few of the many land-based innovations that would be unnecessary in the ocean environment. Marine organisms rely on buoyancy and friction for support, whereas land plants and animals need more complicated support systems. Although there are microbes on land, their total mass is relatively small. Things are exactly the opposite in the ocean, where the overwhelming mass of living creatures are simple and microscopic (Figure 13.1).

ENVIRONMENTAL STABILITY

Marine organisms have a very stable living environment. Deep waters remain unchanged from season to season. Surface waters are a little more variable, but are very steady compared to land areas on Earth. Light intensity changes on a daily cycle, and seasonal changes in light intensity are important at temperate and higher latitudes. But even in surface waters seasonal variations in temperature are typically less than 5°C and salinity variations are typically less than one part per thousand. Because the shallow nearshore waters are the most variable, nearshore marine organisms tend to be more tolerant of small variations in their environment than those found farther out or in the deep ocean environment (Figure 11.11).

REGULATION OF BODY FLUIDS

Because marine organisms are bathed in an environment of nearly constant temperature and salinity, it is of little advantage to them to provide any additional regulation of their body fluids. Consequently, most don't. Due to this environmental stability, the indigenous species within most regions have become rather intolerant of change, and so the distribution of most marine species is sharply defined by the temperature and salinity of the various ocean regions.

For the large fast swimmers that travel long distances, it is clearly advantageous to regulate internal body fluids, so that as these organisms travel from one environment to another, their cells are continually bathed in fluids that ensure optimum performance. Bony fish (but not cartilaginous fish, like sharks and rays) regulate the salinity of their body fluids through membranes in their gills and guts. Mammals can regulate body temperatures as well.

The skin and gut of many larger animals are *semipermeable*—that is, they allow the passage of water molecules more easily than salts.[1] On a molecular scale, there is a very strong electrostatic attraction between salt ions and the water molecule. If fresh water and salt water are separated by a semipermeable membrane, this attraction will pull the water molecules through the membrane toward the salty side.[2] The amount of back pressure that must be applied from the salt side to prevent this flow is called the **osmotic pressure**. When typical sea water is separated from fresh water, the osmotic pressure is equal to 25 atmospheres, or equivalently, the pressure at a depth of 250 meters beneath the ocean surface.

The body fluids of the bony fish are only about half as salty as the ocean they live in. Consequently, there is a strong tendency for the water inside these fish to flow through their membranes and out into the saltier ocean environment. To counteract this tendency, marine fish must drink large quantities of water and remove the salt by secreting it back out through special cells in their gills and concentrated salty urine.

[1]Skins are much thicker than guts, so the passage through skin is much slower.

[2]This is why you can reduce swelling in an injury by soaking it in salt water. More precisely, the water molecules diffuse in both directions through the membrane, but they tend to stay on the salt side, where they experience the strongest attraction.

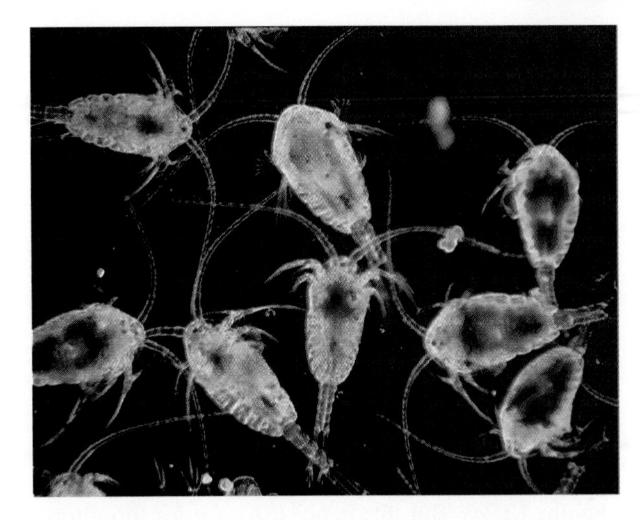

Figure 13.1

Most of the biomass on land is contained in large organisms, whereas most marine biomass is in microscopic organisms. These photos show typical land and ocean grazers: elk feeding on tall grasses and bushes, and microscopic copepods feeding on microscopic particles and microalgae.

DENSITY PROBLEMS

Life-giving sunlight is found in surface waters only. This is of critical importance to plants, of course. However, it is also advantageous for animals to be near the food source, so most animals stay near the surface too. This presents a problem, because organisms contain many materials that tend to make them denser than water. Many have dense mineral skeletons for protection or support, which add to the organism's density problem. How can they remain in or near the surface waters where the life- giving sunlight is found?

Organisms deal with this problem in a variety of ways. One is that marine organisms tend to have a high water content, which helps neutralize their buoyancy. Jellyfish are 95% water, as compared to 65% for typical land animals, such as humans. Among the larger creatures, mammals and reptiles have lungs, whose air content helps hold them near the surface. Many fish have gas-filled **swim bladders** that help neutralize their buoyancy. In some species, the air that fills these bladders is gulped at the surface and passes through the mouth and esophagus, but more commonly, it is taken out of the blood.

Gases are compressible. Consequently, as a fish descends, its bladder is compressed, and conversely it expands as it rises toward the surface.[3] The fish must make appropriate adjustments in the air content of the bladder as it rises or descends. Some deep water fish neutralize their buoyancy with bladders of fat, rather than air. The fat avoids the compression problem associated with gases, but cannot be adjusted quickly.

Some fast swimmers, such as tuna, have no swim bladders and rely on their constant motion to keep them near the surface. Octopus, squid, and cartilaginous fishes, such as sharks, skates, and rays, don't have any sophisticated internal mechanism to neutralize their buoyancy and must either swim or sink (Figure 13.2a). Most of these creatures spend a large portion of their lives resting on the bottom. Some animals, like clams, starfish, and crabs, have no means of overcoming the density problem and are permanently confined to the bottom.

Among the microscopic animals (Figure 13.2b), some have small pouches of secreted wax, oil, or

[3] If a fisherman brings a fish to the surface too fast, its air bladder might expand to the point that it extends into or even out of its mouth.

gases that help neutralize their buoyancy. Many have long thin or feathery appendages or protrusions, and the single-celled organisms may have whiplike *flagella*. These appendages serve two purposes. First, they retard the rate of sinking, much like the fluff on milkweed or cottonwood seeds helps them fall only very slowly through the air. Also, the movement of these appendages and flagella may give the organism sufficient mobility to remain near the surface, and even to change its position in response to daily changes in sunlight.

Of course, it is especially crucial that plants be able to remain near the surface. Large attached algae often have gas-filled bulbs or hollow portions to help keep their photosynthetic parts near the surface (Figure 13.2c). But this is not the case for the microscopic single-celled algae that account for the bulk of the ocean's plant mass. Most of these also secrete protective mineral exoskeletons that add to their density problems (Figures 11.1b and 12.2).

How do these microscopic plants overcome their tendency to sink out of the sunlit surface waters? They conquer this problem in two ways:

1. First, they maximize their friction with the water in order to retard the rate at which they sink.
2. Second, they have high rates of reproduction so that those remaining in the surface waters can reproduce fast enough to replace those lost to the darkness below.

Because friction is proportional to surface area, a man with a parachute falls more slowly than one holding a handkerchief over his head, and fine dust settles out of air much more slowly than large stones. As we have seen, the smaller the object, the greater is its surface area in comparison to its volume. Through small size, the tiny plants have a large surface area relative to their volume, or equivalently, large friction relative to their weight.

Another way that these microscopic plants increase their friction with water is with jagged shapes and long thin appendages. The tiny external skeletons of many are rough and jagged (Figures 11.1b

(a)

(b)

(c)

Figure 13.2
It is absolutely essential for plants to remain near the surface where the sunlight is. It is helpful for most animals to remain near the surface where the food is produced. (a) A shark's body shape, pectoral fins, and tail all provide an upward thrust when swimming. (b) Microscopic plankton, such as these diatoms and the copepods of Figure 13.1, have small sizes and feathery appendages that help retard their sinking rate, and some have tiny flagella that provide some minor mobility. (c) Kelp are large brown algae that have gas-filled bulbs that help hold the leaflike fronds near the surface.

and 12.2). Friction also depends on the viscosity of the water, and warm water is less viscous than cold water. Consequently, microscopic plants in warm water tend to be smaller, and their skeletons tend to have longer needlelike spikes, reflecting their need for increased friction. Finally, many microscopic plants have long thin flagella (Figure 12.2c), which both increase friction and provide minor mobility.

> Because friction is proportional to surface area, a man with a parachute falls more slowly than one holding a handkerchief over his head, and fine dust settles out of air much more slowly than large stones.

Figure 13.3
Warm water animals tend to be smaller and more colorful than their cooler water counterparts.

PREDATION

One aspect of marine life that is difficult for us civilized land mammals to comprehend is the extent of predation in the oceans. We are familiar with large and largely herbivorous land animals, who produce few offspring that they carefully protect. On land, even the first-level carnivores are relatively rare. By contrast, the oceans contain many interwoven layers of carnivores, who produce extremely large numbers of unprotected young and feed ferociously on each other and their own kin. It starts at a microscopic level and proceeds through many trophic levels before arriving at the top carnivores, such as tuna or salmon.

DISTRIBUTION OF ORGANISMS

In Chapter 10, we learned that the intensity of plant activity is determined by the availability of sunlight and nutrients. For animals, the two factors that determine the intensity of their activity are the abundances of food and oxygen. Food is first produced by plants in the sunlit surface waters. Oxygen also enters the ocean at the surface, either as a by-product of photosynthesis, or directly by absorption from the atmosphere. With a few limited exceptions, it is the supply of food, rather than oxygen, that restricts animal life. Where there is high plant productivity at the surface, there will be lots of food and large animal populations. Where plant productivities are low, few animals will be found.

The particular species of plants and animals in any given region depend on the water's temperature and salinity. Most seem to be more sensitive to changes in temperature than salinity, although salinity does present some barriers, especially in estuaries. In many cases,[4] organisms in neighboring regions of the oceans have evolved quite differently, because invisible barriers of temperature or salinity have prevented intermixing. Because deep ocean waters tend to be uniform over much larger distances than those near the surface, there is correspondingly more uniformity of deep water animal species over larger distances.

Water temperature has an interesting effect on organisms, which seems to reflect the fact that the rate of most chemical reactions increases with an increase in temperature. When comparing the activities of warm and cold water species, we find that the warm water organisms tend to metabolize and photosynthesize at a higher rate (if provided with sufficient raw materials). They also tend to grow, reproduce, and age more quickly, and tend to be smaller and more colorful than their cold water counterparts (Figure 13.3). Warm water biological communities also show greater diversity of species, each occupying a narrower ecological niche, or functional role in the community.

In spite of fewer species, cold waters generally support larger numbers of individuals, because cooler waters tend to contain more nutrients and correspondingly higher plant productivities. Cooler waters also tend to contain more dissolved gases so that both photosynthesis and respiration are easier. The schools of large numbers of relatively few species make for better fishing in these cooler waters (Figure 11.6a).

[4]Examples include communities around hydrothermal vents compared to neighboring sea floor communities, or species of similar organisms across the Antarctic Convergence, or even across a large river mouth.

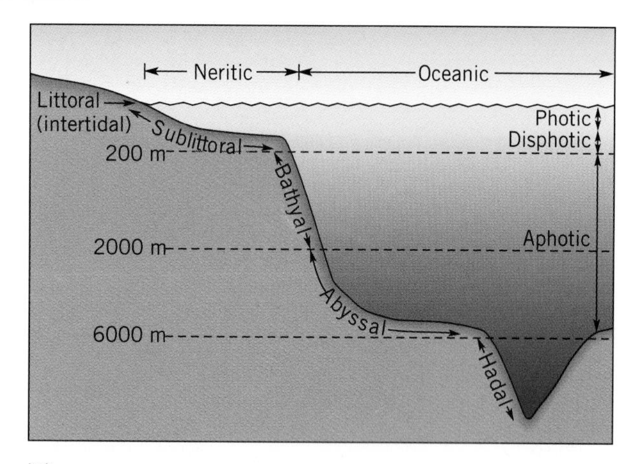

Figure 13.4

Various biozones. The waters over the continental shelf are neritic, and those over the deep ocean oceanic. Vertically, the water is divided into the sunlit photic zone, faintly lighted disphotic zone, and dark aphotic zone. The benthic region is divided according to depth into littoral, sublittoral, bathyal, abyssal, and hadal zones, as indicated.

Polar and subpolar waters are well mixed from top to bottom. As a result, the species of microscopic animals found in surface and deep waters are similar at high latitudes (Figure 7.36). In the tropics, the species found in the two regions are quite different, due to the very different temperatures of the surface and deep water masses. The inhabitants of temperate surface waters change with the seasons, due to migration and spawn of fishes from other areas, and changes in life stages triggered by seasonal changes in these waters.

THE PELAGIC ENVIRONMENT

One way of characterizing marine life is according to where it lives. Organisms that live on the ocean bottom or within the sediment are called **benthic**. Those that live in the water are called **pelagic**. The majority of the ocean's living organisms are pelagic. This is especially true for plant life, because it is necessary for them to remain in the sunlight near the surface. Most marine animals are also pelagic, because it is advantageous for them to remain near the food source.

The pelagic environment is conveniently divided into the upper sunlit **photic** zone and the deeper dark **aphotic** zone (Figure 13.4). Sometimes, an intermediate **disphotic** or *twilight* zone is identified where there is some sunlight, but less than 1% of that at the surface. In clear oceanic water, the photic zone extends down to about 100 meters, the disphotic zone to somewhere around 500 meters, and the aphotic zone below that. In the more turbid coastal waters, photic and disphotic zones are much thinner, because sunlight cannot penetrate as far. Productive plants are found only within the photic zone, although various forms of animal life can be found at all depths.

The penetration of sunlight also influences the colors of organisms. As is illustrated in Figure 7.7, the blue-green travels through clear water best. Reds, on the other hand, are absorbed rather quickly and therefore don't penetrate very far. For this reason, many deep water animals are reddish in color. The color of an object is the color it reflects. If no reds reach these deep water animals, then they reflect no light, and they appear black or invisible. This protects them from predators.

The pelagic environment can also be subdivided into the **neritic** environment, which is the waters over the continental shelves, and the **oceanic** environment, which includes the waters over the deep ocean floor (Figure 13.4). There is much more water in the oceanic region, but the neritic environment usually has more biological activity. (Do you remember why?)

THE BENTHIC ENVIRONMENT

The subdivisions of the benthic environment are based on depth. In order of increasing depth, they are the littoral, sublittoral, bathyal, abyssal, and hadal biozones, as illustrated in Figure 13.4. The **littoral** zone is the intertidal region, and the **sublittoral** zone makes up the rest of the continental shelf. The **bathyal** zone is the intermediate range of depths, encompassing the continental slope, rise, and part of the oceanic ridge system. The abyssal plains and ocean basins comprise the **abyssal** zone, and the narrow, deep ocean trenches make up the **hadal** biozone. The precise depth ranges corresponding to this description are illustrated in the figure.

Where the sunlight reaches the bottom, the benthic community includes plants, something that would not be possible on the deep ocean floor.

Throughout most of the ocean, the benthic environment is dark and isolated from the productive surface waters above. There is a slow and gentle rain of detritus downward from these surface waters—materials that have somehow escaped consumption up there. This is an important connection with the surface, because it is the primary food source that supports the benthic community.

In the shallow coastal waters, however, the two communities merge. Where the sunlight reaches the bottom, the benthic community includes plants, something that would not be possible on the deep ocean floor. Some of this plant life occurs in the form of microscopic algal coatings on rocks or other hard surfaces, and some is in the form of larger, more advanced attached plants. In these shallow productive environments, the benthic community thrives, often being even more heavily populated with plants and animals than the water above (Figure 11.11). This is quite the reverse of the case in deeper offshore waters, where benthic organisms are isolated from the surface sunshine and food production. Although these shallow coastal waters with thriving benthic populations are the most familiar to us land-based creatures, they are only a very minor fraction of the ocean.

13.2 THE SPECTRUM OF LIFESTYLES

There is fascinating diversity among ocean creatures. Each has its own peculiar lifestyle and special characteristics that has made its species successful over the ages. Table 13.1 summarizes some of the common distinctions made on the basis of habitats and lifestyles.

Among the ways we have of describing the various types of organisms is a distinction based on how they get their food. Those that have the ability to produce their own food are called **autotrophic**. Those that cannot are **heterotrophic**. Most plants are autotrophic, and all animals are heterotrophic. For many organisms, especially microscopic ones, this distinction is not very clear. Many microscopic plants, for example, do not produce all of their own needs, requiring some materials produced by other organisms. Many bacteria use inorganic chemical reactions to provide the energy for the synthesis of organic materials, but often produce only a fraction of their total requirement in this manner.

Water inhabitants can also be categorized according to whether they are primarily swimmers or drifters. The swimmers are called **nekton**, and as we learned in Chapter 1, the drifters are called *plankton* (Figure 13.5). Most plankton actually have some minor means of mobility, but are still at the mercy of waves and currents. Plankton include such things as microscopic plants and jellyfish. Examples of nekton are anchovies and whales. The distinction between nekton and plankton is sometimes a matter of size or maturity, rather than breed. There is much more diversity among plankton than nekton. Most nekton are relatively large vertebrates, such as fish, reptiles, and mammals. Most pelagic invertebrates (i.e., nonvertebrates) are planktonic. An exception is squid, which is a fast-swimming invertebrate and close relative of the octopus.

PLANKTON

Plankton include all plants that are not attached to the bottom and most invertebrate animals. They range in size from large jellyfish down to the microscopic organisms that make up the bulk of the ocean's plant and animal communities. Bacteria, spores, eggs, and the juveniles of larger nekton or

Table 13.1 Some Classification Schemes for Marine Organisms

Classified According to	Classes	Characteristics
Mode of nutrition	Autotrophic	Can make their own food
	Heterotrophic	Cannot make their own food
Mobility	Nekton	Swimmers
	Plankton	Drifters
		Phytoplankton (plants)
		Zooplankton (animals)
Habitat	Pelagic	Live in water
		Intertidal
		Neritic
		Oceanic
	Benthic	Live in/on bottom
		Littoral
		Sublittoral
		Bathyal
		Abyssal
		Hadal

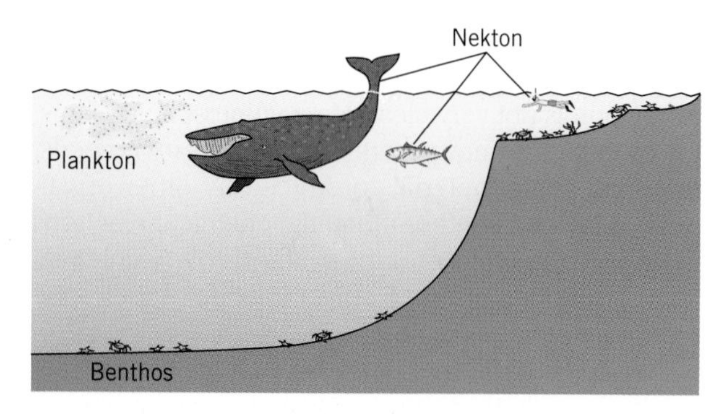

Figure 13.5

Oceanic life can be divided into two classes: benthic organisms, which live on the bottom, and pelagic organisms, which live in the water. The pelagic organisms are further divided into swimmers (nekton) and drifters (plankton). The benthic environment is sparsely populated throughout most of the ocean, because the creatures are far removed from the food source. In shallow coastal waters, however, the two zones merge, food is plentiful, and benthic populations explode. Attached plants are among the benthos in these shallow waters.

Table 13.2 Ways of Characterizing Plankton

Size
Picoplankton (smaller than about 5 μm)
Nanoplankton (between about 5 and 70 μm)
Microplankton (between about 70 μm and 1 mm)
Macroplankton (larger than about 1 mm)
Duration of Planktonic Life
Holoplankton (entire life)
Meroplankton (only part of life)
Ecological Role
Phytoplankton (producers)
Zooplankton (consumers)

benthic organisms are also members of the planktonic community. Planktonic plants are called **phytoplankton**, and planktonic animals are referred to as **zooplankton**.

There is a tendency for the distribution of plankton to be uneven, which is referred to as *plankton patchiness*. For phytoplankton, this probably results from wind- driven surface currents, which concentrate phytoplankton and their life-giving nutrients in some places more than others. The clustering of zooplankton might be caused by the same wind-driven surface currents, or it might reflect the zooplankton's tendency to stay near the food source or avoid predators.

The great diversity among plankton makes further subdivision useful (Table 13.2). Sometimes, scientists distinguish among them on the basis of size. Unfortunately, members of the same species tend to vary in size, which makes it difficult to have precise categories. In fact, there is not yet even complete agreement among marine scientists as to what these categories should be, but the categories in Table 13.2 correspond roughly to the current consensus. Normally, a species is assigned to the category that holds the majority of its members, even though some individuals might cross the line into a neighboring category.

Those plankton whose dimensions are less than about 5 micrometers are called **picoplankton**. Important picoplankters include bacteria, blue-green algae, and coccolithophores, which are important phytoplankton in the central areas of the oceans where nutrients are deficient, and therefore small size and efficiency are important. Those in the range of about 5 to 70 micrometers are called **nanoplankton**. Important members of this group are the diatoms, which are larger single-celled plants found in richer waters of coastal and upwelling areas, where nutrients are more plentiful, and therefore small size and efficiency are not as important. Those plankton in the range of about 70 micrometers to 1 millimeter are called **microplankton**. Most of the ocean's zooplankton falls within this group. Important members of this group include a variety of copepods and other tiny crustaceans, and the larvae and juveniles of larger benthic or pelagic organisms. Those plankton that are larger than about 1 millimeter in size are called **macroplankton**. They include floating seaweed, jellyfish, comb jellies, arrow worms, and again, the larvae and juveniles of larger nekton and benthos.

Plants and animals that are plankton all their lives are called **holoplankton**. These include all the single-celled phytoplankton and a large variety of microscopic animals that are responsible for harvesting this food and detritus. But some species are planktonic only part of their lives and are therefore called **meroplankton**. Included in this group are the eggs, larvae, and juveniles of fish and many benthic organisms, such as barnacles, oysters, worms, starfish, clams, snails, and sea cucumbers (Figure 13.6). In fact, most marine fish and invertebrates lay eggs in open water or on the bottom and have planktonic larvae. Typical spawns include hundreds of thousands or even millions of eggs. Such large spawns are needed because the chances of survival are very slim. Predation is heavy. On the average, only two will survive to replace the parents. But as a result of these heavy spawns, meroplankton make up a significant and diverse portion of the oceans' large community of tiny zooplankton.

(a)

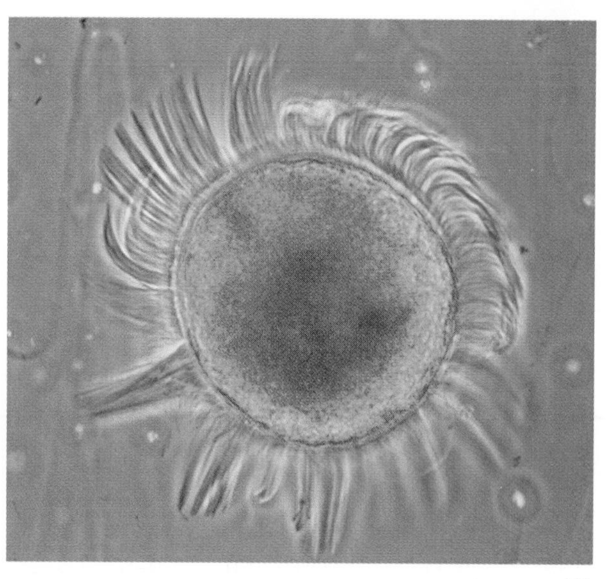

(b)

Figure 13.6
Meroplankton include the tiny lanktonic larvae of organisms that are not planktonic as adults, such as the larvae of a crab (a) and a segmented worm (b).

PHYTOPLANKTON

As we have seen, the overwhelming majority of all marine plant life occurs in the form of microscopic, single-celled phytoplankton. Their small size gives them the following advantages:

1. It gives them a large surface area in comparison to their volume, which increases their efficiency in getting needed raw materials and eliminating wastes.
2. It allows for much quicker transport of materials within the plant.
3. It causes increased friction with the water, which reduces the danger of extinction through sinking out of the photic zone.

All three of these advantages serve to increase their food production efficiency, which is particularly important in the nutrient-deficient marine environment.

One way of demonstrating the relationship between size and efficiency is to take a community of organisms, place them under optimal growth conditions, and see how long it takes them to double their mass, either through growth or reproduction. Typical doubling times for various sizes of organism are displayed in Table 13.3, which clearly demonstrates that smaller size gives greater efficiency. In the marine environment, the dominant planktonic species are smaller in regions where the need for efficiency is greater.

ZOOPLANKTON

The majority of zooplankton are small and microscopic, but multicellular organisms (Figures 11.18, 13.1b, and 13.6). They feed on phytoplankton and other suspended organic particles, and they have adaptations for causing water movement and filtering tiny food particles from the water. Each day, the zooplankton community consumes roughly its own weight in food, on average. Of course, this gets quickly passed on to the organisms that feed on them, so the total zooplankton biomass remains constant, in spite of their heavy feeding.

We have already seen that arthropods are the most abundant animals on Earth. On land, insects are by far the most numerous of all classes of animals, both in terms of numbers of species and numbers of individuals. In the oceans, crustaceans dominate, especially the smaller microscopic species (Figures 11.18 and 13.7). Their jointed legs and external skeletons make it easy to see that they are related to tiny shrimp or insects. Among tiny crustaceans, the smallest are copepods (Figure 11.18). In terms of total biomass, they are the single most important zooplankton. At higher latitudes, especially near Antarctica, their larger cousins, the krill, are dominant (Figure 12.13a). Besides small crustaceans, other im-

Table 13.3 Typical Times Required for Communities of Organisms to Double Their Mass Through Growth and Reproduction (Assume adequate sunlight and material requirements.)

Organisms	Doubling Time
Picoplankton (e.g., bacteria)	4–5 hours
Nanoplankton (e.g., diatoms)	10 hours
Microplankton (e.g., copepods)	4 days
Fish	1 year
Trees	20 years

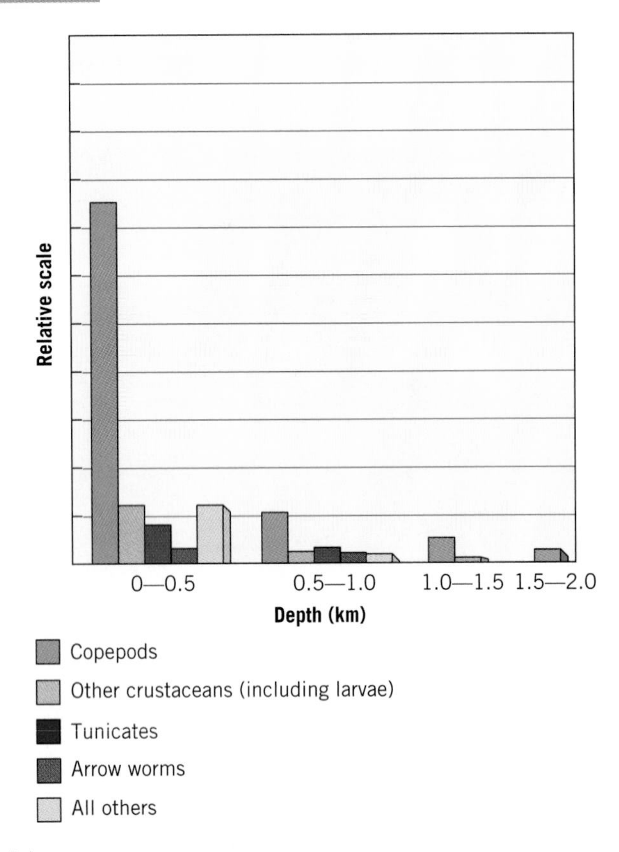

Figure 13.7

Relative abundances of the most common types of zooplankton at various depths in the Central North Atlantic Ocean. (After Deevey and Brooks, 1971, *Limnology and Oceanography* 16:6, 933.)

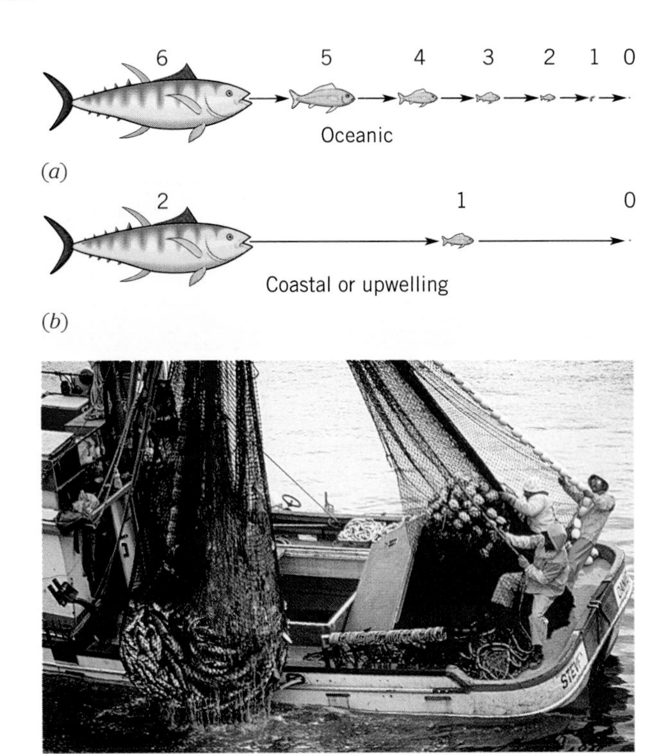

Figure 13.8

(a) In the less productive oceanic waters, emphasis is on efficiency, so the members of each trophic level are as small as possible, consistent with their functions. The predator is just slightly larger than its prey, so there are many extra steps in the food chain before the food reaches the large predators that we harvest. Only a small fraction of the food value at each trophic level gets passed on to the next, so that very little ever reaches large predators. (b) In the more productive coastal and upwelling waters, by contrast, there is less emphasis on efficiency and fewer steps are involved before the food reaches large predators. Both the greater productivity and smaller losses along the food chain combine to make coastal and upwelling waters much richer in the large predators that we harvest (c).

portant kinds of zooplankton include arrow worms, tunicates, comb jellies, and the larvae or juveniles of many larger organisms.

The emphasis on small size and efficiency in most oceanic regions leads to a long food chain (i.e., long succession of trophic levels). Since food is scarce, efficiency demands that each organism be as small as possible, consistent with its particular functions. The predator need be only slightly larger than its prey, so a large number of trophic levels are required before food is passed up to large carnivores, such as anchovies or tuna (Figure 13.8a). We have already seen that of the food consumed at any trophic level, only about 10% gets passed on to the next level, on average. Therefore, not much is left by the time it reaches large predators.

You can see that there are two reasons why large fish are extremely scarce in most oceanic waters. First, there is not much plant productivity over the deep ocean areas, and second, the small amount of food that is produced must pass through many more trophic levels along the way, with heavy losses at each level. Our fishing industry is geared toward the larger predators, and because of these two reasons, very little of our catch comes from oceanic waters (Figure 13.8b).

Most zooplankton migrate vertically in response to light conditions. Using their tiny appendages or by creating small changes in their own density, they can climb or sink at rates of typically 10 to 40 meters per hour. They come to the surface to feed at night and then sink to a depth of several hundred meters by midday (Figure 13.9a), thereby remaining as close to the food producers as possible without exposing themselves in the sunlight to predators. Vertical distances traveled may reach as much as 500 meters in

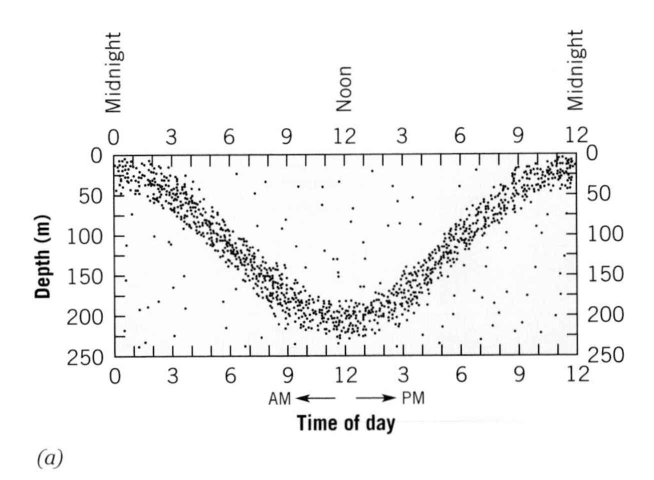

(a)

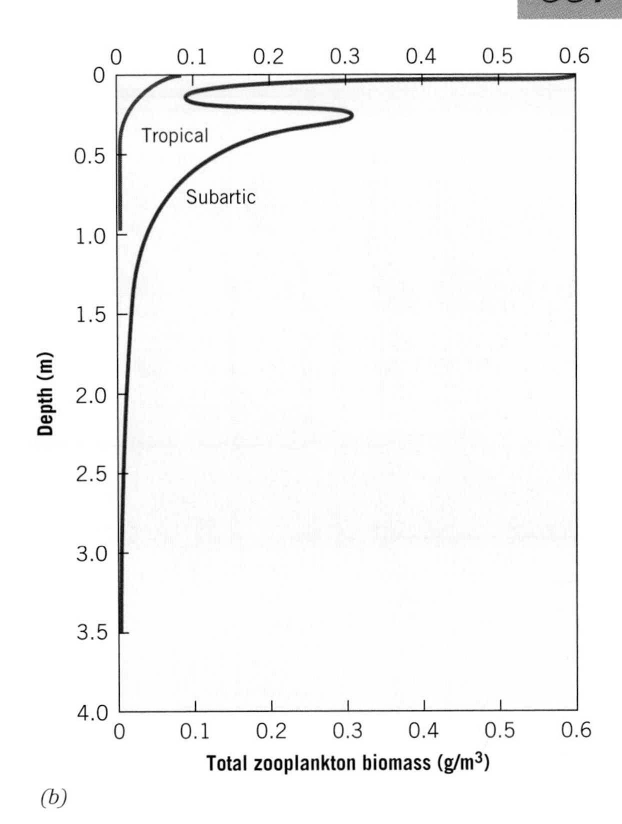

(b)

Figure 13.9

(a) Typical daily migration pattern for the microscopic grazers of the deep scattering layer. (b) Distribution of zooplankton with depth in tropical and subarctic regions of the Pacific Ocean, averaged over many stations. The low abundance in the tropics reflects the small food supply. The deep scattering layer is quite prominent in the subarctic. (Data from Michael J. Kennish, ed., 1989, *Practical Handbook of Marine Science*, CRC Press, Boca Raton, FL, p. 307.)

tropical waters, where the sun is high and penetrates deep, and as little as 10 meters in polar waters, where the sun is low and has low penetration. This daily vertical migration was first observed during World War II, when sonar echo sounders recorded daily changes in the position of this then unexplained **deep scattering layer**, so named because it scattered sonar signals.

In Figure 13.9b, you can see that where food is plentiful (i.e., in the very productive summertime subarctic waters), there are two peaks in the concentrations of zooplankton. Some are at the surface feeding, and others are just below the photic zone resting, or metabolizing that which they recently captured. Equally interesting are the data from tropical waters, where food is scarce and supports fewer grazers and predators. In these waters, the need for food often takes precedence over the need for protection; many zooplankton seem to remain near the surface all the time.

NEKTON

Of the various components of the marine biological community, nekton are the most familiar to us because of their size and visibility. On the whole, however, they make up only about 0.1% of the total live biomass in the ocean. Their distribution in the oceans is similar to that of zooplankton, because their primary food is either zooplankton or other smaller nekton that feed on the zooplankton. This means that much of the ocean's nekton can be found in the deep scattering layer, migrating vertically each day along with zooplankton.

Many species of nekton and zooplankton (such as fish, squid, and tiny crustaceans) tend to form large groups, or *schools* (Figure 13.10), which provides many benefits. Experiments show that members of a school eat more and grow faster than isolated individuals. Perhaps observing others feeding sharpens one's own appetite, and perhaps learning by example is more effective than trial and error. Schooling also has some benefits for reproduction and the rearing of young. Finally, schooling has protection benefits. For detecting danger, hundreds of pairs of eyes are better than one. Also, the motions in large schools often seem to confuse or bewilder a predator, whereas the behavior of an isolated individual prey does not.

One of the protective benefits of schooling is interestingly subtle. You may have noticed that the way to gain weight is to eat constantly. If you eat a little "snack" every few minutes, you consume a lot of food altogether and gain weight quickly. On the other hand, if you eat only one meal every day or two, you will lose weight. Because of the limited size

(a)

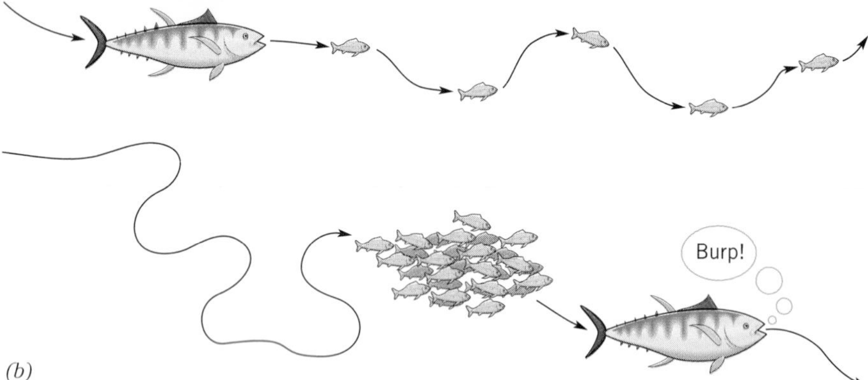

(b)

Figure 13.10

(a) There are several reasons for fish to school, including advantages in feeding, reproduction, rearing of young, and reduction in losses to predators. (b) A predator can consume much more through frequent small snacks than a few infrequent large feasts. (So can we.) Therefore, a species can reduce its losses to predators by schooling.

of your stomach, in addition to shrinkage between meals, there is a limit to how much you can eat at one sitting, no matter how hard you try. That is, through frequent snacks you can consume much more food altogether than you can through infrequent banquets.

When we apply this concept to marine environments, if organisms school, their predators will have a few infrequent banquets as they encounter these schools. But in this fashion, the predators consume less altogether than if the prey were dispersed and the predators could enjoy frequent small snacks (Figure 13.10b).

Although most nekton are found in the surface waters near the primary food source, some species are found in midwater and deep water environments. Because the food supply is meager, infrequent, and mostly detritus from above, these deep sea creatures tend to be small efficient scavengers rather than large predators. Most have small bodies with huge mouths, which allow them to take advantage of infrequent feasts that fall down from surface waters above or result from occasional encounters with less fortunate neighbors (Figure 13.11).

(a)

Figure 13.11

Mid- and deep water fish are small, but usually have huge mouths that allow them to take advantage of infrequent cashes of food that they encounter. (a) Photo of a hatchetfish. (b) Miscellaneous deep water fishes drawn to about ⅔ actual size

BENTHOS

The inhabitants of the ocean bottom are referred to as **benthos**. Throughout most of the ocean, only a very small fraction of the food produced in the surface waters ever makes it down to the ocean bottom. Consequently, with the exception of some shallow coastal waters where sunlight reaches the bottom, the total mass of the benthic community is only a small fraction of that in the surface waters above. Typical average values for the total benthic biomass per square meter of bottom surface range from about 200 grams per square meter beneath the productive

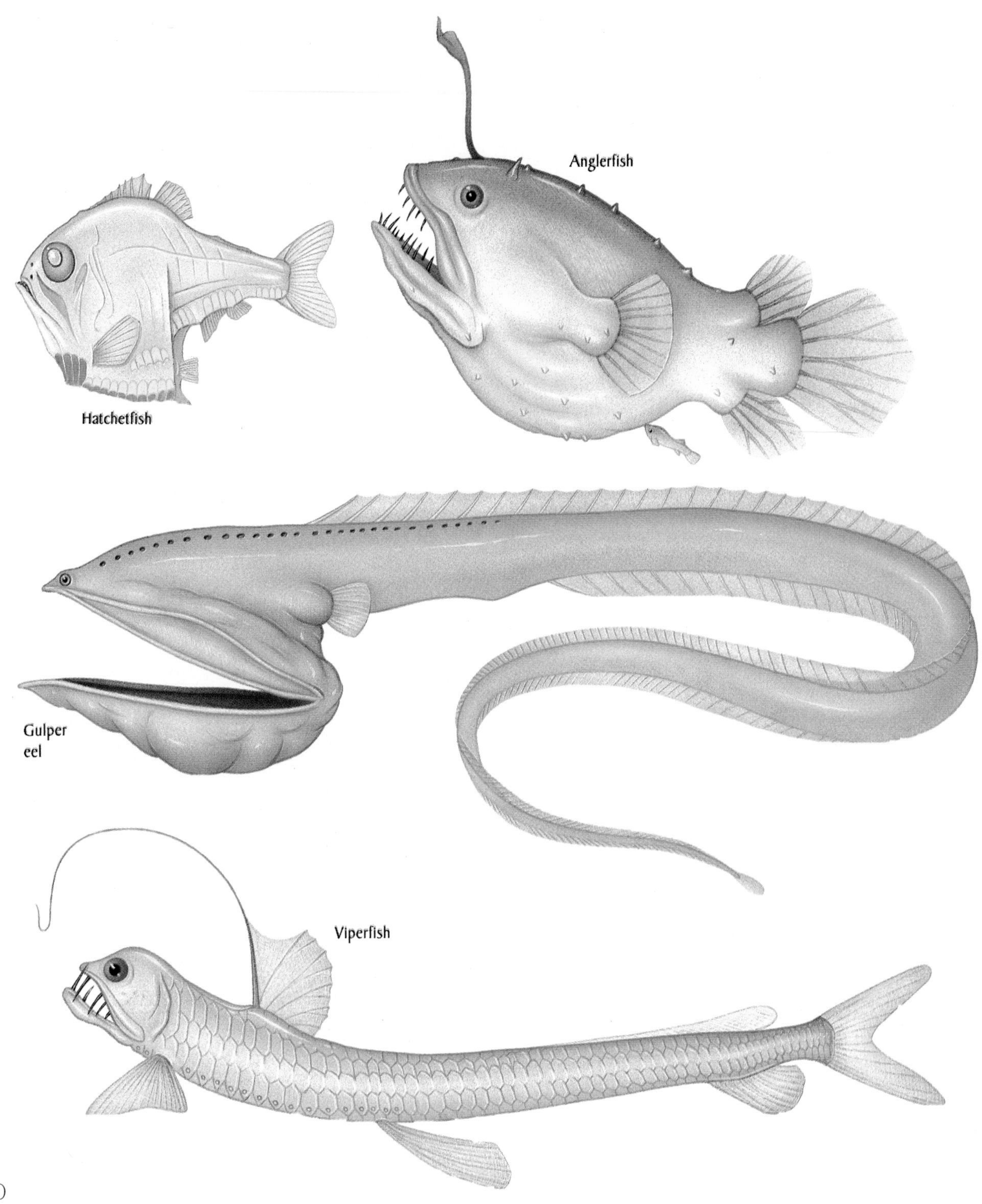

(*b*)

coastal waters, to about 0.2 gram per square meter beneath the central ocean regions.

Although benthos overall represent only a very small fraction of the ocean's total biomass, they come in a wide variety of forms. In fact, 98% of all marine animal species are benthic. In the pelagic environment, we find huge numbers of individuals of relatively few species, whereas benthic environments support relatively few individuals of a very large variety of species. The great amount of diversity among benthic species is a reflection of the wide variety of benthic environments.

Benthos can be placed into various categories according to their mobility (Table 13.4). Those that are attached to the bottom are called *sessile*, and those that can move are said to be *vagrant* or *mobile*. Examples of sessile benthos are barnacles, mussels, kelp, and corals. Most sessile benthos have planktonic larval stages, which explains how they colonize fresh surfaces, such as pilings, ship hulls, and whales. Mobile benthos can be further subdivided into those that live on the bottom but are not confined to it (**nektobenthos**, or **demersal** organisms), such as flounders and rays (Figure 13.12); those that crawl over the bottom, such as starfish, snails, and crabs; and those that burrow within the bottom, such as worms and clams.

Benthos display three very different kinds of feeding behaviors. Some actively search for prey or scavenge, such as octopus, crabs, and starfish. Some are filter feeders (sometimes called *suspension feeders*), such as some clams, corals, and barnacles. Some are deposit feeders, such as some worms, nudibranchs, and sea cucumbers.

Animals that live on the bottom surface are called **epifauna**, and those that live within the sediments are **infauna**. Epifauna that actively crawl over the bottom are generally scavengers or predators, such as crabs or starfish. Those that are attached to the bottom may be predators, such as anemones, or may be filter feeders, such as tube worms, barnacles, oysters, or mussels. Some of the infauna are deposit feeders, such as certain types of segmented worms. But others burrow within the sediments only for protection or stability, gathering their nourishment by filtering it from the water above. These filter-feeding infauna use various mechanisms to get the water to flow through their feathery nets that filter food particles from the water. Some use muscular contraction to pump water through their burrows, some use siphons, and some extend their feathery filters directly into the water above.

Table 13.4 Ways of Characterizing Benthos

Ways of Characterizing Benthos
Sessile (attached)
Vagrant, or mobile
Live on, but not confined to, the bottom
Crawl over the bottom
Burrow in the bottom

Figure 13.12
Nektobenthos, such as this flounder, are capable of fast extended motion, even though they spend most of the time on the ocean bottom.

Deposit feeders and bacteria tend to favor fine-grained muds, because such muds tend to be richer in organic matter. Not only do the finer grains present more surface area for surface-active organic materials to collect on, but also they tend to collect in lower-energy environments where organic detritus can also settle out directly. Consequently, in fine-grained muds, the infauna tend to be dominated by deposit feeders, whereas coarser sediments tend to house more filter feeders, such as clams, sand dollars, and filter-feeding worms. In addition to organic detritus, the bacteria themselves are an important source of lipids and proteins for deposit feeders, although their carbohydrate content is low. The deposit feeders help aerate the sediments, which benefits bacteria and microscopic animals.

SAMPLING

To study plants and animals in the oceans, we must have accurate methods of sampling them, so that we know what organisms live where and in what quantities. Although we touched on this topic in Section 11.3, we now describe in a little more detail some of the sampling methods used, and some of the sources of inaccuracy associated with each.

For nekton, we sometimes use shipboard sonar for locating schools and estimating their species and numbers. We also use fisheries data, trawling gear, purse seines, and line gear (Figures 11.6a and 13.8b). Unfortunately, many species are particularly mobile and skillful at avoiding capture. Furthermore, their tendency to cluster and school creates huge differences among similar measurements in similar regions of the ocean.

For measuring plankton, we often drag plankton nets behind ship (Figure 11.6b). Modern nets have mechanisms to open and close the nets at the appropriate time and depth, and meters for measuring the volume of water passing through. Plankton nets must be towed slowly to avoid tearing them and to minimize the crushing of the microorganisms. Another method for sampling plankton involves pumping water aboard ship and filtering it through a graduated sequence of nets to separate plankton of various sizes. For either method, the samples collected are subsequently examined under a microscope for species identification. Unfortunately, due to plankton patchiness, neighboring samples taken by identical methods often give very different results.

To sample benthos, we drag dredges along the bottom, or bite into the bottom with grab samplers or corers (Figure 5.17). Unfortunately, certain benthic organisms are more mobile than others and can avoid capture. Even among sessile benthos and infauna, the tendency to cluster in communities causes neighboring samples to give quite different results. Other observational methods include lowering appropriate apparatus to the sea floor, which gives us photographs or TV images, climbing into appropriate SCUBA gear, or using submersibles (Figure 1.16b). Unfortunately, intrusions from above scare away the skittish and attract the curious, giving inaccurate measures of the populations of both.

13.3 BENTHOS

In the preceding two sections, we studied marine environments and the lifestyles employed by different organisms to cope in these environments. Now we combine these two topics to examine communities of organisms.

GENERAL

We begin with the ocean floor. The kinds and abundances of organisms in ocean floor community depend on the availability of those materials needed for their livelihood. Important considerations include the abundance of sunlight and nutrients for plants, food and oxygen for animals, and the nature of the bottom sediments for both.

Except for some exposed rock near the crest of active ridges, the deep sea floor is covered almost entirely by very fine muddy sediments, as are the outer continental shelves, slopes, and rises. These contain varying amounts of organic matter, depending on the productivity of the surface waters above and dilution by other sources. The many protected coastal waters, such as those of estuaries or lagoons, also have fine, organic-rich muds. On the other hand, the high-energy environments of exposed coastal waters favor coarser sediments, such as sands, cobbles, or rocks.

Because plants need sunlight, attached seaweeds and benthic microalgae can grow only in the shallow coastal regions where sunlight penetrates to the bottom. Even within this limited region, these plants can only grow where the bottom is stable, such as rocky coastal areas or protected estuarine environments. High-energy coastal sands are not suitable for benthic plants, because the constant motion of the sediments prevents seaweeds from attaching to the bottom and buries the microalgae.

Animals need food and oxygen. Most deep waters have plenty of dissolved oxygen, so the distribution of epifauna reflects the abundance of food. The distribution of infauna decreases rapidly beyond a depth of 10 centimeters or so into the sediments, as the supply of both food and oxygen decreases.

Putting this all together, we expect benthic plants to be found only in shallow coastal waters where the bottom is stable. Benthic animals would be found throughout the oceans, in a pattern that roughly reflects the productivities of the surface waters above (Figure 10.13). That is, they would tend to be most abundant in the shallow coastal waters and decrease in abundance with distance from shore. Deep ocean benthos would be scarce, except for upwelling regions where plant productivities in the surface waters are high. Infauna would be found primarily within the top several centimeters of sediments.

NEARSHORE

Due to the constant shifting of sediments, the sands of exposed coasts are nearly devoid of plants or epifauna. Infauna tend to be predominantly filter feeders. Deposit feeders are scarce because the sands tend to be rather coarse and sterile.

The situation is quite different in more protected coastal waters, such as estuaries. Because sediments are stable in these protected waters, plant life and various epifauna flourish. Organic detritus tends to settle out along with fine sediments, leaving organic-

Figure 13.13

Sand dollars feed on bits of organic matter in the sediments. Sometimes they stand on edge to capture tiny food particles carried by the water motion. The sandy near-shore benthic environment is also home for burrowers, such as some clams and segmented worms, who also filter their food from the water.

Figure 13.14

In and above the intertidal region is a noticeable vertical zonation of organisms, reflecting the varying abilities of the different species to tolerate prolonged periods out of the water.

rich muds on which deposit feeders feast. There is very little water percolation through these fine sediments in quiet waters, so oxygen becomes depleted within these muds, which restricts the activity of animals and aerobic bacteria. In some places, organic sediments collect so rapidly that they cannot fully decompose. (See Section 11.5.)

Where coastal sediments contain organic matter, the deposit feeders include various kinds of segmented worms, nematodes, sea cucumbers, and sand dollars. Common filter feeders within sediments include various kinds of clams and certain segmented worms that either extend feathery particle trapping nets up into the water above, or pump water through their burrows from which they filter their sustenance (Figure 13.13). Burrowing carnivores are rare, although there are some predaceous burrowing shrimp, segmented worms, and a type of burrowing starfish that feeds on buried clams.

Compared to coasts with sands and finer sediments, rocky coasts have considerably more benthic life (Figure 13.14). One reason for this is that there is more food production by plants. Rocky coasts provide stable anchors from which seaweeds can grow, and microalgae grow on surfaces provided by both rocks and seaweeds. In addition to more food production, rocky coasts also provide stability, protection, and anchorage for a variety of epifauna that are not possible in shifting coastal sands. These epifauna include small mobile scavengers and herbivores, such as various small crabs and other crustaceans, snails, limpets, and sea urchins. The epifauna also include carnivores, such as anemones, hydroids, nudibranchs, starfish that feed on clams, and some snails that can bore through bivalve shells and devour the meat inside. Finally, there are also filter feeders, such as clams, barnacles, mussels, and filter-feeding worms.

On rocky coasts, the diversity of species among benthos is particularly evident in the littoral (intertidal) zone, reflecting the change in environments with altitude. We find different types of communities at different levels, corresponding to varying abilities to withstand prolonged periods out of the water as the tides come and go (Figure 13.14). Table 13.5 lists some of the common nearshore benthic organisms and the zones in which they are most likely to be found. The table is for rocky coasts, because rocky coasts have the most populace and visible communities of organisms, and they are more stable, which permits the development of more clearly defined zonation.

Organisms in the spray zone are above the high tide line and never covered by water. On the other hand, organisms in the shallow offshore region are below low tide line and always covered by water. Between these two extremes, the resident organisms vary, depending on their varying abilities to withstand exposure and predation by shore birds. Organisms living in this region have developed various mechanisms to prevent dehydration while the tide is out, such as producing gelatinous coatings or burying themselves in moist sediments.

Table 13.5 Some Common Nearshore Benthic Organisms

Benthic Organism	Spray Zone	High Intertidal	Low Intertidal	Shallow Offshore
Producers				
Lichens	X			
Blue-green algae	X	X	X	X
Green algae (tufts)	X	X		
Green algae (leafy)			X	X
Red algae			X	X
Small brown algae			X	X
Large brown algae				X
Grazers				
Limpets	X	X	X	X
Snails	X	X	X	X
Chitons			X	X
Sea urchins				X
Sea cucumbers				X
Scavengers				
Insects	X			
Rock lice	X			
Sand hoppers	X			
Crabs		X	X	X
Lobsters				X
Filter-feeding epifauna				
Acorn barnacles		X	X	X
Goose barnacles			X	X
Mussels			X	X
Small crustaceans			X	X
Sea cucumbers				X
Oysters				X
Sponges				X
Tunicates				X
Filter-feeding infauna				
Clams, cockles, scallops		X	X	X
Feather-duster worms				X
Burrowing water-pumping worms				X
Deposit feeders				
Nematode worms	X	X	X	X
Segmented worms		X	X	X
Sand dollars			X	X
Predators				
Hydroids			X	X
Anemones			X	X
Corals				X
Boring snails			X	X
Starfish			X	X
Nudibranchs				X

Table 13.5 groups the common nearshore benthos according to whether they are producers, grazers, scavengers, filter-feeding epifauna, filter-feeding infauna, deposit feeders, or predators. This grouping is a little simplistic in that some organisms span more than one of these groups. Within each group, the more prominent members are listed in order of their vertical zonation, with those living in the spray zone appearing on top, and those living only below the low tide line at the bottom. From a quick glance at the table, you can see that the most diverse and heavily populated benthic region is the shallow water just below the low tide level. This and other general patterns are correctly represented by the table. However, some important details are not so easily categorized.

Figure 13.15

Using chemicals, abrasion, or both, some infauna burrow into rocks, pilings, or reef structures for protection. Burrowing organisms have left their marks on these rocks.

We first examine some important details regarding the producers in Table 13.5. Blue-green algae appear in all shallow water environments, but they are dominant nowhere. In the spray zone, they live in symbiotic relation with a fungus to form lichens (Figure 12.5). In the other environments, they grow in thin films on surfaces, such as rocks, seaweeds, or shells. Green algae species that form small stiff tufts seem to withstand exposure better than leafy forms, which we only find in lower regions, where they can be covered by water most or all of the time. Shallow water red algae frequently are filamentous and secrete a hard calcareous skeletal framework, which makes them look and feel like coral. Although some small species of brown algae can withstand exposure for brief periods, large kelp and sargassum can't. Therefore, the larger forms of brown algae grow predominantly below low tide line, where they are covered by water all the time.

Many of the organisms appearing in Table 13.5 have a variety of species with different feeding mechanisms and that live in different environments. Most grazers also feed on any flecks of dead organic matter, so they are scavengers as well. Although many snails are grazers, some are carnivorous. Some also feed on clams, some on worms, and some on coelenterates. Although some sea urchins and sea cucumbers do indeed feed on seaweeds, some filter feed, and some feed on detritus found on the sea floor. Segmented worms come in a variety of forms. Some are deposit feeders. But others display a variety of filter-feeding mechanisms. Although we list anemones, hydroids, and corals as predators, they also ingest any flecks of dead organic matter that fall on their tentacles.

A few of the benthos listed in Table 13.5 need very special environments. Most corals prefer shallow warm tropical waters. Oysters are filter feeders, so they prefer strong currents to bring them their sustenance. Because they are preyed on by some species of starfish, crabs, fish, and boring snails, they prefer brackish waters in which many of these predators cannot survive. Like coral reefs, oyster beds also provide stable platforms on which many of the other shallow offshore benthos thrive.

Although most infauna live within sediments, some barnacles, worms, sponges, and bivalves actually bore into rocks, pilings, shells, or solid reef structures for protection. Some use mechanical abrasion and some chemical dissolution to slowly carve out their homes (Figure 13.15).

CONTINENTAL SHELF

Going outward along the continental shelf, the benthos are similar to those of the shallow offshore region described in Table 13.5, but with two very important modifications. First, benthic producers and grazers thin out and disappear. As the water depth increases, more of the primary production is done by phytoplankton and less and less by seaweeds. The seaweeds disappear altogether in water deeper than about 50 meters, because not enough sunlight reaches the bottom to sustain them. As the benthic plants disappear, so do the grazers. Second, the remaining benthos populations of scavengers, filter feeders, deposit feeders, and predators become sparser with increasing distance from shore. The reasons involve both generally decreasing primary production and increasing distance between the benthos and their ultimate food source in sunlit surface waters.

DEEP OCEAN

Unlike their coastal relatives, deep ocean benthos (Figure 13.16) are far removed from the productive surface waters and must rely on detritus that somehow escaped consumption up there. Furthermore, the surface waters over the deep ocean are usually considerably less productive than coastal waters. For these two reasons, the deep ocean bottom is very sparsely populated. Life on the sea floor is limited by food, and that within the sea floor is limited by oxygen as well.

The sea floor is covered mostly by muds and oozes, but there are some coarser turbidite deposits near the continental margins and some exposed rock near the crest of the oceanic ridge. Sea floor sediments are uniform over huge distances, in sharp contrast to coastal sediments that show large variations over distances of a few tens of meters. It is a cold, dark, and uniform environment. There are no seasons, and benthos reproduce at any time of the year.

The deep sea floor is a cold, dark, and uniform environment. There are no seasons, and benthos reproduce at any time of the year.

The main food source is microscopic detritus falling from above, and about 30 to 40% of this is consumed by bacteria on the sea floor surface. Occasionally, large caches arrive from above, and some benthos, amphipods, and fish have developed the ability to sense these arrivals from remarkable distances, arriving quickly for the feast. The fine organic particulate matter makes the sea floor a good environment for deposit feeders, such as nematodes, segmented worms, and burrowing crustaceans, all of whom ingest benthic bacteria along with organic matter. Common deep ocean epifauna include protozoans, sponges, anemones, sea squirts, snails, brittle stars, sea cucumbers, and tube worms.

Figure 13.16

Photo of the deep ocean floor, showing the typical fine ooze sediments and evidence of a few deposit feeders: a sea cucumber, some worm tracks, and a partially buried brittle star.

Compared to their coastal relatives, deep sea benthos grow more slowly, metabolize more slowly, and live longer. The same is true for bacteria, which was first discovered accidentally and subsequently confirmed in laboratory studies. The research submersible *Alvin* was accidentally dropped from its hoist and settled on the floor at a depth of about 1.5 kilometers. Fortunately, no scientists were on board, but their lunch was. The lost sub was recovered after 11 months, and the food aboard was remarkably well preserved, much better than would be expected if it had been refrigerated at the same temperature. Subsequent studies have shown that it is primarily the high pressure, and not the cold temperature, that seems to slow the bacterial processes.

VENT COMMUNITIES

The vigorous benthic communities surrounding hydrothermal vents (Figure 1.20b) on the oceanic ridges are particularly interesting, because their food is produced locally and without any help from the Sun. The primary producers are chemosynthetic bacteria, which are exceptionally productive. The reason for the high productivities might be related to the fact that most chemical reactions proceed more quickly at higher temperatures.

The water surrounding a vent cools rapidly with distance. The benthic animal community shows corresponding zonation, reflecting varying degrees of tolerance of the elevated temperatures. Vent benthos are primarily filter feeders, such as clams, mussels, crabs, tube worms, and various other worms, including one that looks like spaghetti. A few organisms are found in the mouth of the vent, but most are found in the outer fringes at nearly normal ocean bottom temperatures, relying on turbulence to bring them their bacterial sustenance.

Although the kinds of filter feeders found in vent communities are similar to those on the neighboring sea floor, the particular species differ. In particular, vent benthos tend to have higher metabolic rates, mature more rapidly, and reproduce more quickly. These qualities are clearly advantageous in a temporary vent environment, which may require relocation of their homes as some vents fizzle and others appear.

Figure 13.17
Drawing showing eelgrass and typical infauna in an estuarine community.

13.4 PELAGIC ORGANISMS

NERITIC WATERS

Over the continental shelves, the planktonic community includes representatives of most major classes of marine animals. Phytoplankton are often dominated by diatoms, because nutrients are relatively abundant, and therefore there is no particular need for the efficiency afforded by the smaller species of phytoplankton that live in the central ocean regions. Among zooplankton, holoplankton are mostly small crustaceans, and most of these are copepods, although there are also many amphipods, shrimp, and krill.[5] Meroplankton include the eggs, larvae, and juveniles of many nekton and benthos (Figures 11.18 and 13.6).

Nekton include squid and the major commercial fishes. Herring, menhaden, sardines, and anchovies are small fish that live in large schools and feed directly on plankton. Somewhat larger are rockfish that live individually or in smaller schools near the bottom. These include cod, perch, snapper, and small sharks. Larger fish that often cruise nearer the surface include many types of tuna, mackerel, barracuda, swordfish, and larger sharks. Demersal fish that are sometimes in the water, but more often on the bottom, include flounders, halibut, sole, rays, and skates.

OCEANIC WATERS

Throughout most of the oceanic waters, plant productivities are very low, due to a chronic shortage of nutrients in the photic zone. In oceanic waters, the total biomass is very limited, but diverse. The shortage of food and nutrients places a great premium on small size and efficiency. Therefore, plankton are tiny, and include many trophic levels. The resulting complex interrelationships among microscopic organisms have created considerably more diversity among oceanic plankton than their neritic counterparts, even though the population density of oceanic plankton is much smaller. That is, compared to plankton over the continental shelves, oceanic plankton is sparse but more varied.

The primary oceanic producers are mostly the very tiny coccolithophores and various species of tiny silicoflagellates and dinoflagellates (Figure 12.2) that support a very long food chain. In the very limited areas of upwelling, where nutrients are plentiful, the larger and less efficient diatoms dominate. In these regions, the food chain is similar to that of the continental shelves, discussed above. Consequently, we will not repeat it here, but rather concentrate on the food chains in the much larger nutrient-deficient regions of the central oceans.

The microscopic detritus and food produced by phytoplankton get passed on to tiny herbivorous zooplankton. Prominent among these are the single-celled protozoans, foraminifera and radiolaria. These are preyed on by tiny multicellular animals, especially various kinds of copepods. Next in line are somewhat larger, but still very small, zooplankton, such as arrow worms, planktonic snails (pteropods), and various small crustaceans. At each level, the eggs and larvae of higher levels are also an important food source. Because of the need for small size and efficiency at each level, many levels are required before the remaining food reaches the top predators like squid, jellyfish, and fish.

Below the photic zone, animal life is particularly sparse, consisting mostly of small crustaceans, such as copepods, amphipods, euphasids, and shrimp. Also found are arrow worms, jellyfish, squid, and

[5]Dividing *small crustaceans* into smaller groups like copepods, amphipods, shrimp, and euphasids is analogous to dividing *small insects* into ants, flies, beetles, and so on. That is, each of the smaller groups still contains large numbers of different species.

Figure 13.18
Photo of a salt marsh showing the grass-covered tidal flat and a rather wide tidal channel.

some small fish. Many of these perform daily vertical migrations with the deep scattering layer. Many of the animals at these depths have a reddish coloring that helps make them invisible to predators. Many fish in the disphotic zone have very large eyes, and some in the aphotic zone have none at all. Most deep fish are small, because there is not enough food to support large sizes. Adults are typically from 2 to 20 centimeters long (Figure 13.11). Many have small bioluminescent organs on their surfaces, which are probably used to attract prey.

The water is too deep for attached plants, but floating sargassum collects in the central North Atlantic.[6] These large mats of floating brown algae can produce and reproduce in these central oceanic waters, although their total productivity is still constrained by a scarcity of nutrients. The sargassum is usually not eaten itself, but it provides a platform for other algae and animals to live in and on.

13.5 ESTUARIES

In contrast to exposed coasts, protected waters of estuaries often have fine, organic-rich muds, which provide an excellent habitat for a variety of infauna. In addition, the high productivities also ensure enough organic detritus to support a sizable community of filter feeders, such as mussels, oysters, and clams. Some scavengers, predators, and grazers, such as fiddler crabs, snails, insects, and birds, are also present (Figure 13.17).

The nature of the estuarine benthic community at any particular point in an estuary depends on a number of varying estuarine conditions. These include:

1. The salinity, which changes from fresh at the head to salt at the mouth
2. Currents and turbidity that tend to clog filter feeders, and bury or expose benthos in spite of their efforts to the contrary
3. Oxygen concentration, which tends to diminish quickly with depth in sediments, especially in organic rich muds
4. The nature of the substrate, which could vary from rock to mud

Tidal currents cause considerable sediment movement near the mouth of the estuary, keeping this region relatively free of plant life. However, attached plants do proliferate in the back regions of the estuary where there is much less water and sediment motion. Seaweeds tend to grow in regions that are shallow but always covered by water. Vascular plants tend to grow in the regions that are alternately covered and exposed as the tide comes and goes. Depending on their particular nature, these regions are variously described as tidal flats, mud flats, salt marshes, or swamps. Winding **tidal channels** cut through the beds of seaweed and vascular plants, where these plants help anchor the sediments (Figure 13.18). The currents through these channels keep them clear and deeper than the surrounding plant beds.

The vascular plants in estuaries are often collectively called **halophytes** because of their tolerance of salt. They include marsh grasses as well as mangrove bushes and trees at lower latitudes (Figures 13.18 and 5.11). Although halophytes produce seeds, their primary means of propagation is by sending out rootlike rhizomes, from which new neighboring plants grow.

Estuarine plants are extremely productive, because the shallow water affords them ample access to sunlight, and the freshwater run-off from land brings in plenty of nutrients. In addition to producing food, seaweeds and halophytes provide mechanical support and homes for other smaller organisms, such as tiny algae, bacteria, protozoans, snails, small crustaceans, and other small creatures, which together may have a total mass even greater than that of their host. Most of the production is carried out by large attached plants, although some production is from

[6] Sargassum is carried by surface currents from coastal Caribbean areas to the area of convergence in the central North Atlantic gyre.

small green algae or microscopic blue-green algae, growing either on these larger plants or directly on the sediments. Some food is produced by benthic microalgae—mostly diatoms that live on the muddy bottom. Although some animals graze on the small and microscopic algae, relatively few animals feed directly on the large attached plants. Studies of estuarine filter feeders indicate that the main three components of their diets are organic matter carried into the estuary by rivers or streams, detritus from the estuarine plants, and plankton, with the relative amounts of these depending on their location in the estuary.

In addition to the abundance of nourishment (nutrients for plants and organic matter for animals) in estuaries, there is usually relatively low predation and plenty of dissolved gases. However, relatively few species can tolerate the changes in temperature and salinity that are demanded of estuarine organisms. Therefore, estuaries support large numbers of individuals of relatively few different species.

Plankton are not as important in estuarine communities as they are in the ocean proper, for several reasons. One is that species of both phytoplankton and zooplankton tend to be extremely intolerant of changes in salinity and temperature, both of which vary greatly in estuaries. Second, they tend to get flushed out to sea twice a day with the tides, whereas benthos don't. Third, the attached plants are so productive and the benthic animal community so active that they both completely overshadow the planktonic contributions.

Because of the abundance of food and low predation, estuaries serve as excellent nurseries for the youth of many commercially important fish. Tidal flushing also allows the estuaries to serve the greater ocean community by distributing estuarine food to the neighboring coastal area. For these reasons, estuaries have a much greater influence on marine life than one would suspect from their relatively small size.

The rivers and streams flowing into estuaries are continually supplying sediments, which settle out in the quiet waters and are stabilized by the attached plants. Consequently, estuaries are filling in and are temporary features on geological time scales.

13.6 CORAL REEFS

Coral reefs are built by carbonate-secreting plants and animals. Each generation grows on the remains of its predecessors, building up a solid limestone structure with living organisms on its surface. Besides

(*a*)

(*b*)

Figure 13.19

(a) Closeup photo of coral polyps, showing the tiny tentacles with which they capture particles of food. (b)Underwater photo of some reef corals.

corals, other major contributors to reef structure include various kinds of carbonate-secreting algae, mollusks, and benthic foraminifers.

Calcium carbonate is most easily secreted in warm salty waters, because these are the conditions under which it is least soluble. Warm salty waters are found near the surface in western tropical oceans, because these waters have been driven all the way across the oceans by trade winds, being exposed to solar heating and evaporation all the way. Consequently, reefs are found mostly in western tropical oceans. Reef-building corals are most active in waters that are around 23 to 25°C and within 30 meters of the surface.

INHABITANTS

As we learned in the preceding chapter, corals are small coelenterates (like tiny anemones) that capture zooplankton and other food particles with their tentacles and pull them into their open gut (Figure 13.19). Each individual coral is called a *polyp* and measures typically a few millimeters across. Polyps grow in colonies and secrete calcium carbonate to form large branching structures. These structures support polyps in positions that expose them to water motion and particulate organic matter that is suspended therein. Polyps extend their tentacles through small holes for feeding and withdraw when they feel danger from predators. Since corals must rely on water motion to bring them their food, they grow most vigorously at the outer edges of a reef, where their exposure to wave and water motion is the greatest.

Corals live in close symbiotic relationship with a type of dinoflagellate algae, called zooxanthellae. Zooxanthellae actually live within the tissue of the coral polyp and can sometimes make up more than 50% of the polyp's weight. The coral's structure gives the zooxanthellae protection and good access to sunlight. Furthermore, the coral's waste products provide it with nutrients. In return, the zooxanthellae serve the coral by removing CO_2 and waste materials. They also help the coral secrete calcium carbonate, because the removal of CO_2 lowers the acidity, which decreases the carbonate's solubility.[7] Finally, coral can feed on the zooxanthellae when in need of food.

Coral reefs are found in waters that experience relatively little change over seasons and centuries. Consequently, inhabitants of these reefs have been able to evolve over the ages to fill many varied narrow ecological niches. That is, each reef inhabitant serves a very specialized function in the reef community. This means that there is a large variety of species found, but with relatively few members of each. The relative stability of these waters over long periods of time has also permitted the growth of extensive reefs. The largest of these is the Great Barrier Reef that extends for over 2000 kilometers along the northeast coast of Australia (Figure 13.20).

Although corals provide the most visible and artistic structures in many parts of the reef, the total live biomass of the corals and all other animals combined is usually less than that of the various kinds of algae. On the reefs, as in every other oceanic environment, the animal community is ultimately dependent on plants for their food. The primary producers on a coral reef are zooxanthellae and various forms of red and green attached algae. (Most brown algae prefer cooler waters.) Some of the algae is filamentous and branching, some leafy, some forms mats, and many are hardened by secreting calcium car-

Figure 13.20
Aerial photo of a portion of Australia's Great Barrier Reef. A small island (a "cay" or "key") on the reef is visible.

[7]Dissolved carbon dioxide makes the water acid, as you might know from your experiences with carbonated soft drinks. Calcium carbonate dissolves in acids.

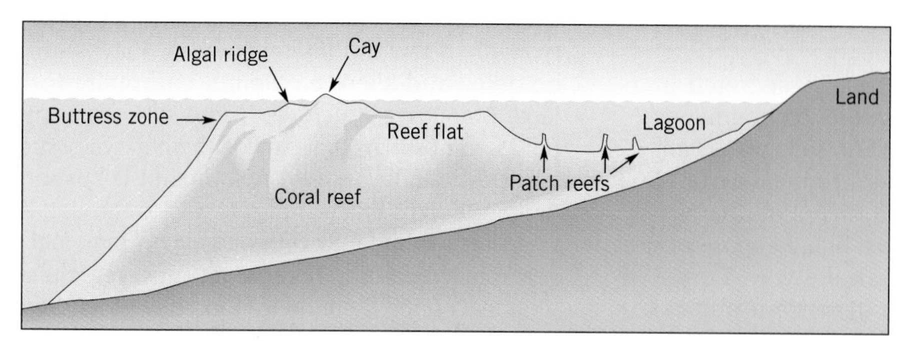

Figure 13.21

Cross section of a typical coral reef.

bonate.[8] However, phytoplankton are rather unimportant, because currents continually flush them off the reef.

Plant productivities are extremely high in reef environments because these shallow tropical waters afford excellent access to sunlight and rapid recycling of nutrients. In addition to nutrients released from the bacterial decay of reef detritus, reefs also often benefit from coastal upwelling and forced upwelling of deeper currents that strike the reef's deeper flanks.

Feeding on these plants are various snails, sea urchins, crustaceans, and grazing fish. Filter feeders include many kinds of mollusks and worms. Carnivores include corals, starfish, sea urchins, anemones, crabs, snails, fish, octopus, and eels. Parrot fish and some worms and mollusks bore into the reefs to feed on the soft parts. Some organisms, such as sponges and clams, bore into the reefs for protection. All these boring organisms tend to weaken the reef structure.

Phytoplankton are rather unimportant, because currents continually flush them off the reef.

Not all corals grow on reefs. Some are solitary. Some species grow in cold and sometimes dark polar waters. Furthermore, not all reefs involve corals. Nevertheless, corals are conspicuous and important members of the world's largest reefs.

STRUCTURE

The surface of a well-developed coral reef is irregular, which is the result of corals competing with each other for space (Figure 13.19b). This irregularity gives more surface area for better feeding. Few corals grow in water deeper than 50 meters, because the water is cold, food scarce, and the sunlight meager. By contrast, the corals in water shallower than about 20 meters are particularly strong, to resist heavy wave motion.

The front edge of the reef, called the **buttress zone**, usually is a series of ridges and grooves facing seaward (Figure 13.21). Incoming wave energy is channeled along these grooves, called **surge channels**, spreading the wave energy over a larger area and exposing more corals to the heavy water motion that they need to bring them their food. Only the sturdiest species grow in this intense environment, such as brain corals, massive branching or encrusting corals, and some calcium-carbonate-secreting algae. At greater depths down the reef flank, the surge channels turn into channels for carrying coralline debris sliding off the reef. At these greater depths, the amount of coralline debris increases, the density of living corals decreases, and the species are more delicate.

Immediately behind the buttress zone is normally an **algal ridge**, formed by calcareous red and green algae. They require sunlight, whereas the corals of the buttress zone require water motion. Therefore, the algal ridge tends to grow upward toward the sun, whereas the coralline buttress zone grows outward toward the incoming waves. Incoming waves pile up coralline debris in some places, sometimes forming small and often temporary islands, called **cays** or **keys**.

Behind the algal ridge is a **reef flat**. Here, the water motion is not as violent as it is in the buttress zone, so the corals needn't be as robust. This allows a wider variety of species to grow on the reef flat. Behind the reef flat is a deeper lagoon, where the water is relatively quiet, therefore not particularly well suited to corals. Corals may grow upward in a few places forming **patch reefs**. Behind the lagoon is land, or in the case of atolls, the other side of the atoll.

[8]These hardened algae are sometimes called *coralline algae* because they look and feel like corals.

Summary

The Marine Environment

The marine environment is very mild compared to land, so most marine organisms have no need of regulating their body fluids or temperatures. A few of the more advanced animals do. Productive plants must stay near the surface, as must many animals that rely on them for food. Consequently, these organisms employ a variety of mechanisms to combat their tendency to sink. For the microscopic organisms, these mechanisms include friction, high reproduction rates, pouches of gas or oils, and flagella. More complex animals might rely on swim bladders or motion.

Surface waters are more variable than deep waters, and coastal waters are even more variable. Coastal organisms are correspondingly more tolerant of environmental changes than creatures of the more stable environments. Plants require sunlight and nutrients; animals require food and oxygen. Throughout most of the ocean, food is what constrains animal activity, but within the bottom sediments, oxygen depletion is also important. Within the ocean, species are constrained by invisible barriers of temperature and salinity. Warm water species tend to be more colorful, varied, and smaller than their cold water counterparts, although cold waters generally support more individuals.

The pelagic environment is divided into photic and aphotic regions according to sunlight, and into neritic and oceanic environments based on the depth of the water. Most marine organisms are pelagic because of the advantages of staying near productive sunlit surface waters.

The Spectrum of Lifestyles

Pelagic organisms are divided according to mobility: Nekton are swimmers and plankton are drifters. Plankton are further subdivided into phytoplankton (planktonic plants) and zooplankton (planktonic animals). Nekton are more familiar to us, but plankton are far more abundant. Phytoplankton are the main producers in the ocean. Most zooplankton remain near the surface, exhibiting daily vertical migrations in response to sunlight. Smaller size yields greater efficiency. In the central oceanic regions where nutrients are scarce, phytoplankton is very tiny, and the food chain is long. The most numerous zooplankton are tiny crustaceans, but also included are protozoans and the planktonic juveniles of larger organisms.

The distribution of nekton is similar to that of zooplankton on which they feed. Both display daily vertical migration in response to changes in sunlight. They tend to school, which has advantages for feeding, reproduction, rearing of young, and minimizing losses to predators. Mid- and deep water nekton tend to be small efficient scavengers.

The benthic environment is divided into the littoral, sublittoral, bathyal, abyssal, and hadal zones. Throughout most of the ocean, the benthic environment is isolated from the productive surface waters. Consequently, the benthic community is relatively sparse, except in the shallow coastal waters. Because it depends on detritus from the surface waters above, the benthic population reflects productivity at the surface.

Benthos

Compared to pelagic organisms, benthos have fewer individuals but display greater diversity of species. Some benthic organisms crawl over the bottom, some are attached to the bottom, and some burrow within the bottom sediments. Some are predators or scavengers, some are filter feeders, and some are deposit feeders. Deposit feeders prefer fine muds. Filter feeders and predators are more common in coarse sediments or rocky environments.

Benthic animals need food and oxygen. Both become depleted beyond the upper several centimeters of sediments, which constrains the infauna. The food supply reflects the productivity of the surface waters above, and this is what most closely controls the distribution of benthos of all species. Benthic plants can be found in shallow coastal waters, where sufficient sunshine reaches the bottom.

Exposed coastal sands are devoid of epifauna and attached plants due to the constantly moving sediments. However, filter-feeding infauna may be present. Stable rocky coasts and protected sediments in estuaries provide much more stable anchorage for attached plants and epifauna, so a much wider and richer collection of benthos is found in these environments. Within the littoral zone, there is a distinct zonation of species according to their ability to withstand exposure. Going outward along the shelf, attached plants and benthic grazers disappear due to lack of sunlight, and benthos generally become sparser, reflecting decreasing productivity of the surface waters.

Except in upwelling areas, deep ocean waters are not very productive, and so deep ocean benthos are sparse. Many processes, including bacterial decay, are slowed at the greater pressures of the deep. Chemosynthetic bacteria provide the food that supports the isolated benthic communities surrounding hydrothermal vents.

Pelagic Organisms

Compared to their relatives in neritic waters, pelagic organisms in oceanic waters tend to be fewer, smaller, more varied, and the food chain is correspondingly longer. Neritic waters are normally more productive, supporting much larger populations of organisms, including most commercial species.

Estuaries

In estuaries, attached halophytes and seaweeds are the main producers, but benthic microalgae also produce. Phytoplankton are relatively unimportant. Abundance of sunlight and nutrients creates very high productivities. The reduced numbers of predators and abundance of food make estuaries into ideal nurseries for many species.

Coral Reefs

Calcium carbonate is least soluble in warm salty waters, so the carbonate-secreting organisms that build coral reefs do best in the warm salty western tropical oceans. These marine environments have been quite stable over the ages, allowing the development of large reefs and diverse populations with intricate interrelationships. Coral lives in close symbiotic relationship with zooxanthellae. Corals need water motion to bring them food. Therefore, they grow outward toward the open ocean, in a rugged reef front, with ridges and grooves. The algal producers on the reef need sunlight. Therefore, the coralline reef front is backed by an algal ridge, growing upward toward the surface. Sometimes, coralline debris collects on this ridge, sometimes forming small low islands, called *keys* or *cays*. Behind the algal ridge is a reef flat, where a wider variety of more delicate corals grow. Behind that is a lagoon of more protected waters, where there is too little water motion for vigorous coral growth.

Key Terms

osmotic pressure
swim bladder
benthic
pelagic
photic
aphotic
disphotic
neritic
oceanic
littoral
sublittoral
bathyal
abyssal
hadal
autotrophic
heterotrophic
nekton
phytoplankton
zooplankton
picoplankton
nanoplankton
microplankton
macroplankton
holoplankton
meroplankton
deep scattering layer
benthos
nektobenthos
demersal
epifauna
infauna
tidal channels
halophytes
buttress zone
surge channels
algal ridge
cay (key)
reef flat
patch reefs

Study Questions

1. How do land and ocean living environments compare? How do the sizes of the dominant land and marine organisms compare? What are some of the advantages of smallness for marine plants? Why don't they need to be as complicated as land plants?
2. Why do so few marine organisms exert much control over the temperature or salinity of their body fluids? Why is it advantageous for those swimmers that can travel large distances to regulate their body fluids?
3. The supply of what two things limits the distribution of marine plants? The supply of what two things limits the distribution of marine animals? Give regions of the ocean where each of these might be scarce.
4. Discuss some of the ways that microscopic marine plants combat their tendency to sink. What are some of the mechanisms employed by other organisms?
5. Compare and contrast the general characteristics of organisms found in warm water to those in cold water.
6. What are the subdivisions of the benthic environment? Of the pelagic environment?
7. Why are most marine plants pelagic, rather than benthic? Why are most of the marine animals found within a few hundred meters of the surface?
8. Explain the daily vertical migration of the deep scattering layer. How does it affect the concentrations of dissolved nutrients and oxygen?
9. Discuss the benefits of schooling.

10. Does the benthic or pelagic animal population have the greatest variety of species? Which has the largest total population?
11. Give some examples of sessile benthos, vagrant benthos, nekton, and plankton.
12. Why are benthos sparse in most deep ocean regions? Why are infauna found only in the upper several centimeters of sediments?
13. Give one example each of a benthic producer, grazer, scavenger, filter feeder, deposit feeder, and predator.
14. Going outward from the coast along the continental shelf, what general changes would we observe in benthos populations, and why?
15. What is the source of food for vent communities? In what general ways do vent benthos differ from their nonvent sea floor relatives?
16. Is the food chain longer in oceanic or neritic regions? Why?
17. Why are estuarine sediments usually rich in organic matter?
18. Why are estuarine plants exceptionally productive? Why are plankton not so prominent in estuaries as in other ocean waters?
19. What are the three different lifestyles displayed by benthos? What are the three different types of feeding behaviors? What type of sediment is favored by deposit feeders, and why?
20. Why do corals grow most vigorously near the outer edges of a reef? What do zooxanthellae gain from their relationship with coral? What does coral get out of it?
21. Why are the plants on reefs exceptionally productive? Are the plants mostly planktonic or benthic?
22. Describe the structure of a typical coral reef.

Critical Thinking Questions

1. Suggest two mechanisms that a planktonic member of the deep scattering layer might use to change its density, so it could rise and sink through the water each day.
2. Discuss some qualitative differences between the benthic community on the deep ocean floor and that in shallow coastal water.
3. Before getting to commercial fish, the food produced by plants must pass through many more trophic levels in oceanic regions than in neritic or upwelling regions. Why? This is one of the reasons why so little of our commercial harvest is taken from oceanic regions. What is another?
4. From a marine animal's point of view, what would be some of the advantages and disadvantages of living in an estuary?
5. Compare and contrast the animal communities found in coastal sediments with those found on rocky coastlines and in protected estuaries. How do they compare with those of the deep ocean floor?

Suggestions for Further Reading

BERTNESS, M. D. 1992. The Ecology of a New England Salt Marsh. *American Scientist* 80:3, 260.

BROWN, W. GREGORY. 1993. Ocean Low Life. *International Wildlife* 23:2, 52.

DENTON, E. 1960. The Buoyancy of Marine Animals. *Scientific American* 203:1, 118.

HOLLOWAY, MARGUERITE. 1994. Diversity Blues: Oceanic Biodiversity Wanes as Scientists Ponder Solutions. *Scientific American* 271:2, 16.

LERMAN, M. 1986. *Marine Biology: Environment, Diversity, and Ecology*. Benjamin-Cummings.

LUTZ, R. A. 1991. The Biology of Deep-Sea Vents. *Oceanus* 34:4, 75.

MISTRY, R. 1992. Lilliputian World of Plankton. *Sea Frontiers* 38:1, 42.

SHINN, E. A. 1989. What Is Really Killing the Corals? *Sea Frontiers* 35:2, 72.

THURSTON, H. 1988. The Little Fish That Feeds the North Atlantic. *Audubon* 90, 52.

WURSIG, B. 1988. The Behavior of Baleen Whales. *Scientific American* 258:4, 102.

If we wish the oceans to play a significant role in reducing the food shortage, we must evolve from hunter to farmer in the oceans as we did on land.

Oil spill cleanup on the Vancouver coast.

Fourteen

COASTAL DEVELOPMENT, POLLUTION, AND FOOD

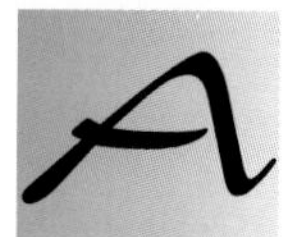

Although we are land animals, our increasing population is having an important impact on the oceans. We are crowding the oceans' shores and polluting their waters. Our search for food is taking an increasing toll on the marine biological community. Furthermore, we are tapping the oceans' energy and mineral resources to satisfy our growing material appetites.

In this chapter, we look into the impacts of coastal construction, pollution, and food harvesting, all of which have particularly strong effects on the marine biological community. In the following chapter, we study the prospects for utilizing the oceans' energy and mineral resources. We also examine the development of the law of the sea, which would decide ownership and extraction rights for various marine resources.

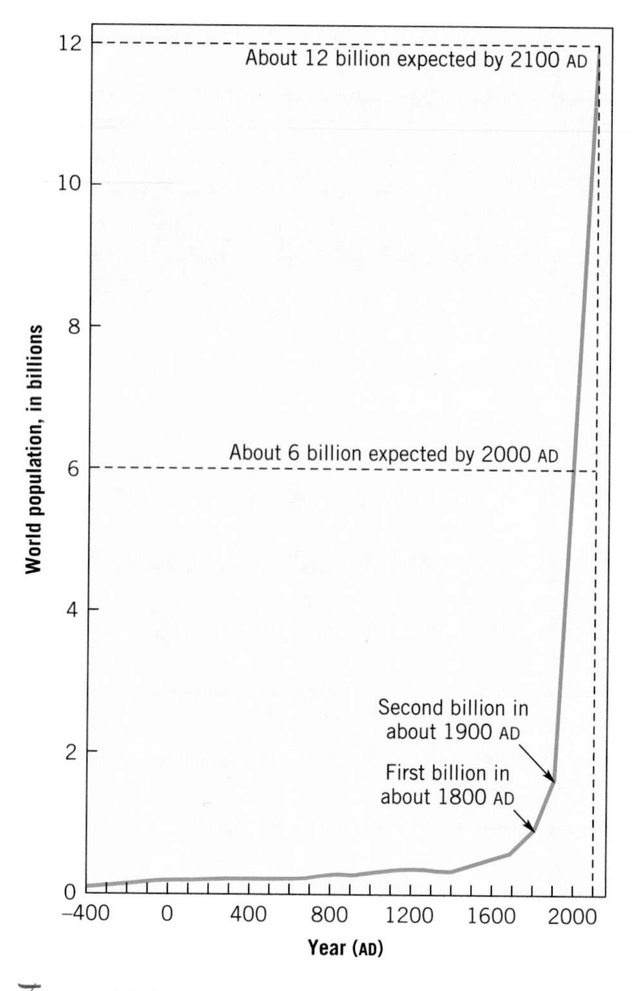

Figure 14.1

World population since 400 BC, including projections to the year 2100. (From A. Hobson, 1994, *Physics and Society* 23:4, 20.)

14.1 THE POPULATION PROBLEM

For the first time in history, the human population is breaching the limits of what the Earth can support. Although humankind is no stranger to environmental destruction, pollution, and starvation, the problems of the past have always been local.

Until recently, nature has had adequate mechanisms to cleanse itself of local and temporal human pollutants, and to rebuild the small areas that humans have altered. But now, our impact is so severe that we are overloading these natural mechanisms. In the past, the world as a whole has had ample food production capacity. The food simply hasn't always been utilized or distributed appropriately. But now we face a global problem, where the total amount of food produced in all parts of the world cannot possibly match our needs, no matter how effectively it is harvested and distributed.

Although birthrates in developed countries have declined, the population of the world as a whole has continued its steady increase—at a rate of nearly 2% per year for the past several decades (Figure 14.1). Being accustomed to growth in all facets of our society, we may not find 2% annual growth very shocking. Nonetheless, this growth has led us into a huge and insurmountable problem. It is a classic problem in *exponential growth*.

EXPONENTIAL GROWTH

Anything that increases at a fixed annual rate will double in a time that is quickly calculated by dividing the annual rate of increase into 70%.[1] This statement is called the **banker's rule of 70**, because bankers use this rule to obtain a rough idea of investment returns. For example, an investment that yields a 10% annual return doubles every 7 years, because 70 divided by 10 equals 7. Each dollar invested would become 2 after 7 years, would double again after 7 more, and so on (Figure 14.2).

Anything that doubles its value in a fixed length of time displays **exponential growth**. Things that grow exponentially can get very large very quickly. For example, suppose that someone were to give you one penny today, two tomorrow, four the next day,

[1] This is not magic. Its mathematical origin is that the natural logarithm of 2 is 0.693, or 69.3%, which becomes 70% when rounded off.

and so on, continuing to double the gift each day. At the end of just 1 month, the daily gift would be 20 million dollars! And even that amount would double on the very next day!

> Suppose that someone were to give you one penny today, two tomorrow, four the next day, and so on, continuing to double the gift each day. At the end of just 1 month, the daily gift would be 20 million dollars!

ZERO POPULATION GROWTH

Applied to the world population, an annual growth rate of 2% means that the population doubles every 35 years (70/2 = 35). This doubling time may not seem too shocking until you realize what happens after just a few doubling times. If this growth rate could continue, the entire surface of the world would be covered shoulder to shoulder with people in just 600 years!

Obviously, such continued growth could not occur. **Zero population growth** will be a reality,

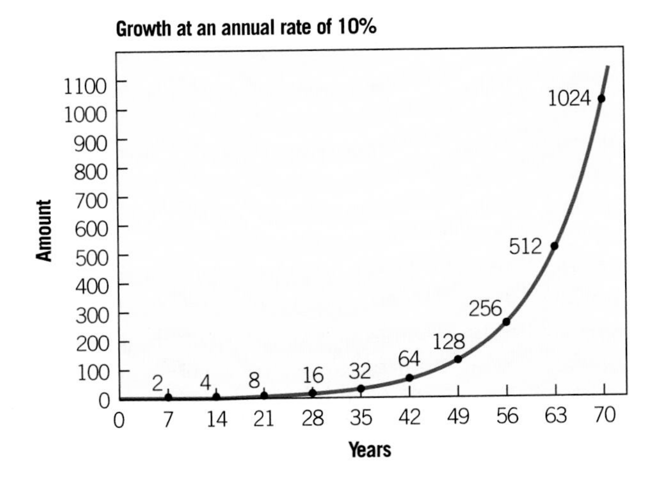

First 7 years

Year	0	1	2	3	4	5	6	7
Compounded annually	1.000	1.100	1.210	1.331	1.464	1.611	1.772	1.949
Compounded continually	1.000	1.105	1.221	1.350	1.492	1.649	1.822	2.014

Figure 14.2
Illustration of exponential growth, starting at 1 and increasing at a rate of 10% per year. The doubling time is 70/7 = 7 years. After 10 doubling times, the amount has reached 1024 times its original value. The explicit numbers showing the doubling in 7 years are given below.

whether we like it or not. It is as certain as sunrise tomorrow. The only option that we as a society have is whether we choose a civilized method of accomplishing this, or allow such events as famine, disease, or wars to accomplish it for us. That is, if we do not deal with the problem ourselves, nature will do it for us, and the remedies nature chooses will not be very pleasant.

POPULATION PROJECTIONS

A world packed shoulder to shoulder is unrealistic. World food production limits, environmental destruction, pollution, disease, and many other factors would halt population growth long before that situation is realized. Through United Nations agencies, we have fairly accurate data on the population and world agricultural capabilities. Since population trends change slowly over many generations and world food production limits are approaching rapidly, it is fairly easy to project future population trends.

We expect that the *rate of increase* in population will start to decline around the year 2000. The population will level off at nearly twice its present number by the mid- to late twenty-first century, with the largest increases in the countries with weaker economies. By the year 2025, over half the world's population will reside in Southern and Eastern Asia and another 20% in Africa (Figure 14.3).

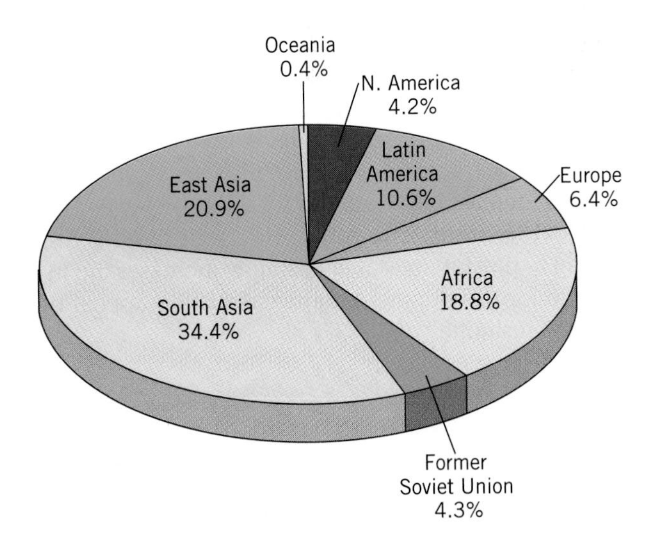

Figure 14.3
Projected distribution of the world population by continent in the year 2025, if we assume adequate food and a total population of 8.2 billion at that time. (Data from *Demographic Indicators of Countries*, 1983, United Nations, New York.)

IMPACT ON THE OCEANS

Now let us examine some of the problems that the growing population is facing. One of these is the development of communities on unstable coastal real estate, creating dangers for residents and causing large amounts of public money to be used for the protection of these private properties. Another is the discharge of pollutants into coastal waters, with associated impacts on both marine biological communities and coastal human populations. In addition, feeding the growing population requires increasing intrusion into the marine biological community and will probably necessitate fundamental changes in the way that marine crops are harvested. Since these problems are likely to continue and even increase in severity, it is perhaps important for us to become more familiar with the nature of these problems, associated natural processes, and possible societal remedies.

Figure 14.4
Barrier islands are temporary and changing features. Developments on them are particularly vulnerable. Barrier islands often remain relatively stable for many years, and then suddenly erode or move overnight during a particularly violent storm.

14.2 COASTAL DEVELOPMENT IN THE UNITED STATES

For reasons that include fisheries, commerce, industry, climate, and scenery, a large fraction of the world's population lives near a coast. In the United States, 35% of the population lives in coastal counties, and 50% lives within 80 kilometers of the coast (including the Great Lakes). Seventy percent of the U.S. coastline is privately owned, and the desirability of this land gives it an expensive price tag. As the high and stable coastal land becomes developed, there is increasing pressure to develop the lower and less stable areas. The high price that people are willing to pay for coastal land simply increases the incentive for development and weakens the appeal of prudent restraint.

THE DANGERS

The coast marks the line of battle between the ocean and the continent, and occasionally this battle becomes extremely violent. Most of the time, the balance between coastal erosion and deposition of sediments is close. As we have seen, deposition often dominates during summer months, and erosion takes over during the winter. But the changes are small. However, exceptionally violent coastal storms can erode large amounts of coastal property overnight. That is, the major modifications of coastal features tend to occur only very occasionally, suddenly, and violently, with long periods of relative stability in between.

This fact just adds to the dangers of coastal development. Any one portion of the coast may go many years or even decades without experiencing a single violent storm. People often acquire the false impression that the coastal area is safe, which is further reinforced by the development of communities complete with schools, shopping centers, and streetlights. Such developments and lack of caution simply increase the amount of devastation when the storm finally arrives (Figure 14.4).

EAST AND GULF COASTS

The ongoing rise in sea level is a particular concern for the U.S. East and Gulf Coasts. Currently, these coasts are retreating at an average rate of about 50 to 80 meters per century. You can see that the average coastal lot on the East or Gulf Coasts does not have a very long life expectancy. Indeed, protection of this coastal real estate is becoming a major concern. Such efforts can at best only delay, but not prevent, the ocean's landward progress.

Barrier islands are particularly vulnerable. These long, low, narrow sandy barriers are extremely dynamic on a time scale of decades, growing, shrinking, moving, disappearing, and reappearing with the passage of years (Section 10.3). As the sea level rises, barrier islands must retreat along with the coast. As is true of most coastal erosion, however, most of this movement will be accomplished during the very brief periods of a few violent coastal storms.

Even where coastal storms are not particularly threatening, rising sea level may still be a major problem. The city of New Orleans is built on the Mississippi River Delta, and the expanse of this delta effectively shields the city from the threat of storm surges. As is typical of rivers, the Mississippi tends to carry larger volumes of water in the spring, which causes it to flood out over the delta region as it nears the Gulf of Mexico. As these waters flood out over the delta, they become calm and the sediments settle out. In this way, deposition by the spring floods continues to build up the delta region.

New Orleans, however, must be surrounded by levees to protect it from flooding. Thus, it receives no new sediments with spring floods, and therefore the land it is built on does not rise along with the neighboring delta. Large portions of the city are already below river level, some as far as 2 meters. Each year it gets worse. To keep the city dry requires 34 large pumps pumping 114,000 cubic meters of water per day. Ground water is so close to the surface that it causes many problems, including drainage, building foundations, and cemeteries.

Figure 14.5

Coastal property is valued but risky. Homes built on sea cliffs risk undercutting by the waves, those built on beach front property risk damage during storms, and those built on land fill risk quicksand conditions during earthquakes.

WEST COAST

Along the rugged West Coast, people pay hefty sums for land with beautiful vistas of the Pacific Ocean. But as waves cut into their base, sea cliffs crumble and mountainsides slide down into the ocean. For this reason, some of the most prized real estate is also in the greatest danger (Figure 14.5).

Another problem on the West Coast is development on extensive regions of marsh lands that have been filled. When an earthquake strikes, these regions experience a *quicksand condition*. The shaking of the ground brings the shallow water table to the surface. Buildings sink and topple over.

14.3 POLLUTION

Not only are we concerned with the impact of the ocean on coastal populations, but we are equally concerned with our own impact on the ocean.

THE GENERAL PROBLEM

Humans are not the only perpetrators of environmental destruction. Storms, fires, floods, volcanic eruptions, glaciers, and a host of other natural causes were creating incidents of wide-spread devastation long before humans were here (Figure 14.6). In the biological realm, excessive populations of rabbits, deer, buffalo, locusts and other insects, and plagues of various kinds have destroyed many ecosystems without any help from humankind.

Significant human contribution to environmental pollution is relatively recent, but has grown rapidly to large proportions. Our population has been growing extremely rapidly over the past few decades. Unfortunately, our planet is not growing with us. As our numbers increase, so does our impact on our environment which is obvious, widespread, often destructive, but sometimes benevolent. Compared to other organisms, we are more aware of our effects and can plan ahead to reduce our impacts. On the other hand, we have material needs that go far be-

Figure 14.6
Some of the world's heaviest pollution comes from non-human sources, such as this environmental disaster from the 1980 eruption of Mt. St. Helens.

yond food production, and these extra needs place extra demands on our environment.

In general, the main problem with pollution from our industrialized society is that it is **acute**. That is, it is produced in large quantities at a few points, rather than being spread out more evenly. Concentrated pollutants pour from a single factory or a single oil tanker that has run aground (Figure 14.7). Although natural processes can usually handle pollutants in diluted form, our acute sources often overload these natural cleansing mechanisms. The inability of nature to handle these concentrated sources gives a bit of credibility to the expression, "The solution to pollution is dilution."

Although natural processes can usually handle pollutants in diluted form, our acute sources often overload these natural cleansing mechanisms.

Figure 14.7
Pollution from human sources is often acute, locally overloading nature's cleansing mechanisms. Here we see oil spilling from a grounded tanker near Brittany.

Everyone agrees that the best ways to reduce our pollution are to:

1. Produce less.
2. Recycle more.

But how best to dispose of the remaining wastes is a question that will remain controversial.

Disposal must involve either the air, the land, or the ocean. Although some wastes can be safely incinerated, disposal in air is often the least satisfactory solution, because we humans must breathe the air and cannot avoid the pollutants that it contains. Although the atmosphere does have ways of cleansing itself of many pollutants, we cannot control where they drift, where they fall, or what waterways they might be washed into. That is, we cannot avoid them when they are in the air and cannot control where they go. Consequently, the atmospheric disposal of toxic materials from industrial smokestacks, or incineration of wastes, is often not an acceptable solution.

Table 14.1 U.S. Ocean Dumping (Numbers are in millions of tons.)

Material	Atlantic	Gulf of Mexico	Pacific	Total	%
Dredge spoils	15.81	15.30	7.32	38.43	80
Industrial waste	3.01	0.70	0.98	4.69	10
Sewage sludge	4.48	0	0	4.48	9
Construction and demolition debris	0.57	0	0	0.57	1
Solid waste	0	0	0.03	0.03	—
Explosives	0.02	0	0	0.02	—
	23.89	16.00	8.33	48.21	100

Source: U.S. Environmental Protection Agency.

This leaves the land or ocean as the two alternatives for the disposal of many wastes and by-products. Both can have severe environmental impacts. But since humans live on land and not in the ocean, disposal in the ocean can sometimes be the lesser of the two evils. The deep ocean floor is the cheapest real estate on Earth, which makes it a leading candidate as a dump site. As our society continues to grow and evolve, the pollution of our oceans will become an increasingly attractive alternative to land disposal. We will be increasingly faced with these kinds of decisions in the future.

Some of our pollutants are inert. Some get quickly recycled into the Earth's greater ecosystem. But some are long-lived and toxic. The United States produces annually about 230 kilograms of hazardous wastes per person. At the moment, most of this is being stored or disposed on land. Although we are disposing annually about 200 kilograms of materials per person in the oceans,[2] most of this is dredged sediments and relatively little is hazardous (Table 14.1).

The marine pollutants of largest concern include:

1. Heavy metals from industrial and municipal discharge
2. Hydrocarbons from shipping, natural seepage, and drilling
3. Nutrients from municipal sewage, industry, agriculture, and freshwater run-off
4. Radioactive wastes from weapons testing, power plants, and hospitals
5. Thermal pollutants from industry and power production
6. Particulate matter from dredging, mining, drilling, and fishery processing

Urban centers are particularly heavy polluters. The amount of material disposed on the continental shelf near New York City in the past century would be enough to cover Manhattan to a depth of greater than six stories.[3] Some of this is toxic. Some depletes the water's dissolved oxygen content. But most is inert, such as demolition rubble and dredge spoils. It is being increasingly suggested that the marine disposal of these inert materials might be used advantageously to create offshore islands for such things as airports, recreation areas, power stations, or fish-processing facilities.

COASTAL VULNERABILITY

Of the various oceanic regions, coastal waters tend to be most vulnerable to pollution for several reasons (Figure 14.8). First, humans tend to live near the coast. Second, the coastal waters are shallow, so pollutants are more concentrated in this thinner layer than they would be in the much deeper waters of the ocean basins. Third, coastal waters receive all the run-off from land, which is the main source of pollutants, both natural and man-made. Finally, the shallow coastal waters are usually the most biologically active and productive, so coastal pollutants have the largest effect on marine ecosystems.

Of course, coastal waters are also particularly vulnerable to natural processes such as weather-induced variations in temperature and salinity, or the coastal discharges of sediments and nutrients by rivers after heavy rains. But human pollutants tend to be much more concentrated than those from natural sources. Many materials that are harmless or even

[2]To give these numbers some perspective, the discharge of sediments into the ocean from rivers amounts to about 2000 kilograms per person per year.

[3]Calculations by M. Grant Gross.

Figure 14.8

Pollutants are more concentrated in coastal waters, both because coastal waters are shallower and because they are closer to the sources of pollution.

beneficial in small quantities become toxic when in excess.

In addition to materials, we also discharge heat into some coastal waters. Thermal discharge may alter metabolic activities, trigger changes in life cycles, reduce the solubility of oxygen or other dissolved gases that are needed by organisms, or exterminate species that cannot tolerate the temperature or variation.

Many of our chemical and organic pollutants are surface-active and stick to solid surfaces such as those of fine suspended sediments. As these sediments settle out, they take the attached pollutants with them. Because coastal waters tend to be turbid, this mechanism is especially effective in coastal waters where both pollutants and suspended sediments are concentrated. Of course, their concentration in the bottom sediments may be harmful to benthic communities, and there is the latent danger that they can be stirred up again by coastal storms, currents, or dredging operations.

ESTUARINE PROBLEMS

The deep ocean is nearly pristine. The concentration of human pollutants is minuscule compared to natural pollutants, and both are small compared to pollution found in coastal waters. Both immense volume and remoteness from human sources ensure that the deep ocean waters will remain clean for a long time.

At the other extreme are the waters of estuaries and semienclosed seas. Because of their shallowness and restricted communication with the open ocean, these waters must endure an especially large build-up of pollutant concentrations. Human populations congregate around these bodies, and rivers bring in additional human and natural pollutants from surrounding regions. The dredging of shipping lanes often just adds to the turbidity of these waters and further imperils an ecosystem that may already be threatened by other pollutants.

The addition of foreign heat or materials to the estuarine environment is not necessarily bad. It is the nature, quantity, or timing of these pollutants that causes problems. Even before humans arrived on Earth, estuaries were places having relatively large seasonal temperature changes, and the makeup of the dissolved and suspended materials also changed considerably with seasonal changes in foliage and run-off from the neighboring land. But over the ages, the surviving organisms have developed the ability to incorporate these changes into their own life cycles. For some, the appropriate changes in life are triggered by these seasonal changes in their environment.

It takes time to develop a stable intricate biological community, with various organisms filling various ecological niches. Altering any component affects the entire community. If the changes were permanent, new communities would eventually be able to develop and adapt. But because our pollution is not constant in time, these extensive stable communities of organisms cannot develop.

There are some natural processes that help cleanse estuaries of pollutants, both natural and man-made. In fact, one important function of estuaries and wetlands is that they cleanse the run-off from land before it flows into the oceans. Although some of the pollutants become entrapped in the estuary's bottom sediments, the cleansing of the water itself depends on flushing the pollutants out to sea. In estimating our impact on an estuary, we must know not only the rate at which pollutants are being dumped into it, but also the rate at which they are being flushed out. This flushing can be accomplished both by:

1. Rivers and streams flowing into the estuary (freshwater flushing)
2. By the ebb and flow of the tides (seawater flushing)

Of course, pollutants flushed out to sea do not disappear. They just go somewhere else. Because of its size, the ocean is able to absorb much more pollution than estuaries. Nonetheless, these pollutants spread through the ocean to other areas, especially along coastlines where they are aided by strong coastal currents. Once in the ocean, pollutants cannot be kept local and are therefore of global concern. Unfortunately, international agreements to try to protect the world's oceans have a few weak links. These are the few coastal countries willing to compromise the quality of their coastal environment for the short-term economic gain offered by those international industries seeking a haven for their polluting operations.

INERT MATERIALS

Roughly 80% of the materials currently disposed of in the oceans are the materials dredged from rivers, harbors, and shipping lanes. The main impact of these operations is the destruction and modification of the community of benthic organisms, which occurs in not only the immediate region where materials are being removed or emplaced, but also the surrounding waters that have increased turbidity.

Mining operations also create turbid waters. In shallow coastal waters, sand and gravel are the materials most heavily mined, but some sediments are mined for their heavy metal content, and some phosphates for fertilizers. In the near future, it is possible that some areas of the deep ocean floor may be mined for manganese nodules, and some areas of the oceanic ridge might be mined for metal sulfides. Although the impact of these operations on local benthic ecosystems might be severe, the alternative of land mining operations is also ecologically destructive.

NATURAL ORGANIC MATERIALS

Organic materials typically decompose 10 to 100 times more slowly in the ocean than in air. The basic molecular framework of organic materials is provided by chains of carbon atoms (Figure 11.4). The overwhelming majority of organic materials on Earth have been produced by living organisms. These "natural" organic materials tend to decay and get recycled through the ecosystems relatively quickly.

By contrast, man-made organic materials tend to resist bacterial decomposition. In fact, durability is often a crucial factor in their manufacture and sale. These "synthetic" organic materials include such things as plastics, fibers, polymers, propellants, fire retardants, fertilizers, solvents, insecticides, disinfectants, and insulators. The very features that cause them to resist bacterial breakdown can also make them toxic to other organisms in relatively small quantities. Symptoms of poisoning by synthetic organics include nausea, jaundice, and sometimes effects on reproduction and deformed offspring.

Although synthetics are only a small component of our total organic pollutants, their tendency to be toxic and resist decomposition makes them especially problematic. Many other chemical pollutants have similar qualities, so we study synthetic organic materials in the succeeding section on chemical pollutants. In this section, we concentrate on the more plentiful, but less toxic and more easily decomposed natural organic pollutants.

The two natural organic pollutants of greatest concern are sewage and petroleum. Sewage enters the ocean primarily from the outfalls of municipal sewage treatment facilities, and petroleum enters mostly from urban drainage and shipping operations, such as flushing of tanker ballasts. Tanker accidents are also a significant source.

To give this problem some perspective, we should note that ocean organisms produce much more organic waste altogether than all the human sewage outflows, and a considerable amount of petroleum would enter the ocean through natural seepage, even if there were no ship accidents or other human discharges. The problem with human pollutants, therefore, is not the total amount, but rather that they enter the ocean in extremely high concentrations at relatively few isolated points, locally overloading the ocean's natural cleansing mechanisms. Furthermore, human pollution occurs in shallow coastal waters where damage to life and the environment is maximum.

Because of health hazards, it is important to keep sewage in places where it does not come in contact with people until after it has decomposed. For that reason, sewage outfalls should be located beneath the permanent thermocline. That is, sewage should be released in the cooler subsurface water. The lighter warm surface layer acts as a lid that prevents the cooler waters from rising to the surface where they can come in contact with people. Sewage outfalls should also be where there are subsurface currents that can both disperse the material and continue to bring in the oxygen needed by the bacteria. Bacterial decomposition releases nutrients, which would then be available for plants whenever this water rises to the surface again.

Major contributors of petroleum pollution are tanker operations, and the most damaging acute sources are accidents (Table 14.2). Accidental spills reached their peak in the late 1970s when about 6% of the world's fleet of 6000 tankers was involved in accidents per year. The amount of oil released into the ocean from these is larger than that from natural seepage up through the ocean floor. Both sources

Table 14.2 Sources of Petroleum Entering the Ocean. Total is about 3.2 Million Metric Tons Per Year. Estimates Vary Greatly, but the Numbers Below are Representative

Source	Percent of Total
Seepage and other natural sources	8
Offshore production	2
Refining	3
Loading, bilge pumping, and other transportation	32
Tanker accidents and other accidental spills	13
Municipal wastes	21
Industrial wastes	6
Urban run-off	3
Rivers	2
Ocean dumping	1
Atmosphere	9
	100

are much larger than the oil released from well blowouts and other accidents from oil platforms.

The oil released in tanker and platform accidents has several fates, the relative importance of which vary from one accident to the next, depending on such factors as wind, sunshine, waves, the oil's composition, and the water's temperature, depth, and turbidity. Typically, about one-third of the oil evaporates, another third joins the bottom (and sometimes shoreline) sediments, and the remaining third is dispersed in the water by waves and winds.

Natural processes eventually clean up the oil spill, but it takes several years. As long as the bacteria have sufficient oxygen, both the oil dispersed in the water and that which joins the bottom sediments eventually decompose. But the decomposition takes typically several years to complete. It goes fastest where there is ample dissolved oxygen for the bacteria and where the oil is broken up into tiny globules, providing more surface area and better access for the bacteria.

In the meantime, severe damage may be done to the local biological community. Such spills are particularly devastating to eggs, larvae, and microscopic animals, which cannot escape, but sometimes are damaging to the larger more mobile animals as well (Figure 14.9). There are some organisms, such as bacteria and certain worms in the bottom sediments, that ingest the petroleum for nourishment.

Because of heavy shipping activities, some estuaries tend to have particularly high concentrations of petroleum pollutants. Because estuaries are shallow and turbid, the surface activity of these pollutants means that relatively large concentrations of petroleum collect in the bottom sediments, where they can reenter the marine environment when disturbed by storm waves or currents.

The organisms in the world today have evolved over a period of several hundred million years and longer. They have evolved to fill niches in an environment that has until recently been free of concentrated human pollutants. For example, although bacteria decompose organic molecules, few have experienced environments as rich in concentrated petroleum and other chemicals as are now found in tanker accidents and municipal sewage outflows. To fill this need, research is now being directed into developing "superbug" bacteria that quickly decompose highly concentrated organic materials and certain other chemical pollutants.

One of the problems produced by high concentrations of organic matter is the depletion of dissolved oxygen, which devastates the marine animal communities in the region. Bacterial decomposition depletes both organic matter and oxygen, but when organic matter is particularly abundant, they will run out of oxygen first. Anaerobic decomposition may continue until the other oxidizing agents are consumed as well. This condition can occur in concentrated sewage outfalls and petroleum spills, and is sometimes observed in stagnant ponds where leaves and other bits of organic matter on the bottom only partially decay. A similar situation occurs when we release an abundance of nutrients into relatively stagnant water. The nutrients stimulate excessive plant growth near the surface. The large amount of organic matter produced by these plants collects on the bottom, where bacterial decomposition depletes the oxygen.

CHEMICALS

Many trace materials are needed by organisms to produce special molecules and tissues that serve special functions. Because these materials are rare, there is the risk that the organism might not happen to run into them at the moment they are needed. Consequently, there is a tendency for the organisms that have survived over the ages to be those that can collect and store these rare materials for such time as when they might be needed. Some organisms can even pass on these stored materials to future generations.

Only very recently in the evolution of life on Earth have we developed an industrial society that produces concentrated quantities of materials that

Figure 14.9
Although we see many photos of the effects of oil spills on larger organisms, in general the larger, more mobile organisms are not as strongly affected as are eggs, larvae, and microscopic organisms.

were formerly very rare and well dispersed.[4] We are finding that many materials that are needed in small quantities become toxic in abundance. The inherited ability to collect and hoard materials that are rare was once advantageous, but is now sometimes detrimental. The tendency of various pollutants to become concentrated in the tissues of organisms is referred to as **biological magnification**. Predators also have the ability to hoard precious or rare materials, so the "magnification" of these materials increases with each trophic level, as is illustrated in Table 14.3.

Among the chemical pollutants that tend to pose the most serious threats are heavy metals, insecticides, disinfectants, and synthetic organic materials. The heavy metals include mercury, lead, cadmium, silver, and nickel. These are already present in our environment from natural sources, but humans tend to release them in concentrated forms. For example, it is estimated that mercury enters the marine environment at a rate of about 5000 tons per year from natural sources and also 5000 tons per year from human sources. But the natural sources are dilute, whereas the human sources are concentrated at certain industrial discharge points.

One famous episode of mercury poisoning occurred in Minamata, a city on the west side of Japan. Mercury discharged from a local industry between 1953 and 1960 contaminated the shellfish in the local bay. Many people who fed on these shellfish were severely poisoned. Over 50 people died, some developed permanent disabilities, and hundreds suffered severe illness. Some of the symptoms of this poisoning were even passed on to the next generation.

Disinfectants and insecticides such as DDT, aldrin, and dieldrin, are purposely manufactured to kill certain microorganisms. This property makes them toxic to larger organisms as well. One study showed that when oysters were in water that had DDT concentrations of 0.1 ppb, the oysters themselves acquired DDT in their tissues in concentrations of 7 ppm, a 70,000-fold magnification. A similar study

[4]For perspective, most chemical pollutants enter the environment in quantities of parts per billion or less, but that may still be millions or billions of times more highly concentrated than their natural abundance.

Table 14.3 Biological Magnification of the Pesticide DDT in an East Coast Estuary

Organism	DDT (ppm)
Water	0.00005
Plankton	0.04
Silverside minnow	0.23
Sheepshead minnow	0.94
Pickerel	1.33
Needlefish	2.07
Heron	3.57
Tern	3.91
Herring gull	6.00
Osprey egg	13.8
Merganser	22.8
Cormorant	26.4

Source: Woodwell et al., 1967, DDT Residues in an East Coast Estuary, *Science* **156**.

of trout in the Great Lakes showed a 100,000- to 1,000,000-fold magnification of a synthetic organic material known as PCB.

Many harmful chemicals released into our atmosphere by transportation and industry eventually find their way into the oceans. These include both toxic materials and those that combine with the moisture in our atmosphere to produce acid rain. Although these are of concern while in our atmosphere and may produce problems in freshwater rivers and lakes, they are generally insignificant in the oceans, because the immense volume of the oceans ensures that they remain extremely diluted. Furthermore, unlike oceanic pollutants with acute sources, airborne pollutants enter the ocean in very diluted form, usually dissolved in rainfall or freshwater run-off from the land. For the particular case of acid pollutants, the ocean's dissolved carbonates neutralize them as well.

NUCLEAR WASTES

As we consume our convenient fossil fuels, our society becomes increasingly reliant on nuclear power production. One of the concerns with this form of energy supply is that the spent fuels contain radioactive materials. Some of these have short half-lives and decay quickly during temporary storage. But some of the radioactive decay products are long-lived, and we must find some way of storing them for extended periods of time spanning thousands of years.

As with other forms of pollution, the major problem here is that the pollution due to nuclear wastes would be concentrated at one spot, so the release of radioactivity and heat would be highly localized. At the moment, radioactivity in the ocean from natural sources is over a thousand times more abundant than that from human sources, and most of the human-created radioactivity is due to bomb fallout, not power production. But if we do rely heavily on nuclear power production, then the activity of the wastes would become a problem, and we must find a suitable place to store them.

The deep ocean floor is both far from population centers and one of the most stable geologic structures. For these reasons, it is quite possible that our society will eventually decide on the sea floor as a permanent repository for our nuclear wastes. Most likely, they would be deposited in appropriate vaults or through drill holes either in the sediments or underlying rock. But there are still many problems to be solved. One of these is the matter of international law. Who has the right to use the deep sea floor as a repository?

14.4 THE FOOD PROBLEM

In addition to producing more pollutants, our growing population is also demanding more food. Just as the ocean is increasingly becoming a repository for our pollutants, it is also being increasingly relied on for food. The two problems may seem incompatible, and of course, they are. But there are

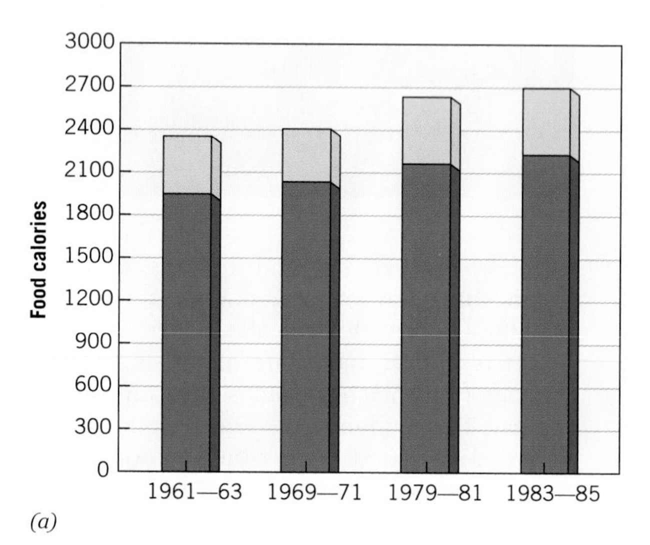

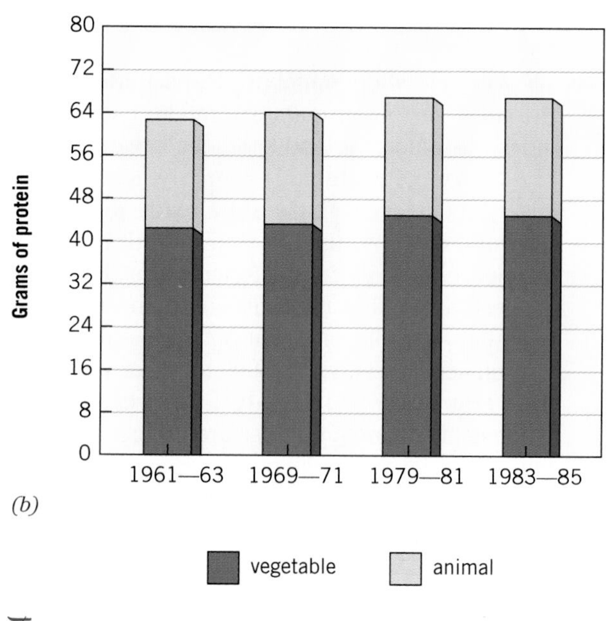

Figure 14.10

World average per capita food supply for the years 1961 to 1985. (a) Calories per person per day. (b) Grams of protein per person per day. About two-thirds of total protein comes from vegetables, and one-third from animals. (From *FAO Product Yearbook*, Vol. **40**, 1986.)

other problems associated with ocean food harvests, which are even more severe, as we will learn in this section.

PRODUCTION LIMITS

Of the various important resources in short supply, the most critical is food. Advances in technology, irrigation projects, and fertilizers have helped increase the world's food harvest in recent decades. This increase has been slightly larger than the population increase, so the per capita food supply has increased slightly (Figure 14.10).

Unfortunately, these production increases cannot last. The Earth is a relatively small planet, and there is a limit to the amount of food it can produce. Right now, the annual world food harvest is about 4.6 billion tons (Table 14.4), and it cannot be increased much beyond this. More than half the tillable land is already under cultivation, and the remaining land is of poorer quality and would be considerably less productive. In fact, most of the remaining land is unsuitable for cultivation at all. Therefore, although it may be possible to increase our agricultural output somewhat, it cannot be doubled. It could not possibly keep up with our population, not even for one more generation.

To make the problem even worse, the parts of the world that will be experiencing the largest population increases are those that are already pushing their agricultural production limits. These regions also tend to have the poorest economies. Consequently, the very places that need food the most will be the least capable of either expanding their own production or importing it from elsewhere.

It is understandable that we should want to turn to the oceans for relief from this impending disaster. Can the oceans save us? If not, can they help us out a little? Or will they be of no use to us at all? There are yet many uncertainties in the prospects for efficient use of our oceans for food production, and considerable research is being directed toward this area. However, from very general considerations we can get a basic understanding of what might be possible, and what is not.

THE DEMAND FOR FOOD

The world population is 5.5 billion. The annual food production of 4.6 billion tons amounts to 0.8 ton per person, or about 2 kilograms (4.5 lb) per person per day. This may seem like ample food, and it is. Of course, a considerable fraction goes into feeding livestock, but still there would be sufficient food to nourish our present society, provided:

1. It were distributed properly.
2. The population would stop growing.

Unfortunately, neither of these two conditions is being met. Consequently, nearly half the world's people are suffering from malnutrition.

Food is one of the few resources that is needed by everyone, and everyone needs roughly equal quantities of it. Comparing the diets of the best-fed societies to those of the most poorly nourished peoples, we find only small differences in food weight, or calories consumed per person. Of course, there are

Table 14.4 World Food Production and Animal Products Breakdown

Food	Billions of Tons	% of Total
Cereal grains	2.03	44
Animal products	0.86	19
Root crops	0.64	14
Vegetables and melons	0.45	10
Fruit	0.35	8
Legumes, nuts, oil, drink beans	0.14	3
Sugar	0.11	2
	4.6	100

Animal Product	Billions of Tons	% of Total Food	% of Animal
Milk	0.56	12.3	65
Meat	0.17	3.6	20
Fisheries	0.10	2.1	12
Eggs	0.03	0.7	3

Source: *FAO Productivities Yearbook*, Vol. 40, 1986; *FAO Yearbook of Fishery Statistics*, Vols. **62** and **63**, 1986.

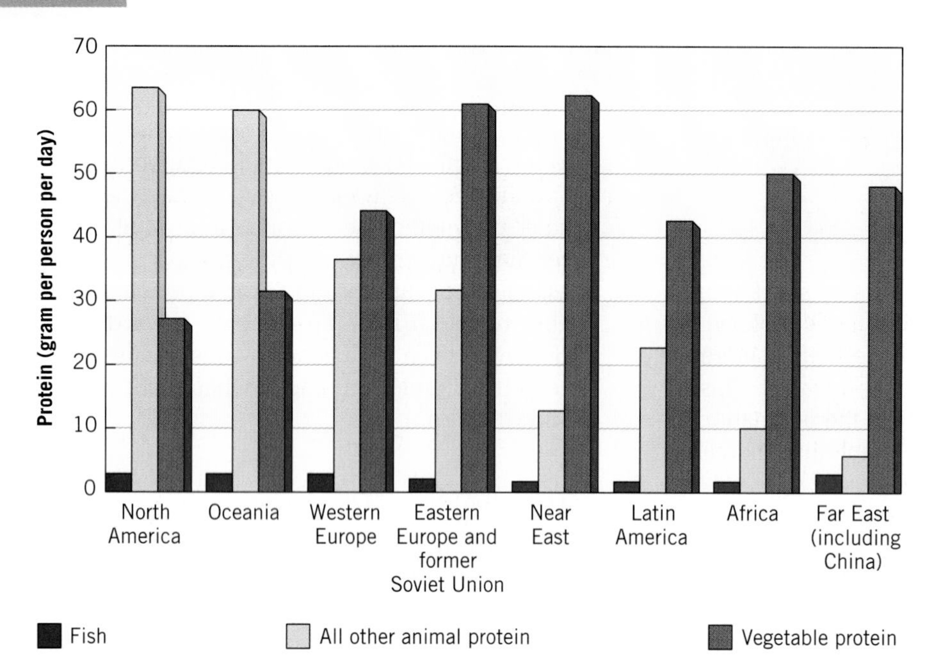

Figure 14.11

The role of fish protein, all other animal protein, and vegetable protein in the diets of various parts of the world. You can see that fish protein is a rather small part of the total protein everywhere. (From S. J. Holt, 1969, The Food Resources of the Ocean, *Scientific American*, **221**:3, 178–194.)

big variations in the *kinds* of foods eaten. For example, the wealthier nations eat much more expensive animal protein, whereas poorer nations rely more heavily on vegetable protein (Figure 14.11). But on the whole, the demand for food supplies does not vary nearly so much from person to person as does the demand for other resources (Figure 14.12).

In the language of economics, the demand for food is **inelastic**. This means that each person's demand for food remains about the same, irrespective of other factors, such as price or supply. If the supply should decrease, we will still demand the same amount of food. One result of this inelasticity is that suppliers can raise their prices considerably during times of shortage and still be assured that the food will sell.

DISTRIBUTION ECONOMICS

The wealthier nations can pay more for food. Consequently, in times of shortage, the available supply tends to flow toward the wealthier nations, and the poor nations suffer. For example, the average American spends 13% of his or her disposable income on food, whereas the average person in India spends between 60 and 90% of his or her income on food. If the price of food should double, it is clear who will suffer most.

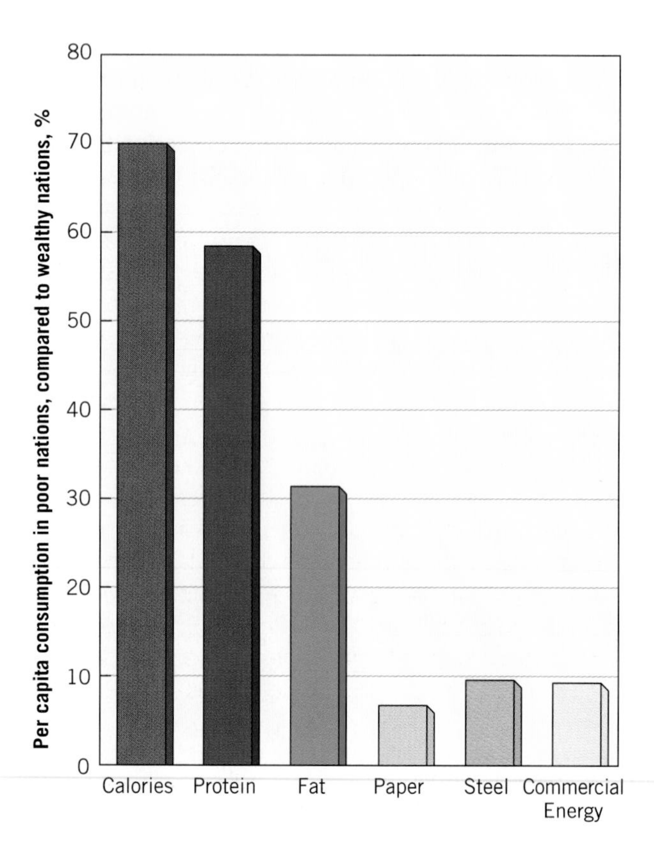

Figure 14.12

Per capita consumption of various resources in developing nations, as a percentage of that in developed nations. You can see that food consumption does not differ as much between rich and poor nations as the consumption of other resources. In the language of economics, the demand for food is more inelastic (i.e., does not change as much) compared to the demand for other resources. (After William D. Ruckelshouse, 1989, *Scientific American* **261**:3, 172.)

> The average American spends 13% of his or her income on food. The average person in India spends 60 to 90% of his or her income on food. If the price of food doubles, it is clear who will suffer most.

You can see that as the population continues to grow and the food shortage becomes more severe, the distribution will become even more inequitable. The economics are such that as the shortage increases, food will tend to go to the wealthier countries who are suffering least, rather than the poorer countries who are suffering most.

STRETCHING THE RESOURCE

One way to make the limited food supply stretch further would be to find some way to ensure that it is distributed equitably. As we have seen, such a system would have to successfully oppose economic forces, which work in the wrong direction. Another important goal to pursue is diet education. Most people are not efficient eaters, overeating in certain nutritive areas in order that their bodies get the minimum requirements in others. Through balanced diets these wastes could be reduced, and a limited food supply could be stretched further.

Even if we could ensure balanced diets and an equitable distribution of food, there is still a limit to the population that the Earth can support. Unfortunately, we have already nearly reached that limit. We are already the victims of our own numbers. To make matters worse, the bulk of the present population is young and haven't yet reached the parenting stage. As they become young parents, the birthrates will continue to grow, and starvation will take even larger tolls.

Table 14.5 World Marine and Freshwater Harvest in 1967, Each as a Percent of Total

Freshwater fish		13.6
Marine fish		77.6
Herring, sardines, anchovies, etc.	32.5	
Cod, hake, haddock, etc.	13.5	
Redfish, bass, congers, etc.	5.2	
Mackerel, billfish, cutlassfish, etc.	4.4	
Jack, mullet, etc.	3.4	
Tuna, bonito, skipjack	2.4	
Flounder, halibut, sole, etc.	2.2	
Sharks, rays, chimaeras	0.7	
Unsorted and unidentified fishes	13.7	
Crustaceans, mollusks, and other marine invertebrates		7.4
Mollusks	5.1	
Crustaceans	2.2	
Sea cucumbers, sea urchins, etc.	0.1	
Aquatic mammals (porpoises, dolphins, seals, etc.)		0
Turtles and frogs		0.1
Plants		1.3

14.5 THE ROLE OF THE OCEANS

THE OVERALL IMPACT

Some characteristics of the world fisheries' harvests are given in Table 14.5 and Figure 14.13. You can see that about 90% of the harvest is fish, and most of the rest is shellfish and crustaceans. The large harvests of Peru and Chile are attributable to the large area of productive waters off the West Coast of South America. The Asian fisheries account for over a third of the total world harvest. The large population is stretching the limits of Asian agricultural resources, giving added incentive for expanding their fishery harvests. Japan and members of the former Soviet Union have invested heavily in modern technology and fishing fleets, which are capable of extended voyages and able to harvest nearly anywhere in the world's oceans.

The annual world fish harvest is nearly 100 million tons. This is only about 2% of the world's total food harvest.[5] Consequently, even phenomenal increases in the ocean food production would have relatively little effect on the overall food problem. For example, if we were somehow able to *double* our ocean harvest, it would represent only a 2% increase in the world's total food harvest. This increase would be completely wiped out by just 1 year's growth in population.

The reasons for the ocean's relatively small impact on our food problem should already be clear. We have seen that scarcity of nutrients in the oceans severely limits the productivity of plants. In addition, because marine plants are microscopic and well dispersed, we can't harvest them. Instead, we must harvest at much higher trophic levels. Since only 10% of the food value at one trophic level gets passed on to

[5] About 30% of this fish harvest is diverted into producing fish meal for enriching the feed of poultry and other livestock.

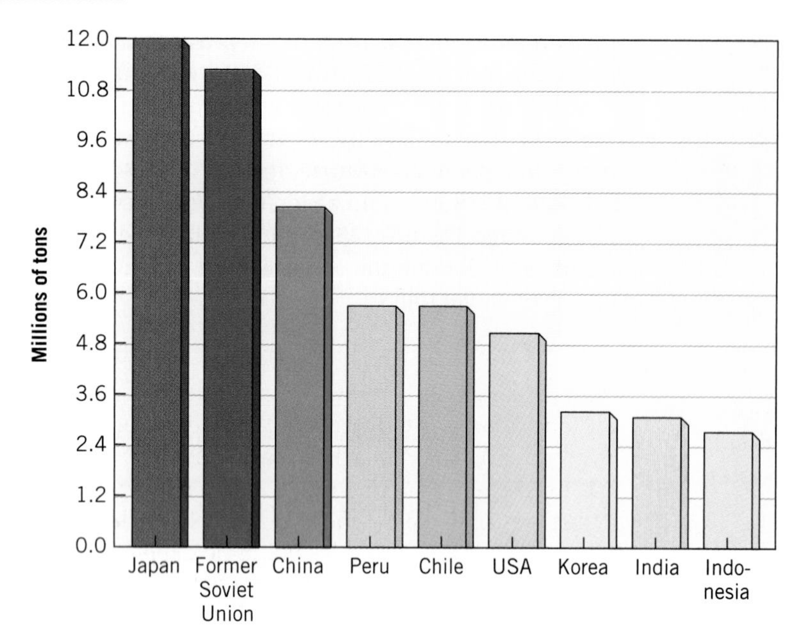

Figure 14.13
Annual fish harvests of the leading nations. (From *FAO Yearbook of Fishery Statistics*, Vols. **62** and **63**, 1986.)

the next, relatively little of the original food value remains by the time it reaches the levels that we harvest.

In summary, the reasons for the relatively small food harvest from the oceans are:

1. The marine plants produce less food altogether than land plants.
2. We must harvest at higher trophic levels, where relatively little of the original food value remains.

SEAFOOD IN OUR DIETS

Our ocean harvest yields almost entirely animal protein. Some kelp is harvested for human consumption, and some research is being undertaken to try to develop grains that can be grown in salt water. But it is not possible for the ocean to ever become a major source of vegetable matter in our diet. The scarcity of nutrients throughout most of the ocean surface waters makes small size and efficiency imperative. The cultivation of large plants for harvest in most ocean regions is therefore not practical.

We can make limited food resources stretch farthest by harvesting the lowest trophic level possible. Land plants are large and can be harvested directly for human consumption, whereas marine plants are microscopic and cannot. Therefore, the most efficient use of our resources would be to use the land almost entirely as a source of vegetable matter in our diets, and the ocean primarily for the production of animal protein.

There are also nutritional reasons for relying more heavily on seafoods as a source of animal protein. Compared to land animals, fish have the following advantages:

1. Their protein is of better quality for humans. They are a better source of vitamin B-12 and the amino acids needed for human tissue production.
2. They have less cholesterol and saturated fats, and more polyunsaturated fats and essential fatty acids—important factors in the reduction of blood cholesterol and heart disease.

Most nutritionists agree that a small amount of animal protein in our diet is advisable, primarily because of our need for balanced amino acids and vitamin B-12. Although not much is needed, a minimum of 7 to 9 kilograms of animal protein per person per year is recommended. This would be an average of about 20 to 25 grams (nearly 1 oz) per day.

14.6 FISHERY MANAGEMENT

THE BASIC IDEAS

With these motivations for increasing the harvests from our ocean fisheries, we now study the ways this might be accomplished and their impact on our food problem. The theory of proper fishery management is fairly simple. Unfortunately, in

Figure 14.14

One way to keep the stock young is to harvest adults only. This can be done by regulating net mesh sizes to ensure that only adults get captured. In this photo, the net mesh is being checked by the U. S. Coast Guard.

practice it has been very difficult to enforce. We need firm decisions, backed up by clear authority and made on the basis of good information. All three of these requirements have been somewhat lacking in the past.

The size of any stock of fish is naturally limited by its food supply. If we harvest some of the fish, more food will be available for the others, and so they will tend to grow and reproduce until they have again reached the numbers limited by their food supply. However, if we fish too heavily, we run the risk of reducing their numbers to the point where the stock has difficulty reproducing itself, and it may take years or decades to recover. That is, if we overfish in an effort to increase this year's yield, we jeopardize our harvests in future years.

The most intelligent use of our fish resources would demand that we harvest only as much this year as will not unduly jeopardize next year's harvest. The maximum amount of any stock that may be harvested year after year is called the **maximum sustainable yield**. We try to determine this number for the various commercially important species, using records of past fish catches. Uncertainties arise, however, due to the influence of several unknown or variable factors, such as the abundances of food and predators, the size of the spawn, and the prevalence of disease.

The solution to the problem of how to maximize our fishery harvests is based on the simple observation that youth are more efficient at putting on weight and increasing their size than adults. That is, if a ton of youth and a ton of adults are given the same amount of food, the young will grow and put on more weight than the adults. The more weight that the stock of fish gains per year, the more tonnage of that stock may be harvested per year. Therefore, efficient management of our fisheries would keep the fish stock young, because the young are more efficient at gaining weight from the available food supply.

KEEPING THE STOCK YOUNG

One way of keeping the stock young would be to harvest adults only, leaving the youth to fatten up for future years. In principal, this is easy to do. For those fish harvested by nets, the net mesh size could be regulated to allow smaller youth to be able to escape, although larger adults would be caught (Figure 14.14). The fish caught on line gear are usually larger and harvested one at a time, so they could be individually inspected as they are landed.

Unfortunately, international negotiations in these areas have been laborious, and when agreement is reached, regulation and enforcement have often proven nearly impossible. Individual greed and the short-sighted desire for immediate gain have often resulted in the breakdown of such regulations, which otherwise would have increased the long-term welfare of all.

Fortunately, there is another way of controlling the age of fish stock. We simply regulate the total tonnage of fish harvested. To see how this works, consider the thoughts of some very perceptive fish. If the fishing were heavy, they would know that their days were numbered, whereas with lighter fishing, they might rightfully expect to reach a ripe old age before dying. As you can see, the more heavily a particular stock is fished, the shorter the average life expectancy, and the younger the average age of the stock.

Although it is advantageous to keep the stock young, some adults must be left in order to carry out reproduction. Therefore, the job of fishery management is to allow enough fishing to ensure that the fish stocks are young, but not so much fishing that too few adults are left to carry out the necessary reproduction and regeneration. This optimum amount of fishing varies from one species to another, but usually it is such that the abundance of the species in the ocean is somewhere between one-third and two-thirds of its original virgin abundance.

APPLICATION OF THESE IDEAS

We can apply this understanding to our fisheries to see that there are two ways to increase our harvests. One is to restrict the harvests from presently overfished stocks, giving them a chance to regenerate. Bottom-dwelling mollusks and crustaceans are currently overfished in many areas because of the high price that they bring. Also overfished are many species of fish in easily reached fishing grounds, such as herring, cod, and perch from the North Atlantic, and anchovies from the Pacific near Peru.

The other way of increasing fishery harvests is taking more from presently underfished stocks. For example, most of the Indian Ocean is underfished, yielding altogether only 5% of the total ocean fish harvests (Figure 14.15). In addition, we could increase our harvest of fish whose taste or smell makes them unpopular. These fish can be processed to produce a tasteless powder called *fish protein concentrate*, or **FPC**. FPC does not spoil, giving it an advantage over other fishery products in storage and transportation. FPC also has very high nutritive value, and it can be used as an additive in many food products.

EFFECT ON THE FOOD PROBLEM

If the problems of regulation and enforcement could be overcome and these principles of proper fishery management applied, the harvests from our oceans could be greatly increased. Estimates vary, but most experts predict that the total food output of our oceans could more than double and perhaps even triple.

How far would this increase go toward reducing the present world food crisis? We know that the present fishery harvests amount to less than 2% of the total world food harvest, so doubling or tripling this would not have a large impact on the total food tonnage (Figure 14.16). However, this 2% is almost exclusively animal protein and, in fact, accounts for 11% of the world's total animal protein consumption (Figure 14.16b). So although increasing the ocean

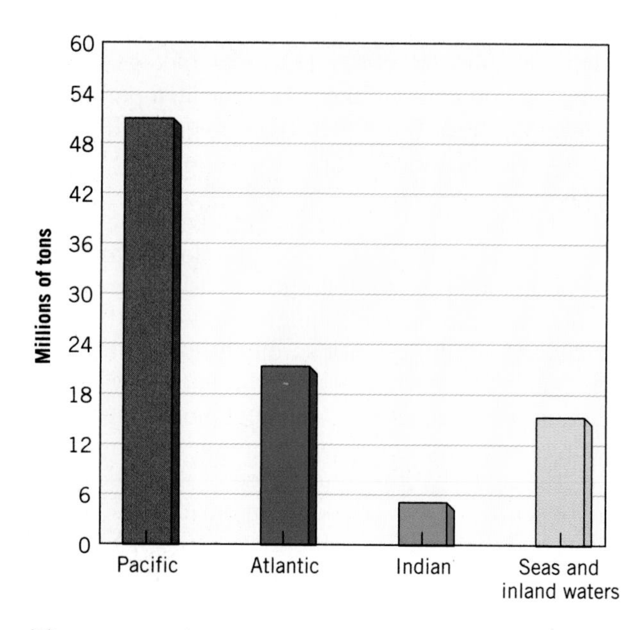

Figure 14.15

World fish harvests by ocean. Notice how little is being taken from the Indian Ocean. (From *FAO Yearbook of Fishery Statistics*, Vols. **62** and **63**, 1986.)

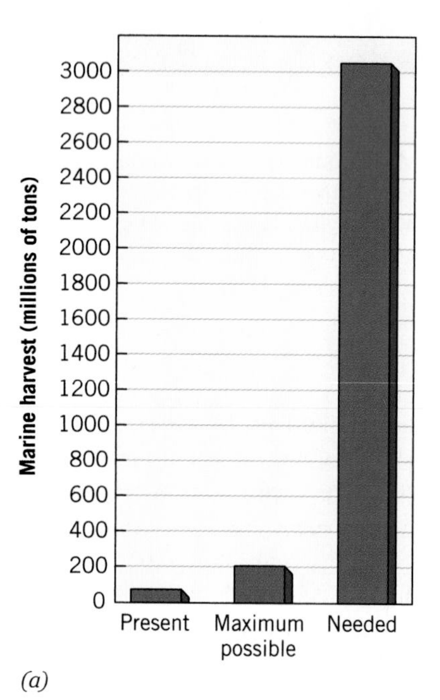

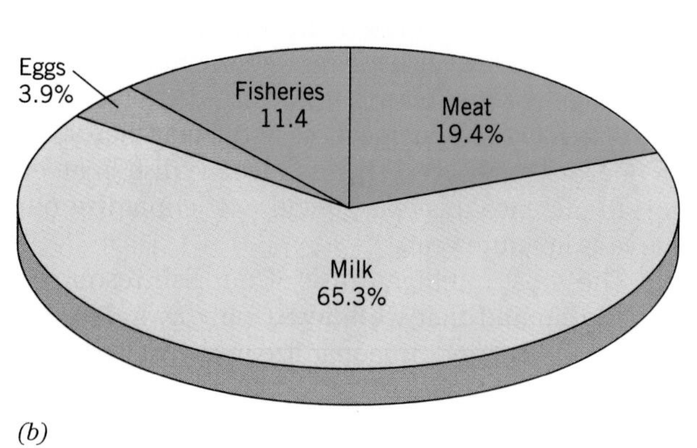

Figure 14.16

(a) Illustration of our present marine harvest, the maximum possible harvest using modern techniques and good management, and the amount that would be needed in three decades if the oceans were to solve the food problem then. You can see that fisheries can have only a very minor impact on the overall problem. (b) In the particular area of animal protein, however, fisheries can have significant impact, as is illustrated in this plot of the annual world harvest of animal protein by raw weight. (From *FAO Productivities Yearbook*, Vol. **40**, 1986; *FAO Yearbook of Fishery Statistics*, Vols. **62** and **63**, 1986.)

harvests would have only a very small influence on the overall problem, it would occur in an important nutritive area.

Increasing the fishery harvests would also have an important secondary impact. It would decrease our reliance on protein from land animals, thereby releasing for human consumption grains that would otherwise have been fed to these animals (Figure 14.17).

BETTER TECHNOLOGY

Hunters can do better by not only hunting more intelligently, but also using better weapons. New and emerging technologies are making it feasible to hunt previously untapped stocks.

Among these stocks, the krill in Antarctic waters are extremely attractive. They come in schools, typically several hundred meters in diameter and 40 meters thick, which would be convenient for harvesting. Furthermore, these schools tend to be segregated according to age, adults in some and juveniles in others. This age segregation would make it easy to harvest adults only, keeping the stock young for more efficient growth.

Estimates of possible krill harvests vary, because the size of the **standing crop** (i.e., the total unharvested amount that remains year after year) is poorly known. Through several methods, including estimates of krill consumed by whales, seals, porpoises, and other natural predators, it appears that the standing crop may be between 1 and 5 billion tons, and an annual human harvest of a half billion tons may be feasible. This harvest alone would be six times greater than our present total ocean harvest and could make our fisheries contribute roughly 15% of our total world food harvest, if these optimistic estimates are correct. Again, this increase would not come close to solving the food problem, but it would be a rich source of protein.

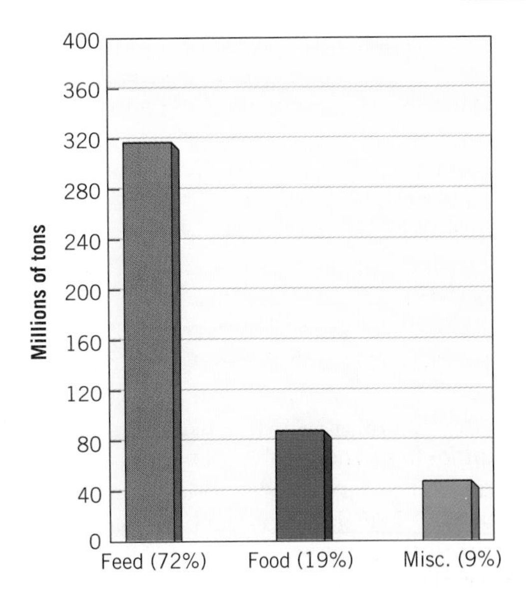

Figure 14.17

Disposition of annual grain harvests in the Western Hemisphere. Notice that the majority are used to feed livestock, and less than one-fifth of the total goes to humans. It would be about 10 times more efficient to eat the grain directly, than to eat the animal that eats the grain. (From *FAO Productivities Yearbook*, Vol. **40**, 1986.)

All successful attempts to increase the world's food harvest, either from agricultural or conventional fishery resources, would be completely negated by just a few years of continued population growth.

If there is to be any significant help from the ocean, then revolutionary changes are needed, not just improvements on what we are already doing. To see if there are any such possibilities, we return to fundamentals and reexamine our use of the ocean from first principles.

14.7 OCEAN FARMING

At this point in the chapter, we have learned that intelligent management of our world fisheries could have an impact on the animal protein component of our diets, but little effect on the overall problem. Our huge population is pushing the limits of the world's food-producing ability. All successful attempts to increase the world's food harvest, either from agricultural or conventional fishery resources, would be completely negated by just a few years of continued population growth.

THE FLOW OF ENERGY

Our bodies need energy to perform their various functions. The Sun is the ultimate source for this energy, which plants transform into stored chemical energy through photosynthesis. Before it reaches our bodies, this energy usually gets channeled through many intermediate steps, and the losses in these intermediate steps are heavy.

Only about 2% of the solar energy incident on the plants is stored and passed on to the herbivores that graze on them. From this point on, the efficiencies improve slightly. At each trophic level, an aver-

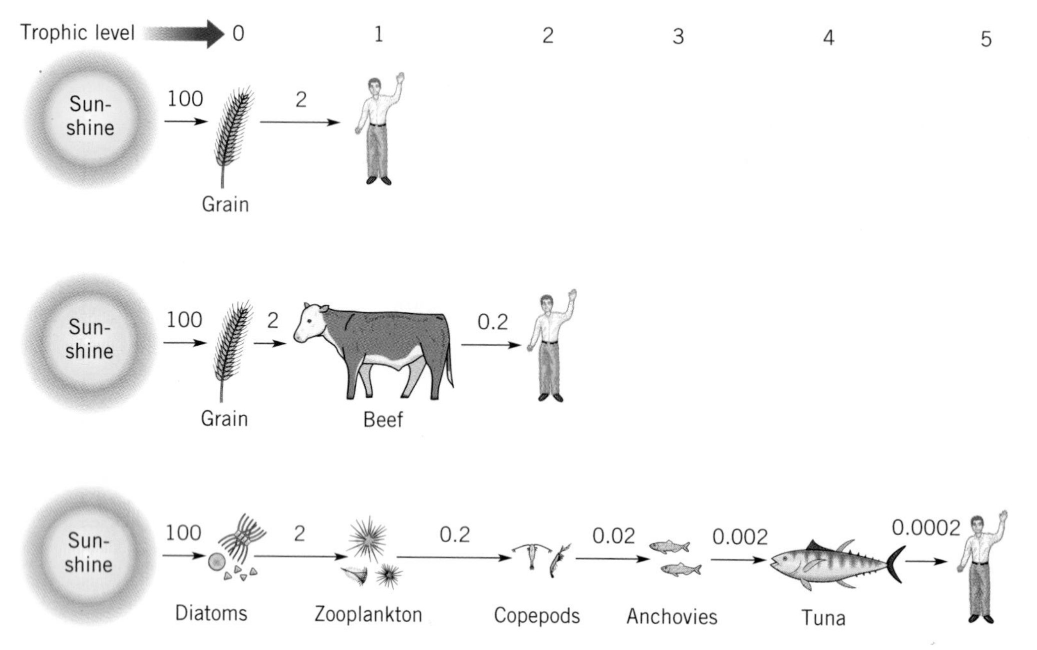

Figure 14.18
Illustration of the amount of energy being passed on to successive trophic levels, starting with 100 calories of solar energy incident on the primary producers. Clearly, it is most efficient to harvest the lowest trophic levels possible.

age of about 10% of the food energy consumed is stored and passed on to the next trophic level (Figure 14.18). The remaining 90% is used by the organisms for their own functions, such as locomotion, reproduction, digestion, production of bone and other nonnutritive material, and so on.

As an example, suppose you eat a secondary carnivore, such as a tuna fish. Of the original solar energy incident on the phytoplankton, 2% of it was stored and passed on to the herbivores, 10% of that to the carnivores, 10% of that to the secondary carnivores, and then 10% of that is passed on to you. You receive 10% of 10% of 10% of 2% (or 0.002%) of the original solar energy available. This is not very efficient use of the original solar energy!

Our harvest from the land is at low trophic levels. For example, cereal grains and other vegetables are the zeroth trophic level. Even beef cattle or hogs are only at the first trophic level, since they feed directly on the hay or grains of the zeroth trophic level. By contrast, our ocean fisheries harvest animal protein at much higher trophic levels. This is the main reason that only 2% of our total food harvest comes from the oceans.

The fact that harvests are smaller at higher trophic levels was demonstrated quite clearly in an analysis of ocean food harvests carried out by J. H. Ryther. His results are summarized in Table 14.6 and Figure 14.19. They show that less than 1% of the ocean harvest comes from the open ocean areas, although more than 80% of total primary productivity takes place there. By contrast, 50% of our fish harvest comes from upwelling areas, although less than 1% of total primary productivity occurs there. The difference is that the open ocean harvest is taken at the fifth trophic level on the average, whereas the harvest in upwelling waters comes from the first and second trophic levels. That is, although much more food is produced by plants in the open ocean, so much is lost at the intermediate trophic levels that by the time it reaches the fish we harvest, there is little left. (Do you recall why the food chain is longer in oceanic waters than upwelling or coastal waters?)

HARVESTING MARINE PLANTS

From this understanding of efficiencies, it is clear that it would be advantageous if we could harvest phytoplankton rather than fish, because this could provide much more food value to us. Unfortunately, we can't. As we have seen, most phytoplankton are microscopic and well dispersed (Figure 11.1b). To obtain significant harvests would require forcing huge volumes of sea water through fine filters. This would require the development of new technology, it would be very slow and tedious, and it would require a great deal more energy and equipment than would be economically feasible.

In addition to phytoplankton's microscopic size and dispersed distribution, another insurmountable

Table 14.6 Summary of Ryther's Results, Showing the Strong Inverse Correlation between Trophic Level Harvested and the Size of the Harvest

Province	Productivity per m^2 ($gC/m^2/yr$)	Percent of Ocean Area	Total Productivity (billion tons of carbon per year)	Percent of Total Ocean Productivity	Percent of Total Fish Production	Trophic Level Harvested
Oceanic	50	90.0	16.3	81.5	<1	5
Coastal	100	9.9	3.6	18.0	50	3
Upwelling	300	0.1	0.1	0.5	50	1.5

Source: J. H. Ryther, 1969, *Science* 166, 72–76. Copyright © 1969 by the American Association for the Advancement of Science.

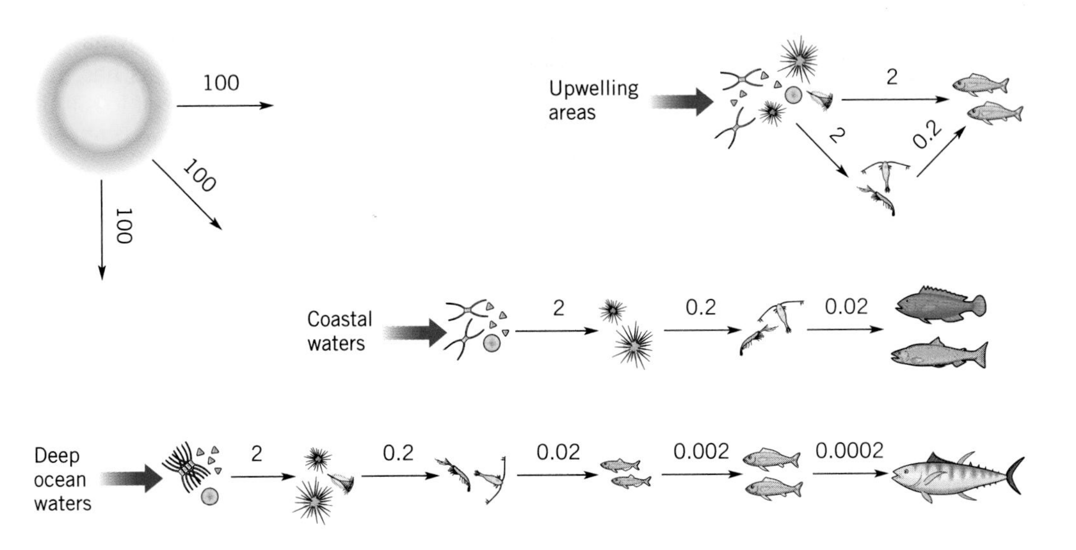

Figure 14.19
Illustration of Ryther's analysis, starting with 100 calories incident on the primary producers, 2% of which is stored in the organic matter produced by the plants, and 10% passed on to each higher trophic level. The longer the food chain, the greater the losses along the way. Although the surface waters of the deep ocean have the largest total productivity, they produce the least food for us, because we must harvest the fifth trophic level there.

difficulty involves the rate at which we would have to conduct these harvests. If we remove half of the phytoplankton today, most will be replaced again by tomorrow, because they reproduce and grow very quickly when nutrients and sunlight are sufficient. So to maximize the ocean's food output, we would have to harvest a good fraction of the phytoplankton in all parts of the ocean each day. Imagine what a supernatural effort that would take! Even if we had the appropriate filtering technology, the sheer volume of the water involved would ensure that we couldn't do it.

It turns out that nature has a harvesting mechanism that is more efficient than anything we could produce. Tiny grazers (Figures 13.1 and 13.6) feed on phytoplankton and then themselves are eaten by larger organisms. Even though there are large losses as the food is passed on up the food ladder to the levels that we harvest, the fact that nature makes these large-scale and daily phytoplankton harvests means that the total food yield is greater than that which we could hope to accomplish by ourselves.

NUTRIENT RECYCLING

In this chapter, we learned that through improved management and technology, the production from our world fisheries can be increased significantly. As welcome as these increases would be, they would still not come close to solving the food problem caused by our expanding population (Figure 14.16a).

If we wish our oceans to have a more significant impact on our food supply, then we would have to make some fundamental changes on the way we use them. In particular, we would have to change from hunters who harvest whatever nature happens to provide, to **ocean farmers** who cultivate and nourish particular crops for harvest. We have already made this transition from hunter to farmer on land.

As we learned in Chapter 11, scarcity of nutrients in sunlit surface waters severely restricts primary productivity in vast oceanic regions. Decaying organic detritus tends to sink into deeper waters, where the released nutrients cannot be used by plants due to lack of sunlight there. By contrast, on land the decaying detritus remains at the surface, where it is readily available for use by the next generation of plants. As a result, the total annual productivity of the oceans is less than that of the lands (Table 11.3), even though the oceans cover 2.4 times more of the Earth's surface and the land environment is much harsher.

Of course, there are some regions of the oceans where nutrients do get recycled quickly, and plant productivities in these regions are high. Examples include continental shelves and regions of upwelling. At high latitudes, lack of sufficient sunlight may restrict plant productivities, even where nutrients are plentiful. However, the vast majority of the ocean waters receive ample sunshine but simply lack nutrients. To make these vast regions more productive would require the pumping of nutrient-rich deeper waters to the surface (Figure 14.20).

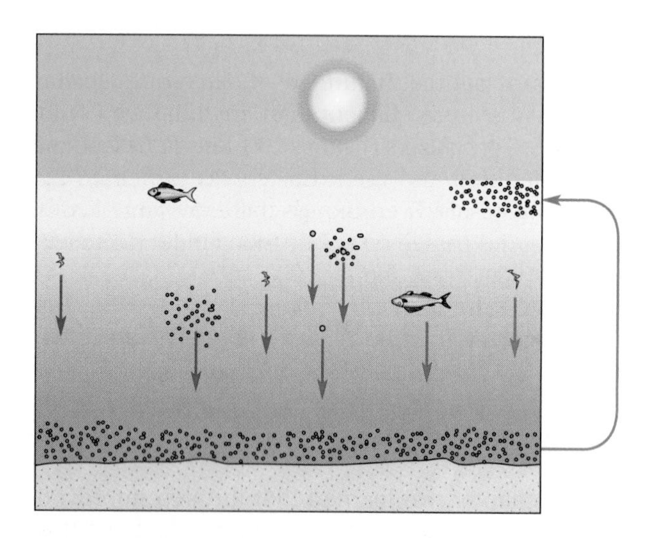

Figure 14.20

The relatively small productivity throughout most of the ocean is a result of its depth, which allows the sinking organic detritus to remove nutrients from the surface waters. If we could bring these nutrient-laden deeper waters to the sunlit surface, we could increase productivity immensely.

If we could bring these nutrients to the surface, plant productivity would increase immensely, supporting greatly increased populations at higher trophic levels for human harvest. It would make many large oceanic regions as productive as tropical rain forests on land. Of course, we would have to harvest higher trophic levels, so the total harvest would still be more restricted than that of comparable land areas. Nonetheless, if this were done over large regions of the ocean, we could experience sizable gains in world food production—much more than would be possible from the improved hunter techniques discussed previously.

If done on a large scale, the pumping of cool deep water to the surface would undoubtedly have a significant impact on our environment, as has the clearing of land for farming and burning of fossil fuels. You may wish to speculate on what these effects might be.

If we wish the oceans to play a significant role in reducing the food shortage, we must evolve from "hunter" to "farmer" in the oceans as we did on land.

In summary, if we wish the oceans to play a significant role in reducing the food shortage, we must evolve from "hunter" to "farmer" in the oceans as we did on land. On land, we learned to plow under last year's organic detritus to make the nutrients available for this year's crop. In the oceans, we would have to do the same, returning nutrients from the organic detritus of previous years to the surface, where they could help life flourish again.

14.8 FIRST STEPS

About 10% of the world's seafood production comes from aquaculture, mostly in Asia (Table 14.7) and mostly involving fish grown in freshwater ponds. Carp and catfish are major products of Asian aquaculture. In the United States, aquaculture accounts for only about 2% of the fishery products, producing about half the commercial catfish and nearly all the trout. Only a small fraction of the world's eligible freshwater ponds are being utilized. In particular, many agricultural ponds could be used

Table 14.7 World Aquaculture Production in 1975, in Millions of Tons

Region	Finfish	Mollusks	Seaweeds	Crustaceans	Total	%
Asia	1.20	0.46	0.75	0.02	2.43	40
China	2.20		0.30		2.50	41
Europe	0.42	0.40			0.82	14
North America	0.02	0.13			0.16	3
Latin America	0.03	0.05			0.07	1
Africa	0.11				0.11	2
Australia		0.01			0.01	0
Total	3.98	1.05	1.05	0.02	6.10	
% of total	65	17	17	0	100	

Source: After T. V. R. Pillay, 1976, in *Advances in Aquaculture*, FAO Technical Conference on Aquaculture, Kyoto, 1976, Unipub, New York.

for growing fish, while simultaneously storing agricultural water. Making optimum use of these ponds could increase the world's freshwater aquaculture production 10-fold.

There is even greater potential for farming the oceans (**mariculture**), as we saw in the previous section. At present, only about a third of the world's aquaculture production projects use sea water, and these are mostly geared toward cultivating shellfish and large algae. Production of crustaceans, such as shrimp and lobsters, has been difficult.

Most fish need to spend part or all of their lives in a natural setting, so although they might be spawned and harvested in artificial ponds, they need to be released into the ocean for the major part of their lives. This is referred to as *ranching*, and it has been done successfully with various Pacific salmon.

The use of artificial upwelling to bring nutrients to the surface has been done successfully in small-scale operations, but has not yet been accomplished on large scales for a variety of reasons, including environmental concerns, the costs of running the pumps, and legal uncertainties over harvesting rights.

Forced upwelling for food production could be combined with other needs. One interesting possibility is using deep water as the coolant in electrical power plants (Chapter 15). Not only are the colder waters better coolants, but also they carry fewer living organisms and therefore reduce fouling of the cooling apparatus. Furthermore, after being released from power plants, the water's elevated temperature and nutrient content would be ideal for mariculture. Food production from these waters would still be very small compared to the world's needs, but they could serve as relatively inexpensive and risk-free tests of the use of artificial upwelling to support mariculture on a much larger scale.

Summary

The Population Problem

The world's population is growing exponentially, presently doubling every 35 years. Eventually, zero population growth will happen, whether we like it or not. Among the problems created by our growing population are the development of unstable coastal real estate, the discharge of pollutants into the coastal waters, and increasing emphasis on ocean fisheries to feed us.

Coastal Development

Coastal erosion and destruction are occasional and violent, separated by long periods of calm. Along most of the U.S. East and Gulf Coasts, the sea level is slowly rising and the shoreline retreating. Many coastal properties are doomed. Developments along barrier islands are particularly risky. Much of the West Coast's seacliffs are vulnerable to wave undercutting, and building on fill risks quicksand conditions during earthquakes.

Pollution

Coastal waters are particularly vulnerable to pollution because they are shallow, close to human populations centers, receive the run-off from land, and are biologically active. Throughout the ages, coastal waters have received natural pollutants of many forms, but human pollutants are becoming a major problem because the sources are acute—that is, large quantities at a single place. Nature's cleansing mechanisms are not able to handle some of these large concentrations. It is clear that we need to produce less and recycle more.

Our wastes must be disposed of somewhere, and the choices are the air, land, or ocean. The air is often the least desirable of these, and the fact that the sea floor is cheap and far from humans makes this an attractive dump site. Most of our pollutants are inert, but some are toxic. Some are short-lived, but some are lasting. All affect the coastal ecosystems.

Bacteria and evaporation can clean up natural organic pollutants, such as sewage and petroleum, although they can be locally intense and temporarily devastating. We still have no good way of handling organic pollutants from tanker operations and accidents. Durability is one of the requirements for synthetic organic materials, such as plastics and pesticides, so they are a much more lasting problem in the marine environment.

Organisms have the inherited ability to store trace materials for later use. This ability means that trace materials become more concentrated in the organisms' tissues than the external environment, and even more concentrated at higher trophic levels. Although needed in small amounts, many of these trace materials become toxic when in excess. Consequently, our acute sources of certain chemical pollutants can devastate local marine communities. We still have found no satisfactory repository for our nuclear wastes, and the sea floor is increasingly being viewed as the best solution.

The Food Problem and the Role of the Oceans

Each person needs a certain amount of food, regardless of the price. This inelasticity in the demand causes large price fluctuations and inequitable distribution in times of shortage. A limited supply may be stretched further through wise distribution and diets.

Nutritionally, fish are a better source of animal protein for humans than land animals. The most efficient use of our food resources would be to harvest the lowest trophic levels possible. In general, land plants are large and harvestable, whereas marine plants are not. Therefore, the wisest use of our food resources would be to use the oceans primarily as a source of animal protein and the lands primarily for vegetable matter.

Fishery Management

Young fish are more efficient at gaining weight from a limited food supply, but some adults are needed to carry out reproduction. Good fishery management controls the annual harvests to ensure that these two constraints are met. By enforcing good fishery management principles, the world ocean harvest could be doubled, or even tripled. If krill could be harvested, the fishery output could be increased even more. Still, the oceans could only provide a relatively small percentage of the total food harvest, although this could be a substantial portion of the total animal protein needed.

We cannot harvest much of the zeroth trophic level in the oceans as we do on land, because the phytoplankton are microscopic, well dispersed, and we would need to make frequent harvests of large oceanic regions. Nature's harvesters do a much better job of collecting and concentrating these materials than we could. As a result, we can only harvest higher trophic levels. The higher the trophic level, the less food energy is left, as was demonstrated in the Ryther analysis.

For ocean fisheries to ever produce substantial portions of the world's food would require us to change from hunter to farmer in the oceans. We would have to see that nutrients get recycled quickly in the oceans as they do on land, by pumping nutrient-rich deep waters back to the surface.

Key Terms

banker's rule of 70
exponential growth
zero population growth
acute
biological magnification
inelastic demand
maximum sustainable yield
FPC
standing crop
ocean farming
mariculture

Study Questions

1. What is the present population of the world? What is the annual rate of growth? Why is zero population growth sure to happen?
2. What are some of the most vulnerable kinds of areas on the U.S. East and West Coasts, respectively, and why?
3. How is the Mississippi River delta getting built up by additional layers of sediments over the years? Why is this not happening to New Orleans?
4. For many kinds of pollutants, nature produces more than humans. However, the human sources tend to be more damaging. Why? What are the two best ways to reduce our pollution?

5. What are the three general repositories for human wastes? Why might the atmosphere sometimes be the least desirable?
6. Why are coasts and estuaries more vulnerable to human pollution than the deep ocean areas?
7. Petroleum has always been in the marine environment through natural seepage, and it decays relatively quickly. Why then are oil spills so disruptive to the local marine ecology?
8. Explain how biological magnification is a problem with chemical pollutants.
9. How much food is harvested in the world annually? What fraction of this comes from the ocean? Why can't the world's agricultural production be increased by several times the present amount?
10. World food prices shoot up much more in response to minor shortages than the prices of other resources. Why?
11. What are the nutritional advantages of fish over land animals as a source of animal protein?
12. Explain why annual fish harvests are larger if the fish stocks are kept young. How could this be done?
13. What happens if we overfish a stock? What are some of the presently overfished stocks? Underfished stocks?
14. Where does the energy for life come from? How does it get to our bodies, and what losses occur along the way?
15. Why can't we harvest the zeroth trophic level in the oceans? Explain how it could stretch our limited food resources if we take our animal protein from the oceans rather than land animals.
16. Explain why the total primary productivity from land is greater than that in the oceans, in spite of the fact that the land area is considerably smaller and has a much harsher environment.
17. What does it mean to change from hunter to farmer in the oceans?

Critical Thinking Questions

1. Suppose the population of a town increased at a rate of 3.5% per year. How long would it take to double? How many times larger would it be after 100 years? If you had invested $1000 at 7% annual return when you were born, what would it be worth when you are 80-years-old?
2. What do you think the coastal nations could do to discourage a few of their members from polluting the coast excessively for short-term financial gain?
3. Discuss ways to make a limited food supply stretch further.
4. What fraction of the present total world food production comes from our fisheries? How much of our animal protein comes from fisheries? Roughly how many years would it take for continued population growth to wipe out any increase in food production that is gained from the enforcement of proper fishery management? From harvesting krill?
5. Discuss the Ryther analysis. In particular, explain why less than 1% of our harvest comes from oceanic regions with 80% of total primary productivity, and why 50% comes from regions with less than 1% of total primary productivity.

Suggestions for Further Reading

ABLESON, P. H. 1987. World Food. *Science* 236: 4797, 9.

BROADUS, JAMES M. 1991. The Oceans and Environmental Security. *Oceanus* 34:2, 14.

BUTNER, J. K. ET AL. 1987. Aquaculture. *BioScience* 37:5, 308.

FOOD AND AGRICULTURE ORGANIZATION OF THE UNITED NATIONS. 1990. *Yearbook of Fisheries Statistics, Catches, and Landings*. p.70.

HAUB, C. 1988. Trial by Numbers. *Sierra* 73:6, 40.

OLSON, R. E. 1989. World Food Production and Problems in Human Nutrition. *Nutrition Today* 24:1, 15.

PAIN, STEPHANIE. 1990. On the Edge of Disaster. *New Scientist* 126:1714, 36.

SATCHELL, MICHAEL. 1992. The Rape of the Oceans. *U.S. News and World Report* 112:24, 64.

U.S. NEWS AND WORLD REPORT. 1992. CAN SHARKS SURVIVE? 112:24, 70.

Unless we can come up with suitable alternatives soon, automobiles and city gas systems will join stage coaches and Roman aqueducts in obsolescence.

Offshore oil drilling platform. Our interest in ocean resources increases as we deplete land-based reserves.

Fifteen

OCEAN RESOURCES AND LAW

The oceans are both large and largely unexplored. Many people hope that the discovery of new ocean treasures will somehow save us from impending disaster, as our mushrooming population and growing material appetites rapidly deplete land-held resources.

Undoubtedly, the oceans will rescue us from some impending shortages, at least temporarily. In other cases, they will be of little help. We have no crystal ball and cannot make detailed predictions about what benefits and disappointments lie ahead in the realm of ocean discoveries and technologies. Nonetheless, from very general considerations we can at least get a rough idea of what might be possible and what is not.

The basic problem is that our population is growing exponentially and our rate of consumption of resources is growing even faster. Earth, however, is not growing at all. The combination of heavy and increasing consumption with fixed resources leads to disaster. If we are to survive, we must find a way to make do with existing resources. We must think of ourselves as voyagers on **spaceship Earth** (Chapter 3 opening photo). Wherever this journey takes us, humankind must finish the trip with the same resources that were on board at the beginning. We will be getting no more.

15.1 ENERGY

One of these important limited resources is energy. Much of our energy appetite results from our choice of lifestyles, but energy is also important in the extraction and production of other needed resources, such as food. As we deplete the land-based energy sources, we will turn more and more toward the ocean for help.

THE PROBLEM

The world presently relies heavily on the energy stored in **fossil fuels** (i.e., those derived from the remains of previous living organisms), because they are cheap and easy to extract. For the most part, coal is found in beds near the surface and can be removed by loading it directly onto railroad cars. Petroleum and natural gas can be removed with little more effort than drilling holes in the ground. Due to both this ease of recovery and their rich energy content, fossil fuel consumption has grown quickly and has replaced other energy sources (Figure 15.1).

Unfortunately, these easy energy sources are irreplaceable. They took millions of years to develop, and we are exhausting them in decades. Their accessibility and low price have exacerbated the problem, because we have become reckless and wasteful in their use. The growth in the consumption of fossil fuels has far outstripped even the growth in population.

Of particular concern are the **portable fuels**, the natural gas and petroleum products. These can be used in transportation and piped through pipelines. They are so convenient that their use has skyrocketed in the past four decades (Figure 15.1). Consequently, the world reserves of these high-quality portable fuels are being rapidly depleted (Table 15.1 and Figure 15.2). In fact, production has already peaked, and we are now on the down side. Unless we can come up with suitable alternatives soon, automobiles and city gas systems will join stage coaches and Roman aqueducts in obsolescence.

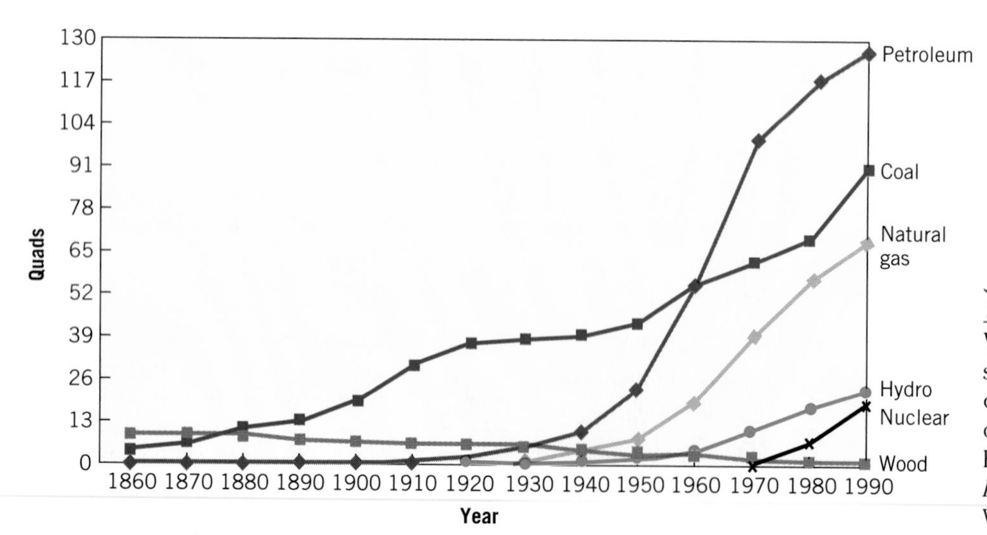

Figure 15.1
World energy use by source, showing the changing reliance on the different energy sources over the past 130 years. (After Richard C. Dorf, 1978, *Energy Resources and Policy*, Addison-Wesley, Reading, MA.)

Table 15.1 List of the World's Remaining Reserves of Nonrenewable Fuels, the Present Rate of Consumption, and a Projection of How Many Years Each Would Last at its Present Rate of Consumption

Fuel	Reserves (quads)	1987 Use (quads)	Years Left
Coal	15,000	90	170
Petroleum	3,500	127	28
Natural gas	2,000	66	30
Uranium ore	800	17	47

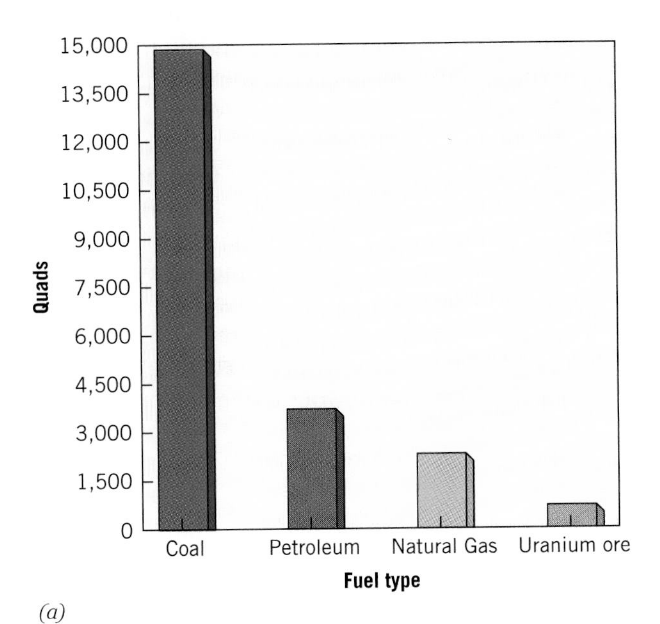

(a)

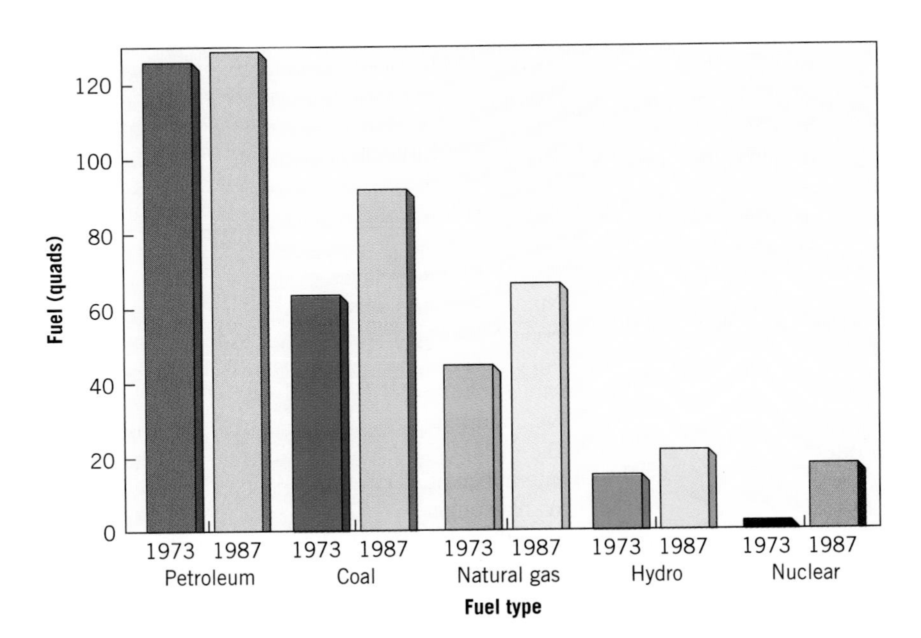

Figure 15.2

(a) Estimated world fuel reserves, in units of quads. (A *quad* is a quadrillion British thermal units of energy, or about 10^{18} J.) For comparison, the present annual world total energy usage is about 300 quads. (b) World energy production in 1973 and 1987 by fuel type. (Data from (a) Richard C. Dorf, 1978, *Energy Resources and Policy*, Addison-Wesley, Reading, MA; (b) Bodansky, 1989, *Physics and Society* **18:3**.)

> Unless we can come up with suitable alternatives soon, automobiles and city gas systems will join stage coaches and Roman aqueducts in obsolescence.

It is clear that within the next few decades we will have to make major changes, resorting to more expensive and lower-grade fossil fuel deposits for the near future, and looking for alternatives to fossil fuels for the longer-term future. The most immediate problem is portable fuels. They will soon be gone. To prolong their lifetime, there is an effort in the world to move away from these high-grade fuels and toward coal and nuclear energy wherever possible (Figures 15.1 and 15.2b). However, these alternatives are also limited, and they also pose serious environmental hazards. They are simply stopgap measures, to be used only until more appropriate long-term solutions can be found. Our search for alternatives will include ocean energy resources.

THE OCEAN AND SOLAR ENERGY

In the search for satisfactory alternatives, one attractive candidate is **solar energy**, which is energy from the Sun. It is a "clean" source, because it doesn't add carbon dioxide, smog, soot, or poisonous gases to the atmosphere like fossil fuels or nuclear reactors, for example.

The disadvantage of solar energy is that it is diffuse. We are accustomed to having our energy delivered at high temperatures in very small regions, such as within the cylinders of our cars, or the furnaces or reactors of our power plants. The Sun's energy, by contrast, arrives at lower temperatures and is spread over much larger areas. Clearly, we are going to have to develop new technologies to harness this. Our ocean is a huge, ready-made collector of solar energy, which it stores in several different ways. Because of the potential importance of this huge resource, we now examine how the ocean might help us meet our energy needs.

In addition to harnessing solar and ocean energy resources, other problems facing us involve:

1. Converting this energy to the form needed
2. Transporting it to the place needed
3. Storing it until the time needed

These are especially critical concerns for ocean energy, because the supply would normally be long distances from the continental population centers. For example, how could ocean energy in the middle of the Pacific right now help you cook your breakfast tomorrow, or drive your car this weekend? Clearly, we must find ways for ocean energy to be stored and transported to the continents without huge losses along the way.

ENERGY CONVERSIONS

In Chapter 2, we learned that energy was the ability to do work, or equivalently, the ability to exert a force over a distance (i.e., to move something). We also learned that it comes in a variety of forms, such as kinetic, potential, thermal, chemical, and electrical energy, and it can be converted from one form to another. We later learned, for example, that plants convert solar thermal energy into chemical energy in the organic matter they produce, and then organisms later convert this chemical energy into other forms as it is needed.

As another example of energy conversions, consider a coal-burning power plant. Chemical energy stored in coal is transformed into thermal energy as the coal is burned. This heat is then used to boil water, so thermal energy is converted into kinetic energy with steam. The steam turns a turbine. The kinetic energy of the turbine is then converted into electrical energy by the electrical generator, which goes though electrical lines to your home. In your home, it

Table 15.2 Efficiencies of Various Power Plants, Engines, and Lightings

For power plants, the efficiencies measure useful energy output compared to energy input, but do not include losses in transformers and transmission lines. For engines, the efficiencies measure useful energy output to energy input, but do not include losses due to friction in transmissions, wheel bearings, tires, wind resistance, and so on, when these engines are in cars. For lighting fixtures, the efficiencies measure light energy output compared to electrical energy input. In all cases, the energy lost leaves as waste heat.

	Efficiency(%)
Power plants	
Fossil fuel power plants	38
Nuclear power plants	31
Solar power plants	28
Solar cells	10
Engines	
Steam turbine	45
Diesel	35
Internal combustion	22
Wankel engine	18
Lighting	
Fluorescent	20
Incandescent	5

is turned back into thermal energy by your stove burner, or into heat and light by your lightbulb.

Such conversions of energy from one form to another are costly in terms of wasted energy. Each time a conversion takes place, only a fraction of the energy goes into the new energy form. Much is lost in the conversion process, and most of these losses end up in the form of unused thermal energy, called *waste heat*. For this reason, we wish to increase the **efficiency** of machines, which is the fraction of the energy input that is turned into the desired energy output:

$$\text{Efficiency} = \frac{\text{useful energy out}}{\text{total energy in}}$$

Higher efficiency means smaller loss. Conversion efficiencies for some familiar machines are listed in Table 15.2. As our present energy sources are exhausted and the remaining energy supply becomes more precious, we will become increasingly interested in the efficiency of conversion from one form to another, in order to reduce the losses.

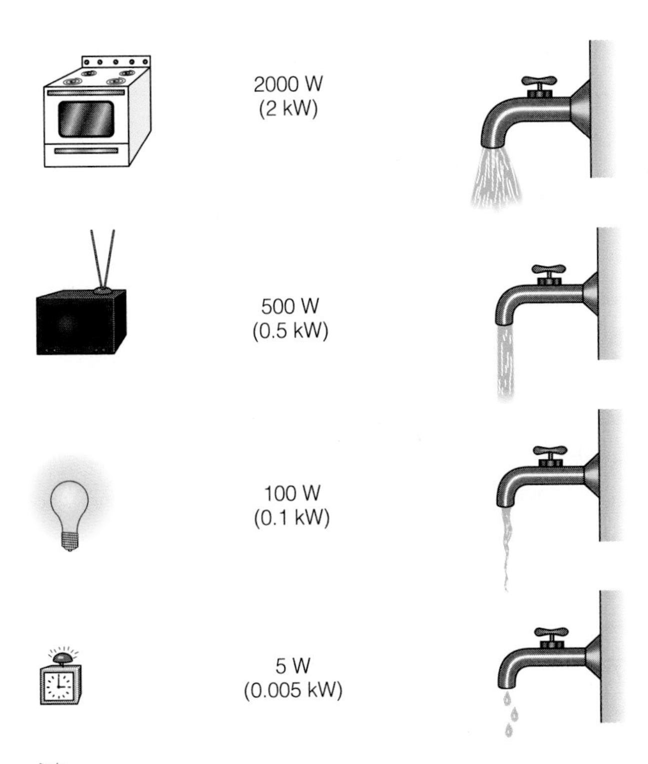

Figure 15.3

The amount of water in your bathtub depends on both the *rate* at which water comes out of the faucet and the amount of *time* the faucet is open. Similarly, the amount of electrical energy used depends on both the *rate* at which it is used and the amount of *time* used. Some analogies for typical appliances are shown above.

Table 15.3 Units for Measuring Energy and Power

	Energy Units
1 calorie	The amount of heat energy required to raise the temperature of 1 g of water by 1°C.
1 BTU (British thermal unit)	The amount of heat energy required to raise the temperature of 1 lb of water by 1°F.
1 joule (J)	The amount of kinetic energy of a 2-kg mass that is moving with a speed of 1 m/sec.
1 kilowatt-hour (kWh)	The amount of energy used, if used at a rate of 1000 W for 1 hour.

Conversions
1 BTU = 252 calories
1 BTU = 1054 J
1 J = 0.24 calories
1 quad = 1 quadrillon BTU
= 1,054,000,000,000,000,000 J

Power Units
1 watt = 1 joule/second
1 kilowatt = 1000 watts
1 horsepower = 746 watts

UNITS

The rate at which energy is used is called **power**. The standard unit of power is the **watt** (W), which is a rate of 1 joule per second. For example, a 60-watt light bulb uses energy at a rate of 60 joules each second. A **kilowatt** (kW) is 1000 watts, or 1000 joules per second. Table 15.3 summarizes the common units for measuring energy and power.

The total energy used is equal to the product of rate times time. To illustrate, compare the use of energy to the use of water in your home (Figure 15.3). Energy comes in a variety of forms, but electrical energy flows through wires and out of wall sockets, just as water flows through pipes and out of faucets. The amount of water you use is the product of the rate at which it comes out of the faucet times how long the faucet is on. For example, if water flows out at a rate of 2 gallons per minute for 10 minutes, you use 20 gallons of water altogether. Similarly, if you use electrical energy at a rate of 2 joules per second (i.e., 2 W) for 10 seconds, you use 20 joules of energy altogether. Recalling that a watt is a joule per second, we have

$$\text{No. of watts} \times \text{no. of seconds} = \text{no. of joules}$$

Table 15.4 Approximate Power Consumption by Various Things When in Use[a]

	Kilowatts
Electric alarm clock	0.01
Electric sewing machine	0.07
Incandescent light bulb	0.1
Average metabolism of an adult	0.1
Black and white television	0.6
Toaster	0.8
Electric iron	1.2
Electric stove burner on "high"	2.0
Automobile	200.0

[a]In terms of the primary fuels that were burned to provide this energy, each number should be multiplied by about 5.

When you pay your electric bill, you pay for the number of *kilowatt-hours* of electrical energy used. Since there are 1000 watts in a kilowatt and 3600 seconds in an hour, a **kilowatt-hour** is equal to (1000 j/sec) × (3600 sec) = 3,600,000 J of energy. You can use 1 kilowatt-hour of energy by using it at a rate of 1000 watts for 1 hour. But you would use the same amount altogether if you used it at half the rate for twice as long (500 W for 2 hours), or one-tenth the rate for 10 times as long (100 W for 10 hours), and so on. The rates of energy consumption for many common appliances are listed in Table 15.4.

Let's take a moment to consider how much energy is in a kilowatt-hour, and how little it costs. It takes about 20 joules of energy to lift a brick from the floor to a table. So with 1 kilowatt-hour of energy, you could lift about 180,000 bricks, or several brick houses, from the floor to the table top. We now pay about 10 cents for this amount of electrical energy and less for energy in other forms. From this example, you can see that at this moment in history, energy is extremely cheap—almost free. But it is sure to get much more costly in the near future. Present luxuries like long hot showers and well-cooked foods will become increasingly unpopular during our lives, due to the rising cost of energy.

Present luxuries like long hot showers and well-cooked foods will become increasingly unpopular during our lives, due to the rising cost of energy.

A unit that is used in some of the graphs and tables in this section dealing with world energy consumption or energy reserves is the **quad**. It is a huge unit, because the world uses huge amounts of energy. It stands for *quadrillion British thermal units*, which is

(a)

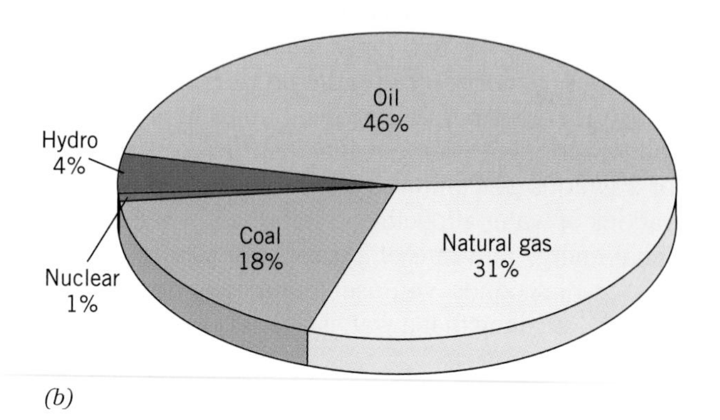

(b)

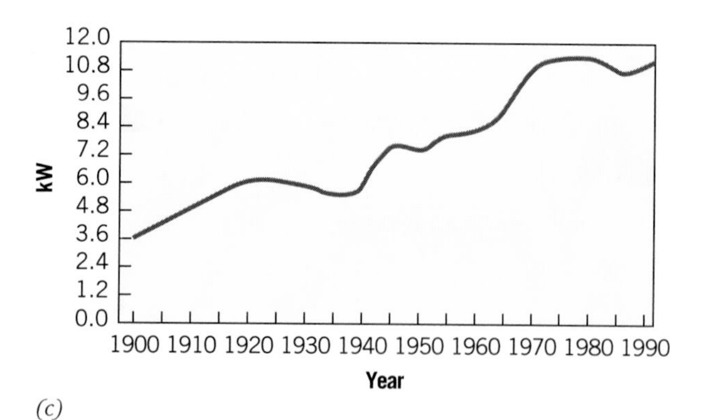

(c)

Figure 15.4
U.S. energy consumption. (a) By sector (b) By source. Fossil fuels account for 95% of all our energy. (c) Rate of per capita consumption since 1900. The steady rise was stemmed in the mid–1970s, when effective energy conservation measures were established. These included mandatory fuel economy in automobiles and insulation standards in new buildings. (After Richard C. Dorf, 1978, *Energy Resources and Policy*, Addison-Wesley, Reading, MA. See also Bodansky, 1989, *Physics and Society* **18:3**.)

Table 15.5 Fraction of World's Resources Used by the United States

Petroleum	32%
Natural gas	57%
Coal	16%
Steel	19%
Aluminum	35%
Copper	27%
Population	6%

Table 15.6 Per Capita Power Consumption by Various Types of Societies[a]

Society	Power Consumption (kW)
Primitive hunter and gatherer	0.1
Advanced hunter and gatherer	0.2
Primitive agricultural	0.6
Advanced agricultural	1.3
Early industrial	3.5
Advanced industrial	10.0
World average today	1.9

[a]These figures include the 0.1 kilowatt needed in human metabolism.

equal to about a quintillion joules, or 300 billion kilowatt-hours (more precisely 1,054,000,000,000,000,000 J, or 293,000,000,000 kW-hours). The world presently uses about 300 quads of energy per year altogether.

ENERGY CONSUMPTION

The average rate of energy usage in the United States is about 11 kilowatts per person. That is, we use energy at a rate equivalent to each of us keeping 110 lightbulbs (0.1 kW each) turned on all the time. Equivalently, we use energy at a rate equal to each of us keeping six stove burners turned on "high" all the time.

About a quarter of this is used in our homes, a quarter for transportation, and the remaining half in the industrial and commercial sectors (Figure 15.4a). That is, on the average, we use energy at a rate of about 3 kilowatts per person for such things as heating and lighting our homes, cooking our meals, and operating our TV sets. On average, we use energy at a similar rate in our cars, trucks, trains, airplanes, and so on. Finally, energy is used at a rate of about 6 kilowatts per person in the industrial and commercial sector for manufacturing or processing the items we buy, or in services we hire.

The United States has less than 6% of the world's population, but consumes about 30% of the world's energy (Table 15.5), most of which is in the form of fossil fuels (Figure 15.4b). This is high even among industrialized societies (Tables 15.6 and 15.7), but enforced energy conservation measures curtailed the rise in the past two decades (Figure 15.4c). The world as a whole uses energy at a rate of 10 billion kilowatts, or 1.9 kilowatts per person, of which nearly 3 billion kilowatts are used by the United States (Figure 15.5).

Table 15.7 Per Capita Power Consumption in 1973 by Country[a]

Country	Kilowatts per Person	Country	Kilowatts per Person
United States	10.8	Lebanon	0.81
Kuwait	9.8	China	0.54
Canada	9.0	Brazil	0.50
Czechoslovakia	6.4	Egypt	0.28
Sweden	5.9	India	0.18
United Kingdom	5.3	Indonesia	0.12
Germany	5.0	Pakistan	0.08
Norway	5.0	Nigeria	0.06
USSR	4.4	Ethiopia	0.04
France	3.8	Afghanistan	0.03
Japan	3.1	Nepal	0.01
Yugoslavia	1.6	World average	1.86
Mexico	1.2	World average without the United States	1.33
Iran	1.0		
Saudi Arabia	0.93		

[a]These figures do *not* include the approximately 0.1 kilowatts per person food energy used in metabolism.

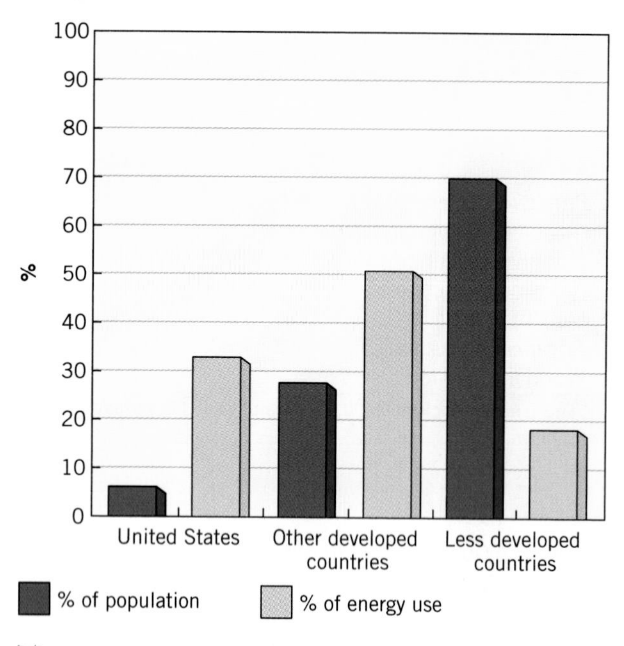

Figure 15.5

Comparison between relative populations and energy usage among the United States, other developed countries, and the less developed nations. (After Richard C. Dorf, 1978, *Energy Resources and Policy*, Addison-Wesley, Reading, MA.)

15.2 OCEAN ENERGY

The ocean stores energy in a variety of forms. The amount of energy available from each is enormous, some being even larger than our society's needs. Harvesting and delivering this energy, however, is an equally enormous task.

THE RENEWABLE SUPPLY

The rate at which energy can be extracted depends on not only the total amount of energy available, but also the **replacement time**. For example, if one resource is large enough to supply our energy needs for 30 years, but it does not get replaced, then after 30 years the resource is gone, and we are again in trouble. This is the predicament we have gotten into through our reliance on fossil fuels.

The long-term welfare of our present society depends on our use of **renewable resources**. These are resources that get replaced as we use them. The maximum rate at which we can continuously extract energy from these resources depends on both the energy available and the rate at which it is replaced.

$$\text{Maximum rate of extraction} = \frac{\text{total energy available}}{\text{replacement time}}$$

For example, if 10 joules of energy were available and it could be replaced every 2 seconds, then the maximum rate at which we could use it would be 5 joules per second (i.e., 5 W). Of course, in the oceans, both the energy available and replacement time are much larger than the numbers in this example.

Recall that energy is the ability to push something over some distance. With greater energy, objects can be pushed more forcefully and over greater distances. This ability can be harnessed in many ways, but among the objects that can be pushed are the blades of a **turbine**. These are like the blades of a fan, and they spin as a fluid passes through them.[1] If they are connected to an electrical generator, the energy of the spinning blades is then converted into electrical energy.

One way of harnessing the ocean's energy is by producing electricity. The electricity could be used directly, or it could be used to produce hydrogen by the electrolysis of water (i.e., pulling the hydrogens off the oxygens in the water molecules). The hydrogen could then be used as a portable fuel, much as petroleum and natural gas are used today. These things will probably be done, but we should look into other possibilities as well.

ENERGY FROM WATER MOTIONS

Water motions in the oceans take the forms of currents and waves. Most currents and waves are driven by the winds, which are caused by the solar heating of the Earth. Thus, they are forms of stored solar energy. Tidal currents are both "lunar" and "solar," since both bodies cause tides.

One way of harnessing the energy of the moving water would be to guide it through turbines that generate electricity. Many designs have been proposed to harness wave energy, one of which is shown in Figure 15.6a. A proposed mechanism to harvest the energy of ocean currents is illustrated in Figure 15.6b.

From numerous studies of waves, currents, and tides, we are able to estimate the total energy available in each of these forms.[2] The replacement times are estimated from simple observations. Because

[1] In conventional power plants, for example, we use the heat from fossil fuels or nuclear reactors to boil water and then shoot the high-pressure steam through turbines.

[2] Kinetic energy can be calculated from the formula $1/2\ mv^2$, where m is the mass of the object, and v its speed. Therefore, to estimate the energy carried by the moving water in waves or currents, we must estimate both the mass of the water involved and its average speed.

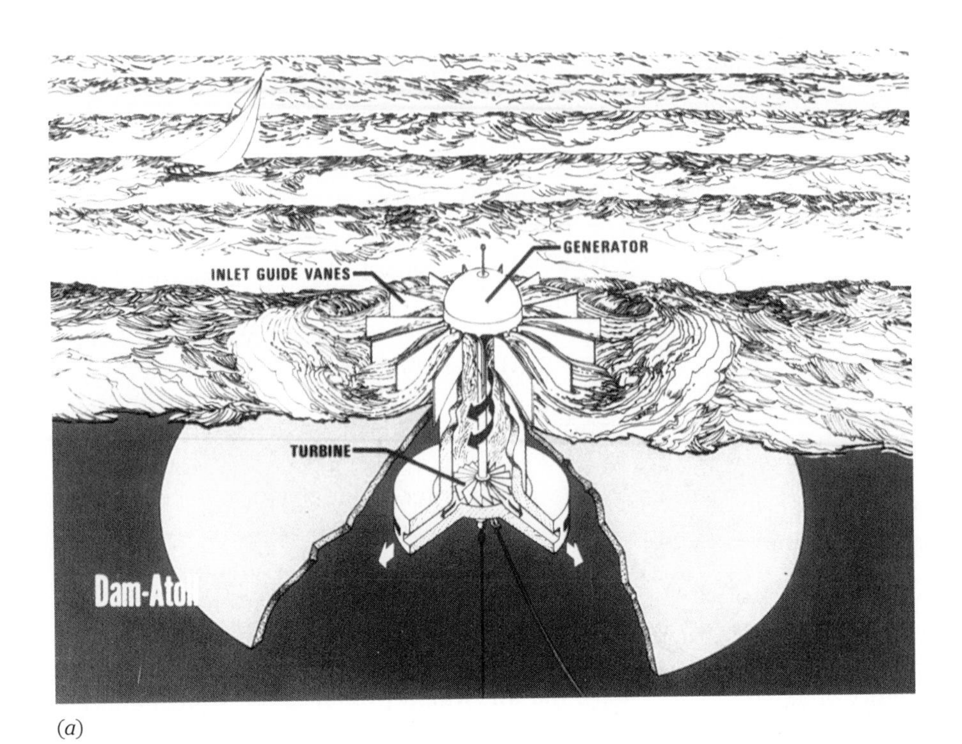

(a)

(b)

Figure 15.6

(a) One possible device for removing wave energy is the *dam-atoll*. In this device, waves refract toward the artificial shoals created by the large underwater shell. The water is then guided through the generator's turbine blades before exiting beneath the shell. (b) Artist's rendition of a large turbine being towed to its mooring. The turbine would be anchored to the ocean bottom and used to generate electricity from ocean currents.

calm seas can turn rough in a matter of days and waves typically last only a week or so before ending their lives on some distant coast, we know that the replacement time for wave energy is a matter of days. That is, if we removed all the energy from waves today, leaving the ocean surface perfectly flat, it would only be a matter of days before the waves were back to normal.

Ocean surface currents undergo seasonal fluctuations, which suggests that if we stopped all the currents today, they would be regenerated in a matter of months. Deeper currents are much slower and have much longer replacement times. Tides rise and fall twice a day, so their replacement time is roughly 12 hours.

With these estimates of available energy and replacement times, we can calculate the maximum rate at which energy could be continuously extracted from each of these sources These numbers are displayed in Table 15.8. You can see that the energy available from both waves and tides is comparable to our society's needs. That available in surface currents is somewhat smaller, and the energy available in deep currents much smaller. Although deep currents

Table 15.8 Various Forms of Energy Stored in the Oceans

For each is given an estimate of the total energy stored in that form and an estimate of its replacement time. By dividing the two, we have an estimate of the maximum rate at which energy could be provided by this source at 100% conversion efficiency.

Energy Source	Total Energy (J)	Replacement Time (sec)	Power (billions of kW)
Thermal energy	3×10^{24}	3×10^{7}	100,000
Wave energy	10^{19}	10^{6}	10
Tidal energy	2×10^{17}	4×10^{4}	5
Surface currents	3×10^{18}	3×10^{7}	0.1
Deep currents	10^{14}	3×10^{7}	0.000003
Present world consumption			10

are massive, they are extremely slow, and this lack of speed is primarily responsible for the reduced energy in them.

Of these ocean energy sources, only tidal energy is being used so far and only to a very small degree. Although available tidal energy is comparable to world needs, to harness this energy on a large scale would be difficult. Huge capital investments would be needed to build the breakwaters and dams, and they would cause serious modifications of coasts and coastal processes. Most promising would be narrow estuaries with large tidal ranges, such as the one in Figure 15.7.

CAPTURED SOLAR ENERGY

Besides the large-scale motions of water masses, solar energy is also stored in more subtle and microscopic ways. In fact, more energy is stored in these ways, because they capture solar energy directly. Currents and waves, by contrast, are secondary effects. The solar energy first heats the Earth, which heats the air, which produces the winds, and the winds in turn produce the water motions. Since large losses are associated with each conversion, less energy is available from these secondary processes than direct absorption of sunlight.

Figure 15.7
The world's largest tidal power station at the mouth of the Rance River Estuary, France.

Ocean Thermal Energy Some of the absorbed solar energy increases the water's temperature. This form of stored solar energy is especially plentiful in warm tropical surface waters. The extraction of this thermal energy may be accomplished through the use of a volatile fluid, such as ammonia or freon, that boils at the temperature of surface waters, but condenses to a liquid at the cooler temperatures of deeper waters. This fluid goes through a closed cycle as follows (Figure 15.8):

1. It is heated to boiling by warm surface waters.
2. The boiled vapors shoot through a turbine, which spins and generates electricity.
3. The vapors are cooled and condensed by cool ocean waters brought up from the depths.
4. The process is repeated.

This cycle is quite similar to that used in conventional power plants, but with a different working fluid. Whereas conventional plants use intense heat sources that boil water, this cycle uses a less intense heat source that boil a more volatile fluid.

This process is referred to as *ocean thermal energy conversion*, or more commonly by the acronym **OTEC**. Small OTEC pilot projects have had moderate success. A great deal of energy is available from the ocean through the OTEC process—many thousands

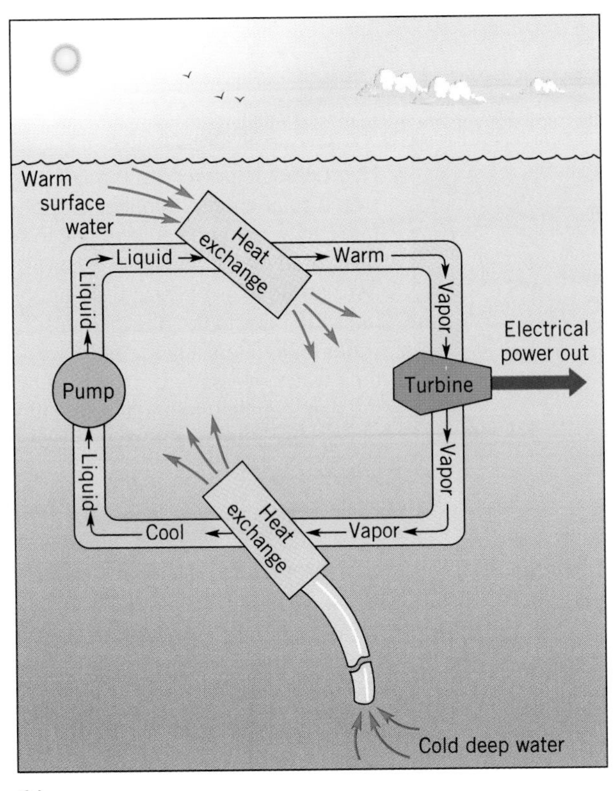

(a)

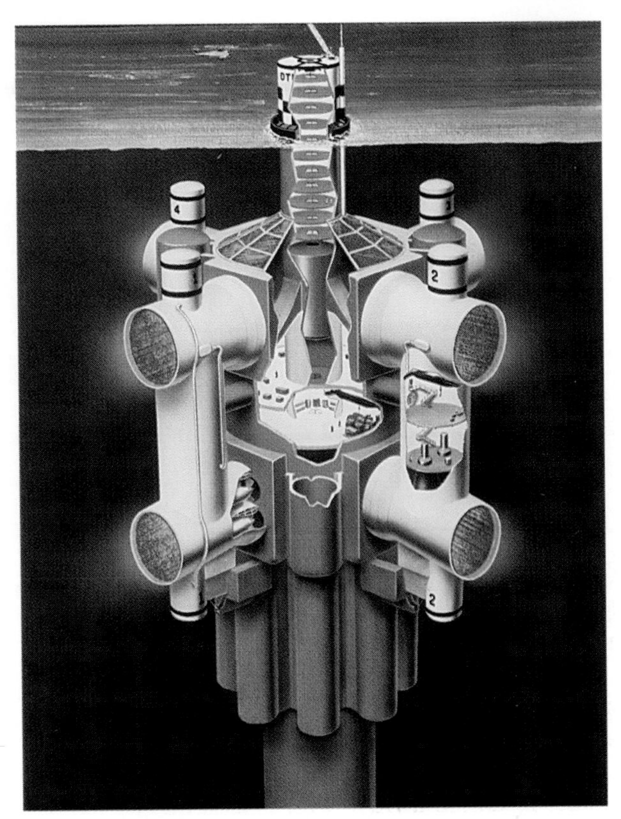

(b)

Figure 15.8

(a) OTEC cycle. The working fluid (perhaps ammonia or freon) is evaporated by the heat of the warm surface waters. These vapors shoot through a turbine that generates electrical power. Then the vapors are cooled and condensed back into liquid form by cooler waters brought up from depth. The cycle then repeats itself. (b) Artist's rendition of an OTEC plant. The tube below brings up cold water from a depth of about 2.5 kilometers.

of times more than is needed by our present society. However, significant technical problems must yet be overcome before OTEC could become commercial. These problems include very low conversion efficiencies (about 2%) and the problem of the fouling of the heat exchangers by marine organisms. Just as the radiator of a car allows heat exchange between the engine coolant and air, OTEC heat exchangers allow heat exchange between the ocean waters and internal volatile working fluid. Marine organisms attach and grow on these heat exchangers, and clog them up.

Hydroelectric Power In addition to heating water, a considerable fraction of the incoming solar energy goes into evaporating water from the ocean surface. Some of this moisture condenses high over the continents and joins rivers and streams on its trip back to the ocean. A dam placed on one of these rivers blocks the flow, and the water collects in a reservoir behind the dam. As the water level rises, the pressure at the bottom increases. Water released from the bottom of the dam exits with a great deal of force, passing through turbines that generate electricity (Figure 15.9). This **hydroelectric power** is already being used for a small but significant amount of electrical energy production in the world.

One interesting suggestion for a source of hydroelectric power is to close off the Red Sea basin and allow the heavy evaporation in this part of the world to remove water until the surface is well below sea level. Then we could allow sea water to flow back into the partially emptied basin at a rate equal to that of evaporation. The inflowing water would turn turbines and generate electrical power. The smaller Dead Sea basin could be used similarly.

Salt Power In addition to thermal motion and evaporation, solar energy is also stored microscopically in a third important way. Recall that the water molecule sticks very tightly to other charged molecules, such as other water molecules, and the salt ions that are dissolved in sea water. When a water molecule evaporates from the ocean, it is torn away

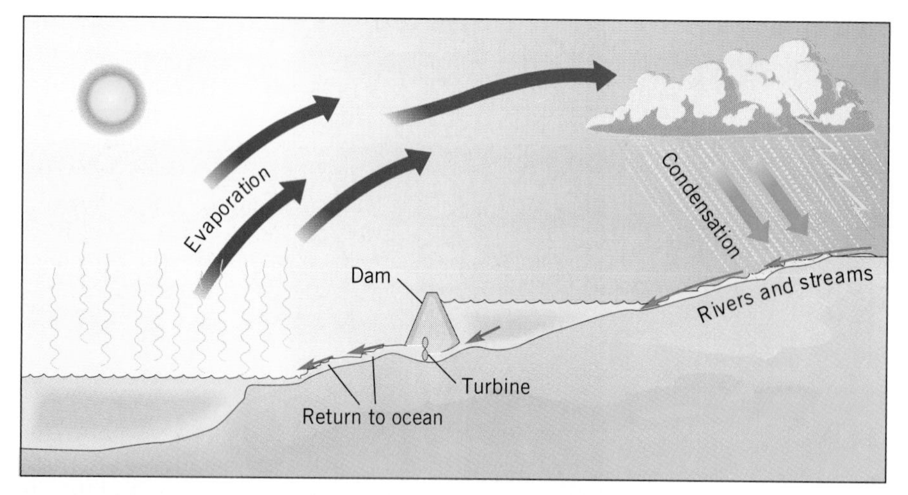

Figure 15.9
Hydroelectric power is a form of stored solar energy. Sunlight evaporates the water and causes some to be carried over land before it condenses and joins rivers and streams in its trip back to the ocean. If this water encounters a dam on the way, then water released from the base of the dam can be directed through turbines, which turn electrical generators.

from these other neighboring molecules. The energy needed to do this is called *latent heat.*

This stored latent heat is released when molecules of water vapor recombine to form droplets. However, this condensed water is fresh, and that which falls on the land remains fresh as it joins rivers and streams leading to the oceans. Although it has recombined with other water molecules, it has not yet recombined with the salt ions of sea water, to which it is also strongly attracted.

Salt power makes use of this strong electrostatic attraction between the electrically polarized water molecules and charged ions of dissolved salts. If fresh water and salt water are separated by a membrane that allows passage of water molecules only, then the salt ions will draw the water molecules through this membrane to the salt water side. The water level on the salt side will rise until the increased pressure stops the flow of water molecules through the membrane. The pressure required to do this is actually very large, amounting to a difference of about 250 meters in water level.

Where fresh water of rivers and streams flows into the ocean, the fresh water could be separated from the salt water by these special membranes and walls, as illustrated in Figure 15.10. The large difference in water level from one side to the other could be used to run hydroelectric power stations. The rate at which energy could be extracted from this source is quite large, in theory—comparable to that from tides or waves. However, we do not yet know how to construct these special membranes that are sufficiently large, rugged, and efficient to handle the large volumes of water involved.

Biomass Conversion Solar energy is also captured and stored in the tissues of plants through photosynthesis. This energy can be released through burning, such as in our fireplaces or the furnaces of electrical power plants. Through known chemical processes, the plant materials can also be converted into portable fuels usable in cars and trucks. The growth and harvest of plants for these energy purposes is called **biomass conversion**. It is not restricted to marine plants alone. Forests on land can be grown and harvested for fuels in the same way.

Although some work on biomass conversion has already been done, it is not expected to have a major impact on our energy problem, for several reasons. First, marine plants are not as productive as are those on land. Second, the vast majority of marine plant life is microscopic. The large harvestable seaweeds grow in only very limited coastal regions of the oceans, and these regions are very small compared to the vast regions of land covered by forests, for example. Third, converting marine biomass to fuel energy has low efficiency. Finally, the cultivation and harvesting of marine biomass are more urgently needed for food than fuel.

IMPLEMENTATION

Before you advocate the adoption of one of these methods as the source for our world's future energy supply, you should first consider some of the problems that must be surmounted to harness them. The enormity of the oceans, problems of energy transportation and storage, inefficiencies of energy conversion, and ecological impact of large-scale modification of the ocean's temperatures or motions pose huge obstacles, both technical and societal. The ocean environment also presents some very difficult problems in engineering and materials, such as corrosion, fouling by organisms, and structural durability.

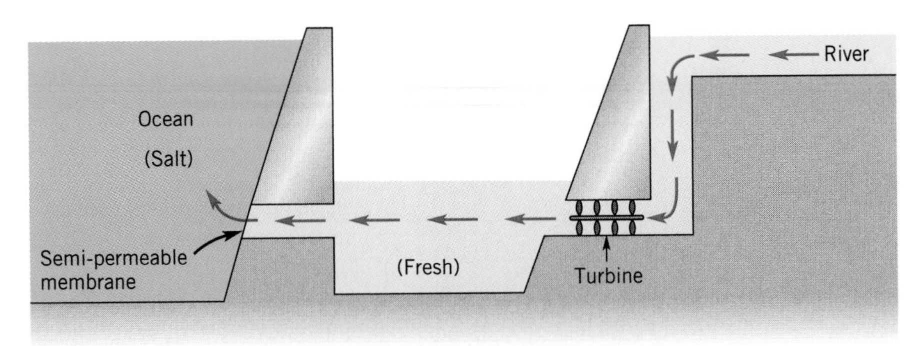

Figure 15.10

One way of using salt power to generate electricity in an estuary. The tendency of fresh water to diffuse through the semipermeable membrane toward the salt water side can be used to maintain a difference in water levels. This difference in water levels can be used to generate hydroelectricity.

It is not feasible to make a gigantic power plant that somehow straddles the ocean and removes a large fraction of the available energy. Any power plant that we could make would be rather small, localized, inefficient, and tap only a small fraction of the available energy. If it proved profitable, then additional and improved plants could be built. In this way, we would gradually build up our use of that particular resource, and the environmental consequences would gradually become apparent.

15.3 MINERAL RESOURCES

In addition to energy, the Earth's most accessible reserves of many other resources are being rapidly depleted by our society's large material appetite. Mineral deposits found on land are usually cheaper and easier to extract than those on the ocean floor, so we exploit them first. But as the consumption of our mineral resources continues, we exhaust the richest deposits and are forced to go after those of poorer quality. Both the increasing demand and decreasing quality of the remaining ores drive up the prices, and these higher prices make it attractive to search for new and alternate sources.

We are becoming increasingly knowledgeable about what may be available from the ocean. However, the area is large and the exploration slow. Searching for sea floor resources from ships is like searching for land-based resources from a blimp that floats 4 or 5 kilometers above the ground and at night. Clearly, progress will be slow. Nonetheless, we are slowly finding and mapping likely sources for various minerals. Although our knowledge is still very sketchy, we have learned enough to begin exploiting some of them.

We are also making progress in our technology for extracting various minerals from the ocean. The initial capital investments are considerably larger than they were for land-based resources. So we will not see exploration by individual prospectors with little more than a shovel and wheelbarrow, as we did on land. Typically, several hundred million dollars are required for ships, platforms, and other hardware that are needed to get started.

Searching for sea floor resources from ships is like searching for land-based resources from a blimp that floats 4 or 5 kilometers above the ground and at night. Clearly, progress will be slow.

The ocean's mineral resources fall into three categories:

1. Those that are removed from the underlaying bedrock
2. Those that are removed from the seafloor sediments
3. Those that are removed from the water.

Of those removed from the underlying bedrock, some are removed through boreholes, and some from mines (Figure 15.11). Of those mined, coal, iron, and tin are the most important. These operations are usually extensions of mine shafts originating on land.

In terms of dollar value (Table 15.9), by far the most important ocean resources are oil and natural gas, which are removed through boreholes. Also removed through boreholes, although of comparatively lesser importance, are sulfur and potassium salts. The resources dredged from the ocean bottom include sand and gravel (for fill and concrete), calcite oozes

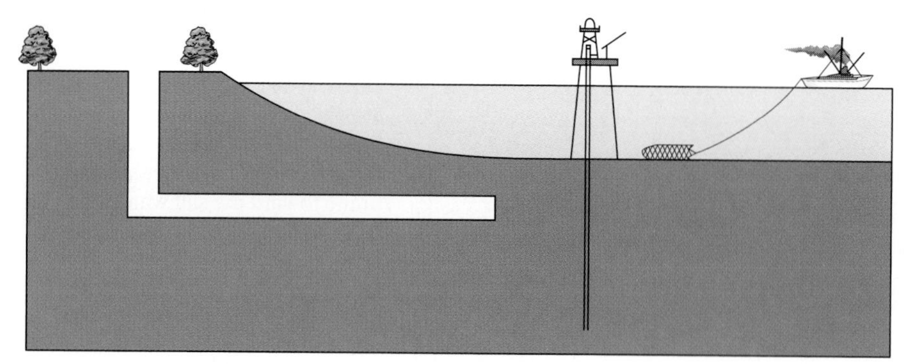

Figure 15.11
Sketch of the three principal means of removing resources from the ocean floor and subfloor: mine shafts, boreholes, and dredge or other means of removing ore from the seafloor surface.

(for cement), and to a lesser extent, heavy-metal enriched sediments and manganese nodules (for metals). Resources removed from the water include various salts and the water itself.

OIL AND NATURAL GAS

The world uses about 20 billion barrels of petroleum annually, or about 4 barrels per person per year. About 20% of this petroleum comes from the ocean, and the fraction is increasing. As with other resources, the easiest and most accessible reserves have been exploited first. Although considerable oil (typically 50% of the original reservoir) remains in spent oil fields on land, it is very expensive to extract the remainder, and so ocean drilling becomes increasingly attractive.

Most marine oil wells have been drilled from fixed platforms in less than 200 meters of water. It is easier and cheaper to drill in shallow water, so the shallow water oil is exploited first (Figure 15.12).

Petroleum comes from the remains of prehistoric organisms, mostly microscopic and marine. The exact chemical processes involved in their transformation are not completely understood, but the general idea is as follows (see Figure 15.13).

Organic detritus falls to the ocean bottom where it begins decomposing, being worked over by bacteria and other small organisms. If this detritus is in a region of heavy sedimentation, or stagnant, oxygen-depleted waters, then it may get buried before being completely oxidized by these organisms. If the organic matter in the sediments is in larger abundance than the oxidizing agents (mostly oxygen and oxygen-containing materials), then it cannot fully decompose and will still contain some stored solar energy in the form of unoxidized organic materials.[3] This stored energy may be released at some later date when it is discovered, brought to the surface, and burned.

These partially decomposed organic materials become transformed into *petroleum*, by the heat and pressure beneath overlying layers of sediment. We know that this transformation takes millions of years and deep burial, because very little petroleum is found in sediments younger than 2 million years.

Over time, the heat cooks the petroleum and breaks the long organic molecules into shorter ones. This process is referred to as **cracking**. As you know, higher temperatures mean greater molecular motion. At higher temperatures, then, the organic molecules shake more violently and tend to break into smaller pieces. If the organic matter is cooked too much, the product is small light molecules of methane gas, or *natural gas*. If not cooked enough, the result is the

Table 15.9 Projected Value of Various Ocean Resources in the Year 2000

Resource	Value (billions of 1973 dollars)
Mineral Resources	
Petroleum	10.50
Natural gas	8.30
Magnesium	0.31
Manganese nodules	0.28
Sulfur	0.04
Fresh water	0.04
Construction materials	0.03
Others	0.02
Biological Resources	
Fish	1.42–4.15
Others	
Energy	3.78–6.03

Source: National Ocean Policy Study, 1974, 93rd Congress, 2nd session.

[3]Remember that the organisms that decompose these materials get their energy by oxidizing (i.e., burning) it. If they run out of oxygen—dissolved for aerobic organisms and in chemical complexes for anaerobic bacteria—then this decomposition stops, and some of the stored energy will remain unused.

Figure 15.12
Offshore oil platforms in the Santa Barbara Channel (California coast).

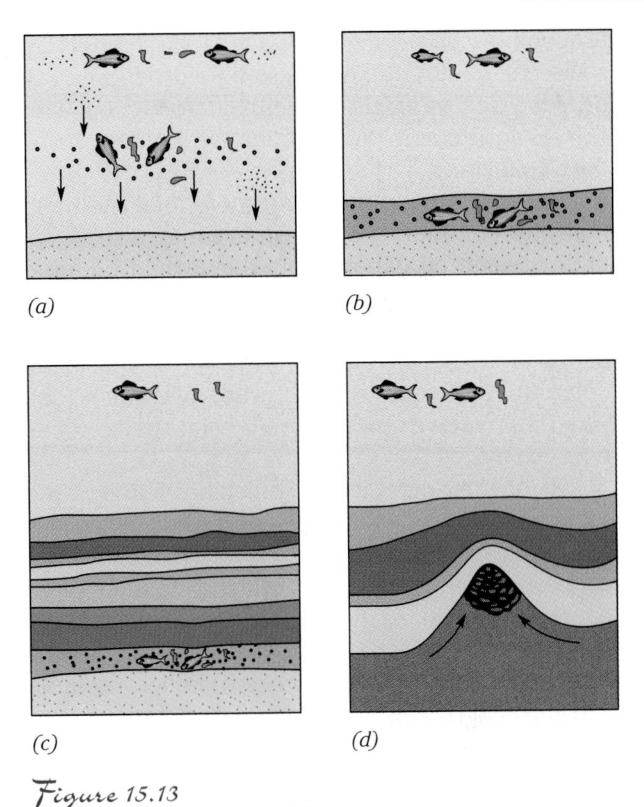

Figure 15.13
Illustration of the steps involved in creating petroleum. (a) Organic detritus sinks to the bottom and (b) must be covered up with sediment before it is completely oxidized. (c) After millions of years, deep within the sediment layers, and being subjected to high temperatures and pressures, it gradually transforms into petroleum. (d) Being lighter than sediments, it rises up into the ridges in the folds in the sediment strata until it encounters an impermeable layer and there it is trapped.

larger molecules of thick tars in oil shale, which we must heat further in order to extract oil. The left-over residues of the natural cracking processes are coal and graphite.

The temperature beneath the Earth's surface increases by about 3°C for every 100-meter depth, on the average. Therefore, the older and deeper deposits are cooked longer and hotter, yielding more natural gas and less oil. At temperatures above 160°C, the end product is mostly natural gas. This temperature is reached at a depth somewhere between 2.4 and 7.6 kilometers, depending on the particular location. Both oil and gas can rise above this depth and might be found all the way up to the surface. But below this depth, only natural gas is found.

Petroleum and natural gas are less dense than the sediments they form in, so they tend to rise. When geological processes wrinkle the originally flat sediment strata, the petroleum or gas rises until it fills the ridges beneath relatively impervious material. Consequently, the ridges and domes in deep sediment strata, immediately beneath impervious layers such as salt or clay deposits, are likely regions to explore for oil or natural gas.

This knowledge of the processes that produce petroleum gives us insight for petroleum exploration. Discoveries would be most likely in places where sediments accumulated so fast that they covered up the detritus before it was completely decomposed. Also, areas of especially high productivity in ancient times should have produced especially high amounts of organic detritus on the ocean bottom for burial and conversion. Ancient continental margins and river deltas are promising places to look, in view of these considerations.

Ancient narrow and enclosed seas are also places where oil deposits are likely to have formed. The Red Sea is a modern example of such a sea. Geologically, it is young, and it is opening up as the seafloor spreads from the ridge that runs down its center. Layers of rock salt up to 5 kilometers thick can be found beneath the Red Sea, with organic muds found beneath them. This indicates there have been long periods of high plant productivities, producing the organic muds, and long periods of heavy evaporation, producing the layers of salt deposits. At some distant future date, these buried deposits will produce rich oil fields. The oil will collect beneath domes in the salt layers, because it cannot rise through them.

The continental rises hold more total sediment than the continental shelves, and may also cover undiscovered oil deposits, but they would be more difficult to exploit due to the increased water depth. There are undoubtedly numerous untapped oil reserves in the ocean, and from the above considerations, we have some ideas of where we might look when our reserves on the continental shelves have been exhausted.

Rich sulfur deposits are sometimes found in the same area as petroleum, because the sulfur deposits form from salt domes (i.e., layers of salt deposits that get pushed upward), beneath which petroleum often becomes trapped. As the salt domes push upward beneath the ocean floor, some is dissolved as it encounters sea water, leaving a caprock of rather insoluble calcium sulfate. This then reacts with petroleum and bacteria to produce pure sulfur. These deposits are particularly common along the U.S. Gulf Coast and in Saudi Arabia.

MINERALS IN THE SEDIMENTS

The granular sediments themselves are an important mineral resource. More than 100 million tons of sand and gravel are dredged from the ocean bottom each year. Removing such a large volume of sediments involves high transportation and environmental costs. Of course, the same is true for sand and gravel from land sources.

Other loose sediments that are exploited commercially include barite, which is a source for barium and is used as a drilling mud, sands enriched in heavy metals, phosphorite, and calcium carbonate. Sea floor deposits of phosphorite (a phosphorus-rich sediment that is used in fertilizers) tend to form where cool nutrient-rich deeper water comes to the surface and is warmed. Relatively little of this has been mined so far, because there is still plenty left in land deposits that are easier and cheaper to mine. The sea floor deposits of calcium carbonate are essentially limitless in comparison to our needs, although cheaper land sources are mined more heavily. Nonetheless, about 20 million tons come from the ocean each year. It is pulverized and heated to drive off CO_2. The result is calcium oxide, or *lime*, that is used in agriculture to treat acid soil and also in cement.

When waves work over the beach sediments, there is a tendency for the sediments enriched in heavier minerals to stay and for the lighter sediments to be removed. When the water level changes, the *relic beaches* that are left behind are enriched in heavy metals. These relic beach deposits are called **placers**, and some of these are mined for their metals.

Manganese nodules are an interesting form of sediment that contain high concentrations of heavy metals (Figure 5.14). The nodules usually range between 1 and 20 centimeters in diameter, but the deposits are also found as coatings on rocks or as hard pavements on the ocean bottom. They are mostly in water deeper than 4 kilometers, although some have been found as shallow as 60 meters in the Great Lakes. The heavy metal content of manganese nodules is typically as follows:

Manganese	15%
Iron	15%
Nickel	0.4%
Cobalt	0.4%
Copper	0.3%

These are average values, and the proportions in any one deposit may vary in either direction by as much as a factor of 2.

Although continental sources of manganese and iron are richer, the ocean's manganese nodules are attracting attention for their content of other metals, such as copper and nickel (Figure 15.14). Mining these nodules requires a heavy initial investment, and the financial risks are high. Several companies have formed international consortia to share the expenses, and several have applied for exploratory licenses in the eastern Pacific region.

Figure 15.14
Some manganese nodules that have been dredged from the ocean floor.

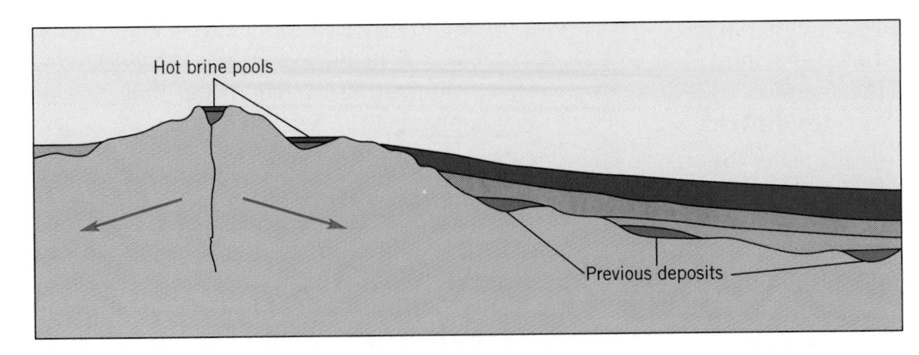

Figure 15.15
Hot brine pools are found filling some depressions in presently active oceanic ridges. Beneath these pools are deposits rich in metal sulfides. There are probably many other similar deposits along the ocean floor, formed earlier when these parts were on the ridge and now covered by more recent sediments.

There has been some excitement in the discovery of hot **brine pools** at a depth of 2 kilometers along the oceanic ridge in the Red Sea. These pools have been found to have salinities up to 250‰, and temperatures up to 56°C. Their formation is related to the heat sources and hydrothermal activity along this active ridge. More than a curiosity, the bottoms of these pools contain sediments that are rich in heavy metals of commercial value. The sizes of these pools have not yet been completely mapped, but it appears that they may be typically a kilometer or so across, and the deposits are 20 to 100 meters thick. Saudi Arabia and Sudan jointly commissioned a German mining company to mine some of these muds. The ores contained a zinc concentration of about 40%, with lesser (but attractive) amounts of silver, copper, and gold. Before such an operation could be done on a commercial scale, however, pumps and plumbing would have to be developed that would withstand the highly corrosive warm brines.

Sediments enriched in metal sulfides are found along all active ridges. The concentration of metals in most of these deposits is too low for commercial interest, but there are some local regions of especially high concentration, perhaps associated with former brine pools like those now found in the Red Sea. We speculate that if we can see these deposits now forming along the active ridges, then it is likely that they have been forming throughout history along all previously active ridges. Since all oceanic crust has originated in these ridges, there may well be such deposits throughout the deep ocean bottom that have been covered up by subsequent sediments (Figure 15.15). Exposed regions of the oceanic ridge, such as Cyprus and Iceland, are attracting increased interest for answers to these questions.

Many of the abyssal clay deposits are enriched in aluminum (up to 9%), iron (up to 6%), and also copper, nickel, cobalt, and titanium. They are not rich enough to yet be competitive with other sources, but their volume ensures that they are virtually an unbounded source for these metals. In fact, most of these metals are accumulating in the ocean bottom sediments faster than our society can use them.

Figure 15.16
A salt recovering operation that uses broad shallow evaporation pools.

ELEMENTS IN SOLUTION

The ocean has been used as a source of salt for thousands of years, and still today salt ranks high in value among the materials extracted from the ocean (Figure 15.16). In addition, magnesium, magnesium salts, and bromine are also extracted commercially. Magnesium is a lightweight metal, like aluminum, but much stronger. It is used in construction where light weight and strength are important. Magnesium is chemically similar to calcium, but it is more soluble. Consequently, magnesium salts are frequently used when solubility is important, such as in medicines. One of the principal uses of bromine has been in antiknock compounds in gasoline.

FRESH WATER

Fresh water is also commercially extracted from the ocean. However, the price paid for fresh water varies so greatly from one part of the world to another that it is difficult to put a value on it. Clearly, water is one of the necessities in life, and people would be willing to pay an extremely high price for it if they had to.

Several methods have been tried to **desalinize** sea water (i.e., to remove the salt from it). One process uses the fact that when water freezes, salts tend to be excluded from the newly forming ice. Consequently, desalinized water can be produced by freezing sea water and then removing and melting the ice. Another process forces sea water through semipermeable membranes that allow the water through, but not salts. Unfortunately, we have not yet been able to manufacture satisfactory membranes for use in large-scale production.

The most common commercial process involves distillation. Water is boiled, and then the vapors are captured and condensed back into liquid water (Figure 15.17). When water evaporates, it leaves salts behind. You can see this process when cooking soup on your stove. The water that condenses on the inside of the pot's lid is pure, having left behind all the salts and other materials in the soup.

Figure 15.17
A fossil fuel desalinization plant.

You can see this process when cooking soup on your stove. The water that condenses on the inside of the pot's lid is pure, having left behind all the salts and other materials in the soup.

An interesting but speculative and untried source of fresh water involves hauling huge icebergs from high latitudes to coastal cities. The huge icebergs would move only extremely slowly under tow, and the trip would probably take more than a year. Melting losses might be reduced by underwrapping the icebergs with some insulating material.

Unfortunately, all desalinization is expensive. Heating, cooling, and towing all use large amounts of energy, which is costly. In the United States, for example, desalinized water costs at least three times as much as water from most city water systems. Solar power is cheap, but not very intense, so solar-powered desalinization plants must be very large if they are to receive as much total heat energy as would be produced in a much smaller fossil fuel plant. For solar desalinization, therefore, what is saved on energy costs is lost to capital expenditures.

The high costs of desalinized water make it too expensive for agricultural use. In order to increase the agricultural output from arid regions of the world, it now seems more promising to try to increase the salt tolerance of plants. If this is successful, brackish or salt water could be used in place of the very expensive desalinized water for agricultural irrigation.

Desalinized water is not too expensive for personal use however. A growing fraction of our population is being attracted to coastal areas, and many of these places do not have sufficient fresh water to support the population. In these areas, desalinization of sea water is increasingly popular, and the populations seem willing to pay the higher price for water in exchange for the coastal climate and lifestyle.

15.4 OCEAN LAW

The increasing interest in ocean resources is a source of international concern. Through the ages, most nations have had rather well-defined territorial boundaries, within which their sovereignty was recognized by others. Sovereignty in the oceans, however, is a different matter. The ocean has no obvious and well-defined boundaries, and there has been little agreement regarding what portions of the ocean should come under any particular nation's jurisdiction and what activities should be permitted. Throughout most of recent history, the deep ocean areas have been assumed to be common territory, and occasional efforts by nations to claim ownership and control over portions of it have been temporary and strongly rebuked.

The result has been a sort of international anarchy regarding the use of the oceans, with accompanying conflicts and misunderstandings. Even relatively innocent use for fishing and transportation has caused much strife. As the world's increasing population places increasing demand on its resources, conflicts of interest are sure to grow more severe.

HISTORICAL PERSPECTIVE

Throughout history, there has never existed a codified, internationally agreed on *law of the sea*. Instead, sailors and nations have relied on *common understanding* rather than "written record." Unfortunately, there has been considerable ambiguity in this common understanding, which has often been interpreted to mean whatever is in a particular nation's interest.

Most western law has its roots in Roman law. The Romans regarded all major waterways as belonging to the international community, therefore not being under the control of any one. The Romans interpreted this principle to cover not only the high seas, but coastal waters as well.

The first major challenge to this point of view came in the thirteenth century from Venice, which at that time had both the power and commercial interest to regulate shipping in the Adriatic Sea. Rather than belonging to "all," it claimed that the seas belonged to "none" and, therefore, could be claimed by any nation with appropriate interest and power. Accordingly, the Venetians controlled passage through this part of the Mediterranean, prohibiting some and extracting tariffs from others.

This proprietary point of view was carried to the extreme in the late fifteenth century, when with the help of the Pope, Portugal and Spain divided all the world's oceans between them. Portugal had explored trade routes around Africa and through the Indian Ocean to the east. Spanish explorers had sailed westward as far as the Americas. Consequently, Portugal was awarded control of the oceans to the south and east, and Spain was awarded control to the west. The dividing line was chosen to be 40°W longitude. Since the coastline of present Brazil extends eastward beyond this line, Portugal colonized Brazil, and Spain explored the rest of the *New World*.

Of course, this division of control was opposed by other interested nations, and it did not last. Countries such as Great Britain, Holland, and the Scandinavian nations based their opposition on the traditional freedom of the high seas, but many still wished to keep certain local waters under their own national jurisdiction. By the early 1700s, the British and Dutch had persuaded many that these local waters should extend a distance of about one cannon shot, or 1 league (3 nautical miles) from shore. Thus, the Venetian influence had become entrenched, and certain local waters were commonly assumed to be exempt from the traditional freedom of the high seas.

The basic beliefs of this compromise position have remained strong for the past three centuries and enjoy nearly universal endorsement. There has been some disagreement on important details, such as the exact width of the local waters over which the coastal nation can exercise jurisdiction, and exactly what activities can and cannot be regulated. The two components of this compromise can be summarized as follows:

1. **High seas.** As was practiced under Roman law and stated in an accepted treatise entitled *Mare Liberum* by a Dutch jurist in 1609, "The high seas should be free for the innocent use and mutual benefit of all."
2. **Territorial waters.** Contiguous to coastal states are territorial waters, over which the coastal state has complete sovereignty, subject only to the rights of others to "innocent passage."

RECENT DEVELOPMENTS

These general ideas were unanimously endorsed by the United Nations General Assembly (108 to 0, with 14 abstentions) in a *Declaration of Principles*, in 1970. Agreement on the details has been much more elusive, however. In particular, there is still disagreement on what exact freedoms are accorded by the first principle, and what particular activities and regions are governed by the second.

Recent interest in the economic exploitation of ocean resources and the control of pollution in coastal waters has given rise to a modification of the *territorial waters* concept. Now there is interest in establishing a third division, called the **exclusive economic zone**, which extends some distance beyond the territorial waters. Within this third zone, the coastal nation has jurisdiction over certain resources.

Unlike the previous two divisions, the exclusive economic zone has no historical precedent.

Increased interest in ocean resources has resulted in additional support for oceanographic research in recent decades. However, along with the increased interest comes increased controversy. Many poorer coastal states see that oceanographic research enables the more developed nations to improve their exploitation of ocean resources. For them, the innocent passage of research vessels no longer seems so "innocent," and they are beginning to restrict the activities of such expeditions in their territorial waters.

Military applications of oceanographic research are also clouding the picture. This difficulty has been exacerbated by the realization that nations sometimes use the disguise of research to carry out military operations. The U.S.S. Pueblo, captured by North Korea in 1968, was supposedly doing oceanographic research, when in fact it was carrying out espionage activities. The United States once outfitted a ship ostensibly for deep sea drilling, which in fact was used to retrieve a sunken Soviet nuclear submarine. For many years, Soviet "trawlers" seemed to appear in U.S. coastal waters whenever events of military interest were happening nearby. No wonder there is growing suspicion of oceanographic research ventures, both among developed and developing nations.

THE NEED FOR CODIFICATION

There has long been a recognized need for a codified law of the seas (Figure 15.18). For one thing, piracy has long been a serious problem for maritime nations. From time to time, such nations have found it necessary to send out naval expeditions to crush piracy in certain troublesome regions. Unfortunately, there are ideological, philosophical, and practical difficulties in prosecuting piracy in the absence of codified international laws or specified jurisdiction. The perpetrators can be perceived either as "villains"

(a)

(b)

Figure 15.18

A codified Law of the Sea would make it easier to identify and prosecute criminal activity without risking international incident. (a) The U. S. Coast Guard arrests a suspected drug smuggler. (b) Piracy is often difficult to define. The scourge of one nation's maritime interests might be a boon to another's. This engraving shows Drake's ship, the Golden Hind (right) capturing the Spanish treasure ship, Nuestra Señora, in the Pacific Ocean near Peru. Although considered a murdering, plundering pirate by the Portuguese and Spanish, Drake was a naval hero to the British, and was even knighted by Queen Elizabeth for his "accomplishments."

or "heroes," engaging in "piracy" or "protecting national interests," depending on one's political or economic interest. One nation's "heroic naval fleet" may be (and often has been) viewed as a "band of outlaw pirates" by another (Figure 15.18b).

Recent economic interest in ocean resources has greatly increased both the desire to achieve international agreement in a codified law of the sea and the difficulty in attaining such an agreement. The exact extent of territorial waters has always been one sore point. For example, one challenge to the traditional views of the extent of territorial waters came in 1945, when President Truman unilaterally proclaimed for the United States the resources of the U.S. continental shelf. Encouraged by this precedent, many other nations soon proclaimed sovereignty over both the shelves and overlying waters to some specified limit. Conflicts quickly arose, which heightened the need for some sort of international agreement.

Another impetus for international accord came from rapid developments in ocean technologies. The proliferation of large fishing fleets, capable of fishing distant waters, and the development of the ability to extract petroleum and other mineral resources from the sea floor caused concern among less developed nations that these resources would soon be exploited by the nations with the technology to do it. The more developed nations, on the other hand, were reluctant to make extensive investments in these technologies, without some guarantees that these investments would be protected and not outlawed.

CONVENTION ON THE LAW OF THE SEA

Attempts to reach international agreements have been frustrating. For example, regarding the use of the high seas, the phrase "for the mutual benefit of all" was interpreted quite differently by coastal and inland nations. Does "all" include inland and nonseafaring nations? Other differences in interpretation arose between developed and less developed nations. Both the different national interests and nature's uneven distribution of oceanic resources nourished disagreements as to the proper definitions territorial waters and exclusive economic zones. Even within one nation, different economic interests favored different definitions.

In spite of all these obstacles and differences, the advantages of some sort of international accord were strong enough that a great deal of effort was directed toward this end. After 11 years of very hard work, a "Draft Convention on the Law of the Sea" was produced in 1982 by the United Nations for the consideration and endorsement of its members.

The United States, the former Soviet Union, and some other developed nations had wanted to handle different aspects of the problem separately. However, they were outvoted by other member nations, who for the most part wanted the law of the sea to be one comprehensive document. The United States made some last-minute demands for changes, only some of which were conceded. Even without U.S. support, the Draft Convention was passed in 1982 by a vote of 130 yes, 4 no, and 17 abstentions. (Joining the United States in opposition were Turkey, Israel, and Venezuela.)

The salient features of this proposed international "law of the sea" are as follows:

1. It defines territorial waters as extending 22 kilometers (12 nautical miles) from shore. Foreign vessels are guaranteed innocent passage and transit passage through straits used for international navigation.
2. It establishes a 370-kilometer (200-nautical mile) exclusive economic zone.[4] It also defines the rights of the coastal nation regarding its sovereignty over natural resources, economic ventures, and environmental protection within this zone. It additionally guarantees certain rights within this zone for the endeavors of landlocked nations.
3. It affirms a coastal nation's rights to the resources of the contiguous continental shelf, but with provisions for sharing the revenues from resources beyond the 370-kilometer limit with the rest of the international community.
4. It asserts the traditional freedoms of the high seas and establishes an "International Seabed Authority" to oversee and regulate the extraction of mineral resources from these regions.
5. It advocates fairness, cooperation, and consideration for others in the various uses of the oceans and outlines the means to settle differences.

The subsequent history of this attempt at a universal law of the sea has been disappointing. In spite of strong approval on the floor of the United Nations, the governments of only 30 of these nations have subsequently ratified it, and it doesn't appear the push will go much farther. As interest in the ocean's re-

[4]This could be extended to 350 nautical miles, if the continental shelf extends that far.

sources becomes more intense, self-interest becomes a stronger motivation than the common good.

Nonetheless, this effort has unprecedented historical significance. It is the first time in history that the majority of the world's nations have even come close to agreement on an explicit codified law of the sea.

Summary

Energy

We are rapidly exhausting our world's cheapest and most accessible energy resources, particularly the portable fuels. We need to find some way of using clean and renewable sources, such as solar energy, some of which is stored in the oceans. Not only are we concerned with conversion to the form needed, but also with the problems of transportation to the place needed and storage until the time needed. Whenever energy is converted from one form to another, some is lost.

The rate at which energy is used is called *power*. It is measured in joules per second, or equivalently, *watts*. The total amount used is the product of rate times time. A kilowatt-hour is the amount of energy consumed if used at a rate of 1000 watts for 1 hour.

The average rate of energy consumption in the United States is about 11 kilowatts per person. About half is used in the industrial and commercial sector, a quarter in our homes, and a quarter in transportation. The world average is less than 2 kilowatts per person.

Ocean Energy

The ocean is a large collector of energy. Some of this energy is stored in the motions of the water, such as waves and currents. A large amount is also available from tidal currents. Solar energy is absorbed directly and stored in several ways. The temperature difference between warm surface and cool deep waters could be used to power machines, such as electrical generators.

Hydroelectric power uses the energy of water flowing down rivers and streams back toward the ocean, completing the cycle that began when solar energy evaporated it from the ocean surface. The electrostatic attraction between water molecules and salt ions can be used to provide energy at places where fresh water runs into the ocean. However, the manufacture of appropriate membranes has been troublesome. Fuels can be produced from biomass, such as seaweeds in the ocean or forests on land. However, the ocean biomass is in even greater demand for food.

Mineral Resources

Interest in ocean mineral resources is increasing for three reasons: economic incentive, increased knowledge of what is available, and improved ocean technology.

Of the ocean mineral resources, by far the most valuable are the oil and natural gas reserves beneath the sea floor. However, these reserves cannot possibly keep up with our demand. Regions where ancient biological productivity was high and decomposition incomplete are the most likely places for petroleum deposits.

Sand, gravel, and calcium carbonate deposits are among the most important of the resources derived from the sediments. Manganese nodules, placers, and special sediments found beneath brine pools and along the ridges are attracting interest for their metal content. Phosphorite could be mined for use in fertilizers, and abyssal clays for metals.

Of the elements in solution, salt has been removed for thousands of years. Also magnesium, magnesium salts, and bromine are extracted. The water itself is removed in desalinization plants. It is an important source of fresh water in parts of the world, but it is too expensive for agricultural use.

Ocean Law

Historically, the law of the sea has relied on common understanding. The Romans held that the seas were the common heritage of all nations and could be ruled by none. This idea was first seriously challenged by the Venetians in the thirteenth century, who maintained that the sea belonged to no one and therefore could be controlled by whoever had the interest and power. These two points of view provided the basis for the present compromise position that the high seas are common heritage, available for the innocent use and mutual benefit of all, whereas territorial waters are under the control of the coastal state and subject to the rights of innocent passage. Recently, increased interest in the ocean's mineral and food resources has led to acceptance of the idea of a third region, called the exclusive economic zone.

Although these principles enjoy universal endorsement, detailed implementation has been difficult, for reasons mostly related to self-interest. Nonetheless, the historical troubles with piracy, recent interest in resources, and recent developments in technology have emphasized the need for agreement and spurred nations to work together to arrive at a detailed written law. The result is the "Draft Convention on the Law of the Sea," approved by a large majority of U.N. members, but then ratified by few.

Key Terms

spaceship Earth
fossil fuels
portable fuels
solar energy
efficiency
power
watt
kilowatt
kilowatt-hour
quad
replacement time
renewable resources
turbine
OTEC
hydroelectric power
salt power
biomass conversion
cracking
placers
brine pools
desalinize
high seas
territorial waters
exclusive economic zone

Study Questions

1. What type of fossil fuels will we run out of first? Why are they in such high demand?

2. What are one advantage and one disadvantage of solar energy as compared to fossil fuels?

3. What are some of the forms that energy comes in? What is meant by conversion efficiency? Give an example of the conversion of energy from one form to another. What is power, and what is the metric unit for measuring it?

4. What is the average per capita rate of energy consumption in the United States, and how does it compare with the world average? How is this divided among the home, transportation, and industrial/commercial sectors? At an average of about 10 cents per kilowatt-hour, what is the dollar value of the per capita energy consumed in the United States per year?

5. What are the three major problems associated with using the ocean as an energy source?

6. Why is replacement time an important consideration in determining the maximum rate at which energy can be used? Give an example.

7. Explain the difference between *direct* and *secondary* effects of solar energy on the oceans. Give an example of each.

8. Of the various renewable ocean energy resources, which is the largest? How might it be converted into electricity?

9. Why do we expect interest in the ocean's various resources to increase over the next few decades?

10. What are some of the seafloor minerals extracted from mine shafts? From boreholes? By dredging?

11. What are some of the mineral resources taken from sediments? What is each used for? Where would you find metal-enriched sediments?

12. What are some of the minerals removed from the water itself?

13. What methods can be used to extract fresh water from sea water? Why are they expensive?

14. What principle in our present understanding of the law of the sea came from Roman law? What part of our present understanding was first introduced by Venice?

15. What are some of the issues that have spurred international interest in achieving a written law of the sea?

16. What are the salient features of the U.N. Convention on the law of the sea?

Critical Thinking Questions

1. Discuss the concept of spaceship Earth.

2. How do you think future generations will judge us for having burned petroleum and other hydrocarbon resources?

3. Why might the energy for an oven, which is used only a few hours per month, cost you more than that for an alarm clock, which is on all the time?

4. Discuss how each of the various forms of energy stored in the oceans is produced by the Sun (including salt power and biomass). Which of these is also produced by the Moon?

5. What would you think might be some of the ecological side effects of extracting energy from waves? Tides? Ocean thermal energy?
6. Describe how petroleum and natural gas deposits are formed. If you were looking for new petroleum deposits, where would you look, and why?

Suggestions for Further Reading

ABLESON, P. H. 1987. Energy Futures. *American Scientist* 75:6, 584.

AKSENOV, VLADIMIR V. 1991. Satellite Oceanography. *Oceanus* 34:2, 69.

BERGHUIZEN, ANDRAE. 1993. Sea of Troubles. *World Press Review* 40:6, 45.

BORGESE, ELISABETH M. 1991. The United Nations Convention on the Law of the Sea. *UNESCO Courier* Aug.–Sept., 53.

BREUER, GEORG. 1991. A Strategy for the Sea Floor. *New Scientist* 132:1790, 34.

BYRNE, T. 1988. The Energy Index. *Science* 242:4886, 1639.

CAPULONG, EDUARDO R. C. 1987. Tapping Ocean Thermal Power. *Popular Science* 231, 97.

CHANGERY, MICHAEL J. 1987. Coastal Wave Energy. *Sea Frontiers* 33, 259.

CONE, JOSEPH. 1988. Prospecting Two Miles Down. *Oceans* 21:1, 16.

DICKINSON, W. W. 1988. Plankton to Petroleum. *Earth Science* 41:4, 21.

LEWIS, R. 1988. Twenty Thousand Drugs Under the Sea. *Discover* 9:5, 62.

PENNY, T. R. AND D. BHARATHAN. 1987. Power from the Sea. *Scientific American* 256:1, 86.

PRIEST, JOSEPH. 1984. *Energy: Principles, Problems, Alternatives*. Addison-Wesley, Reading, MA.

REVKIN, ANDREW C. 1989. Mapping a Wet Frontier. *Discover* 10:9, 30.

RONA, P. A. 1988. Metal Factories of the Deep Sea. *Natural History* 97:1, 52.

ROOT, M. 1988. Underwater Pharmacy. *Sea Frontiers* 34:1, 42.

SASANOW, W. 1989. Wellheads on the Seabed. *New Scientist* 121:1656, 43.

SEA FRONTIERS. 1987. Sea Bottom Road Maps. 33:1, 381.

STEVENSON, JOHN R. 1994. The Future of the United Nations Convention on the Law of the Sea. *American Journal of International Law* 88:3, 488.

YATES, J. ET AL. 1986. Marine Mining: Birth of a New Industry. *Endeavor* 10:1, 44.

GLOSSARY

absolute scale A temperature scale on which absolute zero is 0 K, water freezes at 273 K, and water boils at 373 K. (The degree has the same size as a Celsius degree.)

absolute zero The coldest possible temperature.

abyssal Those regions of the ocean floor that lie at depths between 3.5 and 6 kilometers, which include the bottoms of all ocean basins and most abyssal plains.

abyssal clays Fine-grained inorganic sediments that cover large portions of the deep ocean floor.

abyssal hills All hills on the deep ocean bottom that rise no more than 0.9 kilometers above the floor. Also called seaknolls.

abyssal plains Extensive and very flat regions of the deep ocean floor, usually bordering on continental margins, caused primarily by the deposits of terrestrial sediments in turbidity currents.

acid A solution in which the concentration of the positive hydrogen ion, H^+, is greater than that of the negative hydroxide ion, OH^-, or alternatively, greater than one part in 10^7.

active margins Edges of continents that are geologically active.

acute Very intense or sharp.

adiabatic processes Processes in which changes in temperature are caused by expansion or compression, and not the addition or removal of heat.

algae A common name applied to marine plants, including single-celled plants and the multicellular red, green, and brown algae.

algal ridge The region on a coral reef that lies in quieter waters behind the buttress zone and on which algal growth is particularly intense.

alluvial fans The large fan-shaped deposits of sediments and rock debris found at the base of mountains, especially where ravines empty out into the lowlands.

amino acids Complicated molecular structures that are the building blocks for the construction of the even more complicated protein molecules.

ammonia A material whose molecules are made up of one nitrogen atom and three hydrogens, NH_3.

amorphous Having no recurring order in the arrangement of atomic groups. Not crystalline.

amphidromic motion The large-scale circular motion of water, resulting from Coriolis deflection.

amphipods Insectlike crustaceans with bodies that are flattened from side to side.

amplitude The maximum up or down displacement of a wave measured from the still water level.

anaerobic bacteria Bacteria that can decompose organic matter in the absence of free oxygen, using other oxidizing agents instead.

anemones A common coelenterate animal in tide pools that looks like a flower and has tentacles that it uses to capture food particles and small prey.

Antarctic Bottom Water (ABW) The densest of the major deep water masses. It is formed beneath ice near Antarctica and flows northward along the bottom of the Atlantic Ocean.

Antarctic Circumpolar Current A major ocean current driven eastward around the entire Earth at about 50 to 60°S latitude by the prevailing westerlies.

Antarctic Circumpolar Water (ACW) The water of the Antarctic Circumpolar Current, found at about 50 to 60°S latitude. It extends from the surface to the ocean bottom, except where the Antarctic Bottom Water flows beneath it in the Atlantic.

Antarctic Convergence A region where the surface waters converge at the northern edge of the Antarctic Circumpolar Current, located at about 50°S latitude in all oceans.

Antarctic Divergence The region along the southern edge of the Antarctic Circumpolar Current where surface waters are diverging.

anticyclonic Clockwise in the Northern Hemisphere and counterclockwise in the Southern Hemisphere, as seen from above.

antinode The point on a standing wave where there is maximum vertical displacement.

aphotic Without light. Those regions of the ocean that are completely dark all the time.

Archimedes' principle Any object that is immersed in a fluid receives an upward buoyant force that is equal to the weight of the fluid displaced.

armored beach A beach from which the waves have removed the finer sediments, leaving a rocky surface.

arrow worms Members of the animal phylum Chaetognatha. Small thin worms that have small fins for locomotion and primitive jaws with fine bristles for capturing food particles.

arthropods Members of the phylum Arthropoda, which includes all animals with jointed legs and external skeletons (e.g., crustaceans and insects).

asthenosphere The rather plastic region in the upper mantle and immediately beneath the lithosphere, on which the lithosphere rides.

atmospheric pressure The pressure exerted by our atmosphere due to its weight. At sea level, it is about 14.7 lb/in.^{2}, or 1.01×10^7 N/m^2.

atoll A ring-shaped coral reef that results from coral growth on a sinking volcanic structure.

atom The fundamental constituent of matter, made up of a very small, dense, positively charged nucleus, surrounded by a cloud of negatively charged electrons.

atomic magnets A term referring to the fact that many atoms have small amounts of magnetism, due to the motions of the electrical charges within them.

atomic mass number The total number of protons and nucleons in the nucleus of an atom.

atomic number The number of protons in the nucleus of an atom.

authigenic Sediment that is derived from sea water and precipitates directly onto the ocean floor, as opposed to falling through a column of water first.

autotrophic Able to manufacture its own food.

backarc basins The region between a continent and an offshore subduction zone. These are particularly common in the western Pacific

backshore The portion of a beach above the high tide level.

backwash The water returning back down the beach face toward the surf zone, after a wave has washed up on the beach.

baleen whales One of the two major classes of whales, which includes all those that use baleen to filter tiny organisms from the water.

banker's rule of 70 For anything that grows at a constant rate, the doubling time in years can be calculated by dividing the annual percentage rate of growth into 70%.

bar-built estuary An estuary that is formed primarily by the growth of a baymouth bar or spit to isolate the coastal water from that of the ocean.

barnacle An arthropod that attaches itself to surfaces, forms a hard outer shell, and filters its food from passing water by extending feathery legs out through a hole in its shell.

barrier island A long low thin sandy island that more or less parallels a coast from which it is separated by a shallow lagoon.

barrier reef A reef whose shallowest portion is separated from the mainland by a lagoon.

basalt A fine-grained igneous rock characteristic of oceanic crust.

basins The large deep portions of the oceans extending from continental margins to oceanic ridges, with depths normally in the range of 4 to 5 kilometers.

bathyal Of or pertaining to the portion of the ocean bottom that has intermediate depths, ranging from 0.2 to 3.5 kilometers. Characteristic of continental slopes and oceanic ridges.

baymouth bar Any bar extending across a portion of the mouth of a bay. It usually begins on a headland and extends in the direction of a longshore current.

beach The region of the coast where change occurs from day to day, or month to month. It extends from several meters above the high tide line to several meters below the low tide line.

beach compartment A stretch of beach, typically over 100 kilometers long. The sand in any one compartment neither comes from a neighboring compartment, nor flows into a neighboring compartment.

beach face The surface of the beach. Sometimes used to mean only the beach surface extending from the surf zone to the berm crest.

beach nourishment Adding sediment to a beach to replace that lost to erosion.

Beagle The ship on which Charles Darwin sailed.

Beaufort scale A scale of numbers used to describe the state of the sea. The number 1 represents the calmest conditions and 17 the most violently stormy conditions.

benthic Pertaining to the ocean floor.

benthos Organisms that live on or in the ocean floor.

berm The large reservoir of loose dry sediments on the beach above the high tide line.

berm crest An elevated ridge in the berm.

bicarbonate A radical made up of one hydrogen, one carbon, and three oxygen atoms, HCO_3^-.

biogenous Derived from biological organisms.

biological magnification The tendency for some rare materials to become increasingly concentrated in the tissues of organisms at higher trophic levels.

bioluminescent Giving off light through biological processes.

biomass conversion A method of generating energy for human use by growing plants for fuel.

bivalves A class of mollusks that have hinged shells, like clams and mussels.

bloom A period of rapid increase in life.

blue-green algae Cyanobacteria. Very primitive single-celled algae that have no central cell nucleus.

boiling point The temperature at which water boils.

bony fish Fish that have bony skeletal material (as opposed to cartilage).

borehole A hole drilled into the Earth by a drilling rig.

boundary currents Currents along ocean margins.

brackish Of salinity less than that of sea water, resulting from the mixture of salt water with fresh water.

breakers Breaking waves.

breakwater A man-made barrier built in coastal waters and used for protection from incoming waves.

brine Water having salinity greater than that of sea water.

brine pools Pools of brines found on the deep ocean floor near the axis of the oceanic ridge.

brown algae One of the three common kinds of seaweeds that are usually brown in color, including kelp and sargassum.

BTU British thermal unit, a unit of energy equal to about 1054 joules.

bulkhead A wall within a ship that gives structural support and prevents water from flowing between compartments.

buoyancy The tendency of an object to rise or float due to the forces of the fluid that it is in or on.

buoyant force An upward force exerted by a fluid on a body immersed in it.

buttress zone The rugged outermost face of a coral reef that faces incoming waves.

caballing The process of surface water being forced downward where surface waters converge.

calcareous Made of calcium carbonate.

calcite A crystalline form of calcium carbonate having a hexagonal lattice structure.

calcium carbonate compensation depth The depth in the ocean beneath which calcium carbonate deposits cannot form, because they dissolve.

calorie A unit for measuring energy, which is defined as the amount of energy required to raise the temperature of 1 gram of water by 1°C.

calve The process of icebergs breaking off of glaciers.

Cambrian Period The period of Earth history extending from 600 to 500 million years ago.

capillary waves The small waves on the water surface having wavelengths less than 1.73 centimeters.

carbohydrate A molecule formed from carbon and water (hydrated carbons), including sugars and starches.

carbonate A radical composed of one carbon and three oxygen atoms, CO_3^{2-}.

carbon chain A sequence of carbon atoms that are bound together.

carbon dioxide A material whose molecules are made of one carbon and two oxygen atoms, CO_2.

carbon-14 A radioactive isotope of carbon having 14 nucleons in the nucleus (6 protons and 8 neutrons), as opposed to the normal 12.

carnivora A class of mammals that includes otters, polar bears, and pinnipeds.

carnivore Any meat-eating animal.

cartilaginous fish Fish whose skeletal material is made of cartilage (e.g., sharks and rays).

catalyst A material that aids or facilitates certain chemical interactions.

cay (key) A small temporary reef island made of reef debris.

Celsius scale A temperature scale on which water freezes at 0°C, boils at 100°C, and absolute zero is at −273°C.

cell The microscopic fundamental unit of all living organisms, each cell having a certain set of functions it must perform.

cell efficiency A measure of the ability of a cell to perform its functions expeditiously.

cell membrane A membrane that separates a cell from its outer environment.

cell nucleus A localized region in cells that holds the materials which govern the activities of the cell.

Cenozoic Era The most recent era of Earth history, extending from 65 million years ago to the present.

cephalopods Mollusks with tentacles and a chewing beak, including octopus, squid, and chambered nautilus.

cetaceans Whales.

Challenger The ship that carried a famous British oceanographic exploratory expedition from 1872 to 1876.

chemical energy Energy stored in the chemical bonds within molecules, which can be released through chemical reactions such as burning.

chemosynthesis The production of organic matter using energy from chemical interactions.

chitons Small flat mollusks with eight calcareous plates on their backs.

chordates Animals that have an internal skeletal rod.

clay Sediment with grains having dimensions in the range of 0.25 to 4 micrometers.

coast The strip of land bordering the ocean, including the beach and extending landward as far as the environment is noticeably affected by the ocean.

coastal upwelling The process by which deeper water rises to the surface along a coast, caused by departing surface waters.

coccolithophores A class of very tiny microscopic single-celled marine planktonic plants having exoskeletons made of tiny calcareous plates.

coelenterates A phylum of primitive marine carnivores, each having a gut lined with protoplasm and tentacles for capturing and maneuvering prey into the gut. Examples include jellyfish and anemones.

Common Water A deep water mass formed by the mixing of North Atlantic Deep Water with the Antarctic Circumpolar Water. It is the most voluminous deep water mass in the Indian and Pacific Oceans.

compensation point The depth beyond which plants can no longer produce enough food to sustain themselves.

conduction The transfer of heat through contact, being passed from atom to atom through their mutual interactions.

constructive interference The superposition of two or more waves in such a way that the resulting displacement is larger than that of any of the original component waves.

continental crust The outer layer of the solid Earth beneath the continents, which has a thickness of typically 30 to 50 kilometers, and is slightly less dense than the crust beneath the oceans.

continental margin The edge of a continent, where it borders on the ocean basin.

continental rise The wedge of sediments deposited at the base of the continental slope.

continental shelf The shallow submarine extension of the continent, generally rather flat and characteristically 100 meters deep. It ends at the continental slope.

continental slope The region of the ocean bottom beginning at the outer edge of the continental shelf and plummeting downward at a relatively steep angle to a depth of typically 3 to 4 kilometers.

contour line A line drawn on a chart connecting all points for which a certain variable has the same value. For example, it could connect all points on the ocean floor having the same depth, or it could connect all points on the ocean surface having the same salinity.

convection The transport of heat or material from one place to another via the motion of the material.

convergence A region where surface waters converge.

convergent boundary A boundary between two plates of the lithosphere that are undergoing collision with each other, which is marked by a subduction zone in the ocean or young folded mountains on the continents.

copepods Small arthropods resembling very tiny shrimp or insects that feed heavily on microscopic plants.

coral reef A shallow part of the ocean floor made of the cemented-together skeletons of previous generations of corals and other reef organisms.

core (Earth) The central region of the Earth extending to about 2900 kilometers beneath the surface. It consists of a liquid outer core and solid inner core.

core (sediment) A vertical column of sediment retrieved by certain types of sediment samplers.

corer A sediment sampling device that operates by plunging a hollow tube into the bottom.

Coriolis effect The apparent deflection of moving bodies from their expected straight-line motion caused by the rotation of the observer's reference frame.

cosmogenous Coming from outer space.

cotidal line A line on a chart of the ocean that connects all points where the tide is at the highest point in its cycle at any given time.

covalent bond A bond between atoms where the shared electrons are shared rather evenly.

cracking The refining of petroleum by heating to break up the organic molecules into smaller ones.

crest The highest part of a wave.

crust The outer layer of the Earth's solid surface, extending to depths of roughly 6 kilometers beneath the ocean and 30 to 50 kilometers beneath the continents. The crust is composed of materials that are lighter, cooler, and more brittle than those of the mantle beneath.

crustaceans Marine arthropods, characterized by jointed legs and an external skeleton. Examples include copepods, krill, shrimp, crabs, and lobsters.

crystal structure The periodic arrangement of atomic groups in certain solids.

crystalline Having some periodic recurring order in the arrangement of its constituent atomic groups.

ctenophores Members of a phylum of small luminous animals that are similar to small jellyfish and have rows of hairs along their sides for propulsion.

current gyres Large-scale circular current patterns in the ocean.

cyclonic Counterclockwise in the Northern Hemisphere and clockwise in the Southern Hemisphere, as seen from above.

Dark Ages The period of western European history extending from about 500 to 900 A.D.

decapods Crustaceans with 10 jointed legs, such as crabs, shrimp, and lobsters.

deep An especially deep portion of the ocean bottom.

deep scattering layer The layer of microscopic grazers that show daily vertical migrations in response to sunlight. It derives its name from the scattering of sonar signals.

Deep Sea Drilling Project A research project carried out from 1967 to 1983, which used a deep sea drilling ship, the *Glomar Challenger*, to sample sea floor sediments.

deep water wave A wave in water that is deeper than half its wavelength.

delta A low flat area at the mouth of a river that is formed by the deposition of sediments carried by the river.

demersal Residing on the ocean bottom.

density The ratio of mass to volume, often measured in units of grams per cubic centimeter.

density contours Lines drawn on charts that connect points in the water column of equal density. That is, the water at all points on any one contour has the same density.

deposit feeder An animal that gains its nourishment by ingesting sediments containing organic matter.

desalinize To remove the salt from water.

destructive interference The superposition of two or more waves in such a way that the resulting displacement is less than that of any of the original component waves.

detritus Debris.

diatomaceous earth Sediments composed of the microscopic skeletons of diatoms.

diatoms Single-celled microscopic marine planktonic plants that have siliceous exoskeletons.

diffraction The bending or spreading out of waves as they pass through openings, or as they pass an obstacle, which is not attributable to reflection or refraction.

diffusion The transport of materials from regions of higher to lower concentrations, due to the thermal motion of individual molecules.

dinoflagellates Microscopic single-celled organisms that can propel themselves with the use of tiny whip-like flagella. Some can photosynthesize.

direct techniques Any technique that measures a property directly, as opposed to measuring some effect of the property.

dispersion The sorting out of waves in a group as the longer faster waves take the lead, and the shorter slower waves fall behind.

disphotic The *twilight* region of the ocean beneath the sunlit photic surface waters and above the black aphotic waters. Typically about 100 to 500 meters deep in the open ocean.

dissolved gases Gases that are dissolved in sea water.

diurnal Daily.

divergence A region where surface waters are diverging.

divergent boundary A plate boundary along which the two adjacent plates are spreading away from each other. It is marked by the central axis of the oceanic ridge.

DNA A very long, thin twisted, ribbon-like protein molecule that contains all the information necessary to govern the activities of a cell.

doldrums The region of the Earth near the equator characterized by relatively light breezes, rising air, and frequent rainstorms.

dolomite An evaporite sedimentary rock composed of calcium and magnesium carbonates.

dome A slight rise in the ocean surface caused by the Coriolis deflection of a surface current.

downwelling The process in which surface waters flow downward.

dredge A basketlike apparatus that is dragged along the ocean bottom to retrieve biological or geological specimens.

drowned river valley The former seaward end of a river valley that was flooded as sea level rose.

dune field A region where the sand carried by the littoral drift along the beach accumulates in dunes.

East Pacific Rise The oceanic ridge in the eastern Pacific Ocean.

ebb tide The part of the tidal cycle when water is flowing out of estuaries and into the ocean.

echinoderms Animals belonging to the phylum that contains spiny-skinned radially symmetric bottom dwellers, with internal skeletons (e.g., starfish and sea urchins).

echo sounding Determining the depth and topography of the ocean floor by using echoes.

echolocation The use of echoes to locate objects or determine one's own position.

eddy A small localized current with a circular or whirling motion.

edge waves A wave that travels parallel to a barrier, resulting from the superposition of incoming and reflected waves.

efficiency (energy) The ratio of useful energy out to total energy in.

Ekman spiral The pattern of wind-driven surface currents where the Coriolis deflection causes the water at any depth to flow in a direction slightly to the right (Northern Hemisphere) and more slowly than the water above it.

El Niño A condition that occurs in the eastern Pacific Ocean around Christmas time each year, when warm surface waters arrive. The slackening of trade winds in October causes the surface water that was pushed up against the western ocean margin to begin to flow back eastward across the ocean, arriving at the eastern margin around Christmas.

El Niño condition A condition that occurs every few years in the Pacific Ocean when warm surface waters that arrive around Christmas remain for a prolonged period, usually in excess of a year.

electrical force One of the fundamental forces of nature, involving the repulsion of similar and attraction of dissimilar electrical charges.

electromagnetic waves Waves that travel through space at the speed of light, which are created by accelerating electrical charges. Ordered from long to short waves, they include radio waves, microwaves, infrared waves, visible light, ultraviolet rays, x-rays, and gamma rays.

electron cloud A term applied to the atomic regions where electrons are found.

electron shell A group of electrons on an atom that form a spherically symmetric distribution, or layer in the electron cloud. The first completed shell has two electrons, the second has eight more, and so on.

element Any of the more than 92 chemically different fundamental constituents of matter. The atoms of any one element are characterized by having a certain number of protons in each nucleus.

energy The ability to do work, or equivalently, the ability to exert a force over a distance.

enzymes Organic catalysts that are produced in cells and facilitate certain chemical reactions.

epifauna Animals that live on the surface of the ocean floor.

equation of life The following chemical equation: (carbon dioxide) + (water) + (energy) $\Longleftrightarrow$ (organic matter) + (oxygen).

equatorial bulge The slightly flattened equatorial region of the Earth caused by its spin.

equatorial countercurrents Eastward return flows of waters that have been blown across the oceans by the trade winds.

equatorial currents Surface currents blown westward across the ocean by the trade winds.

Equatorial Divergence The region near the equator where surface currents diverge.

equatorial undercurrents Equatorial countercurrents that flow beneath the surface.

equilibrium theory The model for the production of tides on Earth that does not take into account complications due to such factors as the continents and shallowness of the oceans.

erosion The removal of debris, especially rock debris, from its original location through the action of such things as water, wind, gravity, and animals.

estuary A coastal embayment where salt water from the ocean mixes with fresh water from land.

eucaryotes Any organism whose cells have cell nuclei.

evaporites Sedimentary deposits formed by the evaporation of sea water, which causes the precipitation of the salts that reach saturation.

excess volatiles Those light elements abundant in our atmosphere and oceans whose presence cannot be accounted for by the weathering of the crust.

exclusive economic zone A zone beyond the territorial waters. In this zone, the coastal state has jurisdiction over certain resources, but not complete sovereignty in all affairs.

exoskeletons A skeleton that covers, or partially covers, the exterior of an organism.

exponential growth Something that grows at a fixed rate, such as growing by a certain percent each year.

fast ice Ice that forms adjacent to a land mass.

fault A crack in the Earth's surface along which there has been some displacement.

feeder beach A beach on which

sand is deposited to nourish downstream beaches.

fetch The distance across the ocean's surface over which the wind blows.

field reversals Reversals of the Earth's magnetic field, with the former north magnetic pole becoming a south magnetic pole, and vice versa.

filter feeder An animal that feeds by filtering tiny food particles from the water.

fjord A glacially cut valley that empties into the ocean.

flagella Tiny whiplike appendages on microscopic organisms that give them minor mobility.

flatworms A primitive class of worms. Marine species live on or in the ocean bottom.

flood tide That part of the tidal cycle when water is flowing into estuaries from the ocean.

flooded coastal plain A coastal region where rising sea level has flooded into previous low coastal regions, creating wide shallow lagoons and drowned river valleys.

fluid Any material that flows, usually referring to liquids and gases.

fluke The tail fin of a whale.

food chain The sequence of trophic levels through which food is passed.

foraminifers Single-celled animals belonging to the phylum Protozoa and having calcareous tests.

forced waves Waves that remain under the influence of the forces that generate them, rather than traveling on their own after having been generated.

foreshore The intertidal portion of the beach extending from the low tide line up to the high tide line.

fossil fuel Any fuel, such as coal, petroleum, or natural gas, that is derived from the remains of organisms.

FPC Fish protein concentrate, a dehydrated powder made from fish. It is rich in protein and stores easily without spoiling.

fracture zone Faults oriented roughly perpendicular to the oceanic ridge, which are usually extensions of transform faults that cross the ridge axis.

Fram The ship captained by Fridtjof Nansen when he explored Arctic waters in 1893 to 1896.

free waves Waves that are traveling independent of the force that originally generated them.

frequency The number of oscillations per second.

fringing reef A reef immediately adjacent to land, and not separated from it by a lagoon.

fronds Leaflike photosynthetic parts of marine algae.

fully developed sea A state of wave activity reached when the wind-generated waves have reached their maximum size and will not grow larger or longer even if the wind continues to blow.

gamma rays Electromagnetic waves having the shortest wavelengths and highest energies. Wavelengths are shorter than 10^{-12} meters.

gastropods A class of mollusks that includes snails and slugs.

geostrophic flow A state where the current is flowing in such a way that the Coriolis deflection is exactly balanced by the pressure gradient force, or downslope component of gravity.

grab sampler A device used to retrieve sediments from the ocean bottom. It has jaws that close to take a "bite" out of the sea floor.

graded bedding The sorting of sediment within layers, having the coarsest sediments at the bottom of each layer and successively finer sediments above that until the next layer is reached. Then the pattern repeats itself.

gravimeter A sensitive instrument used to measure the strength of gravity.

gravitational differentiation The gradual separation of materials inside the Earth, caused by the denser materials sinking toward the interior and the lighter materials rising toward the surface.

gravity A force with which two bodies attract each other that is proportional to their masses and inversely proportional to the square of their separation.

gravity waves All waves of wavelength greater than 1.73 centimeters that are driven by gravity and their own inertia, regardless of how they were originally created.

grazers Animals that feed on plants: herbivores.

green algae A phylum of multicellular algae (seaweeds) that grow in shallow water. A common example is sea lettuce.

greenhouse effect The warming effect caused by our atmosphere's transparency to sunlight coming in and its opacity to the Earth's infrared radiation going back out, therefore making it easy for energy to come in and difficult for it to leave.

greenhouse gases Any gases in the atmosphere that tend to absorb infrared light and therefore trap solar heating.

groin A wall extending out into the ocean perpendicular to the beach, which is intended to decrease beach erosion by interrupting the longshore transport of sand.

Gulf Stream An intensified surface current that runs north-eastward along the eastern continental margin of North America and toward Northern Europe.

guyots A flat-topped submarine volcanic mountain.

gypsum An evaporite sediment composed of hydrated calcium sulfate ($CaSO_4 \cdot 2H_2O$).

hadal Of or pertaining to the very deepest portions of the ocean bottom below 6 kilometers in depth (i.e., in trenches).

half-life (radioactive) The time required for half the original nuclei in a sample to decay.

halocline A region where salinity varies quickly with depth.

halophyte A salt-tolerant land plant.

head (estuary) Where the fresh water empties into an estuary from a river or stream.

headland Any point or piece of land sticking out into the ocean from the mainland.

herbivore Any animal that feeds on plants.

heterotrophic An organism that is unable to manufacture its own food.

high seas The region of ocean farther from land than the territorial waters and exclusive economic zone.

highly stratified An estuary where

fresh water is on top, salt water on the bottom, and there is relatively little mixing of the two along the length of the estuary.

holdfasts Rootlike appendages used by seaweeds to anchor themselves to the ocean floor.

holoplankton Organisms that are planktonic their entire lives.

horse latitudes Those regions around 30° to 35°N and S latitudes characterized by light surface winds, sinking air masses, and arid conditions.

hot spot A hot region in the upper mantle that causes volcanism on the sea floor above, leaving a trail of volcanoes as the lithosphere passes overhead.

hurricane A violent storm having wind velocities in excess of 110 kilometers per hour.

Huygens' principle The statement that every point on the present wave crest acts as a point source for generating new wavelets traveling outward in all directions.

hydrate To add water.

hydroelectric power Electrical power generated by capturing water in reservoirs behind dams, and then using the pressure of this water to turn turbines that generate electricity.

hydrogen bond The strong attraction between the positively charged hydrogen end of a water molecule and negatively charged portions of other molecules.

hydrogenous Derived from the materials in solution in sea water.

hydrologic cycle The cycle of water in our hydrosphere during which water in the ocean evaporates, precipitates, and returns to the ocean through any of a variety of routes.

hydrosphere The outer fluid layers of the Earth where water is plentiful, including ocean, atmosphere, ice, lakes, streams, and ground water.

hydrothermal vents Plumes of hot water coming out of the sea floor near the ridge crest.

igneous Rock that solidified from molten magma.

in situ Measured in place, as opposed to being first brought to some other place and then measured.

indirect techniques Techniques in which the measure of a property is inferred from a related property.

inelastic demand When the demand for a product varies little with price.

inert gas Any element whose outermost electron shell is completely filled and therefore does not interact chemically with other atoms. Examples are helium (first shell filled) and neon (first and second shells filled).

inertia The tendency for a resting body to remain at rest and a moving body to maintain its motion in a straight line. A quantitative measure of this property is the mass.

inertial currents Currents following looped trajectories that are characteristic of an object that tries to pursue inertial straight-line motion along the surface of a spinning Earth.

inertial force An imaginary force that seems to cause the deflection of a moving body. The apparent deflection is actually due to the rotation of the observer's reference frame.

infauna Animals that live within the bottom sediments.

infrared waves That portion of the electromagnetic wave spectrum having wavelengths longer than the longest visible waves (red light) but shorter than microwaves.

inlet A gap in a barrier island through which water can flow in or out.

inner core The solid innermost region of the Earth, extending about 1300 kilometers outward from the center.

inner planets Mercury, Venus, Earth, and Mars. (Closer to the Sun than Jupiter.)

internal waves Subsurface waves that travel along the boundary between layers of water of different densities.

intertidal The region of the shore that is between low and high tide levels.

inverse estuary An estuary in which more fresh water is lost to evaporation than enters the estuary from land. The estuarine water is therefore saltier than sea water.

ion Any atom or group of atoms that has a net electrical charge due to excess or deficit electrons.

ionic bond A chemical bond in which the atoms are held together primarily by the electrostatic attraction between ions. The shared electrons are not shared equally, giving strong electrical charge to the binding partners.

ionic salts Salts whose atomic groups are bound by ionic bonds and that tend to separate into charged ions when dissolved in sea water.

isopods Small insectlike crustaceans that are flattened from top to bottom.

isostasy The condition of floating in buoyant equilibrium. The tendency of the lithosphere to float atop the more plastic asthenosphere, with surface elevations reflecting the buoyancy of the lithospheric materials.

isostatic adjustment Upward or downward movement of the lithosphere in order to attain buoyant equilibrium as it rides on the asthenosphere.

jetty A man-made wall or barrier projecting into the ocean along the mouth of a river or stream, in order to guide water flow or prevent the entry of beach sediment.

joule The unit of energy in the metric system that is equal to the kinetic energy of a 2-kilogram mass moving at 1 meter per second and also equal to 0.239 calories.

Kelvin scale The absolute temperature scale, where the degrees are the same size as the Celsius degree, and on which absolute zero is 0 K, water freezes at 273 K, and water boils at 373 K.

kilowatt A rate of energy consumption equal to a thousand joules per second.

kilowatt-hour The amount of energy consumed if used at a rate of a kilowatt for 1 hour. Equivalently, 3,600,000 joules.

kinetic energy Energy of motion.

kingdom The broadest and most general category in the scheme for classifying organisms.

knot A unit of speed equal to 1 nautical mile per hour, or 1.15 miles per hour, or 1.85 kilometers per hour.

krill Arthropods that resemble small shrimp, typically a few centimeters in length. They are very abundant in cooler waters of higher latitudes and are a favorite food for many whales.

lagoon A broad area of shallow water separated from the open ocean by barrier islands or other shallow banks.

land breeze A coastal breeze that blows from the land toward the ocean.

Langmuir cells The circulation pattern formed when strong winds blow across relatively calm waters. The cells are typically a few meters thick and oriented parallel to the wind direction. The water in each cell undergoes corkscrew motion in the direction of the wind.

latent heat of evaporation The amount of heat required to evaporate 1 gram of water, without raising its temperature.

latent heat of fusion The amount of heat needed to melt 1 gram of ice at 0°C.

latitude A measure of north–south position on the Earth relative to the equator, measured in degrees, with the equator being 0° latitude and the poles at 90°N and S latitudes, respectively.

lava The molten rock that flows out of volcanic areas and onto the Earth's surface.

law of gravity The law of nature that requires any two masses to attract each other with a force that is proportional to both masses and inversely proportional to the square of their separation.

leeward The side facing away from the wind.

light–dark bottle system A technique for measuring the rate of primary productivity using water samples in a clear bottle and blackened bottle. In the dark bottle, only respiration occurs, whereas both photosynthesis and respiration occur in the clear bottle.

limestone A sedimentary rock made of calcium carbonate.

limey A colloquial name for a British sailor.

lipids A class of organic materials that includes fats and waxes.

lithogenous Formed from the weathering of rock.

lithosphere The relatively cool and rigid outer region of the Earth, including the crust and the outer 80 to 100 kilometers of the mantle, which is also cool and rigid and rides with the crust in the asthenosphere below.

littoral Moving along the shore, parallel to the shore.

littoral drift The transport of sediments along the beach in the surf zone.

longitude A measure of east–west position on the Earth, measured in degrees east or west of the prime meridian that runs through Greenwich, England.

longshore bars Sand bars located in or just beyond the surf zone, oriented approximately parallel to the shore.

longshore current The net motion of the water in the surf zone along the shore, caused by breaking waves coming in at a slight angle.

longshore transport The transport of sediments along the shore by the longshore current. Same as littoral drift.

low tide terrace A low flat, portion of the beach exposed and damp at low tide.

low velocity zone A less rigid portion of the Earth's upper mantle, extending from about 100 to 200 kilometers below the surface. Seismic waves travel more slowly in this zone than the layers above or below.

lunar tide The portion of the tide that is caused by the Moon's gravity.

macroplankton Plankton large enough to be easily visible to the unaided eye, generally considered to be those having dimensions greater than 1 millimeter.

Magellan Expedition A voyage commanded by Ferdinand Magellan in 1519 to 1522 that was the first to sail around the world.

magma Molten or partially molten rock that is capable of fluid flow.

magnetic anomaly Any local deviation from the normal pattern in the Earth's magnetic field.

magnetic dip The angle that the Earth's magnetic field makes with the horizontal.

magnetic reversals Reversals in the direction of the Earth's magnetic field that occur rather suddenly on geological time scales.

major constituents The most abundant salts in sea water, whose concentrations can appropriately be expressed in parts per thousand.

mammals A class of vertebrates that have hair and give birth to live young which are fed from the mammary glands of the female.

manganese nodules Hard nodular deposits (diameter typically a few centimeters) that are found in some places on the ocean bottom and are enriched in heavy metals.

mangrove swamps Shallow water coastal estuaries where mangrove trees or bushes are abundant.

mantle The interior region of the Earth between the crust and core, extending from roughly 30 to 2900 kilometers beneath the surface.

mariculture Farming the oceans or ponds of ocean water for food.

maturity A measure of how uniform and well rounded the grains of a sediment are.

maximum sustainable yield The maximum weight of any kind of organism that may be harvested year after year.

mediterranean Surrounded by land.

meroplankton Plankton that spend only a portion of their lives as plankton, such as the juveniles of many nekton.

Mesozoic Era The era of Earth history extending from 230 to 65 million years ago.

meteorological equator The Earth's equator from a meteorological point of view, above and below which the climate cools with increasing distance. It is slightly north of the geographical equator, due to the unequal distribution of land.

methane A material whose molecules are made of one carbon and four hydrogen atoms, CH_4.

micrometeors Tiny dust-sized meteors.

micrometer (μm) A millionth of a meter. A micron.

micron A micrometer.

micronutrients Ingredients needed

by organisms in the construction of special materials, but only in extremely trace amounts.

microplankton Plankton whose size ranges from about 70 to 1000 microns.

microwaves Electromagnetic waves with wavelengths between 100 micrometers and a meter. Shorter than radio waves, but longer than infrared.

Mid-Atlantic Ridge The oceanic ridge that runs north–south down the center of the Atlantic Ocean.

Middle Ages The period of western European history extending from about 900 to 1200 AD.

mineral A solid material having both a definite chemical composition and definite crystalline configuration.

ml/l A unit for expressing the concentrations of dissolved gases in terms of milliliters of the gas that are dissolved in a liter of sea water.

Mohorovicic discontinuity (Moho) The boundary between the crust and mantle.

molecular mass number The total of all the neutrons and protons in the atomic constituents of a molecule. The sum of the atomic mass numbers of the molecule's constituent atoms.

molecule Groups of atoms that bond together by sharing electrons.

mollusks Members of an animal phylum, for which each member has a soft "foot" used for locomotion, and most have calcareous protective shells. Examples include clams, snails, slugs, mussels, and squid.

Monera A kingdom of single-celled organisms with no cell nucleus. Examples are bacteria.

mountainous coast Coasts that are mountainous.

mouth Where a river or estuary empties into an ocean.

multicellular Composed of many cells.

nanoplankton Plankton whose dimensions range from 5 to 70 micrometers.

neap tide The two times during the month when tidal variations are smallest.

nektobenthos Animals that live on the ocean bottom, but are able to swim when necessary (e.g., flounders and rays).

nekton Marine organisms that are capable of fast sustained motion, as opposed to plankton, which are not.

neritic Pertaining to those regions of the ocean over the continental shelves.

net mesh size The size of the openings in the weave of fish nets.

nitrate A radical consisting of one nitrogen and three oxygen atoms, NO_3^-.

nitrite A radical consisting of one nitrogen and two oxygen atoms, NO_2^-.

nitrogen-fixing Able to take nitrogen gas N_2 and convert it into a nitrate or nitrite.

nodal points Points along a standing wave where there is minimum vertical motion.

node A nodal point.

North Atlantic Deep Water A deep water mass produced south of Greenland by the convergence of the East Greenland, West Greenland, and Gulf Stream surface currents.

northerlies Winds blowing from the north.

nudibranchs A gastropod that is like a colorful snail without a shell.

nuclear Pertaining to either an atomic nucleus or a cell nucleus.

nucleus (atomic) The tiny dense central portion of an atom containing the neutrons and protons, having a positive charge and dimensions of about 10^{-15} **meters, which is about** 10^5 **times smaller than the dimensions of the surrounding electron cloud.**

nucleus (cell) A separate region within a cell that contains the material which governs the cell's activities.

nutrient An ingredient needed by plants for the synthesis of organic material, whose scarcity may limit productivity. Common nutrients are nitrate and phosphate.

nutrient cycle The cycle of nutrients into and out of organic matter.

ocean basin The large deep portions of the oceans extending from continental margins to oceanic ridges, with depths normally in the range of 4 to 5 kilometers.

ocean farming Mariculture. Cultivation of marine crops.

oceanic Pertaining to the regions beyond the continental shelves where the oceans are deeper.

oceanic crust The type of crust underlying the oceans that is typically 6 kilometers thick and somewhat richer in iron and magnesium minerals, and therefore somewhat denser, than the continental crust.

oceanic ridge A very long undersea mountain range, extending throughout all the major oceans. It is characterized by a central ridge crest from which the sea floor spreads.

ocean trench A trench in the ocean floor marking where one oceanic plate is going down beneath the adjacent plate. Trenches are characteristically several thousand kilometers long and 100 kilometers wide, and extend to a depth of about 10 kilometers below sea level.

offshore Out to sea, beyond the low tide line.

ooze Any sediment whose composition is more than 30% biogenous.

orbital (atomic) The path followed by an electron in the electron cloud.

orbital (wave) A circular path followed by a water molecule as a wave passes overhead.

osmosis The tendency of water molecules to pass through a semipermeable membrane that separates fresh and salt water, going from the fresh water to the salt water side.

osmotic pressure The back pressure required to prevent fresh water from passing through a semipermeable membrane that separates fresh water from salt water.

OTEC Acronym for ocean thermal energy conversion. A process that uses the temperature difference between surface and deep waters to generate electrical power.

outer core The liquid outer portion of the Earth's core, extending from about 1300 to 3500 kilometers from the Earth's center.

outer planets Jupiter, Saturn, Uranus, Neptune, and Pluto. The planets farther from the Sun than Mars.

outgassing The process, including surface volcanism, by which volatiles arrive at the Earth's surface from interior regions.

out of phase Adjective describing oscillations that are not synchronized, so that one oscillating object is either ahead of, or lags behind the others.

oxide A chemical compound formed by combining anything with the O^{2-} ion.

oxidize To combine chemically with oxygen.

oxygen minimum layer The region of the ocean immediately beneath the productive photic zone that is partially depleted of dissolved oxygen because of the respiration of the many animals that reside there.

ozone A material whose molecules are made of three oxygen atoms, O_3.

pack ice Ice that forms seasonally on the surface of the open ocean at high latitudes, generally in the form of large slabs with cracks and sometimes open water between them.

paleomagnetism The study of the magnetism in ancient rocks.

Paleozoic Era The era extending from 600 to 230 million years ago.

Pangaea The supercontinent that existed from about 250 to 180 million years ago, which was broken up as the Atlantic opened.

paralytic shellfish poisoning A potentially fatal sickness caused by eating shellfish that have fed on certain species of dinoflagellates.

partial melting The heating of rock to the point where the minerals with lower melting points melt and lubricate the flow of the crystals of the unmelted materials.

partial tides The various tidal components that are used in making tide tables.

partially mixed Estuaries where the salinity of water increases from head to mouth and top to bottom.

passive margins Continental margins that are geologically quiet.

patch reef A small coral reef growing upward from a lagoon floor on a portion of a coral reef where there is too little wave action and nourishment to sustain vigorous coral growth.

pelagic Pertaining to the waters of the oceans, as opposed to the ocean bottom.

period The time required for one complete oscillation as a wave passes a stationary point.

permeability The ability of a material to allow water (or any fluid) to flow through it.

petroleum A liquid fossil fuel formed from the remains of ancient microscopic organisms.

phase change When a material changes from one phase (i.e., solid, liquid, or gas) to another.

phosphate A radical made of one phosphorus and four oxygen atoms, PO_4^{3-}.

photic zone The upper surface layer of the ocean where there is sufficient sunlight for photosynthesis.

photosynthesis The process of producing organic matter by using carbon dioxide, water, nutrients, and sunlight.

phylum The next to the broadest and most general category in the scheme for classification of organisms. A subdivision of a kingdom.

physical processes Evaporation, precipitation, melting, or thawing—processes that add or remove fresh water on the ocean surface.

phytoplankton Planktonic plants.

picoplankton Plankton with dimensions smaller than 5 microns.

pinnipeds Marine mammals of the order carnivora that have flippers, such as seals, sea lions, and walruses.

placers Former beaches that have been enriched in heavy metal deposits by the winnowing action of the waves.

plankton Marine organisms having little or no means of mobility.

plankton patchiness The tendency of plankton to stay together in clusters.

plate A large piece of the Earth's outer shell (the lithosphere) that moves as a unit and grinds against neighboring plates as it moves.

plate boundaries The outer edges of the plates of the lithosphere.

plate tectonics The motions of the Earth's surface plates.

plateau An elevated region of the sea floor, with composition more like the materials of the continental crust than those of the oceanic crust.

plunging breaker A breaking wave with the crest plunging in front of the wave, as opposed to spilling down its face.

plutonic Formed beneath the Earth's surface by the intrusion of magma into cracks.

pod A group of whales.

polar Pertaining to the region of the Earth near the North or South Pole.

polar easterlies Surface winds that blow from the east at high latitudes.

polar ice Ice cover that remains permanently throughout the year near the North Pole.

polar wandering A change of the positions of the magnetic North Pole in past times, made by measuring the relic magnetism in ancient continental rocks of various ages. The chart is based on the assumption that the continents were stationary and the magnetic North Pole moved, rather than vice versa.

polarization The separation of electrical charges to form one region of net positive charge and another region of net negative charge.

pollutant Any disruptive addition to the environment, usually in the form of either heat or foreign materials.

polymorphism The ability of a single species to appear in two or more very different forms (aside from sex-linked differences).

porosity A measure of what fraction of the volume of a sedimentary deposit is not filled by the sediment.

portable fuels Fluid fuels like natural gas and petroleum products that are easily transported.

potassium-argon dating A method for determining the age of a rock by measuring the amount of radioactive potassium that has decayed into argon within it.

potential energy Energy that is stored for later release.

potential temperature The temperature of a water sample after the pressure has been reduced adiabatically to atmospheric pressure, as it is brought up from depth.

power The rate of energy use.

ppm Parts per million.

‰ The symbol for parts per thousand.

Precambrian Period The period of history occurring earlier than 600 million years ago.

precipitation (atmosphere) Rainfall, snow, hail, sleet, dew, frost, or falling mist.

precipitation (ocean) The process by which dissolved materials are removed from solution and join the bottom sediments.

pressure Force per unit area.

pressure gradient force The force that results from a change in pressure, so that an object experiences a larger pressure on one side than another.

prevailing westerlies The winds that blow from the west in temperate latitudes.

primary producers Organisms, including plants and some bacteria, that produce organic matter.

primary production The production of organic matter.

procaryote An organism whose cells have distinct cell nuclei.

productivity The rate at which organic matter is produced by primary producers.

propagate To travel across or through a medium.

proteins A class of complicated organic molecular structures formed from the chemical alteration of carbohydrates, including the addition of nitrogen, phosphorus, and trace elements.

Protista A kingdom that includes single-celled organisms that have cell nuclei.

Protozoans A class of single-celled animals.

P-wave A type of seismic wave in which the direction of the vibrations parallels the direction of propagation of the wave.

pycnocline A region where the density of water varies quickly with depth.

quad A large unit of energy, equal to a quadrillion BTUs, or 1,054,000,000,000,000,000 joules.

radiation The transport of energy from one region to another via electromagnetic waves.

radical Any common group of atoms that can be found as a negatively charged ion. Examples include carbonate (CO_3^{2-}), nitrate (NO_3^-), and silicate (SiO_4^{4-}).

radioactive decay The decay of unstable nuclei, in which they emit energetic particles.

radio waves The category of electromagnetic waves of longest wavelength, being longer than a meter.

radiolarians Single-celled animals with siliceous exoskeletons.

red algae A phylum of colorful seaweeds, usually with deep reddish colors, that can grow in relatively deep coastal waters.

red tides Times when the water has a reddish tinge due to the presence of a certain type of dinoflagellate. These are also times when shellfish may be poisonous to eat.

reef Any region where the remains of living organisms have made the water shallow.

reef flat A flat region of a coral reef on the side away from the open ocean.

reflection The process by which waves bounce off of surfaces that they run into.

refraction The process by which the direction of propagation of waves changes due to a change in wave speed. For water waves, this is caused by changes in the depth of the water.

relic Left over from ancient times.

remote sensing Using instruments to make measurements where either the item being observed is remote from the instrument, or the instrument is remote from the observer.

Renaissance The period of European history from about 1200 to 1500 AD.

renewable resource Any resource that gets replaced as we use it.

replacement time The time required for a given amount of a resource to be restored once it has been removed.

reptiles A class of cold-blooded vertebrates that includes snakes, lizards, and turtles.

residence time A measure of the average time that a particular component of sea water spends in solution between the time it first enters and the time it is permanently removed from the ocean.

resonance A condition of extra-large wave amplitude obtained when the frequency of an external wave-generating force matches a natural frequency for waves to bounce back and forth along or across an enclosed space, such as a protected harbor or estuary.

resonant frequency The frequency at which waves naturally bounce back and forth across an enclosed region.

respiration The oxidation of organic matter in organisms to release useful energy.

resting spores Dormant cells with the ability to wait out extended periods in inhospitable environments and then spring back to life when conditions improve.

restoring force Any force that tends to restore the water surface to a level surface, such as gravity for large waves, and surface tension for small ones.

revetment Large stones or slabs that are placed directly on a beach to reduce erosion.

rift mountains The rugged heavily faulted mountains that border the rift valley at the crest of an oceanic ridge.

rift valley The long "crack" that runs along the center of the oceanic ridge in some places. The sea floor spreads away from this feature on both sides.

ring of fire A term applied to the perimeter of the Pacific Ocean, being characterized by its volcanic and seismic activity, which results from the subduction of oceanic crust on all its margins.

rip current A swift narrow seaward-flowing current in the surf zone.

RNA Any of a variety of long, thin protein molecules found in biological cells that form the scaffolding on which complicated molecules are constructed.

round worms (nematodes) A phylum of small primitive worms, found mostly in sediments.

ROV Acronym for remote operated vehicle. An unmanned submersible.

rule of constant proportions The

rule stating that the major salts in sea water are always found in the same relative proportions, although the amount of fresh water in the mixture may vary.

Ryther analysis An analysis of plant productivities in various types of ocean regions and of the fish harvests taken from these regions.

salinity A measure of the amount of salt dissolved in the water, measured in grams of salt per kilogram of sea water, or parts per thousand.

salt marsh A shallow marshy region of a coastal estuary.

salt power A method of generating electrical power that makes use of the attraction between salt ions and water molecules. It involves separating fresh and salt water with a semipermeable membrane.

salt wedge A wedge of salt water that intrudes from the ocean into an estuary along the bottom. A class of estuaries that fits this model.

San Andreas Fault A transform fault along the West Coast of North America that separates the Pacific Plate from the North American Plate.

sand Sediments with grain size in the range of 0.06 to 2.0 millimeters.

saturated Holding as much of the material as possible, such as sea water being saturated with a certain salt.

scarp A steep rise or drop-off in the ocean bottom.

schooling The tendency of fish and other marine creatures to form groups.

scurvy A disease that results from a deficiency of vitamin C.

sea (waves) An unsorted mixture of waves.

sea (water) An enclosed or semienclosed body of salt water that is large compared to estuaries and embayment, but small compared to the oceans.

sea breeze A coastal breeze that blows from the ocean toward the coast.

seaknolls Mountains on the ocean bottom that rise less than 0.9 kilometers from the floor. Abyssal hills.

seamount A mountain on the ocean floor, which extends upward more than 0.9 kilometer.

seawall A man-made wall built on the beach parallel to shore to protect the beach from wave erosion.

sedimentary cycle The cycle experienced by some sediments as earlier sedimentary deposits erode and wash into the ocean where they form new deposits. These deposits might subsequently be raised by tectonic processes, exposed, weathered, eroded, and formed into new deposits. The cycle may repeat itself over and over.

sedimentary rock Rock that forms by the cementing together of neighboring sediment grains.

sedimentation The deposition of sediments.

segmented worms A phylum of worms whose bodies are wrinkled in a way that makes them appear to be segmented. Marine species generally either live in tubes and filter feed or live in the sediments and feed on deposits. Annelids.

seiche A resonant sloshing of water in a confinement, such as a harbor or estuary.

seismic echo profiling The use of sound to map subsurface layers of the Earth. Depths are measured by the arrival times of echoes.

seismic wave Vibrations that travel through the Earth.

semidiurnal Twice a day.

semipermeable membrane A membrane that allows water molecules to pass through, but not dissolved salts.

sensible heat Heat that can be detected as a change in temperature (as opposed to latent heat).

sessile Attached.

shallow water wave Any wave traveling on water shallower than one-twentieth of its wavelength.

shelf break The very outer edge of the continental shelf having an abrupt change of slope that indicates the beginning of the continental slope.

shoal To become shallow. A region of shallow water.

shoreline The line of contact between the water surface and land.

sigma-tee A quantitative measure of the density of a water sample, defined as (density-1)×1000, where the density is measured in grams per cubic meter.

significant wave height The average height of the highest one-third of all waves.

silica A mineral made of silicon dioxide, SiO_2. Quartz.

silicate A radical made of one silicon and four oxygen atoms, SiO_4^{4-}.

siliceous Made of silica.

silicoflagellates Any of a variety of tiny single-celled plants having exoskeletons of silica and flagella for minor mobility.

sill Any shallow area or ridge that separates one basin from another.

silt Sediment whose grains have characteristic dimensions ranging from 0.004 to 0.06 millimeters.

simmering The motion in a liquid caused by heating from below.

sine waves A smooth wave form, as is obtained in a plot of the sine of an angle vs. the angle.

sirenia A class of herbivorous marine mammals that includes manatees and dugongs (sea cows).

slack water The time during the tidal cycle when the water is neither entering nor leaving an estuary.

slick A concentration of floating debris on the water surface that results in reduced wave activity and a shiny appearance.

slope water Cool water found between the Gulf Stream and East Coast of North America.

snow line The calcium carbonate compensation depth, named because the light-colored calcium carbonate deposits form above this level, but not below, like the snow on the tops of mountains.

solar energy Energy coming from the Sun.

Solar System The system that includes the Sun and all its planets, moons, asteroids, and comets.

solar tide The portion of the tide that is caused by the Sun's gravity.

solvent The medium in which something is dissolved.

sounding A measurement of water depth.

southern oscillation The weather pattern that normally has warm surface water and storms in the western tropical Pacific with cool surface water and clear skies on the east, but during El Niño has the reverse. That is, the weather patterns switch sides across the tropical Pacific.

spaceship Earth The idea that we must learn to make due with whatever resources are available on Earth, because we will be getting no more. We must make due with what is on board.

Spanish Armada A fleet of 130 Spanish war ships that sailed to England in 1588 in a failed effort to punish England for encouraging raids on Spanish merchant ships.

specific heat The amount of heat energy required to raise the temperature of one gram of a material by one degree.

spilling breaker A breaking wave in which the water spills down the face of the wave.

spit A long low thin strip of sand that extends offshore and usually parallel to shore from a headland or point.

splinter berg A relatively small iceberg broken off a large ice sheet, such as a glacier flowing into the ocean.

sponge A very primitive multicellular animal with little organization among its cells. Structure is provided by tiny spikes. Sponges feed on tiny food particles that get trapped in the body of the sponge as water passes through.

spring tide The two times during the month when tidal variations are greatest, occurring when the moon is either new or full.

stand of tide Any period during the tidal cycle when the water level is neither rising nor falling.

standing crop The population of a stock that remains year in and year out, in spite of harvests, and so on.

standing wave The wave interference pattern that results when two identical wave groups move through each other in opposite directions. The resulting pattern has regions with large up and down motion (antinodes), and other regions with no up and down motion at all (nodes).

still water level The position that the water surface would have if the waves were all flattened.

stipe The long tubular main stalk or trunk of a seaweed.

storm surge A coastal rise in sea level caused by a storm.

stratification Forming horizontal layers.

stream-cut channel A valley, particularly an estuary, that has been cut by the forceful flow of fresh water.

stromatolite A type of rock formation made from layers of sediments that have collected on successive layers of algae.

strong force The fundamental force that is responsible for binding together atomic nuclei.

subduction zone A region where a plate of the lithosphere is sliding downward into the mantle below.

sublittoral Pertaining to those regions of the continental shelf that extend from low tide level all the way to the outer edge of the shelf.

submarine canyons Very large, steep-walled canyons carved in the continental shelves and slopes.

submersible A vehicle that travels beneath the ocean surface.

sulfate A radical made of one sulfur and four oxygen atoms, SO_4^{2-}.

sulfide A molecule made of any element combined with sulfur (S^{2-}).

superposition The placement of one atop another.

surf Coastal breaking waves.

surf beat The periodic change in amplitude of waves coming ashore, with the repeating pattern of a series of larger waves followed by a series of smaller waves, followed by a series of larger waves again, and so on.

surf zone The nearshore region where there are breaking waves.

surface-active Tending to stick to solid surfaces.

surface tension The cohesive force with which the water molecules hold together to form a very thin elastic surface film on the water's surface.

surge channels Grooves in the seaward edge of a coral reef, through which the energy of incoming waves is channeled.

surging breaker Breaking waves that break abruptly on the shore.

swash The water that washes up on shore after an incoming wave has broken.

S-wave A type of seismic wave in which the vibrations are in a direction perpendicular to the direction of propagation of the wave.

swell Long parallel well-sorted waves.

swim bladder A pouch inside a fish that can be filled with gas to adjust the fish's buoyancy.

tablemount A flat-topped submarine volcanic mountain. A guyot.

tabular berg A large flat iceberg, generally formed by breaking away from the ice cover of one of the Antarctic ice shelves.

tectonic-formed estuary An estuary that has been formed by tectonic processes, such as a piece of continental crust on a coast falling beneath sea level, or an offshore piece of crust rising above sea level.

temperate Pertaining to intermediate latitudes, usually considered to be those lying between the tropics (23°N or S) and the Arctic or Antarctic Circle (67°N or S).

temperature A measure of molecular motion.

terrestrial planet A planet with a hard surface and relatively thin atmosphere, such as Mercury, Venus, Earth, or Mars.

terrigenous Pertaining to, or derived from the continents.

territorial waters The strip of water adjacent to a coastal state, over which the state has complete sovereignty, subject to the rights of innocent passage by others.

tests Tiny exoskeletons or skeletal debris of microorganisms.

thermal energy Energy carried in molecular motion.

thermal inertia The tendency for warm things to stay warm, and cold things to stay cold.

thermocline A region where temperature varies rapidly with depth.

thermohaline circulation Ocean circulation caused by differences in water density, such that denser waters flow downhill. The name is derived from the fact that the density of water is determined by its temperature and salinity.

tidal bore A slowly moving, steep-walled front of water created as an incoming tide stacks up in shallow water, especially in rivers and estuaries where there is a slow seaward current.

tidal bulge A region where the ocean surface is elevated due to tidal forces of the Sun and Moon. There are two such regions at any one time, on opposite sides of the Earth.

tidal channels Channels in salt marshes through which water flows as the tides come in and out.

tidal currents Cyclical ocean currents created by the passage of the tides.

tidal flat A flat marshy region of an estuary that is alternately inundated and exposed as the tides rise and fall.

tidal volume (tidal prism) The volume of water contained between high tide and low tide in an estuary, which is the amount of water flowing in and out of the estuary during a tidal cycle.

tidal wave A tide. (Not to be confused with a tsunami.)

tide A large wave that extends halfway around the Earth, which is generated by the gravitational influence of the Sun and Moon.

tide pool A coastal pool of water left above sea level as the tide goes out.

tombolo Any long thin low strip of exposed sand that extends from land out to a small offshore island.

topography Surface features, or relief.

toothed whales One of the two classes of whales: those that have teeth as opposed to baleen.

trace elements Those elements present in sea water in only very small amounts, and whose concentrations are best measured in parts per billion or less.

trade winds The prevailing winds that blow from the east in the tropics.

trajectory The path followed by a moving object.

transform faults A boundary between two adjacent plates that are moving laterally past each other.

transportation (sediments) The means by which sediments are carried over long distances, such as being carried from inland areas to the coast.

trophic level A step in the food chain measured in terms of how many different organisms the organic matter has been in since its original synthesis in plants.

tropical Pertaining to the region of the Earth's surface lying near or between the tropics, at about 23°N and S latitudes.

troposphere The lowest layer of the atmosphere, extending to an elevation of about 12 kilometers.

trough The portion of a wave between crests where the water surface is depressed.

T-S diagram A plot that displays the temperature and salinity at all points in a vertical column of water.

tsunami A wave generated by movement of the ocean bottom, such as an earthquake, underwater landslide, underwater volcanic eruption, and so on. Also called a seismic sea wave.

tunicate A small primitive sack-like chordate.

turbidite A sedimentary deposit resulting from turbidity currents and displaying graded bedding.

turbidity current A muddy slurry of water and sediments that flows downhill along the ocean bottom. Usually, it originates in the head of a submarine canyon and flows down through the canyon and out onto the ocean bottom at the base of the slope.

turbine A machine consisting of a series of fan blades and often connected to an electrical generator. As a fluid passes through the fan blades, the turbine turns and generates electrical power.

turbulence Irregular, unsystematic, and small eddylike motions in a fluid that cause local mixing.

ultraviolet rays Those electromagnetic waves that have wavelengths shorter than the shortest visible light (violet) yet longer than x-rays.

upper mantle The top 650 kilometers of the mantle, which seems to behave differently than the mantle below that.

upwelling The rising of deeper waters to the surface.

vagrant Mobile. Able to move about on the ocean floor.

valence An indication of the number of electrons that an atom or group of atoms will share when forming chemical bonds.

vaporization Conversion from liquid to gaseous form.

vent community A community of organisms found around a hydrothermal vent.

vertebrate An animal with an internal skeleton that includes a spinal column of segments called vertebrae. It also has a central brain, red blood, and two pairs of external appendages.

vertical mixing Mixing between surface and deeper waters.

viscosity The ability of a fluid to resist shear stress, or a measure of the amount of drag exerted on one layer of the fluid by the motion in the layer next to it. For example, syrup has high viscosity and air low viscosity.

visible light Electromagnetic waves with wavelengths in the range of 0.4 to 0.7 micrometers. These have the energies needed to trigger our optic nerves.

volatile Easily vaporized. Having a relatively low boiling point. Also, any volatile substance.

volcanic Pertaining to the extrusion of magmas and other materials onto the Earth's surface from interior regions.

volcanic chain A line of volcanoes on the ocean floor, caused by the passage of the oceanic plate over a hot spot in the upper mantle.

vorticity A measure of the amount of rotary or circular motion in a fluid.

wake The group of surface waves generated by something moving near or on the water surface.

water A material whose molecules are made of two hydrogen and one oxygen atoms, H_2O.

water mass Any large volume of water that has similar characteristics, such as temperature and salinity, and therefore probably has the same origin.

watt A measure of power, or the rate of energy consumption, that is equal to 1 joule per second.

wave amplitude The maximum vertical displacement of the water surface from its average position during the passage of waves, being equal to one-half the wave height.

wave group A group of waves that have roughly the same wavelength, wave height, and wave speed and that move together in the same direction.

wave height The vertical distance between the lowest point in a trough and the highest point on the neighboring crest.

wave period The time it takes for one complete oscillation as a wave passes a stationary point.

wave speed The speed with which an individual wave travels (as opposed to the speed with which a wave group travels).

wave steepness The ratio of wave height to wavelength.

wave of translation The moving mass of water created when a wave breaks, differing from normal gravity waves in that the individual molecules of water no longer have oscillatory orbital motion, but rather move along with the wave.

wavelength The horizontal distance between any point on one wave and the corresponding point on the next wave, such as from one crest to the next.

weathering The processes by which solid rock is broken down into fragments.

well-mixed A type of estuary in which the salinity of the water does not vary significantly with depth at any point in the estuary, although it varies from fresh water at the head to salt water at the mouth. This type of estuary is usually shallow and has good vertical mixing.

westerlies Winds blowing from the west.

West Wind Drift The Antarctic Circumpolar Current that flows eastward around the Earth at about 50 to 60°S latitude.

western intensification The tendency of surface currents along the western edges of all oceans in both hemispheres to be particularly strong, swift, and narrow.

wetlands Shallow coastal marshes.

windows (radiation) Portions of the spectrum of electromagnetic waves to which our atmosphere is transparent.

windward The side facing the wind. (As opposed to leeward.)

work The product of force times distance.

x-rays Electromagnetic waves with wavelengths shorter than ultraviolet, but longer than gamma rays.

zero population growth The state of affairs when the birth rate matches the death rate, so that there is no net increase in population.

zooplankton Planktonic animals.

PHOTO CREDITS

Chapter 1

Opener: Granger Collection. Figure 1.1: Kevin Cullimore/Tony Stone Images/ New York, Inc. Figure 1.2: Courtesy NASA. Figure 1.3: ©Copyright 1961 by California Institute of Technology and Carnegie Institution of Washington. Figure 1.5a: Runk/Schoenberger/Grant Heilman Photography. Figure 1.5b: Sinclair Stammers/Photo Researchers. Figures 1.10a,b: Granger Collection. Figure 1.10c: From Challenger Office Report, Great Britain, 1895. Figure 1.11: Bettmann Archive. Figure 1.12: Courtesy Woods Hole Oceanographic Institution. Figure 1.13: Granger Collection. Figure 1.15a: Courtesy Scripps Institution of Oceanography. Figure 1.15b: Courtesy Ocean Drilling Program, Texas A&M University. Figure 1.16: Courtesy Woods Hole Oceanographic Institution. Figure 1.17: Courtesy Woods Hole Oceanographic Institution, photo by T. Corbett. Figures 1.18 and 1.19: Courtesy NASA. Figure 1.20: Courtesy Woods Hole Oceanographic Institution. Figure 1.21: Courtesy Woods Hole Oceanographic Institution, photo by R.J. Bowen.

Chapter 2

Opener: Jim Corwin/Photo Researchers. Figure 2.2: Malin/Pasachoff/Caltech 1992/Courtesy Anglo-Australian Telescope Board. Figure 2.11c: Courtesy NASA.

Chapter 3

Opener: Courtesy Jet Propulsion Laboratory. Figure 3.3a: Courtesy Astronomical Society of the Pacific. Figures 3.3b and 3.4: Courtesy NASA. Figure 3.7: Harold Sund/The Image Bank. Figure 3.10 : Courtesy Keith Stowe. Figure 3.11b: Dean Lee/The Wildlife Collection.

Chapter 4

Opener: Photo by Cecil W. Stoughton, Courtesy National Park Service. Figure 4.6: Jack S. Grove/Tom Stack & Associates. Figure 4.9b: Barrie Rokeach/The Image Bank. Figure 4.10c: Courtesy Aluminum Company of America. Figure 4.12: Courtesy Woods Hole Oceanographic Institution. Figure 4.14: Courtesy Woods Hole Oceanographic Institution, photo by Dudley Foster. Figures 4.16 and 4.22a: Courtesy NASA. Figure 4.22f: Stephen Frink/WaterHouse. Figure 4.24b: Martin Miller. Figure 4.27: Diagram by Tau Rho Alpha/USGS.

Chapter 5

Opener: David Barnes/Tony Stone Images/New York, Inc. Figure 5.1: Courtesy Richard Buffler, University of Texas. Figure 5.2: Charles Gurche/The Wildlife Collection. Figures 5.3 and 5.4d: Courtesy Keith Stowe. Figure 5.5: Courtesy Ministry of Environment, Lands and Parks Surveys and Resource Mapping Branch, Victoria, British Columbia. Figure 5.6b: Dean Lee/The Wildlife Collection. Figure 5.8: Courtesy Keith Stowe. Figure 5.11: Martin Harvey/The Wildlife Collection. Figure 5.12: Biophoto Associates/Photo Researchers. Figure 5.14a: Institute of Oceanographic Sciences/NERC/Photo Researchers. Figure 5.14b: Courtesy Deepsea Ventures, Inc. Figure 5.15a: Tom Till/The Wildlife Collection. Figure 5.15b: Kim Heacox/Tony Stone Images/ New York, Inc. Figure 5.17: Courtesy Kathleen K. Newell, University of Washington. Figure 5.18a: Courtesy Woods Hole Oceanographic Institution.

Chapter 6

Opener: Kim Westerskov/Tony Stone Images/New York, Inc. Figure 6.3b: Nuridsany and Perennou/Photo Researchers. Figure 6.4: Jean-Gérard Sidaner/Photo Researchers. Figure 6.6: Courtesy NASA. Figure 6.7: James Randkley/Tony Stone Images/New York, Inc. Figure 6.10: Simon Fraser/Photo Researchers.

Chapter 7

Opener: Courtesy NASA. Figure 7.11c: Dave Watts/Tom Stack & Associates. Figure 7.12: Joe Van Os/The Image Bank. Figure 7.23b: Courtesy NASA.

Chapter 8

Opener: Courtesy NASA. Figure 8.19a: Courtesy NOAA. Figure 8.22: George Holton/Photo Researchers. Figure 8.24: Dewitt Jones/Tony Stone Images/New York, Inc.

Chapter 9

Opener: John Callahan/Tony Stone Images/ New York, Inc. Figure 9.1: Joe Devenney/The Image Bank. Figure 9.12e: Courtesy U.S. Army Corps of Engineers, Los Angeles District. Figure 9.19b: Kees van den Berg/Photo Researchers. Figure 9.20: Warren Bolster/Tony Stone Images/New York, Inc. Figure 9.21: Darrell Jones/Tony Stone Images/New York, Inc. Figure 9.34: Andrew J. Martinez/Photo Researchers.

Chapter 10

Opener: G. Büttner/Naturbild/OKAPIA/ Photo Researchers. Figure 10.2a: Harold Sund/The Image Bank. Figure 10.2b: Charles Gurche/The Wildlife Collection. Figure 10.3: Larry Lefever/Grant Heilman Photography. Figure 10.9b: ©John S. Shelton. Figure 10.13a: Courtesy NOAA. Figure 10.13b: Courtesy Pacific Western, Santa Barbara, CA. Figure 10.15: ©John S. Shelton. Figures 10.17 and 10.20: Courtesy NASA. Figure 10.21: Gregory G.

Dimijian/Photo Researchers. Figure 10.22: Gianni Tortoli/Photo Researchers. Figure 10.23: J.H. Robinson/Photo Researchers. Figure 10.25: ©The Port Authority of NY & NJ.

Chapter 11

Opener: M.I.Walker/Photo Researchers. Figure 11.1a: Gary Braasch/Tony Stone Images/New York, Inc. Figure 11.1b: D.P.Wilson/Photo Researchers. Figure 11.1c: Martin Bond/Photo Researchers. Figure 11.3: Ed Pritchard/Tony Stone Images/New York, Inc. Figure 11.6b: Alan Pitcairn/Grant Heilman Photography. Figure 11.6c: Runk/Schoenberger/Grant Heilman Photography. Figure 11.7: Courtesy NASA. Figure 11.11: Randy Morse/Tom Stack & Associates. Figure 11.18: D.P.Wilson/David & Eric Hosking/Photo Researchers. Figure 11.19: Norman R. Lightfoot/Photo Researchers. Figure 11.21b: A.B. Dowsett/Photo Researchers. Figure 11.22: Alfred Pasieka/Photo Researchers.

Chapter 12

Opener: Stephen Frink/Tony Stone Images/ New York, Inc. Figure 12.1a: Vu/© J. Poindexter/Visuals Unlimited. Figure 12.1b: Dr. Tony Brain/Photo Researchers. Figure 12.2a: Photo by Mary Bremer, Courtesy Woods Hole Oceanographic Institution. Figure 12.2b: Jan Hinsch/Photo Researchers. Figure 12.2c: Eric Gravé/Photo Researchers. Figure 12.5: Carr Clifton/Tony Stone Images/New York, Inc. Figure 12.6a: ©Gary R. Robinson/Visuals Unlimited. Figure 12.6b: Lefever/Grushow/Grant Heilman Photography. Figure 12.6c: John D. Cunningham/Visuals Unlimited. Figure 12.7: Jack Dermid/Photo Researchers. Figure 12.8: Stephen Frink/WaterHouse. Figure 12.9a: Stuart Westmorland/Tony Stone Images/New York, Inc. Figure 12.9b: Robert Hermes/Photo Researchers. Figure 12.10: Norbert Wu. Figure 12.11a: A. Kerstitch/Visuals Unlimited. Figure 12.11b: Bruce Iverson. Figure 12.11c: ©Marty Snyderman/Visuals Unlimited. Figure 12.11d: D. Holden Bailey/Tom Stack & Associates. Figure 12.11e: Norbert Wu. Figure 12.12a: Stephen Frink/WaterHouse. Figure 12.12b: Mike Neumann/Photo Researchers. Figure 12.12c: Randy Wells/Tony Stone Images/ New York, Inc. Figure 12.12d: Chris Huss/The Wildlife Collection. Figure 12.13a: V. Siegel/The Wildlife Collection. Figure 12.13b: Runk/Schoenberger/Grant Heilman Photography. Figure 12.13c: Stephen Frink/WaterHouse. Figure 12.14: Fred Bavendam/Peter Arnold, Inc. Figure 12.15: Chris Huss/The Wildlife Collection. Figure 12.16a: Rondi/Tani/Photo Researchers. Figure 12.18a: Stephen Frink/WaterHouse. Figure 12.18b: Mike Bacon/Tom Stack & Associates. Figure 12.18c: Brian Parker/Tom Stack & Associates. Figure 12.19a: Andrew Wood/Photo Researchers. Figure 12.19b: Gary Bell/The Wildlife Collection. Figure 12.20a: John Warden/Tony Stone Images/New York, Inc. Figure 12.20b: Kathy Bushue/Tony Stone Images/ New York, Inc. Figure 12.20c: François Gohier/Photo Researchers. Figure 12.21a: Martin Harvey/The Wildlife Collection. Figure 12.21b: Henry Holdsworth/The Wildlife Collection. Figure 12.21c: Art Wolfe/Tony Stone Images/New York, Inc.

Chapter 13

Opener: Marc Chamberlain/Tony Stone Images/New York, Inc. Figure 13.1(left): John Edwards/Tony Stone Images/New York, Inc. Figure 13.1(right): D.P.Wilson/Photo Researchers. Figure 13.2b: Biophoto Associates/Photo Researchers. Figure 13.2c: Randy Morse/Tom Stack & Associates. Figure 13.3: Stephen Frink/WaterHouse. Figure 13.6a: Courtesy Deborah L. Penry, University of California-Berkeley. Figure 13.6b: Sinclair Stammers/Photo Researchers. Figure 13.8b: Cara Moore/The Image Bank. Figure 13.10a: Darryl Torckler/Tony Stone Images/New York, Inc. Figure 13.11a: Norbert Wu/Tony Stone Images/New York, Inc. Figure 13.12: Gary Bell/The Wildlife Collection. Figure 13.13: Andrew J. Martinez/Photo Researchers. Figure 13.14: Jim Zipp/Photo Researchers. Figure 13.15: Larry Ulrich/Tony Stone Images/New York, Inc. Figure 13.16: Photo by James Broda, courtesy Woods Hole Oceanographic Institution. Figure 13.17: Courtesy Museum of Natural History, NY. Figure 13.18: Brock May/Photo Researchers. Figure 13.19a: Dave Fleetham/Tom Stack & Associates. Figure 13.19b: Stephen Frink/Tony Stone Images/New York, Inc. Figure 13.20: C. Seghers/Photo Researchers.

Chapter 14

Opener: Jurgen Vogt/The Image Bank. Figure 14.4: Matt Bradley/Tom Stack & Associates. Figure 14.5: Will & Deni McIntyre/Tony Stone Images/New York, Inc. Figure 14.6: John Warden/Tony Stone Images/ New York, Inc. Figure 14.7: Martin Rogers/Tony Stone Images/ New York, Inc. Figure 14.8: Lawrence Lowry/Photo Researchers. Figure 14.9: Wilkinson/F.S.P./Gamma Liaison. Figure 14.14: Courtesy U.S. Coast Guard.

Chapter 15

Opener: Bob Thomason/Tony Stone Images/New York, Inc. Figure 15.6a: Courtesy Lockheed Missile and Space Co., Inc. Figure 15.6b: Courtesy AeroVironment. Figure 15.7: ©La Phototheque EDF, photo by Marc Morceau. Figure 15.8b: Courtesy Lockheed Missile and Space Co., Inc. Figure 15.12: Glenn Sport/Gamma Liaison. Figure 15.14: Courtesy Deepsea Ventures, Inc. Figure 15.16: Phillip Hayson/Photo Researchers. Figure 15.17: Gamma Liaison. Figure 15.18a: Courtesy U.S. Coast Guard. Figure 15.18b: Granger Collection.

INDEX